Parasitology Research Monographs

Volume 1

Series Editor:
Heinz Mehlhorn
Department of Zoomorphology
Cell Biology and Parasitology
Heinrich Heine University
Universitätsstrasse 1
40225 Düsseldorf
Germany

For further volumes:
http://www.springer.com/series/8816

Heinz Mehlhorn
Editor

Nature Helps...

How Plants and Other Organisms
Contribute to Solve Health Problems

Springer

Editor
Prof. Dr. Heinz Mehlhorn
Department of Zoomorphology
Cell Biology and Parasitology
Heinrich Heine University
Universitätsstrasse 1
40225 Düsseldorf
Germany
mehlhorn@uni-duesseldorf.de

ISSN 2192-3671   e-ISSN 2192-368X
ISBN 978-3-642-19381-1   e-ISBN 978-3-642-19382-8
DOI 10.1007/978-3-642-19382-8
Springer Heidelberg Dordrecht London New York

Library of Congress Control Number: 2011930797

© Springer-Verlag Berlin Heidelberg 2011
This work is subject to copyright. All rights are reserved, whether the whole or part of the material is concerned, specifically the rights of translation, reprinting, reuse of illustrations, recitation, broadcasting, reproduction on microfilm or in any other way, and storage in data banks. Duplication of this publication or parts thereof is permitted only under the provisions of the German Copyright Law of September 9, 1965, in its current version, and permission for use must always be obtained from Springer. Violations are liable to prosecution under the German Copyright Law.
The use of general descriptive names, registered names, trademarks, etc. in this publication does not imply, even in the absence of a specific statement, that such names are exempt from the relevant protective laws and regulations and therefore free for general use.

Cover design: deblik, berlin

Printed on acid-free paper

Springer is part of Springer Science+Business Media (www.springer.com)

Preface or why such a book

Nature helps ... – of course – at first itself. All living organisms (bacteria, fungi, plants, and animals including prehistoric humans) had and still have to survive the struggle for life, since for millions of years they have been in competition with many individuals of their own species and with even higher numbers of competitors belonging to the rest of the living system, if they fit principally into the conditions given at a particular place on Earth.

This constant need to be always somewhat better / fitter than competitors was and is the motor of *evolution* that led and leads to unbelievable variations in body shape, astonishing physiological adaptations, and many other admirable abilities. Thus, thousands of skilful methods of defense against attacks from the surroundings have been developed. For example, toxic substances and repellents evolved that are used to keep predators away from engorging such individuals. Other species developed medical remedies, which support the wound-healing process or increase the success of reproduction respectively allows a faster growth rate under favorable conditions. Other compounds (e.g., prohibitors of freezing or "overheating") helped survival in poor conditions such as icy winters or extreme heat. These facilities have been developed over millions of years and all species that had not been able to develop such facilities, are today no longer visible as they have returned to the general pool of molecules and elements on Earth. *Thus nature is cruel to all members* that are not flexible in the sense described above, and nature eliminates them for the sake of the survivors.

Of course nature *takes its steps slowly* by testing the chances and benefits of more or less occasional mutations of single individuals in a given biotope under changing environmental conditions.

It is not much longer than 100,000 years ago that humans were thrown as "last-minute man" (on the very late evening of the sixth day of creation) into the battle for survival on Earth. This sending to Earth was not only done by God saying "Go to Earth, rule it and produce successors," but humans were also equipped with a small dose of the divine spirit of recognition – of course some received greater insight, while others received less of this spirit. In any case, the "gifted dwarfs", i.e., humans, developed the skills of observation and mental preservation of important and/or useful events or they recognized helpful facilities of plants and animals in their surroundings from their early beginnings.

Through combination of these empirical observations and by analysis of their background some of these human "skilful dwarfs" used the abilities of plants and animals (later also those of fungi and bacteria) to develop products and measures for survival of diseases or of other critical conditions, which would have killed nonadapted competitors.

Of course plants were first recognized by mankind as beneficial in times of disease. Wise women or magicians – often accused of practicing diabolical witchcraft – have learned over thousands of years to prepare extracts or to use whole plants or their fruits as medicaments or as important contributions to daily food.

At the very beginning – apart from the isolation of extracts – nobody really knew the active compounds in plants, for example nobody knew, why the powder of the South American *Cinchona* tree helped as a powerful remedy against the malaria fever or why extracts of the bark of the willow (*Salix* spp.) suppressed fever and decreased headache. As soon as the methods of chemistry allowed the analysis of details and synthesis processes had been evaluated, medicaments or insecticides etc., based on pure chemicals were developed (e.g., quinine and acetyl salicyl acid instead of bark powders of *Cinchona* or *Salix*). This industrial production of chemotherapeutical remedies still flourishing after 100 years, however, led to the loss of the details of the mostly orally transmitted knowledge on the abilities of many plants and animals, but also led to the discovery of the effects of useful bacteria and fungi as highly effective antibiotics.

Since these chemicals were very pure and often based on a single active ingredient, infectious bacteria, fungi, or parasites developed increasing resistances against these agents, as occurred in the case of MRSA (multiresistant *Staphylococcus aureus* strains) where the efficacy of chemical medicaments became very low or even lost. These events make it understandable, why Jean Jacques Rousseau's original cry "*Back to nature – retour à la nature*" resonates even more loudly today. This stimulates many scientists to test the efficacies of plants and animals against adverse impacts that may affect plants and animals and respectively endanger humans.

The wave of relevant plant-based papers in many scientific journals is growing daily, so that single results can hardly be seen. This is due to the fact that many authors prefer to collect published papers as goodies for their career and forget to develop from their results useful remedies for society.

The present book presents some selected reports on the efficacies of plants used as medicaments, insecticides, and/or parasitizides. It also includes examples of the practical aid given by animals in the fight against pests or describes their use as remedies and to diagnose diseases.

Some chapters show how extracts or particular stages of animals contribute to human or animal health, help in healing wounds, or aid the police in finding the murderer (in case it is not the gardener as usual).

These 15 chapters should stimulate more scientists to think about similar studies hopefully leading to successful products, but they also show the risks that may arise in times of worldwide globalization and regional climate change, when species of fungi, bacteria, plants, and animals are introduced willingly (e.g., as "biofighters" against crop pests) or unwillingly into other environments (as hidden passengers inside the bodies of tourists or inside containers with goods of any kind).

Düsseldorf
May 2011

Heinz Mehlhorn

Contents

1. **Indigenous Traditional Medicine: Plants for the Treatment of Diarrhea** 1
   Clara Lia Costa Brandelli, Raquel Brandt Giordani, Alexandre José Macedo, Geraldo Attilio De Carli, and Tiana Tasca

2. **Efficacies of Medicinal Plant Extracts Against Blood-Sucking Parasites** 19
   A. Abdul Rahuman

3. **Natural Remedies in the Fight Against Insects** 55
   Norbert Becker

4. **The Neem Tree Story: Extracts that Really Work** 77
   Heinz Mehlhorn, Khaled A.S. Al-Rasheid, and Fathy Abdel-Ghaffar

5. **The Efficacy of Extracts from Plants – Especially from Coconut and Onion – Against Tapeworms, Trematodes, and Nematodes** 109
   Heinz Mehlhorn, Gülendem Aksu, Katja Fischer, Bianca Strassen, Fathy Abdel Ghaffar, Khaled A.S. Al-Rasheid, and Sven Klimpel

6. **Curcumin: A Natural Herb Extract with Antiparasitic Properties** 141
   Md. Shahiduzzaman and Arwid Daugschies

7. **Marine Organisms and Their Prospective Use in Therapy of Human Diseases** 153
   Sherif S. Ebada and Peter Proksch

8. **Benefits and Failure of Imported Animals in the Fight Against Pests** 191
   Volker Walldorf

9	**Helminth Therapy to Treat Crohn's and Other Autoimmune Diseases** 211
	Jeff Bolstridge, Bernard Fried, and Aditya Reddy

10	**Insects Help to Solve Crimes** 227
	Jens Amendt

11	**"Living Syringes": Use of Hematophagous Bugs as Blood Samplers from Small and Wild Animals** 243
	André Stadler, Christian Karl Meiser, and Günter A. Schaub

12	**Xenodiagnosis** 273
	Christian Karl Meiser and Günter A. Schaub

13	**Blowfly Strike and Maggot Therapy: From Parasitology to Medical Treatment** 301
	Heike Heuer and Lutz Heuer

14	**Extracts from Fly Maggots and Fly Pupae as a "Wound Healer"** 325
	Heinz Mehlhorn and Falk Gestmann

15	**Living Medication: Overview and Classification into Pharmaceutical Law** 349
	Heike Heuer, Lutz Heuer, and Valentin Saalfrank

Index 369

Contributors

Abdel-Ghaffar, Fathy Department of Zoology, Cairo University, Giza, Egypt, fathyghaffar@yahoo.com

Aksu, Gülendem Department of Parasitology, Heinrich Heine University, Universitätsstr. 1, D-40 225, Düsseldorf, Germany

Al-Rasheid, Khaled A.S. Department of Zoology, King Saud University, Riyadh, Saudi Arabia

Amendt, Jens Institute for Forensic Medicine, Forensic Biology/Entomology, Kennedyallee 104, D-60 596, Frankfurt am Main, Germany, amendt@em.uni-frankfurt.de

Becker, Norbert University of Heidelberg and German Mosquito Control Association (KABS), Ludwigstr. 99, D-67165 Waldsee, Germany, Norbert.Becker@kabs-gfs.de

Bolstridge, Jeff Department of Chemistry, Lafayette College, Easton, PA 18042, USA

Carli, Geraldo A. De Instituto de Geriatria e Gerontologia, Pontifícia Universidade Católica do Rio Grande do Sul, Avenida Ipiranga, 6681, 90610-000, Porto Alegre, RS, Brazil

Costa-Brandelli, Clara L.C. Departmento de Analises, Lab. de Pesquisa em parasitologia, Avenida Iparanga 2752, 90610-000 Porto Allegre, Brazil

Daugschies, Arwid Institute of Parasitology, University of Leipzig, An den Tierkliniken 35, 04 103 Leipzig, Germany, daugschies@vetmed.uni-leipzig.de

Ebada, Sherif S. Institute of Pharmaceutical Biology and Biotechnology, Heinrich-Heine University, Universitaetsstrasse 1, 40225 Duesseldorf, Germany; Department of Pharmacognosy and Phytochemistry, Faculty of Pharmacy, Ain-Shams

University, Organisation of African Unity 1, 11566 Cairo, Egypt, sherif.elsayed@uni-duesseldorf.de

Fischer, Katja Department of Parasitology, Heinrich Heine University, Universitätsstr. 1, D-40 225 Düsseldorf, Germany

Fried, Bernard Department of Biology, Lafayette College, Easton, PA 18042, USA

Gestmann, Falk Department of Parasitology, Heinrich Heine University, Universitätsstr. 1, D-40 225 Düsseldorf, Germany

Giordani, Raquel B. Departmento de Analises, Lab. de Pesquisa em parasitologia, Avenida Iparanga 2752, 90610-000 Porto Allegre, Brazil

Heuer, Heike Fa. Agiltera GmbH, Am Kreuzberg 31, 41 542 Dormagen, Germany, agiltera@t-online.de

Heuer, Lutz Fa. Agiltera GmbH, Am Kreuzberg 31, 41542 Dormagen, Germany

Klimpel, Sven Biodiversity and Climate Change Center (Bik-F), Goethe-University Frankfurt, D-60 325 Frankfurt, Germany

Macedo, Alexandre I. Departmento de Analises, Lab. de Pesquisa em parasitologia, Avenida Iparanga 2752, 90610-000 Porto Allegre, Brazil

Mehlhorn, Heinz Department of Parasitology, Heinrich Heine University, Universitätsstr. 1, D-40 225 Düsseldorf, Germany, mehlhorn@uni-duesseldorf.de

Proksch, Peter Institute of Pharmacological Biology and Biotechnology, Heinrich Heine University, D-40225 Düsseldorf, Germany, proksch@uni-duesseldorf.de

Rahuman, Ahmed Abdul Department of Zoology, C. Abdul Hakeem College, Melvisharam, 632 509, Tamil Nadu, India, abdulrahuman6@hotmail.com

Reddy, Aditya Department of Chemistry, Lafayette College, Easton, PA 18042, USA

Saalfrank, Valentin Fa. Agiltera GmbH, Am Kreuzberg 3, 41 542 Dormagen, Germany

Schaub, Günter Zoology/Parasitology Group, Department of Animal Ecology, Evolution and Biodiversity Group, Universitätsstr. 150, 44780, Bochum, Germany, guenter.schaub@ruhr-uni-bochum.de

Shahiduzzaman, Md. Department of Parasitology, Bangladesh Agricultural University, Mymensingh 2202, Bangladesh, India, szamanpara@yahoo.com

Strassen, Bianca Department of Parasitology, Heinrich Heine University, Universitätsstr. 1, D-40 225 Düsseldorf, Germany

Tasca, Tiana Departmento de Analises, Lab. de Pesquisa em parasitologia, Avenida Iparanga 2752, 90610-000 Porto Allegre, Brazil, tiana.tasca@ufrgs.br

Walldorf, Volker Institute of Zoomorphology and Cell Biology, Heinrich Heine University, Universitätsstr. 1, D-40225 Düsseldorf, Germany, walldorf@uni-duesseldorf.de

Chapter 1
Indigenous Traditional Medicine: Plants for the Treatment of Diarrhea

Clara Lia Costa Brandelli, Raquel Brandt Giordani, Alexandre José Macedo, Geraldo Attilio De Carli, and Tiana Tasca

Abstract Ethnopharmacology can contribute to the exploration of phytotherapeutical resources for use in local contexts and countries of origin. Indigenous people living on their traditional territory largely rely on medicinal plants for healthcare and they are therefore rich in ethnopharmacological knowledge. For public health professionals, such knowledge can help to create the basis for a health system that is more respectful towards local practices and foster better collaborations across medical systems. This chapter presents major studies on the immeasurable and potent traditional knowledge of medicinal plants retained by indigenous communities, relating data within an ethnobotanical and ethnopharmacological framework, and demonstrating its therapeutic merits. Furthermore, the information reported herein could be incorporated into governmental programs of basic healthcare, following recommendations of the World Health Organization (WHO).

1.1 Introduction

Ethnopharmacology and natural product drug discovery remain a significant hope in the current target-rich, lead-poor scenario (Patwardhan, J Ethnopharmacol 100:50–52, 2005). Indigenous people living on their traditional territory largely rely on medicinal plants for healthcare and they are therefore rich in ethnopharmacological knowledge (Uprety et al., J Ethnobiol Ethnomed 6:3, 2010). The South American Indians hold vast knowledge about the physical and chemical properties of medicinal plants, a fact that makes them an important source of new prototype antimicrobial agents (Bueno et al., Acta Bot Bras 19:39–44, 2005). This chapter presents major studies on the immeasurable and potent traditional knowledge of

---

C.L.C. Brandelli, R.B. Giordani, A.J. Macedo, and T. Tasca (✉)
Faculdade de Farmácia, Universidade Federal do Rio Grande do Sul, Avenida Ipiranga, 2752, sala 510, 90610-000 Porto Alegre, RS, Brazil
e-mail: tiana.tasca@ufrgs.br

G.A. De Carli
Instituto de Geriatria e Gerontologia, Pontifícia Universidade Católica do Rio Grande do Sul, Avenida Ipiranga, 6681, 90610-000, Porto Alegre, RS, Brazil

medicinal plants retained by the indigenous communities, relating data within an ethnobotanical and ethnopharmacological framework, and demonstrating its therapeutic merits. Furthermore, the information reported herein could be incorporated into governmental programs of basic healthcare, following recommendations of the World Health Organization (WHO).

1.2 Diarrhea

Infectious diarrhea remains one of the leading causes of childhood morbidity and mortality worldwide. The annual global burden of infectious diarrhea is enormous, involving 3 to 5 billion cases and nearly 2 million deaths, with the latter accounting for almost 20% of all deaths in children younger than 5 years (Boschi-Pinto et al. 2008). Infectious diseases caused 68% of deaths among children worldwide in 2008, with the largest percentages due to pneumonia (18%) and diarrhea (15%) (Black et al. 2010). Diarrhea results from infection of the intestinal tract by a wide range of enteric pathogens that can disrupt intestinal function (Grimwood and Forbes 2009). The resulting symptom complex of diarrhea is characterized by an increased number of loose or watery stools (more than three episodes in 24 h), and therefore fluid loss and dehydration is the cause of death in nearly all patients with diarrhea (Grimwood and Forbes 2009; Santosham et al. 2010).

The annual incidence and the etiologic profile of diarrhea in different populations may vary with several risk factors (Souza et al. 2002). In low-income and middle-income countries, where pathogen transmission occurs mainly through contaminated food or drinking water, bacterial and viral pathogens are responsible for most disease (Santosham et al. 2010). The following factors predispose high incidence rates and bacterial etiology: young age, nutritional deficiencies, inadequate physical and food hygiene, early weaning, densely populated homes and workplaces, lack of basic sanitation in places of stay, access to contaminated water supplies, and a hot season (summer) (Souza et al. 2002; Samie et al. 2009). In industrialized countries the frequency of diarrheal bouts per child is only 0.5–2 episodes per infant/year, whereas in developing regions it may amount to ten episodes/year (Guerrant et al. 1990). However, most episodes are self-limited and rarely result in persistent diarrhea, malnutrition, or death in a previously healthy child (Grimwood and Forbes 2009). In this setting, repeated episodes of enteric infection can contribute to malnutrition by interfering with nutrient absorption, and these episodes usually occur during the first few years of life, a period critical for physical growth and brain development. Moreover, they can be followed by impaired linear growth, intellectual function, and school performance (Grimwood and Forbes 2009; Samie et al. 2009).

The main etiology of diarrhea is related to a wide range of bacteria (such as *Campylobacter jejuni*, *Campylobacter coli*, *Escherichia coli*, *Salmonella* spp., *Shigella* spp., *Vibrio cholerae*, *Yersinia enterocolitica* and *Aeromonas* spp.), enteroparasites (*Giardia lamblia*, *Cryptosporidium* spp., *Entamoeba histolytica*,

Trichuris trichiura), and viruses (adenovirus and rotavirus) (Barreto et al. 2006; Samie et al. 2009). Some bacteria produce secretory enterotoxins that directly invade cells (such as the cholera toxin or the heat-labile or stable enterotoxins produced by *E. coli*). Other bacteria species invade cells or produce cytotoxins (such as those produced by *Shigella*, enteroinvasive *E. coli*, or *Clostridium difficile*) that damage cells or trigger host responses, leading to small or large bowel diseases (enteroaggregative or enteropathogenic *E. coli* or *Salmonella*). Viruses (noroviruses and rotaviruses) and protozoa (*Cryptosporidium, Giardia,* or *E. histolytica*) disrupt cell functions and cause short- or long-term disease (Pawlowski et al. 2009). The diarrhea is treated with antimicrobial drugs, but this treatment is generally ineffective (Obi and Bessong 2002; Nguyen et al. 2005; Samie et al. 2009). The increase in antibiotic resistance among enteric pathogens is becoming a special concern, since the empirical treatment leads to risk of persistent diarrhea (Grimwood and Forbes 2009). This fact has underscored the need of a quick development of antibacterial drugs that will be more effective than those currently in use. In addition, taking into account the intestinal parasitic diseases, large-scale deworming is necessary to reduce the worldwide morbidity of these infections, but without improved water supplies and sanitation this approach cannot be relied on for sustainable reductions in parasite frequency or intensity of infection (Bethony et al. 2006). Strategies including adequate infrastructure, deworming of children in schools, introduction of new anthelminthic, and antimicrobial drugs vaccines and other control tools could result in substantial reductions in the worldwide disease burden.

1.3 Intestinal Parasitic Infections

Parasitic infections are classified as neglected diseases by the World Health Organization (WHO) and they must receive attention due to their high capacity to increase the impact of malnutrition and growth deficiencies, with negative consequences for cognitive function and learning ability (Scolari et al. 2000; Bethony et al. 2006; Grimwood and Forbes 2009; Tanner et al. 2009; Zonta et al. 2010). Even today these diseases are a worldwide public health issue, especially in resource-poor settings (Hurtado-Guerrero et al. 2005; Calzada et al. 2005). Intestinal helminths have a worldwide distribution and infect over one billion of the world's population (Scolari et al. 2000; Zonta et al. 2010). The three soil-transmitted helminthes (*Ascaris lumbricoides, T. trichiura* and hookworms *Necator americanus/Ancylostoma duodenale*) are responsible for an estimated loss of 39 million disability-adjusted life years (DALYs) (Tanner et al. 2009). The most common cestode is *Hymenolepis nana,* affecting 75 million individuals in the world (Cook et al. 2009). *H. nana* infections cause deficiency in vitamin B12 and folate, and high prevalence of stunting in patients (Mohammad and Hegazi 2007).

The clinical features of soil-transmitted helminth infections include acute manifestations associated with larval migration through the skin and viscera, and the acute and chronic manifestations resulting from parasitism of the gastrointestinal tract by adult worms (Bethony et al. 2006). Migrating soil-transmitted helminth larvae provoke reactions in many of the tissues through which they pass. Several cutaneous syndromes result from skin-penetrating larvae, such as those caused by *N. americanus* and *A. duodenale* hookworm third-stage larvae, characterized by itching, a local erythematous and papular rash accompanied by pruritus on the hands and feet (Hotez et al. 2004). *A. lumbricoides* larvae produce eosinophilic granulomas in the liver and verminous pneumonia accompanied by wheezing, dyspnoea, a nonproductive cough, and fever, with blood-tinged sputum produced during heavy infections (Kaplan et al. 2001).

The presence of large numbers of adult *A. lumbricoides* worms in the small intestine can cause abdominal distension and pain. They can also cause lactose intolerance and malabsorption of vitamin A and possibly other nutrients, which might partly cause the nutritional and growth failure (Sharghi et al. 2001). In young children, adult worms can aggregate and cause complete obstruction, leading to bowel infarction and intestinal perforation resulting in fatal peritonitis. In adults, hepatobiliary and pancreatic ascariasis are more common, presumably because the adult biliary tree is large enough to accommodate an adult worm (Khuroo et al. 1990).

The adult *T. trichiura* causes inflammation at the site of attachment from large numbers of whipworms, resulting in colitis. Longstanding colitis produces a clinical disorder that resembles inflammatory bowel disease, including chronic abdominal pain, diarrhea, as well as the anemia of chronic disease. The most serious manifestation of heavy whipworm infection results in chronic diarrhea and rectal prolapse (Bundy and Cooper 1989). Whipworm infection can also exacerbate colitis caused by infection with *C. jejuni* (Shin et al. 2004).

The major pathology of hookworm infection includes eosinophilia and results from intestinal blood loss as a result of adult parasite invasion and attachment to the mucosa and submucosa of the small intestine (Maxwell et al. 1987; Hotez et al. 2004). Hookworm disease occurs when the blood loss exceeds the nutritional reserves of the host, thus resulting in iron-deficiency anemia (Lwambo et al. 1992).

Considering protozoa, the most frequently nonpathogenic amoeba are *Entamoeba coli, Iodameba buetschilii* and *Endolimax nana*, and among the pathogenic protozoa are *E. histolytica* and *G. lamblia* (Ferreira et al. 2000). Amebiasis is caused by *E. histolytica* and is responsible for 100 million annual cases of amebic dysentery, colitis, and liver abscess resulting in 100,000 deaths (World Health Organization 1997a). Among the intestinal forms, acute diarrhea and dysentery account for 90% of cases (Espinosa-Cantellano and Martínez-Palomo 2000). Amebic colitis is often characterized by multiple discrete lesions of varying stages. Symptoms are variable, but most commonly manifest as mild to moderate abdominal discomfort/pain and frequent loose-watery stools containing variable amounts of blood and mucus (Adams and MacLeod 1977a). More unusual manifestations of intestinal amebiasis are amebomas (amebic granulomas), toxic mega colon and fulminating amebic colitis (Ellyson et al. 1986; Mortimer and Chadee 2010). Extraintestinal disease is characterized by amebic infection in the liver resulting in multiple granulomas that expand and coalesce to form large singular abscesses (Adams and MacLeod 1977b). If rupture occurs, it is possible for the parasites to disseminate to other organs, such as the pericardium, lungs, and brain (Adams and MacLeod 1977b; Teramoto et al. 2001).

G. lamblia is one of the most common intestinal parasites worldwide (Moersch et al. 2009). It is estimated that about 300 million people annually are affected by the parasite all around the world (Alizadeh et al. 2006; Ali and Nozaki 2007). Clinical manifestations of giardiosis include acute or chronic diarrhea, abdominal pain, and malabsorption, leading to malnutrition and weight loss, particularly in children (Adam 2001; Pupulin et al. 2004; Ali and Nozaki 2007). It is important to emphasize that *G. lamblia* infections, even asymptomatic, cause a negative impact on children's growth (Prado et al. 2005).

It is well known that enteroparasites may have a negative effect on nutritional status, in particular among children, and the recent decade has significantly strengthened the link between malnutrition and parasitic infection (Stephenson et al. 2000; Saldiva et al. 2002; Matos et al. 2008). The negative impact of the parasitic infections on anthropometric status can be explained by jeopardized food intake, anorexia, and poor absorption of macro and micronutrients (Matos et al. 2008). Parasitic diseases constitute not only a medical but also a social and economic matter, with high morbidity prevalence (Zonta et al. 2010). Malnutrition, in parallel to poverty and inadequate living conditions, is also widespread among indigenous populations (Kühl et al. 2009; Orellana et al. 2009). The nutritional deficiencies are due to reduced diversity of food and poor sanitary conditions (Navone et al. 2006).

Despite progress made in recent years, there is no vaccine against soil-transmitted helminthes and intestinal protozoa (Loukas et al. 2006). The clinical treatment of amebiasis and giardiosis is based on nitroimidazoles, mainly metronidazole. In veterinary treatment, one of the best options against *Giardia* infections is pyrantel pamoate (Hausen et al. 2010). Four anthelminthics are currently on the WHO model list of essential medicines for the treatment and control of soil-transmitted helminthes: albendazole, mebendazole, levamisole, and pyrantel pamoate (World Health Organization 1997b). However, there is considerable concern that large-scale administration of anthelminthics and antiprotozoal in deworming programs might result in the development and spread of drug-resistant nematodes, which is already a significant problem in veterinary medicine (Keiser and Utzinger 2008). Important data from the systematic review and meta-analysis by Keiser and Utzinger (2008) revealed deficiencies regarding the evidence base of current anthelminthic drugs. Well-designed, adequately powered, and rigorously implemented trials should address these issues, not only providing new data regarding the efficacy of antiparasitic drugs, but also aiding in establishing benchmarks for subsequent monitoring of drug resistance. Overall, new anthelminthics and antiprotozoal drugs are urgently needed and plants in widespread use for the treatment of gastrointestinal disorders, such as diarrhea and dysentery, raise the possibility of new alternative therapies.

1.4 Intestinal Bacterial Infections

Diarrhea may be caused by a wide range of bacteria. Diarrheagenic *E. coli* (DEC) is among the most important bacterial enteric pathogens, particularly in developing countries (Moreno et al. 2010). DEC is classified into six categories: enteropathogenic *E. coli* (EPEC), enterohemorrhagic *E. coli* (EHEC), enterotoxigenic *E. coli* (ETEC), enteroinvasive *E. coli* (EIEC), enteroaggregative *E. coli* (EAEC), and diffusely adhering *E. coli* (DAEC) (Souza et al. 2002; Barreto et al. 2006; Moreno et al. 2010). EPEC has been a leading cause of childhood diarrhea in developing countries, but its frequency in industrialized areas has been decreasing (Moreno et al. 2010). The main mechanism of EPEC pathogenesis is the attaching and effacing (AE) lesion, which is characterized by intimate adherence of the bacteria to the intestinal epithelium (Dulguer et al. 2003; Moreno et al. 2010). EAEC is an emerging pathogen, primarily associated with persistent diarrhea in infantiles (Pereira et al. 2010).

Shigellosis is a global human health problem. Four species of *Shigella* – *S. dysenteriae*, *S. flexneri*, *S. boydii* and *S. sonnei* – are able to cause the disease. The symptoms of shigellosis include diarrhea and/or dysentery with frequent mucoid bloody stools, abdominal cramps, and tenesmus. *Shigella* spp. causes dysentery by invading the colonic mucosa. *Shigella* bacteria multiply within colonic epithelial cells, cause cell death and spread laterally to infect and kill adjacent epithelial cells causing mucosal ulceration, inflammation, and bleeding (Niyogi 2005). Antibiotic treatment is indicated for dysentery caused by *Shigella* species, because it can limit both the clinical course of illness and the duration of fecal excretion of the causative organism. However, these bacteria are becoming increasingly resistant to the antibiotics most commonly used in the treatment of diarrhea (Nguyen et al. 2005).

Most of the clinically isolated strains of *Salmonella* belong to the species *Salmonella enterica*, which commonly causes gastroenteritis worldwide. The symptoms often appear 6–48 h after ingestion of contaminated food or water, and are mainly constituted by nausea, vomiting and diarrhea without blood. Fever, abdominal cramps, myalgia, and headache are also frequent. All *Salmonella* species can also cause septicemia mainly in pediatric, geriatric, and immunocompromised patients. *S. typhi* is the causative of typhoid fever; *S. paratyphi* A, *S. schottmuelleri*, and *S. hirschfeldii* produce the mild form of the disease, paratyphoid fever. Ten to 14 days after the ingestion of bacteria, patients gradually experience crescent fever with nonspecific complications: headache, myalgia, malaise, and anorexia. These symptoms persist for 1 week or more and are accompanied by gastroenteritis (Murray et al. 2010).

Campylobacter is a leading bacterial cause of food-borne disease worldwide representing a huge health challenge. This mainly zoonotic genus includes at least 18 species of which *C. jejuni* and *C. coli* are the most common pathogens and account for the majority of diagnosed human *Campylobacter* infections (Silva Quetz et al. 2010). These organisms colonize and disrupt intestinal function, causing malabsorption or diarrhea by mechanisms that involve microbial adherence and localized effacement of the epithelium, production of toxin(s) and direct epithelial cell invasion (Guerrant et al. 1999).

The causative agent of cholera, the Gram-negative bacterium *V. cholerae*, is a facultative pathogen that has both human and environmental stages in its life cycle. It is difficult to gauge the exact morbidity and mortality of cholera because the surveillance systems in many developing countries are rudimentary, and many countries are hesitant to report cholera cases to the WHO because of the potential negative economic impact of the disease on trade and tourism (Nelson et al. 2009). Today, the true burden of cholera is estimated to reach several million cases per year, predominantly in Asia and Africa (Sack et al. 2006). *V. cholerae* is differentiated serologically on the basis of the O antigen of its lipopolysaccharide (LPS) (Stephen 2001). Infection with *V. cholerae* produces a clinical spectrum that ranges from asymptomatic colonization to cholera gravis, the most severe form of the disease. Following host ingestion of contaminated food or water, *V. cholerae* colonizes the small intestine for 12–72 h before symptoms appear. Cholera often begins with stomach cramps and vomiting followed by diarrhea, which may progress to fluid losses of up to 1 liter per hour (Kaper et al. 1995). These losses result in severe fluid volume depletion and metabolic acidosis, which may lead to circulatory collapse and death (Stephen 2001). Rice water stool typically harbors between 10^{10} and 10^{12} vibrios per liter. Symptomatic patients may shed vibrios before the onset of illness and will continue to shed organisms for 1–2 weeks; in contrast, asymptomatic patients shed vibrios for only one day (Kaper et al. 1995).

Since their discovery, antimicrobial drugs have proved remarkably effective for the control of bacterial infections. However, it was soon evident that bacterial pathogens were unlikely to surrender unconditionally, because some pathogens rapidly become resistant to any of the first discovered effective drugs (Obi and Bessong 2002; Barbour et al. 2004; Nguyen et al. 2005; Mandomando et al. 2007; Samie et al. 2009). The vast and widespread empirical use of antimicrobial drugs has increased pathogen resistance and triggered the urgent need of continuous effective drug development (Nguyen et al. 2005; Grimwood and Forbes 2009). New antimicrobial compounds have been isolated from plants that may inhibit bacterial growth by different mechanisms than those currently used by antimicrobials and may have a significant clinical value in treatment of resistant microbial strains (Barbour et al. 2004).

1.5 Ethnopharmacology

The millennium ecosystem assessment examined four future scenarios, all of which require a change from our present deleterious relationship with ecosystems (United Nations 2005). Many factors were considered, including population growth, per capita income, land use change, nutrient loading, and climate change that will result in significant habitat loss, and also will lead to, possibly, 10–15% extinction of plant

Table 1.1 Main plants used in indigenous traditional medicine for diarrhea treatment

Family	Species	Part used
Asteraceae	*Achyrocline satureioides*	Leaves
	Bacharis sp.	Aerial parts
	Cichorium intybus	Roots
	Eupatorium odoratum	Roots
	Guatteria vilosissima	Roots/bark
Myrtaceae	*Campomanesia xanthocarpa*	Leaves
	Eugenia uniflora	Bark
	Psidium guajava	Leaves/bark
	Syzygium cordatum	Whole plant
Apiaceae	*Foeniculum vulgare*	Aerial parts
Rutaceae	*Aegle marmelos*	Leaves/bark
	Casimiroa tetrameria	Leaves
	Esenbeckia sp.	Roots/bark
	Pilocarpus pennatifolius	Whole plant
Solanaceae	*Nicotiana tabacum*	Leaves/bark
Fabaceae	*Bauhinia divaricata*	Leaves
	Piptadenia sp.	Bark
	Piscidia piscipula	Leaves
Celastraceae	*Crossopetalum gaumeri*	Leaves
Euphorbiaceae	*Jatropha gaumeri*	Roots
	Croton gratissimus	Whole plant
Adiantaceae	*Adiantum capillus-veneris*	Aerial parts
	Adiantum peruvianum	Aerial parts
Lamiaceae	*Craniotome furcata*	Aerial parts
	Ocimum micranthum	Leaves
Anacardiaceae	*Ozoroa sphaerocarpa*	Leaves
Punicaceae	*Punica granatum*	Fruits
Sapotaceae	*Manilkara zapota*	Bark
Cannabaceae	*Trema micrantha*	Leaves
Polygonaceae	*Antigonon leptopus*	Leaves
Maliaceae	*Azadirachta indica*	Leaves

species (Cordell and Colvard 2005). In this way it is important to register and catalog the traditional knowledge about medicinal plants to prevent the loss of valuable cultural information (see Table 1.1).

Ethnopharmacology rescues the historical use of plants by people who have great knowledge through family tradition. The information that is known and that can be accumulated on the use of plants in healthcare systems around the world is of inestimable value. Also recognized as valuable to bioprospecting is the ethnobotanical knowledge of the forest-dwelling (or native, tribal, indigenous) people who live in or near tropical forests and possess information on the use of plants for medicinal purposes. For example, of the 120 active compounds isolated from higher plants and used today in Western medicine, 74% have the same therapeutic use as in native societies (Moran 2000). The NAPRALERT database has information on 14,317 species with ethnomedical data, representing 3,703 genera and 272 plant species – 5.2% of all estimated higher plant species (Cordell and Quinn-Beattie 2005). Additionally, for almost 60% of the ethnomedically used plants described in NAPRALERT, no compound has been isolated and no biological work conducted. There is an abundant opportunity for the discovery of new medicinal agents. WHO estimates that approximately 80% of the world's population makes use mainly of traditional remedies for healthcare (Bueno et al. 2005; Kim 2005; Vendruscolo et al. 2005).

All cultural groups use plants as a therapeutic resource and, in urban centers, plants are used as alternative or complementary to official medicine (Kim 2005; Vendruscolo and Mentz 2006). Traditional medicinal plants are being marketed all over the world as phytotherapeuticals, but their quality control is typically very poor or nonexistent. Cordell and Colvard (2005) showed an outline for developing more comprehensive standards for safety and efficacy of plant-based products considering the correct genus and species of plant material is being offered, and that the correct plant part is being used.

In efforts to improve this therapy, ethnobotanical and ethnopharmacological studies disclose a variety of plants used for the treatment of gastrointestinal disorders such as diarrhea and dysentery (Calzada et al. 2005; Barbosa et al. 2007). The indigenous population holds a great knowledge on the use of medicinal plants, making use of some species for the treatment of diarrhea (Bueno et al. 2005). The Convention on Biological Diversity (CBD) instructed sovereign nations to develop plans to catalog and preserve their indigenous knowledge and their biodiversity (Cragg et al. 1997). CBD goals are (1) the conservation of biological diversity, (2) the sustainable use of its components, and (3) the fair and equitable sharing of the benefits arising out of this use. Protecting and compensating local groups for their indigenous knowledge, and for providing access to the biome, is a reasonable expectation for both those who hold the resources and those who are seeking them (Soejarto et al. 2002). The index of cultural significance (ICS) created at the end of the 1980s, aims to register the value of each vegetable species and to disclose its importance for the biological and cultural survival of a traditional community (Stoffle et al. 1990). This index can be an interesting tool for understanding the importance of plants to biological and cultural community survival.

1.6 Indigenous Traditional Medicine

Traditional and millenary cultural habits strongly affect indigenous health. Permanent and indiscriminate contact of indigenous populations with the national society leads to the establishment of an environmental and cultural disequilibrium in the villages (Coimbra and Santos 2000). This fact results in higher susceptibility to local endemic illnesses or to those introduced by the contact, such as enteroparasitic infections (Miranda et al. 1998). There are a reported 300–350 million indigenous people worldwide, 70% of whom reside in Asia. Loss of traditional lands, political and social marginalization, introduction to alcohol and other drugs of addiction, and reduced access to traditional food supplies have resulted in greatly increased susceptibility to disease (Bodeker 2008). The indigenous population is the most vulnerable affected by infectious and parasitic diseases due to lack of instruction, access to drinking water, feeding, healthcare, personal hygiene, and basic sanitation (Miranda et al. 1998; Bóia et al. 2009; Zonta et al. 2010). In fact, indigenous people are likely to be poorer than nonindigenous people (Zonta et al. 2010; Holt et al. 2010). The prevalence of infectious and parasitic diseases is high among Amerindian populations (Scolari et al. 2000; Toledo et al. 2009). The major problem is the absence of infrastructure assigned to the collection of human and domestic animal dejections and the inexistence of drinking water in the villages. Taking into account the dramatic change in lifestyle as a result of acculturation and environmental degradation processes, it is not surprising that intestinal parasites are widely spread (Navone et al. 2006). Moreover, data on the prevalence of intestinal protozoa and helminthes among indigenous groups are very scarce in the literature (Aguiar et al. 2007).

Brazilian indigenous communities have a rich heritage of traditional medicine with an important role in bioprospection of new medicines. Traditions are dynamic entities of unchanging knowledge (Patwardhan 2005), and the medicinal plants are an essential element of the tribal pharmaceutical systems (Leonti et al. 2010). Indigenous people living on their traditional territory largely rely on medicinal plants for healthcare and they are therefore rich in ethnopharmacological knowledge (Uprety et al. 2010).

Few groups have such a vast knowledge on the physical and chemical properties of plants as the South American indigenous populations (Bueno et al. 2005). In Brazil there are 122 indigenous cultures described but only 30% of them were investigated with regard to ethnobotanical aspects (Coutinho et al. 2002). In fact, indigenous people are the keepers of the cumulative knowledge of certain botanical taxa, which has been transmitted over centuries prior to it becoming important in the context of developing novel pharmaceuticals. Importantly, such research would demonstrate the historical development of an intricate relationship between a culture and its environment (Weldegerima 2009). Through family tradition, many plants are used by indigenous people for diarrhea (Coelho-Ferreira 2009).

An example of indigenous knowledge of medicinal plants is given by the ethnic group Mbyá-Guarani, which utilize the leaves and stems of *Achyrocline satureioides* (Asteraceae), barks of *Eugenia uniflora* (Myrtaceae), aerial parts of *Foeniculum vulgare* (Apiaceae), and barks of *Psidium guajava* (Myrtaceae) for the treatment of diarrhea (Brandelli et al. 2009). Quantitative assays of viable *G. lamblia* trophozoites, showed that *A. satureioides* presented the highest cytotoxic effect (93.5%), followed by *P. guajava* (82.2%), and *E. uniflora* (67.3%). The Guarani indigenous group mainly uses *A. satureioides* for the treatment of diarrhea, revealing the conformity with results obtained in vitro (Brandelli et al. 2009; Crivos et al. 2007). Kaiowá and Guarani tribes from Mato Grosso do Sul, Brazil, also described the use of *E. uniflora* and *P. guajava* for diarrhea purposes (Bueno et al. 2005). The Mbyá inhabit the Paranaense Rainforest – among the environmental systems of South America, one of the richest in its biodiversity. This is one of the ethnic groups of the *Tupí-Guaraní* linguistic family which currently inhabit this ecosystem (Crivos et al. 2007). Taking into account that Mbyá populations present high prevalence of diarrhea and parasitic infections and that their strategies for resolution of illness episodes are based on medicinal plants, studies on therapeutic approaches have been justifiable. As a general rule, medicinal plant effectiveness is associated with its organoleptic features, especially its olfactory and gustative characteristics. A strong smell and bitter taste in a plant, for example, is associated with antiprotozoal properties. Plant resources employed in the therapy of diarrhea are usually the same used by the Mbyá-Guarani of Brazil: the aerial parts of *A. satureioides,* leaves of *P. guajava* and *Campomanesia xanthocarpa* (Myrtaceae), roots of *Cichorium intybus* (Asteraceae), and the whole plant of *Pilocarpus pennatifolius* (Rutaceae) (Crivos et al. 2007). Among the different resources used for treatment of diarrhea caused by parasites (called *tachó* in the Mbyá language), this indigenous group uses *Baccharis* sp. (Asteraceae) and for treatment of diarrhea with blood they use leaves and stem bark of *Nicotiana tabacum* (Solanaceae). Indian communities from several tribes in Maranhão, Brazil, reported the use of *Guatteria vilosissima* (Asteraceae) (root and bark), *Piptadenia* sp. (Fabaceae) (bark), *P. guajava,* and *Esenbeckia* sp. (Rutaceae) (root and bark) for antidiarrhea purposes (Coutinho et al. 2002).

The knowledge of medicinal plants was a part of the ancient Maya culture and they are still utilized today by the Yucatec Mayan inhabitants on the Peninsula of Yucatan, Mexico (Ankli et al. 2002). During an ethnobotanical study in three Mayan communities, 360 medicinal plants and 1,828 reports on their uses were documented. The uses of the plants were divided into nine therapeutic groups. Forty-eight species were chosen and evaluated in bioassays relevant to the following groups of illnesses, gastrointestinal disorders; dermatological conditions; women's medicines as well as pain and/or fever. The screening of plant species used for gastrointestinal disorders by the Yucatec Maya showed that the most active extract against *G. lamblia* was *Crossopetalum gaumeri* (Celastraceae). The nonpolar and polar extracts of *Piscidia piscipula* (Fabaceae) and extracts of *Casimiroa tetrameria* (Rutaceae) and *Jatropha gaumeri* (Euphorbiaceae) were also active against *Helicobacter pylori*. The nonpolar root extract of *J. gaumeri* was found to be the most active plant tested against *Bacillus cereus.* The roots of *C. gaumeri* are used orally to combat diarrhea, the leaves of *P. piscipula* are used as a medicine for treating gastrointestinal disorders (especially diarrhea and cramps), *Bauhinia divaricata* (Fabaceae) is used for a variety of illnesses such as gastrointestinal problems, and the roots of *J. gaumeri* are used for treating diarrhea. The antibacterial activities against *H. pylori*, *B. cereus*, and *Staphylococcus epidermidis* observed in this study might be of interest in light of its traditional use as a treatment for diarrhea (Ankli et al. 2002).

In India about 45% of the higher plants are medicinal, being one of the countries that most employ their existing biodiversity. In this context, *Aegle marmelos* (Rutaceae) has been widely used in indigenous systems in India due to its various medicinal properties. Interestingly, the extract obtained from decoction did not present antibacterial activity against enteropathogenic and enterotoxigenic bacteria, but it did inhibit a series of virulence factors, such as bacterial colonization to gut epithelium and production of certain enterotoxins. Moreover, to some extent it could also control giardiosis and rotaviral infections (Brijesh et al. 2009). A study comparing the most common species in India belonging to the family Adiantaceae, proved the indigenous traditional knowledge which reported it as antibacterial, against various bacteria and fungi, including *E. aerogenes*, *E. coli*, and *S. typhimurium*. The methanolic extract from *Adiantum capillus-veneris* (Adiantaceae) reached interesting MICs (minimum inhibitory concentrations) of 0.48 μg/ml against *E. coli* and from *Adiantum peruvianum* (Adiantaceae) a MIC of 15.62 μg/ml against *Enterobacter aerogenes* was obtained (Singh et al. 2008). An Indian research group tested the aerial parts of *Craniotome furcata* (Lamiaceae), reported by indigenous people of Nepal as a treatment for cuts and wounds, against various microorganisms. Extracts obtained with either ethyl acetate or *n*-butanol proved to be effective against *E. coli* and to a lesser extent against *S. typhimurium* (Joshi et al. 2010).

An interesting study has been performed aimed at an evaluation of a synergistic effect of the combination of distinct parts of *Croton gratissimus* (Euphorbiaceae). *C. gratissimus* is known as "koorsbessie" ("koors" = fever) suggesting that the plant is used as a pyrogenic by African indigenous populations. In general, the leaf, bark, and root extract showed higher efficacies than the oils against all the pathogens studied. The combination of leaf and bark extracts brought an enhanced efficacy against *Enterococcus faecalis*. Also a bark and root combination presented a lower MIC for *E. faecalis* and *E. coli* when evaluated independently. The phenomena repeated when leaf and root were combined in a 1:1 ratio, being more effective against most of the pathogens tested, including *E. faecalis*. However, essential oils of *C. gratissimus* presented poor susceptibility against the pathogens tested. The authors highlighted that synergistic interactions clearly indicate that biological activity may be improved through combination therapy, where different complex metabolic pools collectively contribute to the enhanced effect (van Vuuren and Viljoen 2008). In the same way, studies proved the traditional knowledge of the Swazi people against diarrhea, with the combination of *Syzygium cordatum* (Myrtaceae) and *Ozoroa sphaerocarpa* (Anacardiaceae) giving the strongest synergistic interaction (MIC value of 0.33 mg/ml) and a triple combination (1:1:1), including *Breonadia salicina* (Rubiaceae), was also very effective in inhibiting microbial growth (MIC value of 0.44 mg/ml) of *E. coli*. A *S. cordatum* and *B. salicina* combination presented a mildly synergistic effect (MIC of 1.00 mg/ml). Individually *O. sphaerocarpa* was the most potent (MIC of 1.2 mg/ml), followed by *S. cordatum* (MIC of 1.44 mg/ml), and *B. salicina* (MIC of 10.89 mg/ml) (Sibandze et al. 2010).

Nine of the 31 formulae tested used by Mayans were active against bacteria (MIC = 0.5 mg/ml); three formulations were active against *E. coli*, eight against *S. typhi*, and one against *Shigella flexneri* and no activity was found against *Klebsiella pneumoniae*. Moreover, four presented activity against *E. histolytica*, and seven against *G. lamblia* ($IC_{50} \leq 20$ μg/ml). Taking into account only the activity upon the bacteria and no mixture of species of plant fruits, *Punica granatum* (Punicaceae) was the most active including the lowest MIC (0.125 mg/ml) against *S. flexneri*. Also leaves of *Ocimum micranthum* (Lamiaceae), barks of *Manilkara zapota* (Sapotaceae), and leaves of *C. gaumeri* were active against at least one bacterium. A formulation with a mixture of plants containing leaves of *Trema micrantha*

(Cannabaceae), *Antigonon leptopus* (Polygonaeceae), and the root of *Eupatorium odoratum* (Asteraceae) was able to kill two of the four tested bacteria. It is important to note that even though many of the formulations were not active against bacteria, *E. histolytica*, or *G. lamblia*, they might be involved in other mechanisms in the healing process involving immuno-stimulatory and toxin-deactivating activities and molecular changes which occur during the digestive process (Vera-Ku et al. 2010).

Leaves of *Azadirachta indica* (Maliaceae) (locally known as *neem*) are used by indigenous people in different parts of India for curing gastrointestinal disorders such as diarrhea and cholera and its use is widespread. An *A. indica* extract had significant antibacterial activity against multi-drug-resistant *V. cholerae* of serotypes O1, O139 and non-O1, non-O139. Furthermore, *neem* extract showed antisecretory activity on *V. cholera*-induced fluid secretion in mouse intestine. Oral administration of the extract inhibited hemorrhage induced by *V. cholerae* in mouse intestine at a dose ≥ 300 mg/kg. The results obtained in this study give some scientific support to the uses of *neem* by the indigenous people in India for the treatment of diarrhea and the dreadful disease cholera (Thakurta et al. 2007).

1.7 General Conclusions and Comments

Access to good quality traditional medicine is always important, especially where and when it is the only easily accessible treatment. This is particularly so wherever logistical, political, or economic problems make access to modern medicine scarce and difficult. With traditional medicines, users and relatives are in principle protected against the risk of fake drugs when they correctly identify the plants and prepare their own treatments. Ethnopharmacology could decisively contribute to the exploration of phytotherapeutical resources and to the dissemination of the knowledge based on tested effectiveness among locally available traditional medicines. For public health professionals, such knowledge can help create the basis for a health system that is more respectful towards local practices and fosters better collaboration across medical systems.

One goal of this evaluation is to better understand the use of plants by indigenous peoples. In this work we show some correlations between uses of medicinal plants and relevant biological activities. Phytochemical and further pharmacological studies are important tasks for the future in order to better understand the effects of these important pharmaceutical resources. Organizations like WHO and TRAMIL (Central America) encourage the use of remedies provided that they are safe and that scientific evidences for their biological and pharmacological effects exist.

References

Adam RD (2001) Biology of *Giardia lamblia*. Clin Microbiol Rev 14:447–475. doi:10.1128/CMR.14.3.447-475.2001

Adams EB, MacLeod IN (1977a) Invasive amebiasis I. Amebic dysentery and its complications. Medicine 56:315–323

Adams EB, MacLeod IN (1977b) Invasive amebiasis II. Amebic liver abscess and its complications. Medicine 56:325–334

Aguiar JI, Gonçalves AQ, Sodré FC, Pereira Sdos R, Bóia MN, de Lemos ER, Daher RR (2007) Intestinal protozoa and helminths among Terena Indians in the State of Mato Grosso do Sul: high prevalence of *Blastocystis hominis*. Rev Soc Bras Med Trop 40:631–634. doi:10.1590/S0037-86822007000600006

Ali V, Nozaki T (2007) Current therapeutics, their problems, and sulfur-containing-amino-acid metabolism as a novel target against infections by "amitochondriate" protozoan parasites. Clin Microbiol Rev 20:164–187. doi:10.1128/CMR.00019-06

Alizadeh A, Ranjbar M, Kashani KM, Taheri MM, Bodaghi M (2006) Albendazole versus metronidazole in the treatment of patients with giardiasis in the Islamic Republic of Iran. East Mediterr Health J 12:548–554

Ankli A, Heinrich M, Bork P, Wolfram L, Bauerfeind P, Brun R, Schmid C, Weiss C, Bruggisser R, Gertsch J, Wasescha M, Sticher O (2002) Yucatec Mayan medicinal plants: evaluation based on indigenous uses. J Ethnopharmacol 79:43–52

Barbosa E, Calzada F, Campos R (2007) In vivo antigiardial activity of three flavonoids isolated of some medicinal plants used in Mexican traditional medicine for the treatment of diarrhea. J Ethnopharmacol 109:552–554. doi:10.1016/j.jep. 2006.09.009

Barbour EK, Sharif MA, Sagherian VK, Habre AN, Talhouk RS, Talhouk SN (2004) Screening of selected indigenous plants of Lebanon for antimicrobial activity. J Ethnopharmacol 93:1–7. doi:10.1016/j.jep. 2004.02.027

Barreto ML, Milroy CA, Strina A, Prado MS, Leite JP, Ramos EAG, Ribeiro H, Alcântara-Neves NM, Teixeira AG, Rodrigues LC, Ruf H, Guerreiro H, Trabulsi L (2006) Community-based monitoring of diarrhea in urban Brazilian children: incidence and associated pathogens. Trans R Soc Trop Hyg 100:234–242. doi:10.1016/j.trstmh.2005.03.010

Bethony JM, Loukas A, Hotez PJ, Knox DP (2006) Vaccines against blood-feeding nematodes of humans and livestock. Parasitology 133:S63–79. doi:10.1017/S0031182006001818

Black RE, Cousens S, Johnson HL, Lawn JE, Rudan I, Bassani DG, Jha P, Campbell H, Walker CF, Cibulskis R, Eisele T, Liu L, Mathers C, Child Health Epidemiology Reference Group of WHO and UNICEF (2010) Global, regional, and national causes of child mortality in 2008: a systematic analysis. Lancet 375:1969–1987. doi:10.1016/S0140-6736(10)60549-1

Bodeker G (2008) The health care of indigenous peoples. In: Heggenhougen K, Quah S (eds) Nations international encyclopedia of public health, vol 3. San Diego, Academic Press

Bóia MN, Carvalho-Costa FA, Sodré FC, Porras-Pedroza BE, Faria EC, Magalhães GA, Silva IM (2009) Tuberculosis and intestinal parasitism among indigenous people in the Brazilian Amazon region. Rev Saude Publica 43:176–178. doi:10.1590/S0034-89102009000100023

Boschi-Pinto C, Velebit L, Shibuya K (2008) Estimating child mortality due to diarrhoea in developing countries. Bull World Health Organ 86:710–717. doi:10.1016/j.jep. 2004.02.027

Brandelli CLC, Giordani RB, De Carli GA, Tasca T (2009) Indigenous traditional medicine: in vitro anti-giardial activity of plants used in the treatment of diarrhea. Parasitol Res 104:1345–1349. doi:10.1007/s00436-009-1330-3

Brijesh S, Daswani P, Tetali P, Antia N, Birdi T (2009) Studies on the antidiarrhoeal activity of *Aegle marmelos* unripe fruit: validating its traditional usage. BMC Complement Altern Med 9:1–12. doi:10.1186/1472-6882-9-47

Bueno NR, Castilho RO, Costa RB, Pott A, Pott VJ, Scheidt GN, Batista MS (2005) Medicinal plants used by the Kaiowá and Guarani indigenous populations in the Caarapó Reserve, Mato Grosso do Sul, Brazil. Acta Bot Bras 19:39–44

Bundy DAP, Cooper ES (1989) *Trichuris* and trichuriasis in humans. Adv Parasitol 28:107–173

Calzada F, Cervantes-Martínez JA, Yépez-Mulia L (2005) In vitro antiprotozoal activity from the roots of *Geranium mexicanum* and its constituents on *Entamoeba histolytica* and *Giardia lamblia*. J Ethnopharmacol 98:191–193. doi:10.1016/j.jep. 2005.01.019

Coelho-Ferreira M (2009) Medicinal knowledge and plant utilization in an Amazonian coastal community of Marudá, Pará State (Brazil). J Ethnopharmacol 126:159–175. doi:10.1016/j.jep. 2009.07.016

Coimbra CEA Jr, Santos RV (2000) Saúde, minorias e desigualdade: algumas teias de inter-relações, com ênfase nos povos indígenas no Brasil. Cienc Saude Coletiva 5:125–132. doi:10.1590/S1413-81232000000100011

Cook DM, Swanson RC, Eggett DL, Booth GM (2009) A retrospective analysis of prevalence of gastrointestinal parasites among school children in the Palajunoj Valley of Guatemala. J Health Popul Nutr 27:31–40

Cordell GA, Colvard MD (2005) Some thoughts on the future of ethnopharmacology. J Ethnopharmacol 100:5–14. doi:10.1016/j.jep. 2005.05.027

Cordell GA, Quinn-Beattie ML (2005) Unpublished results from the NAPRALERT database. University of Illinois at Chicago

Coutinho DF, Travassos LMA, Amaral FMM (2002) Estudo etnobotânico de plantas medicinais utilizadas em comunidades indígenas no estado do Maranhão – Brasil. Visão Acad 3:7–12

Cragg GM, Baker JT, Borris RP, Carte B, Cordell GA, Soejarto DD, Gupta MP, Iwu MM, Madulid DR, Tyler VE (1997) Interactions with source countries. Guidelines for members of the American Society of Pharmacognosy. J Nat Prod 60:654–655

Crivos M, Martínez MR, Pochettino ML, Remorini C, Sy A, Teves L (2007) Pathways as "signatures in landscape": towards an ethnography of mobility among the *Mbya-Guaraní* (Northeastern Argentina). J Ethnobiol Ethnomed 3:2. doi:10.1186/1746-4269-3-2

Dulguer MV, Fabbricotti SH, Bando SY, Moreira-Filho CA, Fagundes-Neto U, Scaletsky ICA (2003) Atypical enteropathogenic *Escherichia coli* strains: phenotypic and genetic profiling reveals a strong association between Enteroaggregative *E. coli* heat-stable enterotoxin and diarrhea. J Infect Dis 188:1685–1694

Ellyson JH, Bezmalinovic Z, Parks SN, Lewis FR Jr (1986) Necrotizing amebic colitis: a frequently fatal complication. Am J Surg 152:21–26

Espinosa-Cantellano M, Martínez-Palomo A (2000) Pathogenesis of intestinal amebiasis: from molecules to disease. Clin Microbiol Rev 13:318–331

Ferreira MU, Ferreira CS, Monteiro CA (2000) Secular trends in child intestinal parasitic diseases in S. Paulo city, Brazil (1984–1996). Rev Saude Publica 34:73–82

Grimwood K, Forbes DA (2009) Acute and persistent diarrhea. Pediatr Clin N Am 56:1343–1361. doi:10.1016/j.pcl.2009.09.004

Guerrant RL, Hughes JM, Lima NL, Crane J (1990) Diarrhea in developed and developing countries: magnitude, special setting, and etiologies. Rev Infect Dis 12:41S–50S

Guerrant RL, Steiner TS, Lima AAM, Bobak DA (1999) How intestinal bacteria cause disease. J Infect Dis 179:331–337

Hausen MA, Menna-Barreto RF, Lira DC, de Carvalho L, Barbosa HS (2010) Synergic effect of metronidazole and pyrantel pamoate on *Giardia lamblia*. Parasitol Int. doi:10.1016/j.parint.2010.10.003

Holt DC, McCarthy JS, Carapetis JR (2010) Parasitic diseases of remote Indigenous communities in Australia. Int J Parasitol 40:1119–1126. doi:10.1016/j.ijpara.2010.04.002

Hotez PJ, Brooker S, Bethony JM, Bottazzi ME, Loukas A, Xiao S (2004) Hookworm infection. N Engl J Med 351:799–807

Hurtado-Guerrero AF, Alencar FH, Hurtado-Guerrero JC (2005) Ocorrência de enteroparasitas na população geronte de Nova Olinda do Norte Amazonas, Brasil. Acta Amazon 35:487–499. doi:10.1590/S0044-59672005000400013

Joshi RK, Mujawar MHK, Kholkute SD (2010) Antimicrobial activity of the extracts of *Craniotome furcata* (Lamiaceae). J Ethnopharmacol 128:703–704. doi:10.1016/j.jep. 2010.02.001

Kaper JB, Morris JG Jr, Levine MM (1995) Cholera. Clin Microbiol Rev 8:48–86

Kaplan KJ, Goodman ZD, Ishak KG (2001) Eosinophilic granuloma of the liver: a characteristic lesion with relationship to visceral larva migrans. Am J Surg Pathol 25:1316–1321

Keiser J, Utzinger J (2008) Efficacy of current drugs against soil-transmitted helminth infections: systematic review and meta-analysis. JAMA 299:1937–1948

Khuroo MS, Zargar SA, Mahajan R (1990) Hepatobiliary and pancreatic ascariasis in India. Lancet 335:1503–1506

Kim HS (2005) Do not put too much value on conventional medicines. J Ethnopharmacol 100:37–39. doi:10.1016/j.jep.2005.05.030

Kühl AM, Corso ACT, Leite MS, Bastos JL (2009) Perfil nutricional e fatores associados à ocorrência de desnutrição entre crianças indígenas Kaingáng da Terra Indígena de Mangueirinha, Paraná, Brasil. Cad Saude Publica 25:409–420. http://www.scielosp.org/pdf/csp/v25n2/20.pdf

Leonti M, Cabras S, Weckerle CS, Solinas MN, Casu L (2010) The causal dependence of present plant knowledge on herbals – contemporary medicinal plant use in Campania (Italy) compared to Matthioli (1568). J Ethnopharmacol 130:379–391. doi:10.1016/j.jep.2010.05.021

Loukas A, Bethony J, Brooker S, Hotez P (2006) Hookworm vaccines: past, present, and future. Lancet Infect Dis 6:733–741

Lwambo NJ, Bundy DA, Medley GF (1992) A new approach to morbidity risk assessment in hookworm endemic communities. Epidemiol Infect 108:469–481

Mandomando I, Espasa M, Vallès X, Sacarlal J, Sigaúque B, Ruiz J, Alonso P (2007) Antimicrobial resistance of *Vibrio cholerae* O1 serotype Ogawa isolated in Manhiça District Hospital, southern Mozambique. J Antimicrob Chemother 60:662–664. doi:10.1093/jac/dkm257

Matos SMA, Assis AMO, Prado MS, Strina A, Santos LA, Jesus SR, Barreto ML (2008) *Giardia duodenalis* infection and anthropometric status in preschoolers in Salvador, Bahia State, Brazil. Cad Saude Publica 24:1527–1535

Maxwell C, Hussain R, Nutman TB, Poindexter RW, Little MD, Schad GA, Ottesen EA (1987) The clinical and immunologic responses of normal human volunteers to low dose hookworm (*Necator americanus*) infection. Am J Trop Med Hyg 37:126–134

Miranda RA, Xavier FB, Menezes RC (1998) Intestinal parasitism in a Parakanã indigenous community in southwestern Pará State, Brazil. Cad Saude Publica 14:507–511. doi:10.1590/S0102-311X1998000300007

Moersch K, Hanevik K, Rortveit G, Wensaas KA, Langeland N (2009) High rate of fatigue and abdominal symptoms 2 years after an outbreak of giardiasis. Trans R Soc Trop Med Hyg 103:530–532. doi:0.1016/j.trstmh.2009.01.010

Mohammad MA, Hegazi MA (2007) Intestinal permeability in *Hymenolepis nana* as reflected by non invasive lactulose/mannitol dual permeability test and its impaction on nutritional parameters of patients. J Egypt Soc Parasitol 37:877–891

Moran K (2000) Bioprospecting: lessons from benefit-sharing experiences. Int J Biotechnol 2:132–144

Moreno ACR, Filho AF, Gomes TAT, Ramos STS, Montemor LPG, Tavares VC, Filho LS, Irino K, Martinez MB (2010) Etiology of childhood diarrhea in the northeast of Brazil: significant emergent diarrheal pathogens. Diagn Microbiol Infect Dis 66:50–57

Mortimer L, Chadee K (2010) The immunopathogenesis of *Entamoeba histolytica*. Exp Parasitol 126:366–380

Murray PR, Rosenthal KS, Pfaller MA (2010) Enterobacteriaceae. In: Murray PR, Rosenthal KS, Pfaller MA (eds) Medical microbiology, 6th edn. Elsevier, New York

Navone GT, Gamboa MI, Oyhenart EE, Orden AB (2006) Intestinal parasitosis in Mbyá-Guaraní populations from Misiones Province, Argentina: epidemiological and nutritional aspects. Cad Saude Publica 22:1089–1100. doi:10.1590/S0102-311X2006000500022

Nelson EJ, Harris JB, Morris JG Jr, Calderwood SB, Camilli A (2009) Cholera transmission: the host, pathogen and bacteriophage dynamic. Nat Rev Microbiol 7:693–702. doi:10.1038/nrmicro2204

Nguyen TV, Le PV, Le CH, Weintraub A (2005) Antibiotic resistance in diarrheagenic *Escherichia coli* and *Shigella* strains isolated from children in Hanoi, Vietnam. Antimicrob Agents Chemother 49:816–819

Niyogi SK (2005) Shigellosis. J Microbiol 43:133–143

Obi CL, Bessong PO (2002) Diarrhoeagenic bacterial pathogens in HIV-positive patients with diarrhoea in rural communities of Limpopo Province, South Africa. J Health Popul Nutr 20:230–234

Orellana JD, Santos RV, Coimbra CE Jr, Leite MS (2009) Anthropometric evaluation of indigenous Brazilian children under 60 months of age using NCHS/1977 and WHO/2005 growth curves. J Pediatr 85:117–121. doi:10.2223/JPED.1872

Patwardhan B (2005) Ethnopharmacology and drug discovery. J Ethnopharmacol 100:50–52. doi:10.1016/j.jep. 2005.06.006

Pawlowski SW, Warren CA, Guerrant R (2009) Diagnosis and treatment of acute or persistent diarrhea. Gastroenterology 136:1874–1886. doi:10.1053/j.gastro.2009.02.072

Pereira AL, Silva TN, Gomes AC, Araújo AC, Giugliano LG (2010) Diarrhea-associated biofilm formed by enteroaggregative *Escherichia coli* and aggregative *Citrobacter freundii*: a consortium mediated by putative F pili. BMC Microbiol 1:1–18

Prado MS, Cairncross S, Strina A, Barreto ML, Oliveira-Assis AM, Rego S (2005) Asymptomatic giardiasis and growth in young children; a longitudinal study in Salvador, Brazil. Parasitology 131:51–56. doi:10.1017/S0031182005007353

Pupulin ART, Gomes ML, Dias MLGG, Araújo SM, Guilherme ALF, Kuhl JB (2004) Giardíase em creches do município de Maringá, PR. RBAC 36:147–149

Sack DA, Sack RB, Chaignat CL (2006) Getting serious about cholera. N Engl J Med 355:649–651

Saldiva SR, Carvalho HB, Castilho VP, Struchiner CJ, Massad E (2002) Malnutrition and susceptibility to enteroparasites: reinfection rates after mass chemotherapy. Paediatr Perinat Epidemiol 16:166–171. doi:10.1046/j.1365-3016.2002.00402.x

Samie A, Guerrant RL, Barrett L, Bessong PO, Igumbor EO, Obi CL (2009) Prevalence of intestinal parasitic and bacterial pathogens in diarrhoeal and non-diarrhoeal human stools from Vhembe district, South Africa. J Health Popul Nutr 27:739–745

Santosham M, Chandran A, Fitzwater S, Fischer-Walker C, Baqui AH, Black R (2010) Progress and barriers for the control of diarrhoeal disease. Lancet 376:63–67

Scolari C, Torti C, Beltrame A, Matteelli A, Castelli F, Gulletta M, Ribas M, Morana S, Urbani C (2000) Prevalence and distribution of soil-transmitted helminth (STH) infections in urban and indigenous schoolchildren in Ortigueira, State of Paraná, Brasil: implications for control. Trop Med Int Health 5:302–307

Sharghi N, Schantz PM, Caramico L, Ballas K, Teague BA, Hotez PJ (2001) Environmental exposure to *Toxocara* as a possible risk factor for asthma: a clinic-based case-control study. Clin Infect Dis 32:111–116

Shin JL, Gardiner GW, Deitel W, Kandel G (2004) Does whipworm increase the pathogenicity of *Campylobacter jejuni*? A clinical correlate of an experimental observation. Can J Gastroenterol 18:175–177

Sibandze GF, van Zyl RL, vanVuuren SF (2010) Theanti-diarrhoeal properties of *Breonadia salicina*. Syzygium cordatum and Ozoroa sphaerocarpa when used in combination in Swazi traditional medicine. J Ethnopharmacol. doi:10.1016/j.jep. 2010.08.050

Silva Quetz J, Lima IF, Havt A, de Carvalho EB, Lima NL, Soares AM, Mota RM, Guerrant RL, Lima ΛΛ (2010) *Campylobacter jejuni* and *Campylobacter coli* in children from communities in Northeastern Brazil: molecular detection and relation to nutritional status. Diagn Microbiol Infect Dis 67:220–227

Singh M, Singh N, Khare PB, Rawat AKS (2008) Antimicrobial activity of some important *Adiantum* species used traditionally in indigenous systems of medicine. J Ethnopharmacol 115:327–329. doi:10.1016/j.jep. 2007.09.018

Soejarto DD, Tarzian-Sorenson JA, Gyllenhaal C, Cordell GA, Farnsworth NR, Kinghorn AD, Pezzuto JM (2002) In: Stepp JR, Wundham FS, Zarger R (eds) Ethnobiology and biocultural diversity. University of Georgia Press, GA, Athens

Souza EC, Martinez MB, Taddei CR, Mukai L, Gilio AE, Racz ML, Silva L, Ejzenberg B, Okay Y (2002) Etiologic profile of acute diarrhea in children in São Paulo. J Pediatr 78:31–38

Stephen J (2001) Pathogenesis of infectious diarrhea. Can J Gastroenterol 15:669–683
Stephenson LS, Latham MC, Ottesen EA (2000) Malnutrition and parasite infection. Parasitology 121:S23–S38. doi:10.1017/S0031182000006491
Stoffle RW, Evans MJ, Olmsted JE (1990) Calculating the cultural significance of american indian plants: Paiute and Shoshone ethnobotany at Yucca Mountain, Nevada. Am Anthropol 92:416–432
Tanner S, Leonard WR, McDade TW, Reyes-Garcia V, Godoy R, Huanca T (2009) Influence of helminth infections on childhood nutritional status in lowland Bolivia. Am J Hum Biol 21:651–656. doi:10.1002/ajhb.20944
Teramoto K, Yamashita N, Kuwabara M, Hanawa T, Matsui T, Matsubara Y (2001) An amebic lung abscess: report of a case. Surg Today 31:820–822
Thakurta P, Bhowmik P, Mukherjee S, Hajra TK, Patra A, Bag PK (2007) Antibacterial, antisecretory and antihemorrhagic activity of *Azadirachta indica* used to treat cholera and diarrhea in India. J Ethnopharmacol 111:607–612. doi:10.1016/j.jep. 2007.01.022
Toledo MJ, Paludetto AW, Moura Fde T, Nascimento ES, Chaves M, Araújo SM, Mota LT (2009) Evaluation of enteroparasite control activities in a Kaingáng community of Southern Brazil. Rev Saude Publica 43:981–990. doi:10.1590/S0034-89102009005000083
United Nations (2005) Millennium Ecosystem Assessment. Available at http://www.maweb.org
Uprety Y, Asselin H, Boon EK, Yadav S, Shrestha KK (2010) Indigenous use and bio-efficacy of medicinal plants in the Rasuwa District, Central Nepal. J Ethnobiol Ethnomed 6:3. doi:10.1186/1746-4269-6-3
van Vuuren SF, Viljoen AM (2008) In vitro evidence of phyto-synergy for plant part combinations of *Croton gratissimus* (Euphorbiaceae) used in African traditional healing. J Ethnopharmacol 119:700–704
Vendruscolo GS, Mentz LA (2006) Levantamento etnobotânico das plantas utilizadas como medicinais por moradores do bairro Ponta Grossa, Porto Alegre, Rio Grande do Sul, Brasil. Iheringia Sér Bot 61:83–103
Vendruscolo GS, Rates SMK, Mentz LA (2005) Dados químicos e farmacológicos sobre as plantas utilizadas como medicinais pela comunidade do bairro Ponta Grossa, Porto Alegre, Rio Grande do Sul. Rev Bras Farmacogn 15:361–372. doi:10.1590/S0102-695X2005000400018
Vera-Ku M, Méndez-González M, Moo-Pu R, Rosado-Vallado M, Simá-Polanco P, Cedillo-Rivera R, Peraza-Sánchez SR (2010) Medicinal potions used against infectious bowel diseases in Mayan traditional medicine. J Ethnopharmacol 132:303–308. doi:10.1016/j.jep. 2010.08.040
Weldegerima B (2009) Review on the importance of documenting ethnopharmacological information on medicinal plants. Afr J Pharm Pharmacol 3:400–403
World Health Organization (1997a) Amoebiasis. WHO Weekly Epidemiological Record 72:97–100
World Health Organization (1997b) The use of essential drugs: model list of essential drugs (9th list). World Health Organization, Geneva, Switzerland
Zonta ML, Oyhenart EE, Navone GT (2010) Nutritional status, body composition, and intestinal parasitism among the Mbyá-Guaraní communities of Misiones, Argentina. Am J Hum Biol 22:193–200. doi:10.1002/ajhb.20977

Chapter 2
Efficacies of Medicinal Plant Extracts Against Blood-Sucking Parasites

A. Abdul Rahuman

Abstract Mosquito-borne diseases are endemic in more than over 100 countries, causing mortality of nearly two million people every year, and at least one million children die of such diseases each year, leaving as many as 2,100 million people at risk around the world. Mosquitoes are associated with the transmission of malaria, dengue, Japanese encephalitis, filariasis and other viral diseases throughout the globe, apart from being a nuisance insect. Vector control, using agents of chemical origin, continues to be practiced in the control of vector-borne diseases. However, due to some drawbacks including lack of selectivity, environmental contamination, and emergence and spread of vector resistance, development of natural products of plant origin with insecticidal properties have been encouraged in recent years for control of a variety of pest insects and vectors. The work herein is based on activities to determine the efficacies of hexane, chloroform, ethyl acetate, acetone and methanol extracts of medicinal plants tested against blood-sucking parasites.

2.1 Introduction

Vector mosquitoes are capable of transmitting potential pathogens to human beings, and they are responsible for several infectious diseases like malaria, filariasis, Japanese encephalitis, yellow fever, dengue, and chikungunya. It is estimated that every year at least 500 million people in the world suffer from one or other tropical diseases. One to two million deaths are reported annually due to malaria worldwide. Malaria is the world's most important and dreadful tropical disease. It is prevalent in about 100 countries and around 2,400 million people are at risk (Kager 2002). In South East Asia alone, 100 million malaria cases occur every year and 70% of these are reported from India (WHO 2004). Lymphatic filariasis affects at least 120 million people in 73 countries in Africa, India, Southeast Asia, and the

---

A.A. Rahuman
Unit of Nanotechnology and Bioactive Natural Products, Post Graduate and Research Department of Zoology, C. Abdul Hakeem College, Melvisharam, 632 509 Vellore District, Tamil Nadu, India
e-mail: abdulrahuman6@hotmail.com

Pacific Islands. In India, various species of *Aedes*, *Anopheles*, and *Culex* mosquitoes are important insect vectors of human diseases (Pialoux et al. 2007). These diseases not only cause high levels of morbidity and mortality but also inflict great economic loss and social disruption on developing countries such as India, China, etc.

Anopheles stephensi transmits malaria in the plains of rural and urban areas of India. Malaria afflicts 36% of the world's population, i.e., 2,020 million in 107 countries and territories situated in the tropical and subtropical regions. In the South East Asian Region of WHO (World Health Organization), out of about 1.4 billion people living in 11 countries, 1.2 billion (85.7%) are exposed to the risk of malaria, most of whom live in India. Of the 2.5 million reported cases in South East Asia, India alone contributes about 70% of the total cases (Kondrachine 1992). In India, malaria is transmitted by nine anopheline vector species of which six are of primary importance. The primary vectors include *Anopheles culicifacies*, *A. stephensi*, *Anopheles fluatilis*, *Anopheles minimus*, *Anopheles gyrus*, and *Anopheles sundicus*. These species are responsible for transmission in specific ecotypes. Of the six primary vector species, *A. culicifacies* is squarely responsible for transmission of about 60–70% of the malaria in rural plains and peri-urban areas (Raghavendra and Subbarao 2002). The interactive outcome of these disease determinants leads to various combinations of transmission risk factors at local and focal levels. The two million reported cases in the 1980s increased during the 1990s both in terms of morbidity and mortality. In the last 5 years about 40 epidemics including 1,400 malaria deaths have been reported from nine states within our country (Yadav et al. 1999). Although annually India reports about two million cases and 1,000 deaths attributable to malaria, there is an increasing trend in the proportion of *Plasmodium falciparum* as the agent (Dash et al. 2008).

Culex quinquefasciatus, is a vector of lymphatic filariasis which is a widely distributed tropical disease, and there are nearly 1,100 million people living in areas endemic for lymphatic filariasis and exposed to the risk of infection; there are 102 million cases of filariasis, either having patent microfilaraeimeia or chronic filarial disease (Michael et al. 1996). *Wuchereria bancrofti* accounts for approximately 90% of all filariasis cases in the world, followed by *Brugia malayi* and *Brugia timori*. India contributes about 40% of the total global burden of filariasis and accounts for about 50% of the people at risk of infection. Recent estimates have shown that in India, 22 states were found to be endemic for filariasis, and nine states (Andhra Pradesh, Bihar, Gujarat, Kerala, Maharashtra, Orissa, Tamil Nadu, Utter Pradesh, and west Bengal) contributed to about 95% of the total burden of filariasis. A total of 289 districts in India were surveyed for filariasis up to 1995, out of which 257 were found to be endemic. In India a total of 553 million people are at risk of infection and there are approximately 21 million people with symptomatic filariasis and 27 million microfilaria carriers. *W. bancrofti* is the national burden, widely distributed in 17 states and six union territories (ICMR Bulletin 2002), *B. malayi* is restricted in distribution, with decreasing trend. An overview of the traditional endemic foci shows concentration of infection mainly around river basins, and eastern and western coastal parts of India (Sabesan et al. 2000).

The yellow fever mosquito, *Aedes aegypti* is responsible for dengue fever in India, where the number of dengue fever cases has increased significantly in recent years. Dengue infection is endemic in over 100 countries worldwide and causes nearly 100 million cases of dengue fever, 500,000 cases of dengue hemorrhagic fever, and 24,000 deaths each year (Gibbons and Vaughn 2002; Guha-Sapir and Schimmer 2005; Kumaria 2010). Dengue fever is a reemerging disease affecting people in more than 100 countries. Its incidence has increased fourfold since 1970 and nearly half the world's population is now at risk. In 1990, almost 30% of the world population, 1.5 billion people, lived in regions where the estimated risk of dengue transmission was greater than 50% (Derouich and Boutayeb 2006; Hales et al. 2002).

Natural products of plant origin with insecticidal properties have been tried in the recent past for control of a variety of insect pests and vectors. Plants are considered as a rich source of bioactive chemicals and they may be an alternative source of mosquito control agents. Natural products are generally preferred because of their less harmful nature to nontarget organisms and due to their innate biodegradability. In India, over 200 types of vegetable drugs were in use during the *Vedic* period (3700–2000 BC). *Charak Samhita* (600 BC) mentioned 1,270 medicinal plants, while *Sushruta Samhita* (450 BC) and Vagbhatta's *Astangahridaya* (342 BC) mention about 1,100 and 1,150 medicinal plants, respectively (Chadha and Gupta 1995). Many studies on plant extracts against mosquito larvae have been conducted around the world. This chapter reviews the status of medicinal plant extracts and examines the scope for improving their relative contribution to the economy of rural families.

2.2 Background Information of Medicinal Plants

An estimated 14–28% of the 422,000 plants occurring on Earth have been used by human cultures for medicinal purposes at one time or another (Farnsworth and Soejarto 1991). Approximately 80% of the people in developing countries rely even today mainly on traditional medicines for humans (Food and Agricultural Organization (FAO) 1996) as well as domestic animals, a major portion of which are extracts of medicinal plants or their active principles. More than 6,500 species of such medicinal plants have been identified in Asia, 1,900 species in tropical America, and 1,300 species in north-west Amazon (Farnsworth and Soejarto 1991). Global trade in plant-based drugs was estimated at US$ 100 billion, of which traditional medicines using medicinal plants accounted for 60 billion (WHO 2004). In addition, trade in herbal teas, drug adjuncts, dietary foods etc., was estimated at US$ 5 billion in 1997. India has approximately 150,000 practitioners of traditional systems of medicine, and 10,000 licensed pharmacies manufacturing plant-based drugs (WBSICP 1997). The trade in medicinal herbs in India was estimated at US$ 1 billion (EXIM Bank 2003) and the country exports medicinal herbs worth US$ 287 million annually (DGCIS 2004).

2.2.1 Active Principles in Plants

The active principles in medicinal plants are chemical compounds known as secondary plant products. Some secondary products discourage herbivores, others inhibit bacterial or fungal pathogens. Two major categories of these compounds are alkaloids and glycosides.

2.2.1.1 Alkaloids

More than 3,000 alkaloids have been identified in 4,000 plant species; most occur in herbaceous dicots and also in fungi. Alkaloids contain nitrogen, they are usually alkaline (basic), and they have a bitter taste. Their most pronounced actions are on the nervous system, where they can produce physiological and/or psychological results. The difference between a medicinal and a toxic effect of many alkaloids (or any drug) is often a matter of dosage (Levetin and McMahon 2003).

2.2.1.2 Glycosides

Glycosides are so named because a sugar molecule (glyco-) is attached to the active component. Glycosides are generally categorized by the nature of the nonsugar or active component.

2.2.1.3 Cyanogenic Glycosides

The seeds, pits, and bark of many members of the rose family contain amygdalin, the most abundant cyanogenic glycoside. The pits of apricots are a particularly rich source of amygdalin and are ground up in the preparation of laetrile, a controversial cancer treatment; supposedly, hydrogen cyanide is released only in the presence of tumor cells and thus selectively destroys the cells (Levetin and McMahon 2003).

2.2.1.4 Cardioactive Glycosides and Saponins

Both contain a steroid molecule as the active components. Cardioactive glycosides have an effect on the contraction of heart muscle and, in proper doses, some can be used to treat various forms of heart failure. On the other hand, some of the deadliest plants, such as milkweed and oleander, contain toxic levels of cardioactive glycosides. One useful saponin is diosgenin from yams, which can be used as a precursor for the synthesis of various hormones such as progesterone and cortisone (Levetin and McMahon 2003).

2.3 Relative Importance of Selected Medicinal Plants

The use of plant medicines plays an important role in daily healthcare. Local medicines are even preferred to modern medicines. They are of course less expensive, but they are often regarded as being more effective. India possesses a rich and diverse variety of plant resources to meet the growing demand. The following plants have been used in much disease and it is hoped that their study will facilitate selection for further investigation of plants with relatively high levels of potency and a wide range of biological activities. The plants were selected based upon their medicinal and biological activities, which have been reported in the literature. In regard to our literature survey the biological and parasitic activities of the plant species are given herein. *Abutilon indicum* Linn. (Malvaceae), *Acacia arabica* (Lamk.) Willd (Leguminoseae), *Acalypha indica* Linn. (Euphorbiaceae), *Achyranthes aspera* Linn. (Amarantaceae), *Aegle marmelos* Linn. (Rutaceae), *Andrographis paniculata* (Acanthaceae), *Calotropis procera* Linn. (Asclepiadaceae), *Canna indica* L. (Cannaceae), *Cassia auriculata* Linn. (Cesalpinaceae), *Centella asiatica* Linn. (Gentianaceae), *Citrullus colocynthis* Linn. (cucurbitaceae), *Cocculus hirsutus* L. (Menispermaceae), *Ficus racemosa* L. (Moraceae), *Jatropha curcas* L. (Euphorbiaceae), *Jatropha gossypifolia* Linn. (Euphorbiaceae), *Leucas aspera* Willd. (Labiatae), *Mangifera indica* Linn. (Anacardiaceae), *Nicotiana tabacum* Linn. (Solanaceae), *Phyllanthus amarus* L. (Euphorbiaceae), *Ricinus communis* L. (Euphorbiceae), *Rhinacanthus nasutus* KURZ (Acanthaceae), *Solanum torvum* Swartz (Solanaceae), *Tagetes erecta* L. (Compositae), *Vitex negundo* Linn. (Verbenaceae), and *Zingiber officinale* Roscoe (Zingiberaceae) are discussed.

2.3.1 Abutilon indicum *Linn. (Malvaceae)*

Is known commonly as "Thuthi," and is distributed throughout the hotter parts of India (Chopra et al. 1992). It is reputed in the Siddha system of medicine as a remedy for jaundice, piles, ulcer, and leprosy (Yoganarasimhan 2000). The plant is also reported to possess analgesic activity (Ahmed et al. 2000). An approximately 80% ethanol root extract of *A. indicum* showed a toxic effect against *A. aegypti* fourth-instar larvae and guppy fish (Promsiri et al. 2006). The aqueous extract of *A. indicum* was tested for hepatoprotective activity against carbon tetrachloride- and paracetamol-induced hepatotoxicities in rats (Porchezhian and Ansari 2005).

2.3.1.1 Biological Activities

Seven flavonoid compounds were isolated and identified from the flowers of *A. indicum* (Matławska and Sikorska 2002); clomiphene citrate, centchroman, and embelin were isolated from the methanolic extracts of *A. indicum* and studied on uterotropic and uterine peroxidase activities in ovariectomized rats (Johri et al. 1991); the gossypetin 8 and 7 glucosides, cyanidin-3-rutinoside, β-pipene, cincole, farnesol, borneol from oil, eudesmol, geramiol, caryophyllene from flower extract, gallic acid, allantolactone and isoalantolactone were isolated from *A. indicum* (Rastogi and Mehrota 1993, 1995). β-sitosterol has been recognized as the active ingredient of many medicinal plant extracts. β-sitosterol has been isolated from leaf petroleum ether extract of *A. indicum* as a potential new mosquito larvicidal compound against *A. aegypti*, *A. stephensi*, and *C. quinquefasciatus* (Rahuman et al. 2008a).

2.3.2 Acacia arabica *(Lamk.) Willd (Leguminoseae)*

Is a leguminous tree found naturally in the Deccan and other parts of India and tropical Africa. It is an economically valued timber tree and is used for agricultural implements and fuel wood, while the leaves are used for fodder (Anonymous 1985). The gum of *A. arabica* is described as a source of useful medicaments, and it is believed to be of value for treating gingivitis and for reducing plaque (Gazi 1991).

2.3.2.1 Biological Activities

The flowers of *A. raddiana* have been used to attract and catch *Anopheles sergentii* (Müller and Schlein 2006); *A. arabica* was reported as an insecticidal plant which can be used in insecticide preparations (Singh and Saratchandra 2005); the bioefficacies were evaluated for their oviposition inhibition, residual toxicity, and direct toxicity effects on pulse beetle, *Callosobruchus maculatus* (Rahman and Talukder 2006); the ash was used to control the grain weevil, *Sitophilus granarious* (Rahman et al. 2003). Water, hot water, acetone, chloroform, and methanol leaf extracts of *A. arabica* were tested against early fourth-instar larvae of *C. quinquefasciatus* (Rahuman et al. 2009a).

2.3.3 Acalypha indica *Linn. (Euphorbiaceae)*

Is a common annual shrub in Indian gardens, house backyards, and waste areas throughout the plains of India. The leaves possess laxative properties (a substitute for senega) and are used in the form of a powder or decoction. *Acalypha* cures diseases of the teeth and gums, burns, toxins of plant and mixed origin, stomach pain, diseases due to pitha, bleeding piles, irritations, stabbing pain, wheezing, sinusitis, and neutralizes predominance of the Kabha factor (Chopra et al. 1956).

2.3.3.1 Biological Activities

The ethanolic extracts of *A. indica* were evaluated for their wound healing activity in rats (Suresh Reddy et al. 2002). The ethyl acetate, hexane, and methanol extracts of *A. indica* leaves, stem, and roots were investigated against three strains of human pathogenic bacteria, *Bacillus subtilis*, *Staphylococcus aureus*, and *Klebsiella pneumoniae* (Gangadevi et al. 2008). The insecticidal activity of the extract of *A. indica* against *Plutella xylostellai* was evaluated (Grainge et al. 1984); prolonged larval and pupal periods of *C. quinquefasciatus* have been reported by Daniel et al. (1995) while using plant extracts. The acetone, chloroform, ethyl acetate, hexane, and methanol leaf extracts of *A. indica*, were studied against the early fourth-instar larvae of *A. aegypti* and *C. quinquefasciatus* (Bagavan et al. 2008).

2.3.4 Achyranthes aspera *Linn. (Amarantaceae)*

Is widespread in the world as a weed, in Baluchistan, Ceylon, Tropical Asia, Africa, Australia, and America. In the northern part of India it is known as a medicinal plant in different systems of folk medicine.

2.3.4.1 Biological Activities

The saponin isolated from the leaf of *A. aspera* was assessed for cancer chemopreventive activity (Chakraborty et al. 2002); the saponin isolated from the EtOH stem bark of *Maesa lanceolata* exhibited powerful biocidal activity against aquatic adult insects (*Aeschnidae*, *Coenagrionidae*, *Hydrobidae*), mosquitoes (*Anopheles gambiae*, *Anopheles funestus*, *Culex* sp.), snails (*Biomphalaria pfeiffeiri* and *Lymnae natalensis*), *furcocercariae* of *Schistosoma mansoni* and fish (*Haplochromis* sp., *Oreochromis nilotica*, and *Oreochromis macrochi*) (Bagalwa and Chifundera 2007); the soyasaponins and dehydrosoyasaponin I. isolated from *Pisum sativum* showed antifeedant activity against *Sitophilus oryzae* (Taylor et al. 2004a, b). The saponin isolated from *A. aspera* has been reported as a potential mosquito larvicidal compound against *A. aegypti* and *C. quinquefasciatus* (Bagavan et al. 2008).

2.3.5 Aegle marmelos *Linn. (Rutaceae)*

Is commonly known as Bael, an indigenous plant to India and also found in Myanmar, Pakistan, and Bangladesh. It is one of the most useful medicinal plants in India. Its stem, bark, root, leaves, and fruits have medicinal value, and it has a long tradition as a form of herbal medicine. The leaves are widely used to treat diarrhea, dysentery, heart palpitations, and eye diseases (Kirtikar and Basu 1993).

2.3.5.1 Biological Activities

Leaves, fruits, stem, and roots of *A. marmelos* have been used in ethno medicine to exploit its medicinal properties including astringent, antidiarrheal, antidysenteric, demulcent, antipyretic and anti-inflammatory activities. Compounds purified from bael have been proved to be biologically active against several major diseases including cancer, diabetes, and cardiovascular diseases. Preclinical studies indicate the therapeutic potential of crude extracts of *A. marmelos* in the treatment of many microbial diseases, diabetes, and gastric ulcer (Maity et al. 2009).

The methanol extract of *A. marmelos* was assayed for its toxicity against the early fourth-instar larvae of *C. quinquefasciatus* (Rahuman et al. 2008a); evaluated for larvicidal activity and smoke-repellent potential at different concentrations against first- to fourth-instar larvae and pupae of *A. aegypti* (Vineetha and Murugan 2009); the efficacies of leaf hexane, chloroform, ethyl acetate, acetone, and methanol extracts of *A. marmelos* were tested against the adult cattle tick *Haemaphysalis bispinosa*, the larvae of *Rhipicephalus* (*Boophilus*) *microplus*, and sheep fluke *Paramphistomum cervi* (Elango and Rahuman 2010).

2.3.6 Andrographis paniculata *Nees (Acanthaceae)*

Is a medicinal plant widely cultivated in tropical regions of Asia. Traditionally it is used for several applications including as an antidote for snakebite in folk medicine and poisonous stings of some insects and to treat dyspepsia, influenza, dysentry, malaria, cold, fever, laryngitis, and respiratory infection in many Asian countries. The extract of the plant is reported to possess immunological, antibacterial, anti-inflammatory, antithrombotic, hepatoprotective, antihypertensive, and antidiabetic activities (Mishra et al. 2007).

2.3.6.1 Biological Activities

The extract of the plant is a rich source of flavonoids and labdane diterpenoids (Rao et al. 2004; Geethangili et al. 2008). Of the diterpenoids, andrographolide (AP1) and 14-deoxy-11,12-didehydroandrographolide (AP2) were isolated from the 95% alcoholic extract obtained from the aerial parts of *A. paniculata*. Herbal extracts of *A. paniculata* have been known as hepatoprotective and fever-reducing drugs since ancient times and they have been used regularly by the people in the south Asian subcontinent. Methanolic extracts of these plants were tested in vitro on choloroquine sensitive (MRC-pf-20) and resistant (MRC-pf-303) strains of *P. falciparum* for their antimalarial activity (Mishra et al. 2009).

Hyperglycemia and hypertension contribute to diabetic nephropathy; effects of AP1 and AP2 for ameliorating both were reported (Yu et al. 2003; Hsu et al. 2004;

Reyes et al. 2006). Although the constituents of *A. paniculata* are reported to have antidiabetic potency, the precise active compounds responsible for diabetic nephropathic activity of this plant have not been clearly identified (Lee et al. 2010). The leaf acetone, chloroform, ethyl acetate, hexane, and methanol extracts of *A. paniculata* were evaluated for larvicidal activities against *A. subpictus* and *C. tritaeniorhynchus* (Elango et al. 2009a).

2.3.7 Calotropis procera *Linn. (Asclepiadaceae)*

Is a spreading shrub or small tree with a height up to 4 m, exuding copious milky sap when cut or broken; the leaves are gray-green, large up to 15 cm long and 10 cm broad, with a pointed tip, two rounded basal lobes and no leaf stalk; the flowers are waxy white with five petals, purple-tipped inside and with a central purplish crown, carried in stalked clusters at the ends of the branches; fruits are gray-green, inflated, 8–12-cm long, containing numerous seeds with tufts of long silky hairs at one end (Kleinschmidt and Johnson 1977). The native range covers South West Asia (India, Pakistan, Afghanistan, Iran, and Saudi Arabia and Jordan) and Africa (Somalia, Egypt, Libya, south Algeria, Morocco, Mauritania, and Senegal).

2.3.7.1 Biological Activities

Extracts of *C. procera* have insecticidal activity against different insects such as *Sarcophaga haemorrhoidalis* (Moursy 1997), while the latex was used against the third stage larvae of *Musca domestica* (Morsy et al. 2001). The crude latex produced by the green parts of the plant was evaluated for its toxic effects upon egg hatching and larval development (Ramos et al. 2006). They found the whole latex was shown to cause 100% mortality of III instars within 5 min and different aqueous concentrations of this plant affected the gravid female *A. aegypti* mosquitoes and this behavior continued over three gonotrophic cycles (Singhi et al. 2004). The fresh latex extract of *C. procera* was tested against *C. quinquefasciatus* and *A. stephensi* (Shahi et al. 2010).

The root barks methanolic extract of *C. procera* enabled the identification of a novel cardenolide (2″-oxovoruscharin) and it was tested against 57 human cancer cell lines in vitro and in vitro inhibitory influence on the Na^+/K^+-ATPase activity, and in vivo tolerance test (Van Quaquebeke et al. 2005).

2.3.8 Canna indica *L. (Cannaceae)*

Indian shot or Keli is a native of tropical America and is a very popular ornamental and medicinal plant throughout the tropical world. *C. indica* is an upright perennial rhizomatous herb. It has round, shiny black seeds. The plant is used in the treatment of women's complaints. A decoction of the root with fermented rice is used in the treatment of gonorrhea and amenorrhea. The plant is also considered to be demulcent, diaphoretic, and diuretic (Joshi and Pant 2010).

2.3.8.1 Biological Activities

The efficacy of dried, coarsely powdered leaf, flower, rhizome, and seed benzene and methanol extracts of *C. indica* showed significant central, peripheral analgesic activity in mice and anthelmintic activity against *Pheritima posthuma* (Nirmal et al. 2007). The leaf acetone, chloroform, hot water, methanol, petroleum ether (60–80°C), and water extracts of *C. indica* were investigated for larvicidal potential against second- and fourth-instar larvae of the laboratory-reared mosquito species, *C. quinquefasciatus* (Rahuman et al. 2009b).

Four anthocyanin pigments have been isolated from the flowers of *C. indica*; they are (1) cyanidin-3-O-(600-O-α-rhamnopyranosyl)-β-glucopyranoside, (2) cyanidin-3-O-(600-O-α-rhamnopyranosyl)-β-galactopyranoside, (3) cyanidin-3-O-β-glucopyranoside, and (4) cyanidin-O-β-galactopyranoside; the isolated compounds showed good antioxidant activity thus making them suitable for use in food coloration and as a nutraceutical (Srivastava and Vankar 2010).

2.3.9 Cassia auriculata *Linn. (Cesalpinaceae)*

Is a Tanner's *Cassia* commonly found in Asia, and has been widely used in traditional medicine as a cure for rheumatism, conjunctivitis, and diabetes (Joshi 2000). In addition, *C. auriculata* is used by the tribal peoples of the Chittor district of Andhra Pradesh, India for the treatment of skin diseases, asthma, conjunctivitis, and in renal disorders. The dried flowers and leaves of the plants are being used for medical treatment and have been widely used in Ayurvedic medicine as "Avarai Panchaga Choornam" and the main constituent of Kalpa herbal tea has come under extensive study in the light of its antidiabetic effects (Vedavathi et al. 1997). Recently reported was an antiperoxidative effect of *C. auriculata* flowers in streptozotocin diabetic rats (Latha and Pari 2003).

2.3.9.1 Biological Activities

The flower methanol extracts showed larvicidal activity against *A. subpictus* and *C. tritaeniorhynchus* and the hexane and methanol extracts exhibited antimicrobial activity (Kamaraj et al. 2009; Duraipandiyan et al. 2006). The acetone, chloroform, ethyl acetate, hexane, and methanol extracts of dried leaf and flower of *C. auriculata* were tested against larvae of the cattle tick *R.* (*Boophilus*) *microplus*, adult of *H. bispinosa*, hematophagous fly *Hippobosca maculata*, nymph of goat-lice *Damalinia caprae*, and adult sheep parasite *P. cervi* (Kamaraj et al. 2010). Leaf extracts of *C. auriculata* were evaluated for anticancer effects in vitro through cell cycle arrest and induction of apoptosis in human breast and larynx cancer cell lines (Prasanna et al. 2008); the flower hydromethanolic extract of *C. auriculata* was evaluated for antidiabetic activity in alloxan-induced diabetes in rats (Surana et al. 2008); and the ethanol and methanol extracts of flowers were screened for antioxidant activity (Kumaran and Karunakaran 2006). The alcoholic extract of the aerial part of *C. auriculata* displayed potent antioxidant activity and the major antioxidant constituent kaempferol-3-O-rutinoside together with kaempferol, quercetin, and luteolin were isolated from the ethyl acetate fraction (Badaturuge et al. 2011).

2.3.10 Centella asiatica *Linn. (Gentianaceae)*

Is commonly known as "Mandukaparni." In Sri Lanka and Indonesia, it is given the name "Thankuni Sak." In classical Indian Ayurveda literature, it is considered to be one of the "Rasayana" (rejuvenator) drugs.

2.3.10.1 Biological Activities

In common with most traditional phytotherapeutic agents, *C. asiatica* is claimed to possess a wide range of pharmacological effects, being used for human wound healing, mental disorders, atherosclerosis, fungicidal, antibacterial, antioxidant and anticancer purposes (Jayashree et al. 2003). *C. asiatica* has also been reported to be useful in the treatment of inflammation, diarrhea, asthma, tuberculosis, and various skin lesions and ailments like leprosy, lupus, psoriasis, and keloid. In addition, numerous clinical reports verify the ulcer-preventive and antidepressive sedative effects of *C. asiatica* preparations, as well as their ability to improve venous insufficiency and microangiopathy (Zheng and Qin 2007). The *n*-hexane, carbon tetrachloride, and chloroform soluble fractions of the methanol extract of *C. asiatica* were subjected to antioxidant, antimicrobial and brine shrimp lethality bioassays (Obayed Ullah et al. 2009).

The adulticidal and larvicidal effects of leaf hexane, chloroform, ethyl acetate, acetone, and methanol extracts of *C. asiatica* were investigated against the adult cattle tick *H. bispinosa*, sheep fluke *P. cervi*, fourth-instar larvae of the malaria vector *A. subpictus* and Japanese encephalitis vector *C. tritaeniorhynchus* (Bagavan et al. 2009). The ethanolic extract of *C. asiatica* leaves were evaluated for larvicidal and adult emergence inhibition activity against *C. quinquefasciatus* (Rajkumar and Jebanesan 2005). The ethyl acetate extract of *C. asiatica* showed anthelmintic properties and antifilarial effects (Chakraborty et al. 1996). The larvicidal activity of crude acetone, hexane, ethyl acetate, methanol, and petroleum ether extracts of the leaf of *C. asiatica* were assayed for their toxicity against the early fourth-instar larvae of *C. quinquefasciatus* (Rahuman et al. 2008b). Triterpenoid acids, volatile and fatty oils, alkaloids, glycosides, flavonoids, and steroids have been isolated from different parts of the plant (Jayashree et al. 2003).

2.3.11 Citrullus colocynthis *Linn. (Cucurbitaceae)*

Is an annual herb found in the wild as well as cultivated throughout India in warm areas. It is locally known as Makkal in Hindi, bitter apple in English, and Paitummatti in Tamil. In traditional medicine, this plant has been utilized to treat constipation, diabetes, edema, fever, jaundice, bacterial infections as well as cancer. The fruit of this plant is traditionally used as an antidiabetic in the Mediterranean part of the World.

2.3.11.1 Biological Activities

The colocynithin and hydrated colocynithin isolated from the alcoholic extract showed toxicity against cockroaches, adult honey bee, housefly, cotton leaf worm, bed bug, and mosquito (el-Naggar et al. 1989). The petroleum ether and ethyl acetate seed extracts showed antioviposition, F1 adult emergence, and ovicidal and repellent activity against the pulse beetle *Callosobruchus maculatus* (Seenivasan et al. 2004); the crude extracts (70% ethanol) were tested for their mortality, repellency, and the number of eggs laid against the carmine spider mite *Tetranychus cinnabarinus* (Mansour et al. 2004). Colocynth in the form of the solid extract enters into many of the purgative pills of modern pharmacy. It is useful for biliousness, fever, intestinal parasites, constipation, hepatic and abdominal, visceral and cerebral congestions, dropsy, etc. The juice of the fruit mixed with sugar is a house-hold remedy in dropsy (Anonymous 1970). The ethanolic extracts of *C. colocynthis* fruits, leaves, stems, and roots were found to be active against Gram-positive *bacilli*, viz., *Bacillus pumilus* and *S. aureus*, while fruit and root extracts in double strength gave positive results against a Gram-positive bacillus (*B. subtilis*). The leaf petroleum ether extract of *C. colocynthis* showed 100% mortality against *A. aegypti* and *C. quinquefasciatus* (Rahuman and Venkatesan 2008). The fatty acids, oleic acid and linoleic acid were isolated and identified in *C. colocynthis* petroleum ether extract. As mosquito larvicidal compounds oleic and linoleic acids were quite potent against fourth-instar larvae of *A. aegypti*, *A. stephensi*, and *C. quinquefasciatus* (Rahuman et al. 2008b).

The main chemical constituents of *C. colocynthis* reported in the literature are docosan-1-ol acetate, 0, 13-dimethyl-pentadec-13-en-1-al, 11, 14-dimethyl hexadecane, 14-ol-2-one, 10, 14-dimethyl hexadecane 14, ol, 2-one, linoleic acid, oleic acid, carbohydrate, amino acid, organic acid, lipid, sterols, and phenols (Ayoub and Yankov 1981; Basalah et al. 1985; Habs et al. 1984 ; Navot and Zamir 1986).

2.3.12 Cocculus hirsutus *L. (Menispermaceae)*

Is commonly known as Jal-jammi. It is a climber found in tropical and subtropical regions of India. A decoction of the leaves is taken for eczema, dysentery and urinary problems. Leaves and stem are used for treating eye diseases.

2.3.12.1 Biological Activities

Roots and leaves of *C. hirsutus* are given for Sarsaparilla, as a diuretic and in gout (Nadkarni 1982). Aerial parts of the plant are reported to be used as a diuretic and laxative (Ganapathy and Dash 2002) and the root extract showed analgesic and anti-inflammatory effects (Nayak and Singhai 1993). The leaf juice was used in the treatment of eczema (Maasilamani and Shokat 1981). The leaf hexane, chloroform, ethyl acetate, acetone and methanol extracts of *C. hirsutus* were tested against the adult cattle tick *H. bispinosa*, the larvae of *R. (Boophilus) microplus*, sheep fluke *P. cervi*, and for oviposition-deterrent, ovicidal, and repellent activities against fourth-instar larvae and adults of malaria vector *A. subpictus* (Elango et al. 2009b; Elango and Rahuman 2010). The ethanolic extract of the whole plant showed the presence of isoquinoline alkaloids D-trilobine and DL-coclaurine (Jagannadha Rao and Ramachadra Raw 1961), cohirsinine (Viquaruddin and Tahir 1991), jamtinine (Viquaruddin and Iqbal 1992), and cohirsutine (Viquaruddin and Iqbal 1993).

2.3.13 Ficus racemosa *L. (Moraceae)*

Indian fig is an evergreen, lactiferous, deciduous tree with moderate to large spreading, without a prominence of aerial roots, found throughout the greater part of India in moist localities and is often cultivated in villages for its edible fruit (Anonymous 1952). Different parts of *F. racemosa* are traditionally used as fodder, as edible and for ceremonial uses (Manandhar 1972). All parts of this plant (leaves, fruits, bark, latex, and sap of the root) are medicinally important in the traditional system of medicine in India. The leaves powdered and mixed with honey are given in bilious infections (Kirtikar and Basu 1975). The fruits are a good remedy for visceral obstruction and also useful in regulating diarrhea and constipation (Vihari 1995). The astringent nature of the bark has been employed as a mouth wash in spongy gum and also internally in dysentery, menorrhagia, and hemoptysis (Chopra et al 1958). The bark is antiseptic, antipyretic and vermicidal, and the decoction of bark is used in the treatment of various skin diseases, ulcers, and diabetes. It is also used as a poultice in inflammatory swellings/boils and regarded to be effective in the treatment of piles, dysentry, asthma, gonorrhea, gleet, menorrhagia, leucorrhea, hemoptysis, and urinary diseases (Nadkarni et al. 1976). Apart from its usage in traditional medicine, scientific studies indicate *F. racemosa* to possess various biological effects such as hepatoprotective, chemopreventive, antidiabetic, anti-inflammatory, antipyretic, antitussive, and antidiuretic effects (Mandal et al. 1999; Khan and Sultana 2005; Rao et al. 2002; Mandal et al. 2000; Rao et al. 2003; Ratnasooriya et al. 2003).

2.3.13.1 Biological Activities

The tetracyclic triterpene derivative gluanol acetate was isolated from bark acetone extract of *F. racemosa* as a new mosquito larvicidal compound, and it was shown to be a quite potent compound against fourth-instar larvae of *A. aegypti*, *A. stephensi*, and *C. quinquefasciatus* (Rahuman et al. 2008c). Aqueous extracts of the bark were evaluated for anthelmintic activity using adult earthworms, which exhibited a spontaneous motility (paralysis) (Chandrashekhar et al. 2008). Gluanol acetate, β-sitosterol, leucocyanidin-3-O-β-D-glucopyranoside, leucopelargonidin-3-O-β-D-glucopyranoside, leucopelargonidin-3-O-a-L-rhamnopyranoside, lupeol, ceryl behenate, lupeol acetate, and α-amyrin acetate were isolated from the stem–bark of *F. racemosa* (Husain et al. 1992). The compounds gluanone and gluanol were isolated from *Hoppea dichotoma* (Ghosal et al. 1978). The methanol stem–bark extract showed activity against *B. subtilis* (Mahato and Chaudhary 2005); it also showed significant anthelmintic activities (Hansson et al. 1986). The alcoholic and aqueous fruit extracts caused death of microfilariae (Mishra et al. 2005) and racemic acid showed potent anti-inflammatory, cytotoxic, and antioxidant activity; the bark has also been evaluated for cytotoxic effects using 1BR3, Hep G2, HL-60 cell lines and found to be safe and less toxic than aspirin, a commonly consumed anti-inflammatory drug (Li et al. 2004).

2.3.14 Jatropha curcas L. *(Euphorbiaceae)*

Is a shrub/small tree that grows up to 15-ft high. It is known as "Jungli erand" in Hindi and "Katalamanakku" by the local people. It is used in traditional medicine for fevers, venereal diseases, dysentery (Iwu 1993); the seeds of *J. curcas* are a valuable source of biodiesel in Asian countries (Heller 1996).

2.3.14.1 Biological Activities

This plant exhibits bioactive activities for fever, mouth infections, jaundice, guinea worm sores, and joint rheumatism (Irvine 1961; Oliver-Bever 1986). Fagbenro-Beyioku et al. (1998) investigated and reported the antiparasitic activity of the sap and crushed leaves of *J. curcas*. The water extract of the branches also strongly inhibited HIV-induced cytopathic effects with low cytotoxicity (Matsuse et al.

1999). Many *Jatropha* species possess antimicrobial activity (Aiyela-agbe et al. 2000). The extracts showed nematicidal, fungicidal (Sharma and Trivedi 2002), antifeedant (Meshram et al. 1996), molluscicidal (Liu et al. 1997), and abortifacient activities (Goonasekara et al. 1995) against white flies (*Bemisia tabaci*), fourth-instar mosquito (*Ochlerototatus triseriatus*) larvae, and neonates of *Helicoverpa zea* and *Helicoverpa virescens* (Georges et al. 2008), and exhibited insecticidal activities against moths, butterflies, aphids, bugs, beetles, flies, and cockroaches (Wink et al. 1997). The in vitro antimicrobial activity of crude ethanolic, methanolic and water extracts of the stem bark of *J. curcas* were investigated and phytochemical screening revealed the presence of saponin, steroids, tannin, glycosides, alkaloids, and flavonoids in the extracts (Igbinosa et al. 2009). The ethyl acetate, butanol, and petroleum ether extracts of *J. curcas* were tested against early fourth-instar larvae of *A. aegypti* and *C. quinquefasciatus* (Rahuman et al. 2008d); leaf extracts were tested against molluscicide activities of *S. mansoni* and *S. haematobium* (Rug and Ruppel 2000).

2.3.15 Jatropha gossypifolia *Linn. (Euphorbiaceae)*

Is a bushy gregarious shrub, which grows wildly almost throughout India. It possesses significant anticancer and pesticidal activity (Hartwell 1969; Chatterjee et al. 1980). The roots are employed against leprosy, as an antidote for snakebite and in urinary complaints. A decoction of the bark is used as an emmenagogue and the leaves for stomach ache, venereal disease, and as a blood purifier (Kirtikar and Basu 1996; Banerji and Das 1993). *J. gossypifolia* leaves contain histamine, apigenin, vitexin, isovitexin, and tannins. The bark contains the alkaloid "jatrophine" and a lignan "jatrodien" is found in its stems (Matsuse et al. 1999; Omoregbe et al. 1996) The latex of *J. gossypifolia* yielded two cyclic octapeptides, i.e., cyclogossine A and B (Das et al. 1996; Horsten et al. 1996).

2.3.15.1 Biological Activities

This species contains many secondary metabolites like apigenin (Subramanian et al. 1971), cyclogossine A (Horsten et al. 1996), jatrophatrione, and jatropholone (Rahman et al. 1990). The extracts from plant species have biological properties for example antiallergic (Adolf et al. 1984), molluscicidal (Adewunmi et al. 1987), and insect-repellent activity (Areekul et al. 1987). The insecticidal activity of the leaf extract of *J. gossypifolia* was tested against second-instar larvae of *Spodoptera exigua* (Khumrungsee et al. 2009). The larvicidal activity of crude hexane, ethyl acetate, petroleum ether, acetone and methanol extracts of *J. gossypifolia* were assayed for their toxicity against the early fourth-instar larvae of *C. quinquefasciatus* (Rahuman et al. 2008d). The petroleum ether and methanol extracts of *J. gossypifolia* were tested for larvicidal activity against *A. aegypti*, *C. quinquefasciatus*, *A. dirus*, and *Mansonia uniformis* (Komalamisra et al. 2005). Three known flavonoids, vitexin, isovitexin, and apigenin are isolated from leaves of *J. gossypifolia* (Subramanian et al. 1971). The bark contains the alkaloid, jatrophine and a lignin, jatrodien is found in its stems (Matsuse et al. 1999; Omoregbe et al. 1996).

2.3.16 Leucas aspera *Willd. (Labiatae)*

Is commonly called thumbai in India and is found in open, dry, sandy soil, is a weed and locally abundant and used as an indigenous system of medicine (Sadhu et al. 2003). The plant is an erect or diffusely branched, annual herb and leaves are linear, blunt-tipped and the margins are scalloped. *L. aspera* possesses anti-inflammatory activity and is used against cobra venom poisoning and also as an analgesic (Goudgaon et al. 2003).

2.3.16.1 Biological Activities

The acetone, chloroform, ethyl acetate, hexane, and methanol leaf extracts of *L. aspera* were studied against the early fourth-instar larvae of *A. aegypti* and *C. quinquefasciatus* (Bagavan et al. 2008). Murugan and Jeyabalan (1999) reported that *L. aspera* had a strong larvicidal, antiemergence, adult repellency and anti-reproductive activity against *A. stephensi*. Methanol extracts from leaves showed significant larvicidal and growth regulatory activities even at very low concentrations on *C. quinquefasciatus* (Muthukrishnan et al. 1997). Acetone, chloroform, ethyl acetate, hexane, methanol, and petroleum ether extracts of leaf and flower of *L. aspera* extracts were tested against fourth-instar larvae of malaria vector *A. subpictus* and Japanese encephalitis vector *C. tritaeniorhynchus* (Kamaraj et al. 2009). The leaf extract was tested for prostaglandin inhibitory and antioxidant activities (Sadhu et al. 2003); water extracts of *L. aspera* were investigated for in vitro anthelmintic activity against *Haemonchus* sp., *Trichostrongylus* sp., *Oesophagostomum* sp. and *Mecistocirrus* sp., *Bunostomum* sp., *Strongyloides* sp., *Trichuris* sp. and *Capillaria* sp (Amin et al. 2009).

2.3.17 Mangifera indica *Linn (Anacardiaceae)*

Vimang (Mango) grows in the tropical and subtropical regions and its parts are used commonly in folk medicine for a wide variety of conditions. The fruits are known to be an excellent source of many vitamins such as ascorbic acid, thiamine, riboflavin, and niacin, and β-carotene. Recently, much attention has been given to phytochemicals and the distinctive roles they play in anti-inflammatory and anticancer properties related to the consumption of fruits and vegetables.

2.3.17.1 Biological Activities

Several authors have reported pharmacological activities of extracts of *M. indica*, including antispasmodic, analgesic, anti-inflammatory, antipyretic, and antioxidant effects (Das et al. 1989a, b; Coe and Anderson 1996; Awe et al. 1988; Sánchez et al. 2000; Martínez et al. 2000; Garrido et al. 2001); the stem bark aqueous extract of *M. indica* was evaluated for antiallergic and anthelmintic activities of *Trichinella spiralis* (García et al. 2003). The water, hot water, acetone, chloroform, and methanol leaf extracts of *M. indica* were tested against filarial vector *C. quinquefasciatus* (Rahuman et al. 2009a). The aqueous leaf extract of *M. indica* was used to control *Aphis craccivora*, *Mylabris* sp., *Maruca vitrata*, and *Ootheca mutabilis* (Kossou et al. 2001). The major polyphenol component of vimang is mangiferin, a C-glucosylxanthone (1,3,6,7-tetrahydroxyxanthone-C2-β-D-glucoside) with antiviral, antitumor, antidiabetic, and antioxidant activity (Zheng and Lu 1990; Guha et al. 1996; Li et al. 1998; Sánchez et al. 2000; Yoosook et al. 2000). The stem bark components of *M. indica* also present antimicrobial and antiamoebic activities (Das et al. 1989a, b; Tona et al. 2000). Barks, leaves, and seeds of *M. indica* contain tanins (Kerharo and Adam 1974; Watt and Breyer-Brandwijk 1962; Bouquet and Debray 1974; Anonyme 1993) and this may also explain its effectivity against diarrhea (Pousset 1989). According to Pousset (1989), a decoction of 30 g of barks or leaves in 1 l of water is effective against diarrhea. Burkill (1985), Oliver-Bever (1986), Kambu et al. (1989), Das et al. (1989a, b) have reported the anti-inflammatory properties of the plant. The anti-inflammatory properties of *M. indica* may explain its use against toothache.

2.3.18 Nicotiana tabacum *Linn. (Solanaceae)*

Commonly known as tobacco, this is a world-renowned plant used for its narcotic properties. It was originally native to the Americas but is now popularly cultivated in the Indian subcontinent where it is locally known as tambaku. Tobacco has been studied extensively for its effects on biological systems both in animals and humans while a large number of alkaloids have also been identified in the plant. Nicotine is known as the main alkaloid of tobacco, isolated in 1828 from the tobacco leaf, accounting for over 90% of the total alkaloidal content (Bowman and Rand 1980). Apart from nicotine, the other known alkaloids are nicotyrine, nornicotine, anabasine, myosmine, anatabine, nicotelline, and isonicoteine (Kuhn 1965).

2.3.18.1 Biological Activities

Dried leaves, stalks, and the whole herb of tobacco are widely used traditionally in the subcontinent for their antispasmodic, emetic, purgative, sedative, analgesic, and insecticidal properties (Nadkarni 1976; Murray 1989). The leaf hot water, acetone, chloroform, and methanol extracts of *N. tabacum* were tested against the larvae of *C. quinquefasciatus* (Rahuman et al. 2009a). The crude aqueous and methanol extracts of *N. tabacum* were investigated in vitro and in vivo for anthelmintic activity against *Haemonchus contortus* (Iqbal et al. 2006). *N. tabacum* extracts were tested for pesticidal activity against *Tribolium castaneum*, and shown to be very active against *Boophilus microplus* (Williams and Mansingh 1993; Mansingh and Williams 1998).

2.3.19 Phyllanthus amarus L. *(Euphorbiaceae)*

Is an annual, glabrous herb that grows to between 30 and 60 cm; known as "Kilanelli" by local people, it is found in tropical areas, in subtropical regions, and is usually quite scattered in its distribution (Unander et al. 1990). Fresh leaves are ground and mixed with a cup of cow or goat's milk and taken internally to cure jaundice. The aerial parts of the herb have been widely used in folk medicine in India and other tropical countries for the treatment of various diseases and disorders, such as jaundice, diarrhea, ringworm, ulcers, malaria, genitourinary infections, hemorrhoids, and gonorrhea (Unander et al. 1991).

2.3.19.1 Biological Activities

The extracts were found to possess antiviral properties against the hepatitis B virus (Thyagarajan et al. 1988; Yeh et al. 1993) and antitumor, anticarcinogenic, and anti-inflammatory properties (Kiemer et al. 2003; Rajeshkumar et al. 2002). *P. amarus* has reportedly been used to treat jaundice, diabetes, otitis, diarrhea, swelling, skin ulcer, gastrointestinal disturbances, weakness of the male organ, and blocks DNA polymerase in the case of the hepatitis B virus during reproduction (Oluwafemi and Debiri 2008). Root and leaf of *P. amarus* were assessed for antibacterial activity against *Escherichia coli* (Akinjogunla et al. 2010). The ethyl acetate, butanol, and petroleum ether extracts of *P. amarus*, were tested against the early fourth-instar larvae of *A. aegypti* and *C. quinquefasciatus* (Rahuman et al. 2008d); the butanol fraction of the whole plant also exhibited antihepatotoxic activity against CCl4-induced liver damage in rat (Sane et al. 1995). The major lignans phyllanthin and hypophyllanthin have been reported to exhibit antihepatotoxic activity (Sharma et al. 1993).

2.3.20 Ricinus communis L. (Euphorbiceae)

Is a soft wooden small tree, widespread throughout the tropics and warm temperature regions of the world (Ivan 1998). In the Indian system of medicine, the leaf, root, and seed oil of this plant have been used for the treatment of inflammation and liver disorders (Kirtikar and Basu 1991), hypoglycemia (Dhar et al. 1968), and as a laxative (Capasso et al. 1994).

2.3.20.1 Biological Activities

R. communis, the castor bean, has been used to control insect pests in several crops. Aqueous castor-bean leaf extract has been shown to possess insecticidal activity against *Callosobruchus chinensis* (Coleoptera: Bruchidae) (Upasani et al. 2003), *Cosmopolites sordidus* (Coleoptera: Curculionidae) (Tinzaara et al. 2006), *Culex pipiens*, *Aedes caspius*, *Culiseta longiareolata*, and *Anopheles maculipennis* (Diptera: Culicidae) (Aouinty et al. 2006); whereas a methanolic leaf extract had insecticidal activity against *C. chinensis* (Upasani et al. 2003). In addition, both aqueous and acetone leaf extracts had different activity against *Acromyrmex lundi* (Himenoptera: Formicidae) (Caffarini et al. 2008). Castor oil showed insecticidal activity against *Zabrotes subfasciatus* (Coleoptera: Bruchidae) (Mushobozy et al. 2009). Zahir et al. (2009) reported that the acetone, chloroform, ethyl acetate, hexane, and methanol dried leaf and seed extracts of *R. communis* had been tested against the larvae of the cattle tick *R. (Boophilus) microplus*, sheep internal parasite *P. cervi*, fourth-instar larvae of *A. subpictus* Grassi and *C. tritaeniorhynchus*, and also exhibited acaricidal and insecticidal activities against the adult of *H. bispinosa* and *H. maculata* (Zahir et al. 2010).

2.3.21 Rhinacanthus nasutus KURZ (Acanthaceae)

Is commonly called Nagamalli in Tamil. It is a valuable plant that is widely distributed and cultivated in South China, Taiwan, India, and Thailand. *R. nasutus* is well known as a source of flavonoids, steroids, terpenoids, anthraquinones, lignans, and especially naphthoquinone analogs. Various parts of this plant have also been used for the treatment of diseases such as eczema, pulmonary tuberculosis, herpes, hepatitis, diabetes, hypertension, and several skin diseases (Siripong et al. 2006). *R. nasutus* is the best known member of the genus as it has been used as a traditional medicinal plant over thousands of years in the Ayurvedic system of medicine as is practiced on the Indian subcontinent. Thus, traditional uses of *R. nasutus* are well established.

2.3.21.1 Biological Activities

Antifeedant and larvicidal activity of the acetone, chloroform, ethyl acetate, hexane, and methanol leaf extracts of *R. nasutus* were studied against fourth-instar larvae of *Spodoptera litura*, *A. aegypti*, and *C. quinquefasciatus* (Kamaraj et al. 2008); the acetone, chloroform, ethyl acetate, hexane, and methanol dried leaf and flower extracts of *R. nasutus* were tested against malaria vector, *A. subpictus* and Japanese encephalitis vector, *C. tritaeniorhynchus* (Kamaraj et al. 2009), the larvae of cattle tick *R. (Boophilus) microplus*, adult of *H. bispinosa*, hematophagous fly *H. maculata*, nymph of goat-lice *Damalinia caprae*, and adult sheep parasite *P. cervi* (Kamaraj et al. 2010). The dried root powder methanol extract of *R. nasutus* was tested against the larvae of *A. aegypti* and *C. quinquefasciatus* (Rongsriyam et al. 2006); the petroleum ether extract showed larvicidal activity against *A. aegypti*, *C. quinquefasciatus*, *A. dirus*, and *M. uniformis* (Komalamisra et al. 2005); the methanol extracts from *Calophyllum inophyllum* and *R. nasutus* seeds and leaves showed significant larvicidal and growth-regulatory activities even at very low concentrations on the juveniles of *C. quinquefasciatus*, *A. stephensi*, and *A. aegypti* (Pushpalatha and Muthukrishnan 1999). The leaves and stems of this plant are often used for the treatment of hepatitis, diabetes, hypertension, and skin disease (Wu et al. 1998). The leaf petroleum ether fraction shows significant mosquitocidal activity (Shaalan et al 2005). It has been reported that rhinacanthin-C, rhinacanthin-D, and rhinacanthin-N isolated from *R. nasutus* possessed antifungal, antibacterial, antiviral, anti-inflammatory, antiallergic, and antihemorrhoidal activity, with additionally activity against various types of cancer (Puttarak et al. 2010; Gotoh et al 2004). *R. nasutus* root afforded a 1,4-naphthoquinone ester, rhinacanthin-Q, accompanied by 24 known compounds which were isolated and showed cytotoxicity and antiplatelet effects (Wu et al. 1998).

2.3.22 Solanum torvum *Swartz (Solanaceae)*

Is a prickly, tomentose, erect shrub, 1.5–3 m high, with leaves without prickles, white bell-shaped flowers, and lobed fruits seated on the calyx. It is a common plant found throughout the Indian subcontinent. It is locally known as tit begoon, gota begoon, or hat begoon in Bengali and in Tamil commonly known as Sundakai, turkey berry, susumber, gully-bean Thai eggplant, or devil's fig. Common people especially members of the tribes use the fruit of *S. torvum* as a vegetable in their daily diet (Ghani 1998a, b).

2.3.22.1 Biological Activities

S. torvum was evaluated for cytotoxic effect in vitro against a panel of human cancer cell lines (Lu et al. 2009). Pharmacological studies revealed that *S. torvum* possesses antimicrobial (Chah et al. 2000; Wiart et al. 2004), antiviral (Arthan et al. 2002), antioxidant (Sivapriya and Srinivas 2007), analgesic, anti-inflammatory (Ndebia et al. 2007) activity, and the methanolic extracts of *Solanum* species have been tested for molluscicidal activity against *Biomphalaria glabrata* (Silva et al. 2005). The acetone, chloroform, ethyl acetate, hexane, methanol, and petroleum ether extracts of leaf and seed of *S. torvum* were tested against fourth-instar larvae of malaria vector *A. subpictus* and Japanese encephalitis vector *C. tritaeniorhynchus* (Kamaraj et al. 2009) and for parasitic activity against larvae of cattle tick *R. (Boophilus) microplus*, adult of *H. bispinosa*, hematophagous fly *H. maculata*, nymph of goat-lice *D. caprae*, and adult sheep parasite *P. cervi* (Kamaraj et al. 2010). Chloroform and methanol extract of leaves, stem, and roots of *S. torvum* were evaluated for antibacterial and antifungal affects against human pathogenic bacteria and fungi (Bari et al. 2010). Leaves have been reported to contain the steroidal gluco-alkaloid, solasonine. In addition, they contain steroidal sapogenins, neochlorogenin, neosolaspigein, and solaspigenine. They have also been found to contain triacontanol, tetratriacontanic acid, z-tritriacontanone, sitosterol, stigmasterol, and campesterol. Fruits also contain the gluco-alkaloid, solasonine, sterolin (sitesterol-D-glucoside), protein, fat, and minerals (Yuanyuan et al. 2009).

2.3.23 Tagetes erecta L. *(Compositae)*

Is a stout, branching herb, native of Mexico and other warmer parts of America and naturalized elsewhere in the tropics and subtropics including Bangladesh and India. Different parts of this plant including flower are used in folk medicine to cure various diseases. Leaves are used as an antiseptic and in kidney troubles, muscular pain, piles, and applied to boils and carbuncles. The flower is useful in fevers, epileptic fits (Ayurveda), astringent, carminative, stomachic, scabies and liver complaints and is also employed in diseases of the eyes. They are said to purify blood and flower juice is given as a remedy for bleeding piles and is also used in rheumatism, colds, and bronchitis.

2.3.23.1 Biological Activities

The plant *T. erecta* has been shown to contain quercetagetin, a glucoside of quercetagetin, phenolics, syringic acid, methyl-3,5-dihydroxy-4-methoxy benzoate, quercetin, thienyl and ethyl gallat (Kirtikar and Basu 1987; Ghani 1998a, b). Phytochemical studies of its different parts have resulted in the isolation of various chemical constituents such as thiophenes, flavonoids, carotenoids, and triterpeniods (Faizi and Naz 2004). It is very popular as a garden plant and yields a strongly aromatic essential oil (tagetes oil), which is mainly used for the compounding of high-grade perfumes (Manjunath 1969). The flowers of *T. erecta* have antibacterial, antifungal, cytotoxic (against brine shrimp nauplii), and insecticidal activity (against *Tribolium castaneum* and *C. quinquefasciatus*); the potency of the chloroform fraction was higher than that of the ethanol extract or petroleum ether fraction of the flower of *T. erecta* (Nikkon et al. 2002).

Tagetes species, popularly known as marigold, are grown as ornamental plants and thrive in varied agroclimates. *T. erecta* known as Chendu hoovu is used traditionally as leaf paste applied externally to infected feet, wounds and for worms in cattle (Rajakumar and Shivanna 2009). *T. erecta* extract was tested against fourth-instar larvae of malaria vector *A. subpictus* and Japanese encephalitis vector *C. tritaeniorhynchus* (Kuppusamy and Murugan 2006), and as an insect repellent (Prakash and Rao 1997). *T. erecta* methanol and dichloromethane extracts showed a significant pesticidal activity against *S. oryzae* (Broussalis et al. 1999). The acetone and methanol extracts were evaluated for in vitro anthelmintic activity against the fourth larval stage of *Haemonchus contortus* (Aguilar et al. 2008).

2.3.24 Vitex negundo *Linn. (Verbenaceae)*

Is a small aromatic plant that flourishes abundantly in waste lands and is widely distributed in tropical to temperate regions, being a native of South Asia, China, Indonesia, and the Philippines (Dharmasiri et al. 2003). *V. negundo* is a woody, aromatic shrub growing to a small tree. It commonly bears tri- or penta-foliate leaves on quadrangular branches, which give rise to bluish-purple colored flowers in branched tomentose cymes. It thrives in humid places or along water courses in wastelands and mixed open forests and has been reported to occur in Afghanistan, India, Pakistan, Sri Lanka, Thailand, Malaysia, eastern Africa, and Madagascar. It is grown commercially as a crop in parts of Asia, Europe, North America, and the West Indies (de Padua et al. 1999). Traditional medicine mainly comprises uses in Indian Ayurveda, Arabic Unani medicine, and traditional Chinese medicine. In Asia and Latin America, populations continue to use traditional medicine as a result of historical circumstances and cultural beliefs. Traditional medicine accounts for around 40% of all healthcare delivered in China. Up to 80% of the population in Africa uses traditional medicine to help meet their healthcare needs (WHO 2002). Plants are known to produce a variety of compounds which have evolved as defense compounds against microbes and herbivores (Wink 2004).

2.3.24.1 Biological Activities

Petroleum ether (60–80°C) extracts of the leaves of *V. negundo* acted as a promising repellent against *C. tritaeniorhynchus* (Karunamoorthi et al. 2008); the methanol leaf extracts of *V. negundo* were tested against the early fourth-instar larvae of *C. quinquefasciatus* (Kannathasan et al. 2007). The leaf hexane extract of *V. negundo* was tested against the larvae of *A. subpictus* and *C. tritaeniorhynchus* (Kamaraj et al. 2009). The petroleum ether and ethyl acetate 3:1 fraction of

V. negundo, inflicted considerable larval mortality and interfered with pupal–adult metamorphosis against different instars of *C. quinquefasciatus* and *A. stephensi* (Pushpalatha and Muthukrishnan 1995). In spite of several advancements in the field of synthetic drug chemistry and antibiotics, plants continue to be one of the major raw materials for drugs treating various ailments of humans. Clinical and pharmaceutical investigations have in fact elevated the status of medicinal plants by identifying the role of active principles present in them and elaborating on their mode of action in human and animal systems (Dutta 1973).

2.3.25 Zingiber officinale *Roscoe (Zingiberaceae)*

Is one of the most widely used species of the ginger family (*Zingiberaceae*) and is a common condiment for various foods and beverages. Ginger has a long history of medicinal use dating back 2,500 years in China and India for conditions such as headaches, nausea, rheumatism, and colds (Grant and Lutz 2000). *Z. officinale* is a well-known and widely used herb, especially in Asia, where it has been widely used as a spice and condiment in different societies. Besides its food-additive functions, ginger has a long history of medicinal use for the treatment of a variety of human ailments including common colds, fever, rheumatic disorders, gastrointestinal complications, motion sickness, diabetes, cancer, etc. Ginger contains several nonvolatile pungent principles viz. gingerols, shogaols, paradols, and zingerone, which account for many of its health beneficial effects (Kundu et al. 2009).

2.3.25.1 Biological Activities

The larvicidal activity of a petroleum ether extract of *Z. officinale* was evaluated against *A. aegypti* and *C. quinquefasciatus*; bioassay-guided fractionation led to the isolation of (a) 4-gingerol, (b) (6)-dehydrogingerdione, and (c) (6)-dihydrogingerdione which were tested against mosquitoes (Rahuman et al. 2008e). There are several reports of the insect activity of *Z. officinale* extracts (Sahayaraj 1998; Shelly et al. 2003; Shelly and McInnis 2001). This was observed earlier for other isolated compounds from *Z. officinale*, for example [6]-gingerol and [6]-dehydroshogaol exhibited maximum insect growth regulatory (IGR) and antifeedant activity against *Spilosoma oblique* (Agarwal et al. 2001). Prajapati et al. (2005) reported that the essential oils of *Z. officinale* and *Rosmarinus officinalis* were found to be ovicidal and repellent against *A. stephensi, A. aegypti*, and *C. quinquefasciatus*. Ginger contains a number of pungent constituents and active ingredients. Steam distillation of powdered ginger produces ginger oil, which contains a high proportion of sesquiterpene hydrocarbons, predominantly zingiberene (Govindarajan 1982a). The major pungent compounds in ginger, from studies of the lipophilic rhizome extracts, have yielded potentially active gingerols, which can be converted to shogaols, zingerone, and paradol (Govindarajan 1982b).

2.4 Conclusions

In the study medicinal plants were observed to be important elements in the maintenance of a diversity of useful resources, although the use of these plants appears to be little influenced by cultural traditions. One important aspect that must be considered in future comparisons of surveys is the concept of medicinal plants for depending on the context, these categories may be interpreted in significantly different manners. Among the different listed above mentioned plant parts used for the preparation of medicine, the leaves were found to be the most frequently used plant parts in the preparation of remedies and the majority of the remedies. Accurate knowledge of the plants and their medicinal properties is held by only a few individuals in this community. Some of them have a strong tendency towards keeping their knowledge secret. The wealth of Kadar, Muduvar, Malai Malasar and Pulayar tribal knowledge of medicinal plants points to a great potential for research and the discovery of new drugs to fight diseases, obtaining new foods, and other new uses. So, further scientific assessment of these medicines in phytochemical, biological, and clinical studies is greatly needed.

References

Adewunmi CO, Ariwodola JO, Olubunmi PA (1987) Systemic effects of water extract of *Tetrapleura tetraptera*, a Nigerian plant molluscicide used in *Schistosomiasis* control. Pharmac Biol 8:7–14

Adolf W, Opferkuch HJ, Hecker E (1984) Irritant phorbol derivatives from 4 Jatropha species. Phytochemistry 23:129–132

Agarwal M, Walia S, Dhingra S, Khambay BP (2001) Insect growth inhibition, antifeedant and antifungal activity of compounds isolated/derived from *Zingiber officinale* Roscoe (ginger) rhizomes. Pest Manag Sci 57:289–300

Aguilar HH, de Gives PM, Sánchez DO, Arellano ME, Hernández EL, Aroche UL, Valladares-Cisneros G (2008) In vitro nematocidal activity of plant extracts of Mexican flora against *Haemonchus contortus* fourth larval stage. Ann NY Acad Sci 1149:158–160

Ahmed M, Amin S, Islam M, Takahashi M, Okuyama E, Hossain CF (2000) Analgesic principle from *Abutilon indicum*. Pharmazie 55:314

Aiyela-agbe OO, Adesogan EK, Ekunday O, Adeniyi BA (2000) The antimicrobial activity of roots of *Jateopha podagrica* Hook. Phytother Res 14:60–62

Akinjogunla OJ, Eghafona NO, Enabulele IO, Mboto CI, Ogbemudia FO (2010) Antibacterial activity of ethanolic extracts of *Phyllanthus amarus* against extended spectrum β- lactamase producing *Escherichia coli* isolated from stool samples of HIV sero-positive patients with or without diarrhea. Afr J Pharm Pharmacol 4(6):402–407

Amin MR, Mostofa M, Hoque ME, Sayed MA (2009) In vitro anthelmintic efficacy of some indigenous medicinal plants against gastrointestinal nematodes of cattle. J Bangladesh Agric Univ 7(1):57–61

Anonyme (1993) Fiche espèce sur *Mangifera indica* L. Revue de Médecines et Pharmacopées Africaines 6(2):119–124

Anonymous (1952) The wealth of India. Council of Scientific and Industrial Research, New Delhi, India, pp 35–36

Anonymous (1970) Hamdard pharmacopoeia of eastern medicine, 2nd Impression. Hamdard National Foundation, Pakistan, p 373

Anonymous (1985) The wealth of India: a dictionary of Indian raw material and industrial products, vol 1, Revised Edition. CSIR, New Delhi, pp 37–47

Aouinty B, Outara S, Mellouki F, Mahari S (2006) Évaluation préliminaire de activité larvicide des extraits aqueux des feuilles du ricin (*Ricinus communis* L.) et du bois de thuya (*Tetraclinis articulata* (Vahl) Mast.) sur les larves de quatre moustiques culicidés: *Culex pipiens* (Linné), *Aedes caspius* (Pallas), *Culiseta longiareolata* (Aitken) et *Anopheles maculipennis* (Meigen). Biotechnol Agron Soc Environ 10(2):67–71

Areekul S, Sinchaisri P, Tigvatananon S (1987) Effect of Thai plant extracts on the oriental fruit fly. I.Toxicity test. Kasetsart J 21:395–470

Arthan D, Svasti J, Kittakoop P, Pittayakhachonwut D, Tanticharoen M, Thebtaranonth Y (2002) Antiviral isoflavonoid sulfate and steroidal glycosides from the fruits of *Solanum torvum*. Phytochemistry 59:459–463

Awe SO, Olajide OA, Oladiran OO (1988) Antiplasmodial and antipyretic screening of *Mangifera indica* extract. Phytother Res 12:437–438

Ayoub SMH, Yankov LK (1981) On the constituents of the peel of *Citrullus colcynthis*. Part-2. Fitoterpia 52(1):13–16

Badaturuge MJ, Habtemariam S, Thomas MJK (2011) Antioxidant compounds from a South Asian beverage and medicinal plant, *Cassia auriculata*. Food Chem 125:221–225

Bagalwa M, Chifundera K (2007) Environmental impact evaluation of the stem bark extract of *Maesa lanceolata* used in Democratic Republic of Congo. J Ethnopharmacol 114:281–284

Bagavan A, Rahuman AA, Kamaraj C, Geetha K (2008) Larvicidal activity of saponin from *Achyranthes aspera* against *Aedes aegypti* and *Culex quinquefasciatus* (Diptera: Culicidae). Parasitol Res 103:223–229

Bagavan A, Kamaraj C, Elango G, Zahir AA, Rahuman AA (2009) Adulticidal and larvicidal efficacy of some medicinal plant extracts against tick, fluke and mosquitoes. Vet Parasitol 166:286–292

Banerji J, Das B (1993) MAPA, vol 15. Department of Chemistry, University College of Science, Calcutta, India, pp 1002–1017

EXIM Bank (2003) Export potential of Indian medicinal plants and products. Publication No. OP 98. Export and Import Bank of India (EXIM Bank), Mumbai, India (see also http://www.eximbankindia.com/publications. Accessed 9 Oct 2005)

Bari MA, Islam W, Khan AR, Mandal A (2010) Antibacterial and antifungal activity of *Solanum torvum* (Solanaceae). Int J Agric Biol 12(3):386–390

Basalah MO, Ali Whaibi MH, Sher M (1985) Comparative study of some metabolites of *Citrullus colocynthis* Schrad and *Cucumis prophetarum* L. J Biol Sci Res 16(1):105–123

Bouquet A, Debray M (1974) Plantes médicinales de la Côte d'Ivoire. Mémoire ORSTOM, Paris

Bowman WC, Rand MJ (1980) Textbook of pharmacology. Blackwell Scientific, Oxford, pp 42.29–42.31

Broussalis AM, Ferraro GE, Martino VS, Pinzón R, Coussio JD, Alvarez JC (1999) Argentine plants as potential source of insecticidal compounds. J Ethnopharmacol 67:219–223

ICMR Bulletin (2002) Prospects of elimination of lymphatic filariasis in India. ICMR Bull 32(5–6)

Burkill HM (1985) The useful plants of West Tropical Africa, families A–D. Royal Botanic Gardens, Kew, 1

Caffarini P, Carrizo P, Pelicano A, Rogggero P, Pacheco J (2008) Effects of acetonic and water extracts of *Ricinus communis*, *Melia azedarach* and *Trichillia glauca* on black common cutting ant (*Acromyrmex lundi*). IDESIA 26(1):59–64

Capasso F, Mascolo N, Izzo AA, Gaginella TS (1994) Dissociation of castor oil induced diarrhea and intestinal mucosal injury in rat: effect of NG-nitro-L- arginine methyl ester. Br J Pharmacol 113:1127–1130

Chadha KL, Gupta R (1995) Medicinal and aromatic plants. In: Advances in horticulture, vol 11. Malhotra Publishing House, New Delhi, 932 p

Chah KF, Muko KN, Oboegbulem SI (2000) Antimicrobial activity of methanolic extract of *Solanum torvum* fruit. Fitoterapia 71:187–189

Chakraborty T, Babu SPS, Sukul NC, Babu SPS (1996) Preliminary evidence of antifilarial effect of *Centella asiatica* on Canine dirofilariasis. Fitoterapia 67(2):110–112

Chakraborty A, Brantner A, Mukainaka T, Nobukuni Y, Kuchide M, Konoshima T, Tokuda H, Nishino H (2002) Cancer chemopreventive activity of *Achyranthes aspera* leaves on Epstein–Barr virus activation and two-stage mouse skin carcinogenesis. Cancer Lett 177:1–5

Chandrashekhar CH, Latha KP, Vagdevi HM, Vaidya VP (2008) Anthelmintic activity of the crude extracts of *Ficus racemosa*. Int J Green Pharmacy 2(2):100–103

Chatterjee A, Das B, Adityachaudhary N, Dabkirtaniya S (1980) Note on the insecticidal properties of the seeds of *Jatropha gossypifolia* Linn. Indian J Agric Sci 50:637–638

Chopra RN, Nayar SL, Chopra IC (1956) Glossary of Indian medical plants. CSIR, New Delhi

Chopra RN, Chopra IC, Handa KL, Kapur LD (1958) Indigenous drugs of India, 2nd edn. Academic, Calcutta, pp 508–674

Chopra RN, Nayar SL, Chopra IC (1992) Glossary of Indian medicinal plants, vol 2. Council for Scientific and Industrial Research, New Delhi, p 12

Coe FG, Anderson GJ (1996) Screening of medical plants used by Garifuna of Eastern Nicaragua for bioactive compounds. J Ethnopharmacol 53:29–50

Daniel T, Umarani S, Sakthivadivel M (1995) Insecticidal action of *Ervatamia divaricata* L. and *Acalypha indica* L. against *Culex quinquefasciatus* Say. Geobios New Rep 14(2):95–98

Das PC, Das A, Mandal S (1989a) Anti-inflammatory and antimicrobial activities of the seed kernel of *Mangifera indica* extract. Phytother Res 60:235–240

Das PC, Das A, Mandal S, Islam CN, Dutta MK, Patra B (1989b) Antiinflammatory and antimicrobial activities of the seed kernel Mangifera indica. Fitoterapia 6(3):235–241

Das B, Rao SP, Srinivas K, Das R (1996) Jatrodien, a lignan from stems of *Jatropha gossypifolia*. Phytochemistry 41:985–987

Dash AP, Valecha N, Anvikar AR, Kumar A (2008) Malaria in India: challenges and opportunities. J Biosci 33:583–592

de Padua LS, Bunyapraphatsara N, Lemmens RHMJ (1999) Medicinal and poisonous plants, plant resources of South East Asia. Backhuys, Leiden

Derouich M, Boutayeb A (2006) Dengue fever: mathematical modelling and computer simulation. Appl Math Comput 177:528–544

DGCIS (2004) Monthly statistics of foreign trade of India. Annual report for 2003 – 2004(Vol. 1). Exports including re-exports. Directorate General of Commercial Intelligence and Statistics, Ministry of Commerce, Kolkata

Dhar ML, Dhar MM, Dhawan BN, Mehrotra BN, Ray C (1968) Screening of Indian plants for biological activity. Part I. Indian J Exp Biol 6:232–247

Dharmasiri MG, Jayakody JR, Galhena G, Liyanage SS, Ratnasooriya WD (2003) Anti-inflammatory and analgesic activities of mature fresh leaves of *Vitex negundo*. J Ethnopharmacol 87(2–3):199–206

Duraipandiyan V, Ayyanar M, Ignacimuthu S (2006) Antimicrobial activity of some ethnomedicinal plants used by Paliyar tribe from Tamil Nadu, India. BMC Comp Alt Med 6:35–42

Dutta SC (1973) Medicinal plants. National Council for Education Research and Training, New Delhi

Elango G, Rahuman AA (2010) Evaluation of medicinal plant extracts against ticks and fluke. Parasitol Res. doi:10.1007/s00436-010-2090-9

Elango G, Bagavan A, Kamaraj C, Zahir AA, Rahuman AA (2009a) Oviposition-deterrent, ovicidal, and repellent activities of indigenous plant extracts against *Anopheles subpictus* Grassi (Diptera: Culicidae). Parasitol Res 105(6):1567–1576

Elango G, Rahuman AA, Bagavan A, Kamaraj C, Zahir AA, Venkatesan C (2009b) Laboratory study on larvicidal activity of indigenous plant extracts against *Anopheles subpictus* and *Culex tritaeniorhynchus*. Parasitol Res 104(6):1381–1388 el-Naggar ME, Abdel-Sattar MM, Mosallam SS (1989) Toxicity of colocynithin and hydrated colocynithin from alcoholic extract of *Citrullus colocynthis* pulp. J Egypt Soc Parasitol 19 (1):179–185

Fagbenro-Beyioku AF, Oyibo WA, Anuforom BC (1998) Disinfectant/ antiparasitic activities of Jatropha curcas. East Afr Med J 75:508–511

Faizi S, Naz A (2004) Palmitoleate (9Z-Hexadeca-9-enoate) esters of oleanane triterpenoids from the golden flowers of Tagetes erecta: isolation and autoxidation products. Helv Chim Acta 87:46–56

Farnsworth NR, Soejarto DD (1991) Global importance of medicinal plants. In: Akerele O, Heywood V, Synge H (eds) The conservation of medicinal plants. Cambridge University Press, Cambridge, pp 25–51

Food and Agricultural Organization (FAO) (1996) Forests, food and health. http://www.fao.org/forestry/site/28813/em. Accessed Oct 10 2005

Ganapathy S, Dash GK (2002) Diuretic and laxative activity of Cocculus hirsuts. Fitotherapia 73 (1):28–31

Gangadevi V, Yogeswari S, Kamalraj S, Rani G, Muthumary J (2008) The antibacterial activity of *Acalypha indica* L. Indian J Sci Technol 1(6):1–5

García D, Escalante M, Delgado R, Ubeira FM, Leiro J (2003) Anthelminthic and antiallergic activities of *Mangifera indica* L. stem bark components Vimang and mangiferin. Phytother Res 17(10):1203–8

Garrido G, González D, Delporte C (2001) Analgesic and anti-inflammatory effects of *Mangifera indica* L. extract (Vimang). Phytother Res 15:18–21

Gazi MI (1991) The finding of antiplaque features in *Acacia Arabica* type of chewing gum. J Clin Periodontol 18(1):75–77

Geethangili M, Rao YK, Fang SH, Tzeng YM (2008) Cytotoxic constituents from *Andrographis paniculata* induces cell cycle arrest in jurkat cells. Phytother Res 22:1336–1341

Georges K, Jayaprakasam B, Dalavoy SS, Nair MG (2008) Pestmanaging activities of plant extracts and anthraquinones from *Cassia nigricans* from Burkina Faso. Bioresour Technol 99(6):2037–2045

Ghani A (1998) Medicinal plants of Bangladesh. Chemical constituents and uses, 2nd edn. Asiatic Society of Bangladesh, Dhaka, pp 301–302

Ghani A (1998b) Medicinal plants of Bangladesh: chemical constituents and uses. Asiatic Society of Bangladesh, Dhaka, Bangladesh

Ghosal S, Jaiswal DK, Biswas K (1978) New glycoxanthones and flavanone glycosides of *Hoppea dichotoma*. Phytochemistry 17:2119–2123

Gibbons RV, Vaughn DW (2002) Dengue: an escalating problem. BMJ 324:1563–1566

Goonasekara MM, Gunawardhana VK, Jayaseana K, Mohammed SG, Balasubramaniam S (1995) Pregnancy terminating effect of *Jatropha curcas* in rats. J Ethnopharmacol 47:117–123

Gotoh A, Sakaeda T, Kimura T, Shirakawa T, Wada Y, Wada A, Kimachi T, Takemoto Y, Iida A, Iwakawa S, Hirai M, Tomita H, Okamura N, Nakamura T, Okumura K (2004) Antiproliferative activity of *Rhinacanthus nasutus* (L.) Kurz extracts and the active moiety, rhinacanthin C. Biol Pharm Bull 27:1070–1074

Goudgaon NM, Basavaraj NR, Vijayalaxmi A (2003) Antiinflammatory activity of different fractions of *Leucas aspera* Spreng. Indian J Pharmacol 35:397–398

Govindarajan VS (1982a) Ginger – chemistry, technology, and quality evaluation: part 1. Crit Rev Food Sci Nutr 17:1–96

Govindarajan VS (1982b) Ginger – chemistry, technology, and quality evaluation: part 2. Crit Rev Food Sci Nutr 17:189–258

Grainge M, Ahmed S, Mitchell WC, Hylin JW (1984) Plant species reportedly possessing pest-control properties: a database. Resource Systems Institute, East-West Center, Honolulu, HI, USA, p 240

Grant KL, Lutz RB (2000) Ginger. Am J Health Syst Pharm 57:945–947

Guha S, Ghosal S, Chattopadhyay U (1996) Antitumor, immunomodulatory and anti-HIV effect of mangiferin, a naturally glucosylxanthone. Chemotherapy 42:443–451

Guha-Sapir D, Schimmer B (2005) Dengue fever: new paradigms for a changing epidemiology. Emerg Themes Epidemiol 2(1):1

Habs M, Jahn SAA, Schmaehl D (1984) Carcinogenic activity of condensate from colquint seeds (*Citrullus colcynthis*) after chronic eipcutaneous administration to mice. J Cancer Res Clin Oncol 108(1):154–156

Hales S, Wet ND, Maindonald J, Woodward A (2002) Potential effect of population and climate changes on global distribution of dengue fever: an empirical model. Lancet 360:830–834

Hansson A, Veliz G, Naquira C, Amren M, Arroyo M, Arevalo G (1986) Preclinical and clinical studies with latex from *Ficus glabrata* HBK, a traditional intestinal antihelminthic in the Amazonian area. J Ethnopharmacol 2:105–138

Hartwell JL (1969) Plants used against cancer, a survey. Lloydia 32:153–205

Heller J (1996) Promoting the conservation and use of under utilized and neglected crops. 1. Physic nut: *Jatropha curcas* L. International Plant Genetic Resources Institute, Rome

Horsten SFAJ, Van den Berg AJJ, Kettenes-van den Bosch JJ, Leeflang BR, Labadie RP (1996) Cyclogossine A: a novel cyclic heptapeptide isolated from the latex of *Jatropha gossypifolia*. Planta Med 62:46–50

Hsu JH, Liou SS, Yu BC, Cheng JT, Wu YC (2004) Activation of −1Aadrenoceptor by andrographolide to increase glucose uptake in cultured myoblast C2C12 cells. Planta Med 70:1230–1233

Husain A, Virmani OP, Popli SP, Misra LN, Gupta MM, Srivastava GN, Abraham Z, Singh AK (1992) Dictionary of Indian medicinal plants. CIMAP, Lucknow, p 546

Igbinosa OO, Igbinosa EO, Aiyegoro OA (2009) Antimicrobial activity and phytochemical screening of stem bark extracts from *Jatropha curcas* (Linn). Afr J Pharm Pharmacol 3(2):058–062

Iqbal Z, Lateef M, Jabbar A, Ghayur MN, Gilani AH (2006) In vitro and In vivo anthelmintic activity of Nicotiana tabacum L. leaves against gastrointestinal nematodes of sheep. Phytother Res 20:46–48

Irvine FR (1961) Woody plants of Ghana (with special reference to their uses), 2nd edn. OUP, London, pp 233–237

Ivan A (1998) Chemical constituents, traditional and modern uses. In: Medicine plants of the world. Ross Humana, Totowa, NJ, pp 375–395

Iwu MM (1993) Handbook of African medicinal plants. CRC, Boca Raton, FL, pp 24–33

Jagannadha Rao KV, Ramachandra RL (1961) Chemical examination of *Cocculus hirsutus* (Linn) Diels. J Sci Ind Res 20(B):125–126

Jayashree G, Kurup M, Sudarslal S, Jacob VB (2003) Anti-oxidant activity of *Centella asiatica* on lymphoma-bearing mice. Fitoterapia 74:431–434

Johri RK, Pahwa GS, Sharma SC, Zutshi U (1991) Determination of estrogenic/antiestrogenic potential of antifertility substances using rat uterine peroxidase assay. Contraception 44(5):549–557

Joshi SG (2000) Cesalpinaceae – *Cassia auriculata*. Text book of medicinal plants. India Book House, Bangalore

Joshi SC, Pant SC (2010) Effect of H2SO4 on seed germination and viability of *Canna indica* L, medicinal plant. J Am Sci 6:6

Kager PA (2002) Malaria control: constraints and opportunities. Trop Med Int Health 7:1042–1046

Kamaraj C, Rahuman AA, Bagavan A (2008) Antifeedant and larvicidal effects of plant extracts against *Spodoptera litura* (F.), *Aedes aegypti* L. and *Culex quinquefasciatus* Say. Parasitol Res 103(2):325–331

Kamaraj C, Bagavan A, Rahuman AA, Zahir AA, Elango G, Pandiyan G (2009) Larvicidal potential of medicinal plant extracts against *Anopheles subpictus* Grassi and *Culex tritaeniorhynchus* Giles (Diptera: Culicidae). Parasitol Res 104(5):1163–1171

Kamaraj C, Rahuman AA, Bagavan A, Elango G, Rajakumar G, Zahir AA, Marimuthu S, Santhoshkumar T, Jayaseelan C (2010) Evaluation of medicinal plant extracts against blood-sucking parasites. Parasitol Res 106:1403–1412

Kambu K, Tona L, Luki N, Cimaga K, Makuba W (1989) Evaluation de l'activité antimicrobienne de quelques préparations traditionnelles antidiarrhéiques utilisées dans la ville de Kinshasa-Zaïre. Bull de Méde Tradi Pharmacop 3(1):15–24

Kannathasan K, Senthilkumar A, Chandrasekaran M, Venkatesalu V (2007) Differential larvicidal efficacy of four species of Vitex against *Culex quinquefasciatus* larvae. Parasitol Res 101 (6):1721–1723

Karunamoorthi K, Ramanujam S, Rathinasamy R (2008) Evaluation of leaf extracts of *Vitex negundo* L. (Family: Verbenaceae) against larvae of *Culex tritaeniorhynchus* and repellent activity on adult vector mosquitoes. Parasitol Res 103:545–550

Kerharo J, Adam JG (1974) Pharmacopée sénégalaise traditionnelle. Plantes médicinales et toxiques. Edition Vigot – Frères, Paris

Khan N, Sultana S (2005) Chemomodulatory effect of *Ficus racemosa* extract against chemically induced renal carcinogenesis and oxidative damage response in Wistar rats. Life Sci 29:1194–1210

Khumrungsee N, Bullangpoti V, Pluempanupat W (2009) Efficiency of *Jatropha gossypifolia* L. (Euphorbiaceae) against *Spodoptera exigua* HÜbner (Lepidoptera: Noctuidae): toxicity and its detoxifying enzyme activities. KKU Sci J (Suppl) 37:50–55

Kiemer AK, Hartung T, Huber C, Vollmar AM (2003) *Phyllanthus amarus* has anti-inflammatory potential by inhibition of iNOS, COX-2, and cytokines via the NF-kB pathway. J Hepatol 38:289–297

Kirtikar KR, Basu BD (1975) Indian medicinal plants, vol 3, 2nd edn. International Book Distributors, Dehra Dun, pp 2327–2328

Kirtikar KR, Basu BD (1987) Indian medicinal plants. Lalit mohan Basu, Allahabad, India, pp 1385–1386

Kirtikar KR, Basu BA (1991) Indian medicinal plants, vol 3. Periodical Experts Book Agency, New Delhi, pp 2274–2277

Kirtikar KR, Basu BD (1993) Indian medicinal plants, 2nd edn. Periodical Experts Books Agency, New Delhi, pp 499–505

Kirtikar KR, Basu BD (1996) Indian medicinal plants, vol 3. International Book Distributors, Allahabad, 2247

Kleinschmidt HE, Johnson RW (1977) Weeds of Queensland. – 147 s. (zitiert nach pier.)

Komalamisra N, Trongtokit Y, Rongsriyam Y, Apiwathnasorn C (2005) Screening for larvicidal activity in some thai plants against four mosquito vector species. Southeast Asian J Trop Med Public Health 36:1412–1422

Kondrachine AV (1992) Malaria in WHO Southeast Asia region. Indian J Mal Res 29:129–160

Kossou DK, Gbèhounou G, Ahanchédé A, Ahohuendo B, Bouraïma Y, Huis AV (2001) Indigenous cowpea production and protection practices in Benin. Insect Sci Appl 21 (2):123–132

Kuhn H (1965) Tobacco alkaloids and their pyrolysis products in the smoke. In: von-Euler US (ed) Tobacco alkaloids and related compounds. Pergamon, Oxford, pp 37–49

Kumaran A, Karunakaran RJ (2006) Antioxidant activity of *Cassia auriculata* flowers. Fitoterapia 78(1):46–47

Kumaria R (2010) Correlation of disease spectrum among four Dengue serotypes: a five years hospital based study from India. Braz J Infect Dis 14(2):141–146

Kundu JK, Na HK, Surh YJ (2009) Ginger-derived phenolic substances with cancer preventive and therapeutic potential. Forum Nutr 61:182–189

Kuppusamy C, Murugan K (2006) Mosquitocidal effect of ethanolic extracts of *Andrographis paniculata* Nees on filarial vector *Culex quinqufasciatus* Say (Diptera: Culicidae). In: International conference on diversity of insects: challenging issues in management and conservation, Tamil Nadu, India, 30 Jan–3 Feb 2006, pp 194

Latha M, Pari L (2003) Preventive effects of *Cassia auriculata* L. flowers on brain lipid peroxidation in rats treated with streptozotocin. Mol Cell Biochem 243(1–2):23–28

Lee MJ, Rao YK, Chen K, Lee YC, Chung YS, Tzeng YM (2010) Andrographolide and 14-deoxy-11, 12-didehydroandrographolide from Andrographis paniculata attenuate high glucose-induced fibrosis and apoptosis in murine renal mesangeal cell lines. J Ethnopharmacol 132:497–505

Levetin E, McMahon TK (2003) Plants and society, 3rd edn. McGraw-Hill, Dubuque, Iowa

Li H, Miyahara T, Tezuka Y (1998) The effect of Kampo formulae on bone resorption in vitro and in vivo. I.Active constituents of Tsu-kan-gan. Biol Pharm Bull 21:1322–1326

Li RW, Leach DN, Myers SP, Lin GD, Leach GJ, Waterman PG (2004) A new anti-inflammatory glucoside from *Ficus racemosa* L. Planta Med 70:421–426

Liu SY, Sporer F, Wink M, Jourdane J, Henning R, Li YL, Ruppel A (1997) Anthraquinones in *Rheum palmatum* and *Rumex dentatus* (Polygonaceae), and phorbol esters in *Jatropha curcas* (Euphorbiaceae) with molluscicidal activity against the schistosome vector snails Oncomelania, Biomphalaria, and Bulinus. TM IH Trop Med Int Health 2:179–188

Lu Y, Luo J, Huang X, Kong L (2009) Four new steroidal glycosides from *Solanum torvum* and their cytotoxic activities. Steroids 74:95–101

Maasilamani G, Shokat A (1981) J Res Ayurveda Siddha 2:109

Mahato RB, Chaudhary RP (2005) Ethnomedicinal study and antibacterial activities of selected plants of Palpa District, Nepal. Sci World 3:26–31

Maity P, Hansda D, Bandyopadhyay U, Mishra DK (2009) Biological activities of crude extracts and chemical constituents of Bael, *Aegle marmelos* (L.) Corr. Indian J Exp Biol 47 (11):849–861

Manandhar NP (1972) Fodder trees. The Rising Nepal 7:1–2

Mandal SC, Tapan K, Maity J, Das M, Pal M, Saha BP (1999) Hepatoprotective activity of Ficus racemosa leaf extract on liver damage caused by carbon tetrachloride in rats. Phytother Res 13:430–432

Mandal SC, Saha BP, Pal M (2000) Studies on bacterial activity of *Ficus racemosa* leaf extract. Phytother Res 14:278–280

Manjunath MBL (1969) The wealth of India. PID CSIR, New Delhi, pp 109–110

Mansingh A, Williams LAD (1998) Pesticidal potential of tropical plants – II. Acaricidal activity of crude extracts of several Jamaican plants. Insect Sci Appl 18:149–155

Mansour F, Azaizeh H, Saad B, Tadmor Y, Abo-Moch F, Said O (2004) The potential of middle eastern flora as a source of new safe bio-acaricides to control *Tetranychus cinnabarinus*, the carmine spider mite. Phytoparasitica 32(1):66–72

Martínez G, Delgado R, Pérez G (2000) Evaluations of the in vitro antioxidant activity of *Mangifera indica* L. extract (Vimang). Phytother Res 14:424–427

Matławska I, Sikorska M (2002) Flavonoid compounds in the flowers of *Abutilon indicum* (L.) Sweet (Malvaceae). Acta Pol Pharm 59(3):227–229

Matsuse TI, Lim YA, Hattori M, Correa M, Gupta MP (1999) A search for anti-viral properties in Panamanian medicinal plants – the effect on HIV and essential enzymes. J Ethnopharmacol 64:15–22

Meshram PB, Kulkarni N, Joshi KC (1996) Antifeedant activity of *Azadirachta indica* and *Jatropha curcas* against *Papilio demoleus* L. J Environ Biol 17:295–298

Michael E, Bundy DAP, Grenfel BT (1996) Re-assessing the global prevalence and distribution of lymphatic filariasis. Parasitology 122:409

Mishra V, Khan NU, Singhal KC (2005) Potential antifilarial activity of fruit extracts of *Ficus racemosa* Linn. against *Setaria cervi* in vitro. Indian J Exp Biol 43:346–350

Mishra SK, Sangwan NS, Sangwan RS (2007) *Andrographis paniculata* (Kalmegh): a review. Pharmacogn Rev 1:283–298

Mishra K, Dash AP, Swain BK, Dey N (2009) Anti-malarial activities of *Andrographis paniculata* and *Hedyotis corymbosa* extracts and their combination with curcumin. Malar J 12:8–26

Morsy TA, Rahem MA, Allam KA (2001) Control of *Musca domestica* third instar larvae by the latex of Calotropis procera (Family: Asclepiadaceae). J Egypt Soc Parasitol 31(1):107–110

Moursy LE (1997) Insecticidal activity of *Calotropis procera* extracts on the flesh fly, Sarcophaga *haemorrhoidalis* Fallen. J Egypt Soc Parasitol 2:505–514

Müller G, Schlein Y (2006) Sugar questing mosquitoes in arid areas gather on scarce blossoms that can be used for control. Int J Parasitol 36(10–11):1077–1080

Murray JA (1989) Plants and drugs of Sind. Indus, Karachi, pp 154–155

Murugan K, Jeyabalan D (1999) Mosquitocidal effect of certain plants extracts on *Anophels stephensi*. Curr Sci 76:631–633

Mushobozy DMK, Nganilevanu G, Ruheza S, Swella GB (2009) Plant oils as common bean (*Phaseolus vulgaris* L.) seed protectants against infestations by the Mexican bean weevil *Zabrotes subfascistus* (Boh.). J Plant Prot Res 49(1):35–39

Muthukrishnan J, Pushpalatha E, Kasthuribai A (1997) Biological effects of four plant extracts on *Culex quinquefasciatus* Say larval stages. Insect Sci Appl 17:389–394

Nadkarni KM (1976) Indian materia medica. Popular Prakashan, Bombay, pp 850–885

Nadkarni AC (1982) Indian material medica, vol 1, 3rd edn. Popular prakashan, Bombay

Nadkarni KM, Nadkarni AK, Chopra RN (1976) Indian materia medica, vol 1. Popular Prakashan, Bombay, pp 548–550

Navot N, Zamir D (1986) Linkage relationships of 19 protein-coding genes in atermelons. Theor Appl Genet 72(2):274–278

Nayak SK, Singhai AK (1993) Anti-inflammatory and analgesic activity of roots of *Cocculus hirsutus*. Indian J Nat Prod 9:12–14

Ndebia EJ, Kamga R, Nchunga-Anye Nkeh B (2007) Analgesic and anti-inflammatory properties of aqueous extract from leaves of *Solanum torvum* (Solanaceae). Afr J Tradit Complement Altern Med 4:240–244

Nikkon F, Saud ZA, Rahman MM, Haque ME (2002) Biological activity of the extracts of the flower of *Tagetes erecta* Linn. J Biosci 10:117–119

Nirmal SA, Shelke SM, Gagare PB, Jadhav PR, Dethe PM (2007) Antinociceptive and anthelmintic activity of *Canna indica*. Nat Prod Res 21(12):1042–1047

Obayed Ullah M, Sultana S, Haque A, Tasmin S (2009) Antimicrobial, cytotoxic and antioxidant activity of *Centella asiatica*. Eur J Sci Res 30:260–264

Oliver-Bever B (1986) Medicinal plants in tropical West Africa. Cambridge University Press, London

Oluwafemi F, Debiri F (2008) Antimicrobial effect of *Phyllanthus amarus* and *Parquetina nigrescens* on Salmonella typhi. Afr J Biomed Res 11:215–219

Omoregbe RE, Ikuebe OM, Ihimire IG (1996) Antimicrobial activity of some medicinal plants extracts on *Escherichia coli*, *Salmonella paratyphi* and *Shigella dysenteriae*. Afr J Med Med Sci 25:373–375

Pialoux G, Gaüzère M, Jauréguiberry S, Strobel M (2007) Chikungunya, an epidemic arbovirus. Lancet Infect Dis 7:319–327

Porchezhian E, Ansari SH (2005) Hepatoprotective activity of *Abutilon indicum* on experimental liver damage in rats. Phytomedicine 12:62–64

Pousset JL (1989) Plantes médicinales africaines. Utilisation pratique. ACCT, Paris

Prajapati V, Tripathi AK, Aggarwal KK, Khanuja SP (2005) Insecticidal, repellent and oviposition-deterrent activity of selected essential oils against *Anopheles stephensi*, *Aedes aegypti* and *Culex quinquefasciatus*. Bioresour Technol 96:1749–1757

Prakash A, Rao J (1997) Botanical pesticides in agriculture. CRC, Boca Raton, FL

Prasanna R, Harish CC, Pichai R, Sakthisekaran D, Gunasekaran P (2008) *Cassia auriculata* leaf extract (CALE) was evaluated in human breast adenocarcinoma MCF-7 and human larynx carcinoma Hep-2 cell lines. Cell Biol Int 33(2):127–134

Promsiri S, Naksathit A, Kruatrachue M, Tharava U (2006) Evaluations of larvicidal activity of medicinal plant extracts to *Aedes aegypti* (Diptera: Culicidae) and other effects on a non target fish. Insect Sci 13:179–188

Pushpalatha E, Muthukrishnan J (1995) Larvicidal activity of a few plant extracts against Culex quinquefasciatus and Anopheles stephensi. Indian J Malariol 32(1):14–23

Pushpalatha E, Muthukrishnan J (1999) Efficacy of two tropical plant extracts for the control of mosquitoes. J Appl Entomol 123:369–373

Puttarak P, Charoonratana T, Panichayupakarananta P (2010) Antimicrobial activity and stability of rhinacanthins-rich Rhinacanthus nasutus extract. Phytomedicine 17:323–327

Raghavendra K, Subbarao SK (2002) Case studies on insecticide resistance and its anagement. In: Frederick G (ed) Proceedings of Mekong Malaria Forum, RMCP-EC, pp 17–21

Rahman A, Talukder FA (2006) Bioefficacy of some plant derivatives that protect grain against the pulse beetle, *Callosobruchus maculatusi*. J Insect Sci 6(3):1–10

Rahman M, Ismail HM, Yin LT (1990) Jatropholone a and Jatrophatrione two diterpenes from *Jatropha-Gossypifolia*. Pertanika 13:405–408

Rahman MA, Taleb MA, Biswas MM (2003) Evaluation of botanical product as grain protectant against grain weevil, *Sitophilus granarious* (L.) on wheat. Asian J Plant Sci 2(6):501–504

Rahuman AA, Venkatesan P (2008) Larvicidal efficacy of five cucurbitaceous plant leaf extracts against mosquito species. Parasitol Res 103:133–139

Rahuman AA, Gopalakrishnan G, Venkatesan P, Geetha K (2008a) Isolation and identification of mosquito larvicidal compound from *Abutilon indicum* (Linn.) Sweet. Parasitol Res 102 (5):981–988

Rahuman AA, Venkatesan P, Gopalakrishnan G (2008b) Mosquito larvicidal activity of oleic and linoleic acids isolated from *Citrullus colocynthis* Linn. Schrad. Parasitol Res 1036:1383–1390

Rahuman AA, Venkatesan P, Geetha K, Gopalakrishnan G, Bagavan A, Kamaraj C (2008c) Mosquito larvicidal activity of gluanol acetate, a tetracyclic triterpenes derived from *Ficus racemosa* Linn. Parasitol Res 103(2):333–339

Rahuman AA, Gopalakrishnan G, Venkatesan P, Geetha K (2008d) Larvicidal activity of some Euphorbiaceae plant extracts against *Aedes aegypti* and *Culex quinquefasciatus* (Diptera: Culicidae). Parasitol Res 102(5):867–873

Rahuman AA, Gopalakrishnan G, Venkatesan P, Geetha K, Bagavan A (2008e) Mosquito larvicidal activity of isolated compounds from the rhizome of *Zingiber officinale*. Phytother Res 22 (8):1035–1039

Rahuman AA, Bagavan A, Kamaraj C, Vadivelu M, Zahir AA, Elango G, Pandiyan G (2009a) Evaluation of indigenous plant extracts against larvae of *Culex quinquefasciatus* Say (Diptera: Culicidae). Parasitol Res 104(3):637–643

Rahuman AA, Bagavan A, Kamaraj C, Saravanan E, Zahir AA, Elango G (2009b) Efficacy of larvicidal botanical extracts against *Culex quinquefasciatus* Say (Diptera: Culicidae). Parasitol Res 104(6):1365–1372

Rajakumar N, Shivanna MB (2009) Ethno-medicinal application of plants in the eastern region of Shimoga district, Karnataka, India. J Ethnopharmacol 126(1):64–73

Rajeshkumar NV, Joy KL, Kuttan G, Ramsewak RS, NairMG KR (2002) Antitumour and anticarcinogenic activity of *Phyllanthus amarus* extract. J Ethnopharmacol 81:17–22

Rajkumar S, Jebanesan A (2005) Larvicidal and adult emergence inhibition effect of *Centella asiatica* Brahmi (Umbelliferae) against mosquito *Culex quinquefasciatus* Say (Diptera: Culicidae). Afr J Biomed Res 8:31–33

Ramos MV, Bandeira GP, de Freitas CDT, Nogueira NAP, Alencar NMN, de Sousa PAS, Carvalho AFU (2006) Latex constituents from *Calotropis procera* (R. Br.) displaytoxicity upon egg hatching and larvae of *Aedes aegypti* (Linn). Mem Instit Oswaldo Cruz 101(5): 503–510

Rao BR, Murugesan T, Sinha S, Saha BP, Pal M, Mandal SC (2002) Glucose lowering efficacy of *Ficus racemosa* bark extract in normal and alloxan diabetic rats. Phytother Res 16:590–592

Rao BR, Murugesan T, Pal M, Saha BP, Mandal SC (2003) Antitussive potential of methanol extract of stem bark of *Ficus racemosa* Linn. PhytotherRes 17:1117–1118

Rao YK, Vimalamma G, Rao CV, Tzeng YM (2004) Flavonoids and andrographolides from *Andrographis paniculata*. Phytochemistry 65:2317–2321

Rastogi RP, Mehrota BN (1993) Compendium of Indian medicinal plants, vols 2 and 3. Central Drug Research Institute and Publication and Information Directorate, Lucknow and New Delhi, pp 2, 174, 185, 295, 320

Rastogi RP, Mehrotra BN (1995) Compendium of Indian medicinal plants, vols 1 and 4. Central Drug Research Institute and Publication and Information Directorate, , Lucknow and New-Delhi, pp 2, 188, 321

Ratnasooriya WD, Jayakody JR, Nadarajah T (2003) Antidiuretic activity of aqueous bark extract of Sri Lankan *Ficus racemosa* in rats. Acta Biol Hung 54:357–363

Reyes BA, Bautista ND, Tanquilut NC, Anunciado RV, Leung AB, Sanchez GC, Magtoto RL, Castronuevo P, Tsukamura H, Maeda KI (2006) Anti-diabetic potentials of *Momordica charantia* and *Andrographis paniculata* and their effects on estrous cyclicity of alloxan-induced diabetic rats. J Ethnopharmacol 105:196–200

Rongsriyam Y, Trongtokit Y, Komalamisra N, Sinchaipanich N, Apiwathnasorn C, Mitrejet A (2006) Formulation of tablets from the crude extract of *Rhinacanthus nasutus* (Thai local plant) against *Aedes aegypti* and *Culex quinquefasciatus* larvae: a preliminary study. Southeast Asian J Trop Med Public Health 37(2):265–271

Rug M, Ruppel A (2000) Toxic activities of the plant *Jatropha curcas* against intermediate snail hosts and larvae of *schistosomes*. Trop Med Inter Health 5:423–430

Sabesan S, Planiyandi M, Das PK (2000) Mapping lymphatic filariasis. Ann Trop Med Parasitol 94:591

Sadhu SK, Okuyama E, Fujimoto H, Ishibashi M (2003) Separation of *Leucas aspera*, a medicinal plant of Bangladesh, guided by prostaglandin inhibitory and antioxidant activities. Chem Pharm Bull 51(5):595–598

Sahayaraj K (1998) Antifeedant effect of some plant extracts on the Asian armyworm, *Spodoptera litura* (Fabricius). Curr Sci 74:523–525

Sánchez GM, Delgado R, Pérez G (2000) Evaluation of the in vitro antioxidant activity of *Mangifera indica* L. extract (Vimang). Phytother Res 14:424–427

Sane RT, Kuber VV, Chalissery MS, Menon S (1995) Hepatoprotection by *Phyllanthus amarus* and *Phyllanthus debili* in Ccl4 induced liver dysfunction. Curr Sci 68(12):1243–1246

Seenivasan SP, Jayakumar M, Raja N, Ignacimuthu S (2004) Effect of bitter apple, *Citrullus colocynthis* (L.) Schrad seed extracts against pulse beetle, *Callosobruchus maculatus* Fab. (Coleoptera: Bruchidae). Entomon 29(1):81–84

Shaalan EAS, Canyon D, Younes MWF, Abdel-Wahab H, Mansour AH (2005) A review of botanical phytochemicals with mosquitocidal potential. Environ Int 31:1149–1166

Shahi M, Hanafi-Bojdb AA, Iranshahi M, Vatandoost H, Hanafi-Bojdd MY (2010) Larvicidal efficacy of latex and extract of *Calotropis procera* (Gentianales: Asclepiadaceae) against *Culex quinquefasciatus* and *Anopheles stephensi* (Diptera: Culicidae). J Vector Borne Dis 47:185–188

Sharma N, Trivedi PC (2002) Screening of leaf extracts of some plants for their nematicidal and fungicidal properties against *Meloidogyne incognita* and *Fusarium oxysporum*. Asian J Exp Sci 16:21–28

Sharma A, Singh RT, Handa SS (1993) Estimation of phyllanthin and hypophyllanthin by high performance liquid chromatography in *Phyllanthus amarus*. Phytochem Anal 4:226–229

Shelly TE, McInnis DO (2001) Exposure to ginger root oil enhances mating success of irradiated, mass-reared males of Mediterranean fruit fly (Diptera: Tephritidae). J Econ Entomol 94:1413–1418

Shelly TE, Rendon P, Hernandez E, Salgado S, McInnis D, Villalobos E, Liedo P (2003) Effects of diet, ginger root oil, and elevation on the mating competitiveness of male Mediterranean fruit flies (Diptera: Tephritidae) from a mass-reared, genetic sexing strain in Guatemala. J Econ Entomol 96:1132–1141

Silva TM, Batista MM, Camara CA, Agra MF (2005) Molluscicidal activity of some Brazilian *Solanum* spp. (Solanaceae) against *Biomphalaria glabrata*. Ann Trop Med Parasitol 99:419–425

Singh RN, Saratchandra B (2005) The development of botanical products with special reference to seri-ecosystem. Caspian J Environ Sci 3(1):1–8

Singhi M, Joshi V, Sharma RC, Sharma K (2004) Ovipositioning behaviour of *Aedes aegypti* in different concentrations of latex of *Calotropis procera*: studies on refractory behaviour and its sustenance across gonotrophic cycles. Dengue Bull 28:184–188

Siripong P (2006) Induction of apoptosis in tumor cells by three naphthoquinone esters isolated from Thai medicinal plant: *Rhinacanthus nasutus* Kurz. Biol Pharm Bull 29 (10):2070–2076

Siripong P, Kanokmedakul K, Piyaviriyagul S, Yahuafai J, Ruchirawat S, Ruchirawat S, Oku N (2006) Antiproliferative naphthoquinone esters from *Rhinacanthus nasutus* Kurz. roots on various cancer cells. J Trad Med 23:166–172

Sivapriya M, Srinivas L (2007) Isolation and purification of a novel antioxidant protein from thewater extract of Sundakai (*Solanum torvum*) seeds. Food Chem 104:510–517

Srivastava J, Vankar PS (2010) *Canna indica* flower: new source of anthocyanins. Plant Physiol Biochem 48:1015–1019

Subramanian SS, Nagarajan S, Sulochana N (1971) Flavonoids of the Leaves of *Jatropha Gossypifolia*-D (Oxford). Phytochemistry 10:2548–2549

Surana SJ, Gokhale SB, Jadhav RB, Sawant RL, Wadekar JB (2008) Antihyperglycemic activity of various fractions of *Cassia auriculata* Linn. in alloxan diabetic rats. Indian J Pharm Sci 70 (2):227–229

Suresh Reddy J, Rajeswara Rao P, Reddy MS (2002) Wound healing effects of *Heliotropium indicum*, *Plumbago zeylanicum* and *Acalypha indica* in rats. J Ethnopharmacol 79:249–251

Taylor WG, Fields PG, Elder JL (2004a) Insecticidal components from field pea extracts: isolation and separation of peptide mixtures related to pea albumin 1b. J Agric Food Chem 52 (25):7491–7498

Taylor WG, Fields PG, Sutherland DH (2004b) Insecticidal components from field pea extracts: soyasaponins and lysolecithins. J Agric Food Chem 52(25):7484–7490

Thyagarajan SP, Subramanian S, Thirunalasundari T, Venkateswaran PS, Blumberg BS (1988) Effect of *Phyllanthus amarus* on chronic carriers of hepatitis B virus. Lancet 2:764–766

Tinzaara W, Tushemereirwe W, Nankinga CK, Gold CS, Kashaija I (2006) The potential of using botanical insecticides for the control of the banana weevil, *Cosmopolites sordidus* (Coleoptera: Curculionidae). Afr J Biotechnol 5(20):1994–1998

Tona L, Kambu K, Ngimbi N (2000) Antiamoebic and spasmolytic activities of extracts from some antidiarrhoeal traditional preparations used in Kinshasa, Congo. Phytomedicina 7:31–38

Unander DW, Venkateswaran PS, Millman I, Bryan HH, Blumberg BS (1990) *Phyllanthus* species: sources of new antiviral compounds. In: Janick J, Simon JE (eds) Advances in new crops. Timber, Portland, USA, pp 518–521

Unander DW, Webster GL, Blumberg BS (1991) Uses and bioassays in *Phyllanthus* (Euphorbiaceae): a compilation II. The subgenus Phyllanthus. J Ethnopharmacol 34:97–133

Upasani SM, Kotkar HM, Mendki PS, Maheshwari VL (2003) Partial characterization and insecticidal properties of *Ricinus communis* L. foliage flavonoids. Pest Manage Sci 59:1349–1354

Van Quaquebeke E, Simon G, André A, Dewelle J, El Yazidi M, Bruyneel F, Tuti J, Nacoulma O, Guissou P, Decaestecker C, Braekman JC, Kiss R, Darro F (2005) Identification of a novel cardenolide (2″-oxovoruscharin) from *Calotropis procera* and the hemisynthesis of novel derivatives displaying potent in vitro antitumor activities and high in vivo tolerance: structure-activity relationship analyses. J Med Chem 48(3):849–856

Vedavathi S, Mrudula V, Sudhakar A (1997) Tribal medicine of Chittoor district, AndhraPradesh, India, vols. 48–49. Herbal Folklore Research Center, Tirupati

Vihari V (1995) Ethnobotany of cosmetics of Indo-Nepal border. Ethnobotany 7:89–94

Vineetha A, Murugan K (2009) Larvicidal and smoke repellency effect of *Toddalia asiatica* and *Aegle marmelos* against the dengue vector, *Aedes aegypti* (Insecta: Diptera: Culicidae). Entomol Res 39:61–65

Viquaruddin A, Iqbal S (1992) Cohirsutin, a new iso-quinoline alkaloid from *Cocculus hirsutus*. Fitoterapia 63:308–310

Viquaruddin A, Iqbal S (1993) Jamtinine, an alkaloid from *Cocculus hirsutus*. Phytochemistry 33:735–736

Viquaruddin A, Tahir R (1991) Cohirsinine, a new alkaloid from *Cocculus hirsutus*. Phytochemistry 30:1350–1351

Watt JM, Breyer-Brandwijk MG (1962) The medicinal and poisonous plants of Southern and Eastern Africa, vol 1, 2nd edn. Churchill, London

WBSICP (1997) World Bank stresses importance of coming phytomedicines. Newsl Asian Netw Med Aromatic Plants 23:5–6

WHO (2002) WHO Traditional Medicine Strategy 2002–2005. World Health Organization, Geneva

WHO (2004) First Meeting of the Regional Technical Advisory Group on Malaria, Manesar, Haryana, India. SEA-MAL 239:1–38

Wiart C, Mogana S, Califa S, Mahan M, Ismael S, Bucle M, Narayana AK, Sulaiman M (2004) Antimicrobial screening of plants used for traditional medicine in the state of Perak, Peninsular Malaysia. Fitoterapia 75:68–73

Williams LAD, Mansingh A (1993) Pesticidal potentials of tropical plants – I. Insecticidal activity in leaf extracts of sixty plants. Insect Sci Appl 14:697–700

Wink M (2004) Phytochemical diversity of secondary metabolites. Encyclopedia of plant and crop science. Taylor and Francis, Amsterdam, pp 915–919

Wink M, Koschmieder C, Sauerwein M, Sporer F (1997) Phorbol esters of *Jatropha curcas* – biological activities and potential applications. In: Gübitz GM, Mittelbach M, Trabi M (eds) Biofuel and industrial products from *Jatropha curcas*. Dbv- Verlag Univ, Graz

Wu TS, Hsu HC, Wu PL, Teng CM, Wu YC (1998) Rhinacanthin-Q, a naphthoquinone from *Rhinacanthus nasutus* and its biological activity. Phytochemistry 49:2001–2003

Yadav RL, Lal S, Kaul SM (1999) Malaria epidemic and its control in India. Family Med 3:39–41

Yeh SF, Hong CY, Huang YL, Liu TY, Choo KB, Chou CK (1993) Effect of an extract from *Phyllanthus amarus* on hepatitis B surface antigen gene expression in human hepatoma cells. Antivir Res 20:185–192

Yoganarasimhan SN (2000) Medicinal plants of India, vol 2. Cyber Media, Bangalore, p 10

Yoosook C, Bunyapraphatsara N, Boonyakiat Y, Kantasuk C (2000) Anti-Herpex simplex virus activities of crude water extracts of Thai medicinal plants. Phytomedicine 6:411–419

Yu BC, Hung CR, Chen WC, Cheng JT (2003) Antihyperglycemic effect of andrographolide in streptozotocin-induced diabetic rats. Planta Med 69:1075–1079

Yuanyuan LU, Jianguang L, Xuefeng H, Lingyi K (2009) Four steroidal glycosides from *Solanum torvum* and their cytotoxic activities. Steroids 74:95–101

Zahir AA, Rahuman AA, Kamaraj C, Bagavan A, Elango G, Sangaran A, Kumar BS (2009) Laboratory determination of efficacy of indigenous plant extracts for parasites control. Parasitol Res 105:453–461

Zahir AA, Rahuman AA, Bagavan A, Santhoshkumar T, Mohamed RR, Kamaraj C, Rajakumar G, Elango G, Jayaseelan C, Marimuthu S (2010) Evaluation of botanical extracts against *Haemaphysalis bispinosa* Neumann and Hippobosca maculata Leach. Parasitol Res 107 (3):585–592

Zheng MS, Lu ZY (1990) Antiviral effect of mangiferin and isomangiferin on *Herpex simplex* virus. Chin Med J 103:160–165

Zheng CJ, Qin LP (2007) Chemical components of *Centella asiatica* and their bioactivities. Chin Integr Med/Zhong Xi Yi Jie He Xue Bao 5(3):348–351

Chapter 3
Natural Remedies in the Fight Against Insects

Norbert Becker

Abstract With the discovery and large-scale use of synthetic insecticides in the 1940s and 1950s, the control of insects was mainly based on chemicals until the publication of the book "Silent Spring" by Rachel Carson. This new "green" movement was the driving force to search for new environmental compatible tools in the fight against pest and vector insects.

There are many bacteria known which are causing insect diseases and used in insect control programs. Typical for these bacteria is the production of protein crystals the so-called δ-endotoxins which are produced by the bacilli during sporulation.

Two groups of bacteria are of importance: a) *Bacillus thuringiensis* (*B.t.*) and *B. sphaericus*.

B.t. products are most widely used in insect control programs. Thousands of *B.t.* isolates are known which are grouped into three major pathotypes:

- Pathotype A: lepidopteran specific *Bacillus thuringiensis* strains such as *B.t.* H-3a/3b (*B. t. kurstaki*);
- Pathotype B: nematoceran specific strains such as *B.t.* H-14. (*B. t. israelensis*), which kill especially mosquito and blackfly larvae and some members of the suborder nematocera;
- Pathotype C: coleopteran specific strain *B.t.* H-8a/8b (*B. t. tenebrionis*).

The use of insect-specific toxins from *Bacillus thuringiensis* and *B. sphaericus* is forming an increasingly component of biological control strategies against nuisance, pest or vector species. The discovery of these microbial control agents marked the breakthrough in biological control, because of the special abilities of these microbial agents. Their protein crystals are highly toxic to target organisms and extremely environmentally safe. In Germany for instance, over 1,000 km^2 of breeding sites have been treated with *Bacillus thuringiensis israelensis* and *B. sphaericus* resulting in

---

N. Becker
University of Heidelberg and German Mosquito Control Association (KABS), Ludwigstr. 99, 67165 Waldsee, Germany
e-mail: norbertfbecker@web.de; kabs-gfs@t-online.de

a reduction of mosquito population year by year more by than 90% and without evidence of any harmful impact on the environment.

3.1 Introduction

With the discovery and large-scale use of synthetic insecticides in the 1940s and 1950s, the control of insects was mainly based on chemicals until the publication of the book "Silent Spring" by Rachel Carson made the public aware of the risk of using unspecific insecticides. This initiated a new "green" movement and was the driving force in the search for new environmentally compatible tools in the fight against pest and vector insects. In the 1970s pathogens like arboviruses, fungi, microsporidia, bacteria and parasites like nematodes were screened for their use against pest and vector insects.

3.2 Bacterial Control Agents

Amongst all pathogens/parasites bacteria have been proven to be the most promising group of organisms to combat pest and vector insects. They can be easily mass produced in fermenters and their entopathogenic toxins can be formulated to highly efficient microbial control agents which can be easily stored and handled in routine operations (Becker et al. 2003; Becker et al. 2010).

Many bacteria are known which cause insect diseases and nowadays, some are successfully used in insect control programs. Typical for these bacteria is the production of protein crystals with the so-called δ-endotoxins which are produced by the bacilli during sporulation. The evolutionary advantage of these bacilli compared to other soil bacteria is that they can more easily mass develop. The dead larval cadaver of the target organism serves as an ideal "fermenter" because contrary to normal soil the cadavers are full of nutrition for bacterial development. This advantage was obviously the driving force for entomopathogenic bacteria to spend energy in the production of protein crystals as endotoxins.

In Europe two groups of bacteria are of importance: (a) *Bacillus thuringiensis* (*B. t.*) and (b) *B. sphaericus* (*B.s*). The bacteria are predominantly characterized by their serological features, usually described as serotypes according to the antigen character of the flagella or serovarieties (serovar) according to their serological features. However, for practical reasons it is easier to distinguish them as pathotypes according to their spectrum of activity.

3.2.1 Bacillus thuringiensis

B. t. products are most widely used in insect control programs. In Europe, *B. thuringiensis* (serotype H-1) was isolated for the first time in 1911 from flour moths (*Ephestia kuehniella*) derived from a mill in Thuringia, Germany (Berliner 1915). Some years earlier it was described as silkworm disease in Japan by Ishiwata (1901). These isolates kill only larvae of Lepidoptera. The Gram-positive, endo-spore-forming soil bacterium *B. t.* ssp. is closely related to the soil bacteria *B. cereus*, but *B. t.* differs from the latter by the production of protein crystals (the so-called δ-endotoxins) during sporulation (Langenbruch et al. 2005). The genetic information for the toxins is located in one or more plasmids (seldom in the chromosomes). Nowadays, thousands of *B. t.* isolates are known which are grouped into three major pathotypes:

- Pathotype A: lepidopteran specific *B. t.* strains such as *B. t.* H-3a/3b (*B. t. kurstaki*)
- Pathotype B: nematoceran specific strains such as *B. t.* H-14. (*B. t. israelensis*), which kill especially mosquito and blackfly larvae and those of some members of the suborder nematocera
- Pathotype C: coleopteran specific strain *B. t.* H-8a/8b (*B. t. tenebrionis*)

The protein crystals (protoxins) contain usually several toxins of various molecular weights that are highly effective and specific for the target organisms. According to their insecticidal activity and molecular relationship four major classes of endotoxins are characterized (Höfte and Whiteley 1989):

1. Cry I toxins – active against larvae of lepidopterans
2. Cry II toxins – active against larvae of lepidopterans and dipterans
3. Cry III toxins – active against larvae of coleopterans (Chrysomelidae) and
4. Cry IV toxins – active against larvae of nematoceran flies

The toxins can also be further categorized in subclasses such as CryIA etc. Because of the intensive search for new isolates the nomenclature of the toxins is steadily updated (Langenbruch et al. 2005). Additional toxins can be involved in the mode of action as for example, Cyt-toxins in strains toxic to larvae of nematoceran flies and coleopterans.

The genes of the toxins are named as the toxins but without capital letters and in italic (e.g., *cryIV*).

The selectivity of the *Bacillus* derives from a variety of factors.

The target organism must:

1. Ingest the protein crystal (inactive protoxin), and this depends on its feeding habit;
2. The protoxin has to be solubilized in the midgut milieu of the target insect;
3. Possess suitable proteases to convert the protoxin into biologically active toxins;
4. Possess surface receptors (glycoproteins) on the midgut epithelial cells to which the toxins can bind.

This process disturbs the osmo-regulatory mechanisms of the cell membrane, thereby swelling and bursting the midgut cells (Fig. 3.1). Nontarget organisms do not activate the protoxin into the toxin because of their unsuitable gut milieu, or remain undamaged because of the lack of specific receptors on their intestinal cells

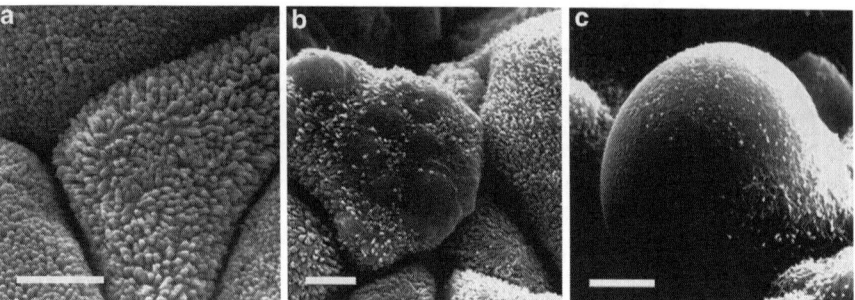

Fig. 3.1 (a) Midgut epithelium of a healthy *Ae. aegypti* larva; (b) 30 min after ingestion of the *B. t. i.* protein crystals, swelling of the midgut cells and reduction of microvilli; (c) 1 h after ingestion cell is about to lyse (micrographs courtesy of J.-F. Charles, Pasteur Institute, Paris)

(Charles and Nielsen-LeRoux 2000; Lüthy and Wolfersberger 2000; Boisvert 2005; Becker et al. 2010).

The protoxin structure contains three domains. Domain I consists of a bundle of seven alpha-helices, which can insert into the gut cell membrane, creating a pore through which ions can pass freely. Domain II contains three antiparallel beta-sheets, which are thought to bind to receptors in the midgut. Domain III is a beta-sandwich which obviously prevents further cleavage of the active toxin by gut proteases (Li et al. 1991).

Formulations

A basic requirement for the successful use of bacterial control agents is the development of effective formulations suited to the biology and habitats of the target organisms. Formulations are available as water-dispersible granules, wettable powders, fluid concentrates, corn-cob or ice granules, pellets, tablets or briquettes containing the bacilli toxins, spores, and inert materials. After harvesting the raw product of fermentation, the fermentation mud is spray-dried to primary powder and then formulated. According to the yield of the fermenter process and the content of toxins the primary powder is then mixed with inert material in order to achieve a standardized product.

Standardization

Standardized methods have been developed to determine the LC_{50} values for microbial larvicides using standard formulations (e.g., IPS 82 for *B. t. israelensis*) as reference products (Langenbruch et al. 2005; Becker et al. 2010). The activity of the standard formulation has been assigned and expressed in "international toxic units" (ITUs or IUs) (e.g., 15,000 ITUs/mg for *B. t. i.* and 16,000 IU *T. ni*/mg for *B. t.* H-1). In a standardized series of bioassays the mortality rates (activity) of the standard product and the product in question are assessed against test organisms

3 Natural Remedies in the Fight Against Insects

($B. t. i.$ = *Aedes aegypti* larvae; $B. t.$ H-1 = larvae of *Trichoplusia ni*). Then the LC_{50} of the series treated with the standard and with the test preparations are read and the potency (titer) of the test material determined by the following formula:

$$\frac{\text{Activity of the standard}(ITU) \times LC_{50} \text{ of standard}}{LC_{50} \text{ of test preparation}}.$$

The potency or titer of the product is then expressed in ITUs or IUs.

Environmental Safety

The exceptional environmental safety of bacterial control agents has been confirmed in numerous laboratory and field tests. The U.S. Environmental Protection Agency (USEPA) categorizes the risk posed by $B. t.$ strains to nontarget organisms as minimal to nonexistent. After more than 50 years of use of $B. t.$ products it can be stated that microbial control agents belong to the most environmentally safe products, which kill target organisms and usually do not harm nontarget organisms such as beneficial insects, plants, or humans.

Ease of Handling

No special equipment is required for the application of bacterial control agents. Generally, simple knapsack sprayers are adequate for accessible breeding sites. Standard ULV (ultra-low volume), airblast or mist blower equipment may also be used. In areas with dense vegetation or wide-spread breeding sites, aerial applications should be preferred. Rotary seeders, nozzle atomizers, or pressurized air sprayers are suitable for the application of granules. Safety precautions are minimal compared to the use of toxic agents. Because of the rapid knock-down effect and the high level of efficiency, the success of the treatment can generally be monitored within a day or two after application.

Lack of Potential for Resistance Development

The development of resistance to chemical insecticides represents a serious problem. Bacterial control agents, however, appear less likely to provoke resistance because their mode of action is more complex (Davidson 1990; Wirth et al. 2005). However, in comparison, resistance in the stored grain pest, *Plodia interpunctella*, to the Lepidoptera specific $B. t. kurstaki$ has been demonstrated in the laboratory (McGaughey 1985). Studies have shown that the commercial use of $B. t.$ preparations in agriculture can lead to resistance within a few years. For example, the diamondback moth, *Plutella xylostella,* which was repeatedly treated with $B. t. k.$ on farms in Hawaii, was found to be 41 times more resistant than populations that were only minimally exposed to *B. t. k.* (Tabashnik et al. 1990). Such resistance phenomena have not yet been observed with *B. t. i.* Resistance studies have been carried out by KABS (German Mosquito Control Association/Kommunale Aktionsgemeinschaft zur Bekämpfung der Stechmückenplage-KABS) with populations of *Ae. vexans* which were constantly exposed to *B. t. i.* over a period of more than 25 years and were therefore subjected to constant and intense selection pressure. The complex mode of action of *B. t. i.* partly explains the relative absence of resistance. The lethal changes in the midgut cells are induced only by the synergistic effects of the different toxin proteins present in the parasporal body of *B. t. i.* This combination reduces the likelihood of resistance. On the other hand, when the gene encoding a single toxin protein was cloned into a microorganism and then fed to larval mosquitoes, resistance was induced within a few generations (Georghiou and Wirth 1997).

In the following paragraphs the different pathotypes of *B. t.* and their practical use will be described.

3.2.1.1 Pathotype A – Lepidopteran Specific Strains

Since the discovery of the first lepidopteran specific strain of *B. t.* (serovar *thuringiensis*; serotype H-1) in 1911 many *B. t.* serovarieties have been isolated which belong to pathotype A. Typical for them is the production of bipyramidal protein crystals that contain several toxins mainly belonging to toxin type Cry1, seldom they contain also Cry2 toxins. The typical protoxins of the Cry 1 proteins have a molecular weight of ~130 KDa; when this parasporal crystal protein is ingested by the lepidopteran larva it gets activated in the alkaline gut milieu of the larva by proteolytic enzymes and converted into the active toxin with a molecular weight of 68 KDa.

In Germany, two *B. t.* serovar are of commercial importance to control pest caterpillars, namely *B. t. kurstaki* (product: Dipel) and *B. t. aizawai* (products: Turex and XenTari).

Dipel is one of the oldest pathotype A (*B. t. kurstaki*) products used against lepidopteran larvae in Germany. It was already registered in 1972. It can be used as a wettable powder or as an oily emulsion (Dipel ES). According to the sensibility of the target organisms, the larval stage, number of larvae, weather conditions, development of the host plants and application equipment, a dosage of 300–3,000 ml of the emulsion mixed with a certain amount of water has to be applied. Turex and XenTari based on *B. t. aizawai* is especially effective against noctuids in cabbage cultures. For ground application in forests usually 600 l of water suspension are used, and when helicopters are used with a nozzle atomizer 50 l of suspension are applied to 1 ha. The application has to be done according to the prescription and only when the weather conditions are suitable, for example when the temperature is higher than 15°C.

The main target organisms for products of pathotype A are:

- The gypsi moth (*Lymantria dispar*) and the oak processionary moth (*Thaumetopoea processionea*) – pest species especially on oak trees. Usually Dipel ES is applied with a helicopter at a rate of 50 l of water suspension containing 2 l of Dipel ES. The winter moth (*Operophtera brumata*), the cabbage white butterfly (*Pieris brassicae*), the diamondback moth (*Plutella xylostella*), the garden pebble moth (*Evergestis forficalis*), the cabbage moth (*Mamestra brassicae*) and other moths as agricultural pest species can be successfully controlled with *B. t.* products by ground and aerial application.

– Storage pest species such as the tobacco moth (*Ephestia elutella*) and the Indian meal moth (*Plodia interpunctella*) or the bee moth (*Galleria mellonella*) are further target organisms.

In Germany the use of *B. t.* products is steadily increasing due to serious problems with pest caterpillars whose development is encouraged by the increasing temperatures over the last 10 years. So far no resistance has occurred against *B. t.* products in Germany.

3.2.1.2 Pathotype B – Nematoceran Specific Strains

Mosquitocidal bacteria have been known since the early 1960s when the first strains of *Bacillus sphaericus* with larvicidal activity were discovered (Kellen and Meyers 1964). However, these strains were not sufficiently toxic to merit commercial development. The discovery of the Gram-positive, endospore-forming soil bacterium *Bacillus thuringiensis. israelensis* (*B. t. i.*) in the Negev desert of Israel in 1976 (Goldberg and Margalit 1977) has inaugurated a new chapter in the control of mosquitoes and blackflies (Singer 1973; Weiser 1984; Becker and Margalit 1993).

Bacillus thuringiensis israelensis

The protein toxins of *B. t. i.* are highly toxic to mosquito and blackfly larvae. The insecticidal effect emanates from the more or less spherical protein crystal or parasporal body (PSB), which contains four major toxin proteins of different molecular weight, referred to as Cry4A (125 kDa), Cry4B (135 kDa), Cry10A (58 kDa), and Cry11A (68 kDa) (Delecluse et al. 1996). These toxins bind to specific glycoprotein receptors on the larval midgut brush border (Charles and Nielsen- LeRoux 1996).

A fifth toxin, called the CytA protein (27 kDa), binds to lipids and does not exhibit the specific binding mechanism which the Cry proteins do (Höfte and Whiteley 1989; Federici et al. 1990; Priest 1992) (Fig. 3.2).

It has been shown that plasmids with a molecular weight of 60–94 MDa play an essential role in the crystal toxin production. Cloning and characterization of the different toxin genes have been accomplished (Bourgouin et al. 1986; Thorne et al. 1986; Ward and Ellar 1988). These results open the possibility of cloning the various toxin genes into host organisms to increase the persistence or the mosquitocidal properties.

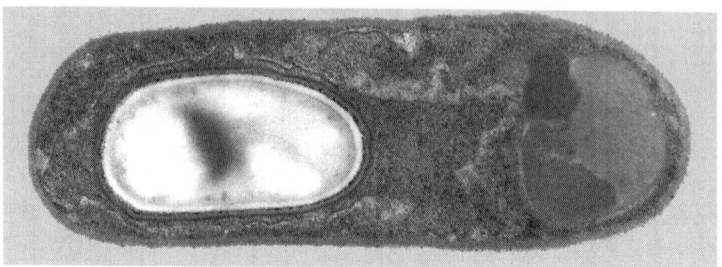

Fig. 3.2 *B. t. i.* with spore (*left*) and protein crystal (*right*) (micrograph courtesy of J.-F. Charles, Pasteur Institute, Paris)

Neither the spore nor the living bacilli are involved in the insecticidal process. The more or less spherical protein crystal is formed at the end of sporulation and consists of three types of protein inclusions separated by thin layers (Ibarra and Federici 1986; Federici et al. 1990).

The mosquitocidal properties of each single solubilized and purified protein have been evaluated in many studies (Beltrao and Silva-Filha 2007; Alam et al. 2008). All tests have shown that each type of protein is mosquitocidal, but none is nearly as toxic as the intact PSB. This high toxicity of the protein crystal is caused by a synergistic interaction of the 25-kDa protein (split from the 27-kDa protein) with one or more of the higher molecular weight proteins (Ibarra and Federici 1986; Chilcott and Ellar 1988; Chang et al. 1993). It is assumed that CytA may function as a Cry11A receptor in the mosquito midgut (Perez et al. 2005). It is thought that the synergism in the mode of action among the proteins reduces the probability of resistance.

The high toxicity of the protein crystal to a great variety of mosquito and blackfly species is the most remarkable property of *B. t. i.* Only at significantly higher dosages are certain other nematocera species affected, but no other organisms are harmed.

B. t. i. was rapidly developed with the support of industry, universities, and national and international organizations such as the World Health Organization (WHO). The WHO Pesticide Evaluation Scheme (WHOPES) which promotes and coordinates the testing and evaluation of pesticides for public health, have evaluated microbial control agents as safe and efficient pesticides for public health programs (WHO/CDS/WHOPES/2004.8). Following extensive safety tests and environmental impact studies, the bacilli were quickly put into use. This rapid exploitation was aided by a series of useful properties of the bacterial control agents. In addition to the relative ease with which they can be mass-produced, bacterial control agents are highly efficient, environmentally safe, easy to handle, stable when stored, cost-effective, and suitable for integrated mosquito management (IMM) programs based on community participation. Furthermore, the costs for development and registration of these agents are many times lower than those for a conventional chemical insecticide. The risk of resistance, especially when *B. t. i.* is used, is much lower than when conventional insecticides are used.

Environmental Safety

The exceptional environmental safety of *B. t. i.* has been confirmed in numerous laboratory and field tests. USEPA categorizes the risk posed by *B. t.* strains to nontarget organisms as minimal to nonexistent and approved the use of *B. t. i.* as early as 1981. In safety tests on representative aquatic organisms, it was shown that in addition to plants and mammals none of the taxa tested such as Cnidaria, Turbellaria, Rotatoria, Mollusca, Annelida, Acari, Crustacea, Ephemeroptera, Odonata, Heteroptera, Coleoptera, Trichoptera, Pisces, and Amphibia appeared to be affected when exposed in water containing large amounts of bacterial preparations (Becker and Margalit 1993; Boisvert and Boisvert 2000; Becker et al. 2010).

Even within the Diptera, the toxicity of *B. t. i.* is restricted to mosquitoes and a few nematocerous families (Garcia et al. 1981; Molloy and Jamnback 1981; Margalit and Dean 1985; Mulla et al. 1982; WHO/IPCS 1999). In addition to mosquito and blackfly larvae, only those of the closely related Dixidae are sensitive to *B. t. i.* Larvae of Psychodidae, Chironomidae, Sciaridae, and Tipulidae are generally far less sensitive than those of mosquitoes or blackflies.

Toxicological tests were carried out using various mammals. *B. t. i.* when given orally, sub- and percutaneously, intraperitoneally, ocularly, through inhalation and scarification appeared to be innocuous even at high dosages of 10^8 bacteria/animal (WHO 1999).

Another important aspect is the widespread occurrence of both bacilli in the soil. They are natural components of the soil microecosystem and not an artificial man-made product where toxic residues may remain after application against nuisance/vector species.

Ease of Handling

No special equipment is required for the application of *B. t.* products. Generally, simple knapsack sprayers are adequate for accessible breeding sites. Standard ULV, airblast or mist blower equipment may also be used. Rotary seeders or pressurized air sprayers are suitable for the application of granules.

Cost Effectiveness

Compared to conventional insecticides, the application of *B. t. i.* is cost-effective when integrated control strategies are designed. For instance, the German Mosquito Control Association (KABS), mosquito abatement project in Germany effectively suppresses mosquitoes originating from a catchment area of more than 600 km^2 involving an average 100 km^2 of actual breeding grounds/year. The total annual budget is about €3 million. More than three million residents of the area are protected from a serious nuisance. In a cost–benefit analysis, balancing public opinion and evaluation of the control measures with the program's costs, it was found that the willingness to pay for the campaign/person (€3.50) is 3.8 times higher than the cost of the campaign/person (€0.92). The consumer surplus, the difference between the total value of €8.365 million and the actual costs of €3 million, is approximately €5.365 million. This amount is KABS' average contribution to societal welfare in this region during the appraisal (Hirsch and Becker 2009).

Environmental considerations, which cannot be expressed in monetary terms, should also be included in these economic calculations.

Formulations

B. t. i. formulations are available as water-dispersible granules (e.g., Vectobac WDG and Vectolex WDG), wettable powders, fluid concentrates (e.g., Vectobac 12AS and Aquabac), corn-cob (Vectobac G and Vectolex G), pellets, tablets (e.g., Vectobac DT/Culinex tablets), briquettes, and ice granules containing the bacilli toxins.

A few hundred grams or even less of powder, 0.5–2 l of liquid concentrate or a few kilograms of granules/ha, are usually enough to kill all mosquito larvae. In some situations, a long-term effect can be achieved if larger amounts are used (Becker and Margalit 1993; Becker and Rettich 1994; Russell et al. 2003; Rydzanicz et al. 2009). With the production of tablet, water-soluble pouches or briquette formulations, progress has been made towards achieving long-term effects (Kahindi et al. 2008; Su 2008).

Tablet formulations, based on *B. t. i.* sterilized by γ-radiation to prevent contamination of drinking water with spores, are successfully used for control of container-breeding mosquitoes such as *Cx. p. pipiens* or *Ae. aegypti* or even against *An. gambiae* (Becker et al. 1991; Kroeger et al. 1995; Mahilum et al. 2005).

In addition to commercially available granules based on ground corn-cobs, sand granules can also serve as a carrier for wettable powder formulations: 50-kg fire-dried quartz sand (grain size 1–2 mm) with 0.8–1.4 l of vegetable oil (as a binding material) and 1.8 kg of *B. t. i.* powder (activity >5,000 ITU/mg) could be mixed in a cement mixer. This mixture is sufficient to treat 2–3 ha. Recently, more cost-effective granules were developed in the form of ice pellets (Becker 2003). Ice granules can be easily produced when water suspensions containing the bacterial toxins are frozen into small ice cubes or pearls (3–5 mm) and kept in cold storage rooms until used. The advantages of using ice granules are:

1. The toxins are bound in the ice pellet, so loss of active material by friction during application is avoided.
2. As the specific weight of ice is less than that of water, the ice pellets remain in the upper water layer where they release the toxins into the feeding zone of mosquito larvae as they melt.
3. The ice pellets penetrate dense vegetation and do not stick to leaves even when it is raining and are less sensitive to wind.

4. Increased swath because the friction is reduced due to the physical properties of ice.
5. The production is cost-effective and
6. the "carrier" is water.

The amount of active material/ha can thus be significantly reduced when compared with granules based on sand.

When appropriately stored, most preparations based on bacterial toxins can be kept for a long period of time without losing activity. Experience has shown that powder or corn-cob formulations lose no or only little of their activity even after many years in storage.

Factors Influencing the Efficacy of Bacterial Control Agents

In addition to the different susceptibility of various mosquito species to bacterial toxins, a variety of factors can influence their efficacy (Becker et al. 1993; Ludwig et al. 1994; Puchi 2005; Sorensen et al. 2007). This efficacy depends upon the developmental stage of the target organisms, their feeding behavior, organic content of the water, the filtration effect of target larvae as well as that of other nontarget organisms, photosensitivity and other abiotic factors such as water temperature and depth, the sedimentation rate as well as the shelf-life of $B.\ t.\ i.$ and $B.\ s.$ (*Bacillus sphaericus*) formulations (Mulla et al. 1990; Becker et al. 1992). The long-term effect is also strongly influenced by the recycling capacity of the agent (Aly 1985; Becker et al. 1995).

Use of Microbial Control Agents in Integrated Mosquito and Vector Management Programs

Many control programs aim at integrated biological control (IBC) strategies which collectively affect the protection of humans against mosquitoes and the conservation of biodiversity with minimal impact on the environment. When the ecosystem is compared with a web and each group of organisms represents one mesh, the strategy is to effectively reduce it to one single mesh (representing mosquitoes) without affecting other meshes in the "food web."

This goal could only be achieved to an optimum, when microbial control agents are used as a component in the overall IBC strategy. The conservation and encouragement of predators are important components of IBC. Therefore, microbial agents and biological methods are integrated with environmental management (e.g., improving ditch systems for regulating water levels and creating permanent habitats for aquatic predators such as fish). IBC programs have been successfully implemented in Quebec (Canada), Switzerland, Sweden and particularly in Germany.

For the successful implementation and use of microbial control agents the following prerequisites are necessary: entomological investigations, precise mapping and logging data on all major breeding sites, assessment of effective dosages in bioassays and conducting field tests, adaptations of application techniques compatible to the requirements in the field, design of the control strategy as well as training of the field staff and finally, dealing with governmental bureaucracies.

Mosquitoes breed in a great variety of habitats. Almost everywhere where stagnant waters are present, the probabilities of finding mosquito larvae will be high. Therefore, formulations and dosages under consideration must be compatible with the ecology of breeding sites and factors affecting the efficacy of microbial control agents.

Two different modes of application of *B. t. i.* should be considered: aerial and ground applications. Fixed-wing aircraft and/or helicopters equipped with conventional buckets, booms and nozzles or rotary mist atomizers are commonly used to treat densely vegetated or/and large surface areas. For ground treatments, the knapsack sprayers or motorized backpack blowers are the conventional equipment used with diluted material or granular formulations.

Various commercial products of *B. t. i.* are used in many countries, on almost all continents (Becker and Margalit 1993; Russell et al. 2003; Puchi 2005; Boisvert 2005; Kahindi et al. 2008). In North America, microbial control agents have been successfully used for ~25 years in nuisance insect control programs and since the beginning of the new millennium, to control mosquitoes transmitting West Nile virus (WNV). In Quebec, each year more than 50 tons of formulations of *B. t. i.* are used against spring species (*Aedes* and *Ochlerotatus*) and summer species (*Culex, Coquillettidia, Anopheles*) to reduce populations of potential vectors of WNV. Contrary to other Canadian provinces adulticides are seldom used in Québec to control mosquito populations.

For the past three decades in Germany, *B. t. i.* have been successfully used against floodwater mosquitoes (*Ae. vexans*) and *Culex* mosquitoes (*Cx. p. pipiens* biotype *molestus*). Over 3,000 km^2 of breeding sites have been treated with *B. t. i.* in the past three decades, resulting in a reduction of the mosquito population every year by >90%. The flood plains of the Rhine are usually inundated several times each summer. The extent of the flooding depends on the snow-melt in the Alps and on rainfall, and it is necessary to constantly monitor the water flow in the Rhine and in the flood plains. During flooding, *Ae. vexans* and other floodwater species hatch within minutes or hours at temperatures exceeding 8°C. Prior to control measures, the larval density and the larval stages are checked by means of sample scoops at representative breeding sites to justify actions taken and to establish correct dosages and the appropriate formulations applied. One day after the application, spot sample scoops are taken to check mosquito density to establish treatment efficacy.

Depending upon the extent of the flooding, 20–30% of the potential breeding sites (~600 km^2) are treated routinely by the 300 collaborators of KABS. 200–400 g of Vectobac WDG (3,000 ITU/mg) or 0.5–1 l of the liquid concentrate Vectobac 12AS (1,200 ITU/mg) are dissolved in 9–10 l of filtered pond water for each hectare treated and applied by knapsack sprayers (Fig. 3.3). For polluted breeding sites or when late instar larvae are present, the dosage has to be increased.

During the most severe floods, usually a third of the area has to be treated with *B. t. i.* granules which are dispensed with the aid of a helicopter (dosages: 10–20 kg/ha).

Fig. 3.3 Ground application of microbial control agents by means of knapsack sprayers

From 1981 to 2009, 100 tons of *B. t. i.* powder and fluid concentrate and almost 1,000 tons of *B. t. i.* ice and sand granules mixed with quartz sand, vegetable oil and *B. t. i.* powder, were used to treat thousands of hectares of floodwater mosquito breeding sites by ground application or by helicopters (Fig. 3.4).

Control of urban mosquito species is mainly carried out by householders or inhabitants. To assist with this, KABS provides information on the biology of *Cx. p. pipiens* biotype *molestus* and the appropriate control measures. Vectobac® DT/Culinex® tablets which contain toxins of one or both bacilli have been particularly successful. They kill *Culex* larvae in water containers over a period of several weeks. In drainage systems and large cesspools with eutrophic water bodies, *B. s.* as a liquid or powder formulation is applied against *Culex* larvae. Each year about one million Culinex–*B. t. i./B. s.* tablets are successfully utilized by residents against *Cx. p. pipiens*, especially in rainwater containers.

Control of Blackflies

In Germany approximately 50 Simuliidae species are described. Especially, the females of the anthropophilic species like *Simulium erythrocephalum, S. ornatum, S. equinum, S. lineatum, S. reptans,* and *S. trifasciatum* are serious pests.

Considering the fact that Simuliidae are vectors of human and animal diseases their control is crucial but should be carried out without harming the environment. The effectiveness and environmental compatibility of VectoBac 12 AS have been proved for years. For an efficient and reasonable treatment some decisive factors have to be considered. First, an optimal proportion between concentration and time of exposure must be given. According to the larval stage and larval density this proportion varies. The temperature also effects the development of the larvae. With

Fig. 3.4 Aerial application of *B. t. i.* granules

rising temperatures larval growth accelerates and ingestion increases. At the same time, the pupation rate rises.

These factors have been considered in laboratory tests and the optimum dosages have been determined under our field conditions. The concentration of 20 ppm of Vectobac applied by a knapsack sprayer for 15 min at 12°C, showed a mortality rate of 91%.

Control of Other Nematoceran Flies

Beside the control of larvae of mosquitoes and blackflies some other members of the suborder nematocera can be controlled by *B. t. i.* However, usually much higher dosages are needed. For instance, chironomid midges can be effectively controlled with *B. t. i.* granules such as Chirex. This granule is based on organic inert material and 10% of Vectobac WDG. The granules sink to the bottom of the water bodies where the midge larvae are feeding and developing. The granules can be applied from a boat or by helicopter with a rotary seeder. At a rate of 60 kg/ha of Chirex (containing 6 kg of Vectobac WDG) mortality rates of more than 80% against species such as *Chironomus thummi* can be achieved.

In grass lands such as golf courses, soccer fields, or meadows tipulid larvae, for example larvae of *Tipula paludosa*, can cause tremendous damage. They can be successfully controlled when 1,000 l of a 2% water suspension containing Vectobac WDG are applied to 1 ha by means of a nozzle atomizer on a truck or by knapsack sprayers when smaller areas are affected. The smaller larvae are more sensitive than older ones. Larvae undergoing pupation or pupae cannot be controlled because they stop feeding on *B. t. i.* toxins.

In garden centers or mushroom cultures sciarids can be serious pest species causing damage to the roots of plants (e.g., larvae of *Bradysia* spp.) The sciarid larvae can be effectively controlled when *B. t. i.* (e.g., Vectobac 12AS) is added to the irrigation water or straight into the watering can. The dosage has to be chosen according to the infestation or application system. Usually less than 1 ml of Vectobac 12AS per watering can of 5 l is enough.

Filter flies such as *Psychoda alternata* – usually associated with filter beds in sewage plants, e.g., with volcanic slag, can be a significant nuisance or even cause allergic reaction in and close to the sewage plants. The application of Vectobac 12AS at a rate of 10 ppm mixed into the sewage water for about 2 h once a month from April to August reduces the number of psychodid larvae by about 80%. Nontarget organisms in the filter bed are not harmed and contribute to the purification of the sewage water.

3.2.1.3 Pathotype C – Coleopteran Specific Strains

For almost 30 years *B. t.* strains have been known that kill leaf beetles (Chrysomelidae), however, with the discovery of *B. t.* strain BI256-82 in the mealworm beetle *Tenebrio molitor* (Tenebronidae) an effective strain could be rapidly developed for the control of some pest beetles such as the Colorado potato beetle *Leptinotarsa decemlineata*. This *B. t.* strain is called *B. t. tenebrionis* according to the original host organism in which it was first found. This strain was first isolated at the Federal Institute for Biological Control, Darmstadt, Germany in 1982 and the efficacy against Colorado beetles was determined at the University of Heidelberg (Krieg 1986; Langenbruch et al. 2005).

B. t. t. produces plate-like protein crystals, mostly two in one cell. The protein crystal (protoxin) has a molecular weight of 73 kDa. In the acidic gut milieu of some beetle larvae the protoxin is activated by proteases and cleaved into the active toxins with a weight of 55 kDa. Several toxins which belong to the Cry3 group are known (e.g., Cry3Aa, 3Ba, 3Bb, or 3Ca) and a Cyt protein (Cyt 2B) can also be involved in the mode of action.

B. t. t. was soon after its discovery produced and formulated for the wide-scale control of pest leaf beetles. In Germany Novodor FC was the first product registered for the control of the Colorado potato beetle and other chrysomelids such as the alder leaf beetle (*Agelastica alni*).

B. t. t. can be applied as a water suspension at a rate of 3-l Novodor/ha against first and second instars of the Colorado beetle or as a 1% suspension against the first instar larvae of the alder leaf beetle. Fourth larval instars of the target organisms need a much higher dosage (Langenbruch et al. 2005).

3.2.2 Bacillus sphaericus

In addition to *B. t. i.*, a second spore-forming bacterium, *Bacillus sphaericus* (*B. s.*) has become increasingly important for mosquito control programs in recent years. The high potential of *B. s.* as a bacterial control agent lies in its spectrum of efficacy and its ability to recycle or to persist in nature under certain conditions, which means that long-term control can be achieved (Hertlein et al. 1979; Mulligan et al. 1980; Lacey 1990; Ludwig et al. 1994; Silva-Filha et al. 2001). The time span between retreatments can thus be extended and personnel costs reduced. This opens up the possibility of a successful and cost-effective control of *Culex* species, particularly of *Cx. pipiens* as a vector of West Nile viruses and *Cx. p. quinquefasciatus* which is the most important vector of lymphatic filariasis, usually breeding in highly polluted water bodies in urban areas.

B. sphaericus can easily be identified by its round spore located terminally in a swollen sporangium (Fig. 3.5). *B. s.* kills only mosquito larvae and, when higher dosages are applied, larvae of Psychodidae also. Certain mosquito species, such as *Cx. p. quinquefasciatus* and *An. gambiae*, are highly susceptible, whereas *Ae. aegypti* larvae are more than 100 times less susceptible. Blackfly larvae, other insects, mammals, and other nontarget organisms are not susceptible to *B.s.* Appropriate formulations have shown a significant residual activity against larvae of *Culex p. pipiens* and *Cx. p. quinquefasciatus* in highly polluted breeding habitats (Davidson and Yousten 1990; Lacey 1990; Becker et al. 1995).

Its mosquitocidal efficacy is based on two different types of toxins: (a) parasporal protein inclusions called binary toxin (Bin) and (b) mosquitocidal toxins (Mtx1: 100 kDa; Mtx2: 31 kDa; Mtx3: 36 kDa). The parasporal protein inclusions are located

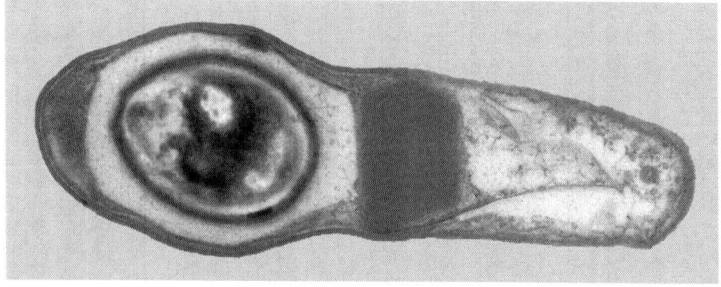

Fig. 3.5 *B. sphaericus* with round spore and protein crystal (dark structure) on the right side of the spore (micrograph courtesy of J.-F. Charles, Pasteur Institute, Paris)

in a coated "spore crystal complex" and consist of proteins of two different molecular weights, a 42-kDa toxin (BinA) and a 51-kDa binding domain (BinB). Both are required for a high level of mosquitocidal activity and are crystallized during sporulation (Broadwell et al. 1990; Baumann et al. 1991; Berry et al. 1991; Priest 1992; Davidson and Becker 1996). The Mtx toxins do not form crystals and degrade quickly upon synthesis during the vegetative stage (Wirth et al. 2007). The mode of action and receptor-binding is similar to that in *B. t. i.* (Davidson 1988; Davidson and Yousten 1990; Charles et al. 1996; Charles and Nielsen-LeRoux 2000).

A particularly attractive feature of *B. s.* is its high efficacy against *Culex* species and its potential to persist and recycle under certain field conditions (Des Rochers and Garcia 1984). In laboratory tests it was shown that the presence of mosquito larval cadavers in the water contributes to the maintenance of toxic levels of *B. s.* Larval cadavers seem to contain all the nutrients necessary, both for vegetative multiplication of the bacteria and for toxin synthesis associated with the sporulation process. Aly (1985) was also able to demonstrate experimentally, the germination of *B. t. i.* in the gut of Aedini larvae.

The application of *B. s.* products is conducted in the same manner as *B. t. i.* products. In most programs *B. s.* is specially used to control *Culex* populations and some anopheline species because of its high efficacy against these species. In some cases resistance was observed against *B. s.*, therefore rotation of *B. s.* with *B. t. i.* products is advised (Sinegre 1984). A recent combined formulation VectoMax® (Valent BioSciences, USA) contains toxins of both bacilli to enlarge the target spectrum and to avoid the onset of resistance against *B. s.*

3.3 The Role of Microbial Control Agents in Malaria Control Programs

The recently initiated global campaign against malaria, particularly in Africa, lead by major international organizations such as Roll Back Malaria (RBM), WHO, President's Malaria Initiative (PMI), Bill and Melinda Gates Foundation (BMGF), Global Fund to Fight AIDS, Tuberculosis and Malaria (GFATM), aims to reduce malaria-related mortality significantly after full implementation of the program. The cornerstones of these strategies include the extension of the facilities for rapid case detection and disease treatment, prophylactics, personal protection such as long-lasting insecticide-treated nets (LLINs) against adult vector-mosquitoes and indoor residual spraying (IRS) (RBM 2005; Makundi et al. 2007; Protopopoff et al. 2007a, b; Aregawi et al. 2008).

The emergence of resistance to both, pesticides and drugs is the most dangerous and difficult to solve problem (Etang et al. 2004; N'Guessan et al. 2007). Furthermore, the use of treated bed-nets and indoor spraying focus only on the control of endophagic and endophilic anophelines, exophagic and exophilic species could still transmit malarial pathogens.

Therefore, research on the development of locally viable integrated vector management (IVM) strategies including larval control and development, testing and/or implementation of sustainable, environmentally sound and cost-effective alternative strategies to DDT use, was strongly encouraged (WHO 2004). Specifically, the use of *B. t. i.* and *B. s.* for mosquito-larval control has been demonstrated to be highly effective for the control of malaria vector-mosquitoes in Africa (Killeen et al. 2002a, b; Fillinger et al. 2003; Yohannes et al. 2005; Fillinger and Lindsay 2006; Majambere et al. 2007; Walker and Lynch 2007; Fillinger et al. 2008).

References

Alam KA, Khan SA, Seheli K, Huda N, Wadud A, Reza SH, Ali E, Mandal C, Salam A (2008) Mosquitocidal activity of Bti Producing Cry Protein against *Aedes aegypti* mosquito. Res J Environ Sci 2:46–51

Aly C (1985) Germination of *Bacillus thuringiensis* var. *israelensis* spores in the gut of *Aedes* larvae (Diptera: Culicidae). J Invertebr Pathol 45:1–8

Aregawi M, Williams R, Dye C, Cibulskis R, Otten M (2008) World Malaria Report 2008. WHO, Geneva, p 118

Baumann PM, Clark A, Baumann L, Broadwell AH (1991) *Bacillus sphaericus* as a mosquito pathogen: properties of the organism and its toxins. Microbiol Rev 55:425–436

Becker N (2003) Ice granules containing endotoxins of microbial control agents for the control of mosquito larvae: a new application technique. J Am Mosq Control Assoc 19:63–66

Becker N, Margalit J (1993) Control of Dipteran pests by *Bacillus thuringiensis* in:*Bacillus thuringiensis:* its uses and future as a biological insecticide. Wiley, Sussex, England

Becker N, Rettich F (1994) Protocol for the introduction of new *Bacillus thuringiensis israelensis* products into the routine mosquito control program in Germany. J Am Mosq Control Assoc 10(4):527–533

Becker N, Djakaria S, Kaiser A, Zulhasril O, Ludwig HW (1991) Efficacy of a new tablet formulation of an asporogenous strain of *Bacillus thuringiensis israelensis* against larvae of *Aedes aegypti*. Bull Soc Vector Ecol 16(1):176–182

Becker N, Zgomba M, Ludwig M, Petric D, Rettich F (1992) Factors influencing the activity of *Bacillus thuringiensis* var. *israelensis* treatments. J Am Mosq Control Assoc 8(3):285–289

Becker N, Ludwig M, Beck M, Zgomba M (1993) The impact of environmental factors on the efficacy of *Bacillus sphaericus* against *Culex pipiens*. Bull Soc Vector Ecol 18(1):61–66

Becker N, Zgomba M, Petric D, Beck M, Ludwig M (1995) Role of larval cadavers in recycling processes of *Bacillus sphaericus*. J Am Mosq Control Assoc 11(3):329–334

Becker N, Petrić D, Zgomba M, Boase C, Dahl C, Lane J, Kaiser A (2003) Mosquitoes and their control. Kluwer Academic/Plenum, New York, p 498

Becker N, Petrić D, Zgomba M, Boase C, Madon M, Dahl C, Kaiser A (2010) Mosquitoes and their control. Springer, Heidelberg, p 577

Beltrao HM, Silva-Filha MH (2007) Interaction of *Bacillus thuringiensis* var. *israelensis* Cry toxins with binding sites from *Aedes aegypti* (Diptera: Culicidae) larvae midgut. FEMS Microbiol Lett 266(2):163–169

Berliner E (1915) Über die Schlaffsucht der Mehlmottenraupen (*Ephestia kühniella* Zell.) und ihren Erreger *Bacillus thuringiensis* n. sp. Zeitschrift für Angewandte Entomologie, 2(1):29–56

Berry C, Hindley J, Oei C (1991) The *Bacillus sphaericus* toxins and their potential for biotechnological development. In: Maramorosch K (ed) Biotechnology for biological control of pests and vectors. CRC, Boca Raton, FL, pp 35–51

Boisvert M (2005) Utilization of *Bacillus thuringiensis* var. *israelensis* (*Bti*)-based formulations for the biological control of mosquitoes in Canada. In: 6th Pacific Rim Conference on the biotechnology of *Bacillus thuringiensis* and its environmental impact, Victoria BC, pp 87–93

Boisvert M, Boisvert J (2000) Effects of *Bacillus thuringiensis* var. *israelensis* on target and non-target organisms: a review of laboratory and field experiments. Biocontrol Sci Tech 10:517–561

Bourgouin C, Klier A, Rapoport G (1986) Characterization of the genes encoding the haemolytic toxin and the mosquitocidal delta-endotoxin of *Bacillus thuringiensis israelensis*. Mol Gen Genet 205:390–397

Broadwell AH, Baumann L, Baumann P (1990) Larvicidal properties of the 42 and 51 kilodalton *Bacillus sphaericus* proteins expressed in different bacterial hosts:Evidence for a binary toxin. Curr Microbiol 21:361–366

Chang C, Yu YM, Dai SM, Law SK, Gill SS (1993) High-level cryIVD and cytA gene expression in *Bacillus thuringiensis* does not require the 20 kilodalton protein and the coexpressed gene products are synergistic in their toxicity of mosquitoes. Appl Environ Microbiol 59:815–821

Charles JF, Nielsen-LeRoux C (1996) Les bacteries entomopathogenes:mode d'action sur les larves de moustiques et phenomenes de resistance. Ann Inst Pasteur Actualites 7(4):233–245

Charles JF, Nielsen-LeRoux C (2000) Mosquitocidal bacterial toxins:diversity, mode of action and resistance phenomena. Mem Inst Oswaldo Cruz 95:201–206

Charles JF, Nielsen-LeRoux C, Delécluse A (1996) *Bacillus sphaericus* toxins: molecular biology and mode of action. Annu Rev Entomol 41:451–472

Chilcott CN, Ellar DJ (1988) Comparative toxicity of *Bacillus thuringiensis* var. *israelensis* crystal proteins in vivo and in vitro. J Gen Microbiol 134:2551–2558

Davidson EW (1988) Binding of the *Bacillus sphaericus* toxin to midgut cells of mosquito larvae: relationship to host range. J Med Entomol 25:151–157

Davidson EW (1990) Development of insect resistance to biopesticides. In: Proceedings of second symposium on biocontrol, Brasilia, pp 19

Davidson EW, Becker N (1996) Microbial control of vectors. In: Beaty BJ, Marquardt WC (eds) The biology of disease vectors. University Press of Colorado, Colorado, pp 549–563

Davidson EW, Yousten AA (1990) The mosquito larval toxin of *Bacillus sphaericus*. In: de Barjac H, Sutherland D (eds) Bacterial control of mosquitoes and black flies: biochemistry, genetics and applications of *Bacillus thuringiensis israelensis* and *Bacillus sphaericus*. Rutgers University Press, New Brunswick, NJ, pp 237–255

Delecluse A, Barloy F, Rosso ML (1996) Les bacteries pathogenes des larves de dipteres: structure et specificite des toxines. Ann Inst Pasteur Actualites 7(4):217–231

Des Rochers B, Garcia R (1984) Evidence for persistence and recycling of *Bacillus sphaericus*. Mosq. News 44:160–165

Etang J, Chandre F, Guillet P, Manga L (2004) Reduced bio-efficacy of permethrin EC impregnated bednets against an *Anopheles gambiae* strain with oxidase-based pyrethroid tolerance. Malar J 3:46–46

Federici BA, Lüthy P, Ibarra JE (1990) Parasporal body of *Bacillus thuringiensis israelensis*: structure, protein composition, and toxicity. In: de Barjac H, Sutherland D (eds) Bacterial control of mosquitoes and blackflies: biochemistry, genetics and applications of *Bacillus thuringiensis israelensis* and *Bacillus sphaericus*. Rutgers University Press, New Brunswick, NJ, pp 45–65

Fillinger U, Lindsay SW (2006) Suppression of exposure to malaria vectors by an order of magnitude using microbial larvicides in rural Kenya. Trop Med Int Health 11:1629

Fillinger U, Knols BGJ, Becker N (2003) Efficacy and efficiency of new *Bacillus thuringiensis* var. *israelensis* and *Bacillus sphaericus* formulations against afrotropical anophelines in western Kenya. Trop Med Int Health 8(1):37–47

Fillinger U, Kannady K, William G, Vanek MJ, Dongus S, Nyika D, Geissbuehler Y, Chaki PP, Govella NJ, Mathenge EM, Singer BH, Mshinda H, Lindsay SW, Tanner M, Mtasiwa D, de Castro MC, Killeen GF (2008) A tool box for operational mosquito larval control: preliminary results and early lessons from the Urban Malaria Control Programme in Dar es Salaam. Tanz Malariol J 7:20

Garcia R, Des Rochers B, Tozer W (1981) Studies on *Bacillus thuringiensis* var. *israelensis* against mosquito larvae and other organisms. Proc Calif Mosq Vector Control Assoc 49:25–29

Georghiou GP, Wirth M (1997) The influence of single vs multiple toxins of *Bacillus thuringiensis* subsp. *israelensis* on the development of resistance in *Culex quinquefasciatus* (Diptera: Culicidae). Appl Environ Microbiol 63(3–4):1095–1101

Goldberg LH, Margalit J (1977) A bacterial spore demonstrating rapid larvicidal activity against *Anopheles sergenti, Uranotaenia unguiculata, Culex univittatus, Aedes aegypti* and *Culex pipiens*. Mosq News 37:355–358

Hertlein BC, Levy R, Miller TW Jr (1979) Recycling potential and selective retrieval of *Bacillus sphaericus* from soil in a mosquito habitat. J Invertebr Pathol 33:217–221

Hirsch HD, Becker N (2009) Cost-benefit analysis of mosquito control operations based on microbial control agents in the Upper Rhine Valley (Germany). Eur Mosq Bull 27:47–55

Höfte H, Whiteley HR (1989) Insecticidal crystal proteins of *Bacillus thuringiensis*. Microbiol Rev 53:242–255

Ibarra JE, Federici BA (1986) Isolation of a relatively nontoxic 65-kilodalton protein inclusion from the parasporal body of *Bacillus thuringiensis* subsp. *israelensis*. J Bacteriol 165 (2):527–533

Kahindi SC, Midega JT, Mwangangi JM, Kibe LW, Nzovu J, Luethy P, Githure J, Mbogo C (2008) Efficacy of vectobac DT and culinexcombi against mosquito larvae in unused swimming pools in malindi, Kenya. J Am Mosq Control Assoc 24:538–542

Kellen WR, Meyers CM (1964) *Bacillus sphaericus* Neide as a pathogen of mosquitoes. J Invert Pathol 7:442–448

Killeen GF, Fillinger U, Kiche I, Gouagna LC, Knols BGJ (2002a) Eradication of *Anopheles gambiae* from Brazil: lessons for malaria control in Africa? Lancet Infect Dis 2:618–627

Killeen GF, Fillinger U, Knols BGJ (2002b) Advantages of larval control for African malaria vectors: low mobility and behavioural responsiveness of immature mosquito stages allow high effective coverage. Malariol J 1:1–7

Krieg A (1986) Bacillus thuringiensis, ein mikrobielles Insektizid. Acta Phytomed 10:191

Kroeger A, Dehlinger U, Burkhardt G, Anaya H, Becker N (1995) Community based dengue control in Columbia:people's knowledge and practice and the potential contribution of the biological larvicide *B. thuringiensis israelensis* (*Bacillus thuringiensis israelensis*). Trop Med Parasitol 46:241–246

Lacey LA (1990) Persistence and formulation of *Bacillus sphaericus*. In: de Barjac H, Sutherland D (eds) Bacterial control of mosquitoes and blackflies: biochemistry, genetics and applications of *Bacillus thuringiensis israelensis* and *Bacillus sphaericus*. Rutgers University Press, New Brunswick, NJ, pp 284–294

Langenbruch GA, Hommel B, Becker N (2005) Bakterienpräparate. In: H. Schmutterer, Huber J (eds) Natürliche Schädlingsbekämpfungsmittel. Eugen Ulmer GmbH, Stuttgart, pp 29–86

Li J, Carroll J, Ellar DJ (1991) Crystal structure of insecticidal delta- endotoxin from *Bacillus thuringiensis* at 2.5 A resolution. Nature 353:815–821

Ludwig M, Beck M, Zgomba M, Becker N (1994) The impact of water quality on the persistance of *Bacillus sphaericus*. Bull Soc Vector Ecol 19(1):43–48

Lüthy P, Wolfersberger MG (2000) Pathogenesis of *Bacillus thuringiensis* toxins. In: Entopathogenic bacteria form laboratory to field application. Kluwer Academic, Dordrecht, Boston, London, pp 524

Mahilum MM, Ludwig M, Madon MB, Becker N (2005) Evaluation of the present dengue situation and control strategies against *Aedes aegypti* in Cebu City, Philippines. J Vector Ecol 30:277–283

Majambere S, Lindsay SW, Green C, Kandeh B, Fillinger U (2007) Microbial larvicides for malaria control in The Gambia. Malariol J 6:76

Makundi EA, Mboera LEG, Malebo HM, Kitua AY (2007) Priority setting on malaria interventions in Tanzania: strategies and challenges to mitigate against the intolerable burden. Am J Trop Med Hyg 77:106–111

Margalit J, Dean D (1985) The story of *Bacillus thuringiensis israelensis* (B.t.i.). J Am Mosq Control Assoc 1:1–7

McGaughey WH (1985) Insect resistance to the biological insecticide *Bacillus thuringiensis*. Science 229:193–195

Molloy D, Jamnback H (1981) Field evaluation on *Bacillus thuringeinsis* var. *israelensis* as a blackfly biocontrol agent and its effect on non target stream insects. J Econ Ent 74:314–318

Mulla MS, Federici BA, Darwazeh HA (1982) Larvicidal efficacy of *Bacillus thuringiensis* serotype H-14 against stagnant water mosquitoes and its effects on non target-organisms. Environ Ent 11:788–795

Mulla MS, Darwazeh HA, Zgomba M (1990) Effect of some environmental factors on the efficacy of *Bacillus sphaericus* 2362 and *Bacillus thuringiensis* (H-14) against mosquitoes. Bull Soc Vector Ecol 15:166–175

Mulligan FS III, Schaefer CH, Wilder WH (1980) Efficacy and persistence of *Bacillus sphaericus* and *B. thuringiensis* H-14 against mosquitoes under laboratory and field conditions. J Econ Ent 73:684–688

N'Guessan R, Corbel V, Akogbeto M, Rowland M (2007) Reduced efficacy of insecticide treated nets and indoor residual spraying for malaria control in pyrethroid resistance area, Benin. Emerg Infect Dis 13:199–206

Perez C, Fernandez LE, Sun J, Folch JL, Gill SS, Soberon M, Bravo A (2005) *Bacillus thuringiensis* subsp. *israelensis* Cyt1Aa synergizes Cry11Aa toxin by functioning as a membrane-bound receptor. Proc Nat Acad Sci USA 102:18303–18308

Priest FG (1992) Biological control of mosquitoes and other biting flies by *Bacillus sphaericus* and *Bacillus thuringiensis*. J Appl Bacteriol 72:357–369

Protopopoff N, Bortel van Marcotty WT, van Herp M, Maes P, Baza D, Alessandro U, Coosemans M (2007a) Spatial targeted vector control in the highlands of Burundi and its impact on malaria transmission. Malariol J 6:158

Protopopoff N, van Herp M, Maes P, Reid T, Baza D, d'Alessandro U, Coosemans M (2007b) Vector control in a malaria epidemic occurring within a complex emergency situation in Burundi: a case study. Malariol J 6:93

Puchi ND (2005) Factors affecting the efficency and persistance of *Bacillus thuringiensis* var. *israelensis* on *Anopheles aquasalis* Curry (Diptera:Culicidae), a malaria vector in Venezuela. Entomotropica 20:213–233

RBM (2005) World Malaria report 2005. WHO, Geneva, S 293

Russell TL, Brown MD, Purdie DM, Ryan PA, Brian H, Kay BH (2003) Efficacy of VectoBac (*Bacillus thuringiensis* variety *israelensis*) formulations for mosquito control in Australia. J Econ Ent 96:1786–1791

Rydzanicz K, Lonc E, Kiewra D, De Chant P, Krause S, Becker N (2009) Evaluation of two application techniques of three microbial larvicide formulations against *Culex p. pipiens* in irrigation fields in Wroclaw, Poland. J Am Mosq Control Assoc 25:140–148

Silva-Filha MH, Regis L, Oliveira CMF, Furtado AF (2001) Impact of a 26-month *Bacillus sphaericus* trial on the preimaginal density of *Culex quinquefasciatus* in an urban area of Recife, Brazil. J Am Mosq Control Assoc 17:45–50

Sinegre G (1984) La résistance des Diptères Culicides en France in:Colleque sur la réduction d'efficacoté des traitements insecticides et acaricides et problèmes de résistance, Paris, pp 47–57

Singer S (1973) Insecticidal activity of recent bacterial isolates and their toxins against mosquito larvae. Nature 244:110–111

Sorensen MA, Walton WE, Trumble JT (2007) Impact of inorganic pollutants perchlorate and hexavalent chromium on efficacy of *Bacillus sphaericus* and *Bacillus thuringiensis* subsp. *israelensis* against *Culex quinquefasciatus* (Diptera:Culicidae). J Med Entomol 44:811–816

Su TS (2008) Evaluation of water-soluble pouches of *Bacillus sphaericus* applied as prehatch treatment against *Culex* mosquitoes in simulated catch basins. J Am Mosq Control Assoc 24:54–60

Tabashnik BE, Cushing NL, Finson N, Johnson MW (1990) Development of resistance to *Bacillus thuringiensis* in field populations of *Plutella xylostella* in Hawaii. J Econ Ent 83:1671–1676

Thorne L, Garduno F, Thompson T (1986) Structural similarity between the Lepidoptera and Diptera-specific insecticidal endotoxin genes of *Bacillus thuringiensis* subsp. *kurstaki* and *israelensis*. J Bacteriol 166:801–811

Walker K, Lynch M (2007) Contributions of *Anopheles* larval control to malaria suppression in tropical Africa: review of achievements and potential. J Med Vet Entomol 21:2–21

Ward ES, Ellar DJ (1988) Cloning and expression of two homologous genes of *Bacillus thuringiensis* subsp. *israelensis* which encode 130-kilodalton mosquitocidal proteins. J Bacteriol 170:727–735

Weiser J (1984) A mosquito-virulent *Bacillus sphaericus* in adult *Simulium damnosum* from Northern Nigeria. Zbl Mikrobiol 139:57–60

WHO (1999) *Bacillus thuringiensis*, Environmental Health Criteria. IPCS, pp 217

WHO (2004) Review of VectoBac WG, PermaNet and Gokilaht 5EC. WHO/CDS/WHOPES/2004.8, Geneva

Wirth MC, Park HW, Walton WE, Federici BA (2005) Cyt1A of *Bacillus thuringiensis* delays evolution of resistance to Cry11A in the mosquito *Culex quinquefasciatus*. Appl Environ Microbiol 71:185–189

Wirth M, Yang Y, Walton WE, Federici BA, Berry C (2007) Mtx Toxins Synergize Bacillus sphaericus and Cry11Aa against Susceptible and Insecticide-Resistant Culex quinquefasciatus Larvae. Applied and Environmental Microbiology 73(19):6066–6071

Yohannes M, Haile M, Ghebreyesus TA, Witten KH, Getachew A, Byass P, Lindsay SW (2005) Can source reduction of mosquito larval habitat reduce malaria transmission in Tigray, Ethiopia? Trop Med Int Hlth 10:1274–1285

Chapter 4
The Neem Tree Story: Extracts that Really Work

Heinz Mehlhorn, Khaled A.S. Al-Rasheid, and Fathy Abdel-Ghaffar

Abstract Traditional or natural medicine is as old as human development of cultural activities. The insight into real or believed activities of plants or minerals was surely obtained by chance as mankind searched for their daily food as nomads. The following chapter reports the insecticidal and acqricidal effects of extracts of the seeds of the neem/tree, which was and is used worldwide for many centuries for a huge number of purposes. The described effects—obtained in clear experiments with 10 species of insects, 3 species of mites and 2 species of ticks—made it possible to develop and lounch safe and effective market products proving that intense research on plants may lead to very good, useful and safe products at other low prices.

4.1 Introduction

Traditional or natural medicine is as old as human development of cultural activities. The insight into real or believed activities of plants or minerals was surely obtained by chance as mankind searched for their daily food as nomads.

Some open-minded individuals of the wandering groups of *Homo erectus* and later the dominant *Homo sapiens* – modern man dating back 100,000 years – noted by experience that eating a particular plant or fruit had good or bad effects stretching from amelioration of diseases to death due to lethal toxicity.

At first this knowledge was only orally transmitted among groups of nomadic humans and later – after development of writing in the early high cultures (e.g., the Babylonians,

H. Mehlhorn (✉)
Department of Zoology and Parasitology, Heinrich Heine University, Universitätsstr. 1, 40 225 Düsseldorf, Germany
e-mail: mehlhorn@uni-duesseldorf.de

K.A.S. Al-Rasheid
Department of Zoology, King Saud University, Riyadh, Saudi Arabia

F. Abdel-Ghaffar
Department of Zoology, Cairo University, Giza, Egypt
e-mail: fathyghaffar@yahoo.com

Table 4.1 Some famous medicinal plants

Name	Ingredients	Users	Targets
False helebore *Veratrum album*	Rodenticides, steroid alkaloids	Romans, Europe, Asia	Rats
Chrysanthemes *Chrysanthemum* species	Pyrethrum	Romans, Persians, today	Insects, pests
Artemisia species	Artemisinin	Today and Old Chinese	Malaria control
Tobacco *Nicotiana tabacum*	Nicotin	Persians, Arabians, today	Pests
China bark trees *Cinchona* spp.	Quinine	Indians of America, today	Malaria, fever
Coffee bush *Coffea* spp.	Coffein	Indians of America, Africans, today	Blood pressure
Tea bush *Camellia sinensis*	Thein	Chinese, today	Blood pressure
Cacao tree *Theobroma cacao*	Theobromin	Indians of America, today	Happy maker, blood pressure
Worm fern *Dryopteris filix-mas*	Filix acid, Aspidinol	Europe, Asia, America	Tapeworm control
Camphora tree *Cinnamonum, C. ceylandicum*	Eugenol	Old Chinese, Asia, Greek, Romans	Wound healing, antiworm activity

the Egyptians, or Greeks) – fixed in early textbooks of Herodot, Hippocrates or Plinius, so that the early physicians could rely on a broad basis of knowledge. Remnants of this orally transmitted knowledge are even today – in times of skilful medical techniques and tremendous chemical medicaments – present in many peoples on Earth, so that there exist in the literature many lists of regional plants that may be used for therapeutic purposes (von Bingen 1974; Tagboto and Townson 2001; Rates 2001; Hussain et al. 2008; Xiao and Fu 1986; Leung 1985; Mulla and Su 1999; Mehlhorn et al. 2005, 2006; Semmler et al. 2009; Schmahl et al. 2010; Mumcuoglu et al. 1996; Oliveira et al. 2009).

Some of these plants have obtained worldwide importance and are used today either still in traditional formulations or in highly sophisticated pure synthetic recombinations. Examples include quinine (found in the tree *Cinchona* sp.), which is used against malaria, heart glycosids = digoxin = digitalis (found originally in plant species of the genus *Digitalis*) used to increase heartbeat or to stimulate other physiological functions (Table 4.1), oils of male farn (*Dryopteris filix-mas*) used against worms, chrysanthemes (*Chrysanthemum* syn. *Pyrethrum* species) producing the insecticide pyrethrum, or artemisinin extracted from *Artemisia* species used for example against malaria. One of the recently most used plants is the so-called neem tree, which has many proven efficacies, but also many, which belong to the "kingdom of believe" (Schmutterer 2002; Costa et al. 2006; Kraus 2002; Kim et al. 2007; Leung 1985; Mehlhorn et al. 2005, 2006; Senthil-Natan et al. 2005a, b; Mulla and Su 1999; Mitchell et al. 1997; Lundh et al. 2005).

4.2 The Neem Tree

This fast-growing tree was first described by the French botanist Adrian Henri Laurent de Jussieu in the year 1830 as *Azadirachta indica* referring to its main

4 The Neem Tree Story: Extracts that Really Work

homeland the subcontinent of India. It belongs to the mahogany family Meliaceae. A second species *A. siamensis* occurs in Asia, which, however, may form hybrids with *A. indica*.

4.2.1 Systematic Position

Kingdom: Plantae
Division: Magnoliophyta
Order: Sapindales (formerly: Rutales)
Family: Meliaceae
Genus: Azadirachta
Species: A. indica (synonyms: *Melia azadirachta*, *Melia indica*, and *Antelaea azadirachta*).

In analyses in other publications, however, *A. indica* was erroneously mixed with the so-called chinaberry *Melia azedarach*, so that several efficacies described might not be related to the neem tree.

4.2.2 Vernacular Names

Since this tree is now found on practically all continents with suitable climates, many local names exist for the neem tree:

Europe:

Germany	Indischer Zedrach, Indischer Flieder, Niem, Nim
France	Azadira d'Inde, lilas d'Indes, Margousier
Portugal	Margosa
Spain	Nim, margosa, paraiso
UK	Indian lilac, neem
America	Neem, nim
Africa	Neem, Nimo, Nimy, Mwarobaini, Ganye, Marrango, Aboodee, Goo gay

Asia/Australia:

India	Limba, limbo, neem, nim, nimb, Vembu, Vepa etc.
Sri Lanka	Kokomba
Pakistan	Nimmi
Malaysia	Mambu
Indonesia	Nimba, imba
Australia	Neem

4.2.3 Size and Appearance

The evergreen neem tree can reach in general a height of 15–20 m [Fig. 4.1 (for figure see end of this chapter)] with crowns of up to 20 m in diameter (if freestanding). The trunk is mostly straight with diameters (in old examples) of up to 1 m. The opposite, pinnate leaves are about 25–30 cm long and extend about 20–30 medium to dark green leaflets with a length of about 5–8 cm [Fig. 4.2 (for figure see end of this chapter)]. Young leaflets appear reddish. The composite flowers appear white and fragrant and reach a length of up to 25 cm, while a single flower is only 5–6 mm long and 8–9 mm wide. Protandrous bisexual flowers and male flowers are found on the same tree. The fruits of the neem tree [Figs. 4.2 and 4.3 (for figures see end of this chapter)] have an olive shape and reach (when ripe) a size of about 2.3 × 1.3 cm.

The regions and climates preferred by neem trees are areas with subarid to subhumid conditions with yearly rainfalls between 400 and 1,200 mm, although *A. indica* is noted also for its drought resistance. The neem trees may grow in many types of soil. Thus, this tree was easily exported to many countries with favorable conditions –being often planted in fruit-tree farms or the tree itself invades new regions due to globalized trade activities.

4.2.4 Use of the Tree

Besides its famous "all healing" efficacies (see below) neem trees are used worldwide in fruit farms to repel pests from monocultures, they are eaten as vegetables, sold as components of cosmetics, and occur in Ayurveda products and Ayurveda activities. Thus, a worldwide industry has been developed around the rearing of this tree that is known in India also as "Holy tree" or "Nature's drugstore."

4.2.4.1 Active Ingredients

The neem tree and its fruits contain numerous ingredients that can become eluted by different methods (acetone, aqueous, methanolic, ethanolic, chloroform, acetonitrile extracts etc.) which deliver different amounts and different components of the typical contents of this plant. Furthermore, the neem trees may differ worldwide, as was shown by Biswas et al. (2002) and Ermel (2002). In such studies it was found that the percentage of oil and azadirachtin contents might vary considerably not only between different regions in the world, but also within one country, if the soil at the various planting sites was different. For example, the seeds of the neem tree might contain 54 ± 1.5% oil in Mali, or only 38 ± 1.5% in Madagascar, while seeds from India and Sri Lanka contained 47.6 ± 5% respectively 50.1 ± 5%.

Table 4.2 Important chemical components inside the seeds/leaves of the neem tree

Name	Chemical group	Main claimed activity
Azadirachtol	Triterpenoids	Antifeedant
Melantriol	Triterpenoids	Antifeedant
Azadirone group	Tetranortriterpenoids	Antifungal, insecticidal, antifeedant
Gedunin group	Tetranortriterpenoids	Antifeedant, antimalarial
Vilasinin group	Tetranortriterpenoids	Antifeedant
Nimbin group	Tetranortriterpenoids	Antifeedant
Salannin group	Tetranortriterpenoids	Antifeedant, growth reducer
Azadirachtin and analogs	Tetranortriterpenoids	Antifeedant, growth inhibition of insects
Nimbione, Margalone	Diterpenoids	Antibiotic activity
Flavonoids, tannins	Nonterpenoids	Larvicidal activities, insecticidal

A similar variation was noted when the contents of azadirachtin were considered. In Togo the highest contents were noted reaching 5.4 ± 1.0 mg azadirachtin/g, while the lowest concentrations were seen in Mali (2.05 ± 0.5 mg/g) and in Madagascar (2.2 ± 0.5 mg/g). Therefore, reliable and comparable results in efficacy tests can only be obtained, if the methods of elution and the origins of the plant material are kept as stable as possible. In other parts of the tree (roots, leaves, stem) the contents of different ingredients vary over even wider ranges. The most important active components of the neem tree are listed in Table 4.2.

Azadirachtin is the most common component and is considered to have the most wide-spread activities, although only a few single tests have been carried out on the efficacy of other less common ingredients of this plant (about 100 were found!) (Koul et al. 1996, 2004).

Azadirachtin belongs to the group of limonoid precursors as a secondary metabolite and represents a highly oxidized triterpenoid (Kraus 2002). Its first total synthesis was accomplished by Gemma et al. in 2007, although many components had already been known since the early 1970s (Schmutterer 2002). The formation of azadirachtin inside the plant follows an elaborate biosynthetic pathway postulating that the steroid component tirucallol is used as a precursor of the neem triterpenoid secondary metabolites. This complicated way of synthesis may explain why it is extremely difficult (and not fully successful till today) to produce azadirachtin de novo by laboratory synthesis. The synthesis described by Gemma et al. (2007) started from decalin which was isolated from the plant itself. The chemical properties and some other characteristics are listed in Table 4.3 and described in many papers (Koul et al. 1996, 2004; Niemann et al. 2002; Kraus 2002).

4.2.4.2 Efficacies of Neem Formulations

The book by Schmutterer (2002) includes reviews and experiences of more than 55 scientists reporting that oils of seeds, crushed leaves, different extracts of leaves, bark pieces, and/or seed have an extremely wide range of activities (Table 4.4), which probably led in traditional medicine to the admiring descriptions of the neem tree as the "wonder tree," "holy tree," or "all-can-treat-tree."

Table 4.3 Chemical properties and characteristics of azadirachtin (according to Niemann et al. 2002)

Name	Azadirachtin
Molecular formula	$C_{35}H_{44}O_{16}$
Molecular mass	720.71 g/mol
CAS number	11141-17-6
PubChem	5281303
Chemical class	Tetranortriterpenoids (structurally similar to insect hormones (ecdysones)
Biological activity	Probably as ecdysone-blocker thus disturbing insect growth
Acute toxicity NeemAzal (=34% Azadirachtin A)	Oral in rats (LD 50): >5,000 mg/kg bw
	Inhalation in rats LD 50): 0.72 mg/L (4 h)
	Skin toxicity rats: >2,000 mg/kg bw
	Skin irritation: rabbit: slightly
	Eye irritation: rabbit: moderately
	Skin sensation: guinea pig: positive
Acute toxicity NeemAzal-T-S (1% Azadirachtin A)	Oral rats: >5,000 mg/kg bw
	Inhalation rats: 5.4 mg/L air/4 h
	Skin toxicity rats: >2,000 mg/kg bw
	Eye irritation rabbit: moderately or not
	Skin sensation: guinea pig: none
Acute toxicity NeemAzal-F (5% Azadirachtin)	Oral in rats: 765 mg/kg bw
	Dermal tox in rats: >5,000 mg/kg bw
	Eye irritation: rabbit: severely
	Skin irritation rabbit: low
Chronic toxicity (90 days)	Even a 90-day oral feeding of rats with 10,000 ppm did not show effects
Reproductive effects (of aqueous leaf extract)	6 weeks daily feeding led to infertility of rats at 66.7%, for 9 weeks: 100%
Teratogenic effects	No informations found
Mutagenic effects	UDS and mouse lymphoma studies: no effects
Carcinogenic effects	No information found
Ecological effects	(a) Birds and other wild life: none
	(b) Aquatic organisms: No significant, since azadirachtin breaks down in water within 50–100 h
	(c) Non-target insects (bees, spiders, butterflies): harmless
Environmental effects	(a) Soil: no accumulations
	(b) Water, groundwater: azadirachtin breaks down in 50–100 h
	(c) Plants: no phytotoxicity and accumulation seen in plants

Of course the neem tree did not develop these efficacies just for humans or other organisms, but for its own survival during evolution. Thus, the antifeedant properties of leaves and seeds were exclusively developed by the tree for its own advantage when attacked by insects feeding on leaves, seeds and/or phloem respectively xylem fluids. In the case of isolation of these substances from the neem tree by humans they might be used for other purposes. The same is true for substances that repel approaching insects, mites and/or worms, as they look for a good meal. Since trees like the neem tree cannot change their location, if the environmental conditions undergo alteration, they must overcome the situation. Therefore, many of them developed substances that had antiviral, antibacterial, or

Table 4.4 Some activities attributed to different neem components (according to Biswas et al. 2002 and Schmutterer 2002)

Antifungal activity	Nimbidin, gedúnin, cyclic trisulfids
Antibacterial	Margolone, machmoodin, nimbolide, nimbidin
Antiviral	Flavonoids
Antifeedant	Azadirachtin
Antimalarial	Nimbolide, azadirachtin
Nematocides	Aqueous extracts of seeds, leaves and roots; nimbidin
Molluscicides	Aqueous extracts
Antimite	Different extracts
Insecticides (more than 500 species susceptible)	Azadirachtin, tannin
Antifeedant	Azadirachtin
Growth reduction of insects	Azadirachtin salannin
Anti-inflammatory	Nimbidin, nimbidale, catechin, polysaccharids
Spermicidal	Nimbidin, Nimbin
Diuretic, hypoglucaemic, antiarthritic	Nimbidin
Immunomodulatory	NB-II-peptidoglycan, gallic acid

Table 4.5 Medical use of special portions of the neem tree in Ayurveda activities (according to Biswas et al. 2002)

Part of the tree	Medical use
Oil of seeds	Leprosy, intestinal worm control
Leaves	Leprosy, eye problems, intestinal worm control, anorexia, skin ulcers
Bark	Fever, analgesia
Flower	Bile suppression, intestinal worm control, phlegm
Fruit	Intestinal worm control, urinary disorder, epitaxis, diabetes, wounds, leprosy
Twig	Cough, asthma, intestinal worm control, diabetes, urinary disorders
All together	Blood problems, itching, leprosy, skin ulcer, intestinal worms

antifungal activities which offered shelter against approaching plants. Therefore, it is not surprising that such efficacies are found in many of the survivors of this evolution today. Plants that were not able to develop the different ingredients cited above or similar ones did not survive in the struggle for life and have disappeared from the Earth.

It is however astonishing that the neem tree had developed these efficacies to such a high extent that they were recognized by humans and thus put to use, as practically all plants must have similar activities, but to different extents. Tables 4.4 and 4.5 summarize some of these important effects compiled by Schmutterer (2002) and Biswas et al. (2002).

When looking at Tables 4.1–4.5 it is obvious that there are numerous studies available dealing with the effects of the neem tree in relation to the different forms of application and the different forms of extraction. It is difficult to compare grades of efficacies because several experiments or sometimes only tales (oral reports) are not very concise or are based on different modes of use of the neem extract.

Thus, it is necessary to define the mode of obtaining the neem extracts and the application doses, if a product is to be successful on the market as an insecticide, repellent, and/or as a medicament. Furthermore, it is necessary to keep in mind that azadirachtin and some other neem components become degraded in water within 50–100 h after first contact. Thus, only water-free versions can be stored in shops and/or pharmacies without loss of activity. In the following a neem seed extract containing MelAza in a water-free composition is presented with its very broad activity against pests in households or against blood-sucking insects, mites, and/or ticks (Locher 2009; Locher et al. 2010; Semmler et al. 2009; Schmahl et al. 2010; Mehlhorn 2009).

4.3 From Research to Products Against Blood-Sucking Orthopods

4.3.1 General Aspects

Before 1900, when the later Nobel Prize winner in medicine (1908) Paul Ehrlich (1854–1915) and other international colleagues started to introduce chemotherapeutics, all medicaments and medicinal remedies had been developed from plant extracts (Brown 1996; Leung 1985). Some of those plants had been used for many centuries, and even today many of them are known and used as "medicinal plants" (Schmutterer 2002; Fajimi and Taiwo 2005; Bäumler 2007). Since long ago, such extracts have also been used against endoparasites (e.g., worms) and ectoparasites (such as mosquitoes, ticks, mites, flea, bugs etc.) of humans or animals (Athanasiadou et al. 2007; Amer and Mehlhorn 2006). With respect to skin penetration by blood-sucking ectoparasites, many plant extracts have been screened either for a defined biocidal (killing efficacy) or for strong repellency activity (prohibiting arthropods from landing on a host). In increasing numbers, papers appear daily that describe hundreds of plant extracts, which are more or less well characterized. However, in most cases, this knowledge remains at a rather theoretical level, since the authors do not try to develop a product from their findings (Oladimeji et al. 2000; Michaelakis et al. 2009).

Such a process is often blocked just for legal reasons:

1. Research results cannot become the basis for a patent, had their publication appeared before a patent was submitted. Exclusivity, however, is needed since no trade company will develop and distribute a product that can be produced easily by another company.
2. The regulations of the European Community have placed repellents onto list 19 of biocides and, in addition, limited the number of compounds on this list. Therefore, it is rather difficult and very costly for newly detected plant ingredients to be produced and officially registered among those already known compounds within a reasonable time. On the other hand list 18 of the European

Community includes compounds that act as insecticides or acaricides and are characterized as biocides, too; their number is also reduced.

However, intense research activities on the efficacy of some plant extracts showed that there are legal possibilities to develop plant extracts for use as biocides or medicinal products against a broad spectrum of ectoparasites of humans and house animals. Some examples of the transfer of scientific results into ready-to-use products are presented in this review. Of course, only such compounds were used, which had proven their activity in a dose-dependent manner and which were nontoxic and did not harm the user's skin. The results presented herein underline the fact that it is worthwhile to develop natural products, since arthropods build up resistances much more easily against chemical compounds with a single active ingredient than against natural products with multiple ingredients.

The tree *Azadirachta* (syn. *Melia indica*) (trivially called neem, niem, or margosa) that grows up to 20 m high originates from India but is now found worldwide, and is often placed into fruit plantations due to its efficacy at repelling insects that might harm the fruits [Figs. 4.2 and 4.3 (for figures see end of this chapter)]. In India, the seeds, leaves, and many different extracts have been used in natural medicine for thousands of years (Schmutterer 2002). Many claimed properties turned out to be false when studied by modern methods. However, many others were found to be true (Schmutterer 2002). The ovoid, about 15-mm-long seeds of this tree appear nut-like [Figs. 4.2 and 4.3 (for figures see end of this chapter)]. Patented di-basic-ester (DBE) when added to MelAza extract has been proven in several in vivo and in vitro tests to eliminate infestations with head lice (*P. humanus capitis*; Heukelbach et al. 2006; Abdel-Ghaffar and Semmler 2007), scabies mites (*Sarcopotes scabiei;* Abdel-Ghaffar et al. 2008a, b, 2009; Locher et al. 2010; Semmler et al. 2009; Schmahl et al. 2010).

4.3.2 Antilice Products

On the basis of these experiments two products were developed and are distributed as "Picksan Lice Stop" (Fa. OTC Pharma, Gorinchem, The Netherlands) and as "Wash-Away Louse" (Fa. Alpha-Biocare, Düsseldorf, Germany and Fa. DEEF, Saudi Arabia). Another formulation follows this year.

Lice efficacy tests of the products were carried out in vitro (immediately after combing children). Then the lice were incubated for 10 or 20 min in the neem seed extracts ("Picksan, Licener" or "Wash-Away Louse") or exposed in combinations to other plant extracts, respectively. Then the lice were washed for 2 min with clear tap water before being placed onto dry filter paper and being observed for up to 24 h (at intervals of 5, 10, 20, 30, 60 min etc.). In vivo tests were also performed while incubating lice-containing hair of children for 10–20 min with the different neem-based shampoos. This washing was repeated after 10 days.

It was seen that lice [Figs. 4.4–4.6a–c (for figures see end of this chapter)] when covered only for seconds with the neem-containing shampoo became detached from the hair apparently due to the blocking of the openings (stigma) of the interior channel system that transports air/oxygen finally to the cells. Since these lice are detached from the hair they will be washed off as soon as the hair is washed with clear tap water at the end of the treatment period. If lice were incubated for at least three minutes in vitro or for about 10 min inside wet, shampoo-containing hair the lice died and no survivors remained. This death was due to suffocation, since the neem extract-containing shampoo entered the tracheole system and blocked at the terminal end the aqueous layer on the cell membranes that constantly require oxygen. Thus, this biocidal activity is a pure physical-mechanistic blocking of the oxygen uptake as is diagrammatically depicted in Fig. 4.6 (for figure see end of this chapter). This way of killing avoids the development of resistances by lice, since every new generation always needs high amounts of oxygen and they cannot adapt to a low supply. Since the eggs of lice are covered by a thick shell (nit), within which the larvae develop within 5–7 days, and need only low amounts of oxygen, the neem extract should be applied a second time about 9–10 days after the first treatment thus eliminating the larvae that hatched during this period.

These neem-based antilice shampoos have been proven to be very effective (Abdel-Ghaffar et al. 2010) and do not endanger the users as was shown while testing products that contain alcohol and/or silicon oils, which were highly inflammable or which may cover the lung epithelia.

4.3.3 Antimite Products

When freshly diluting the neem seed extract (MiteStop®) 1:33 or 1:40 with tap water a ready-to-use product is available. The concentrated extract is sold under the name MiteStop® by Fa. Alpha-Biocare, Düsseldorf, Germany).

4.3.3.1 *Dermanyssus gallinae*

Red chicken mites [*D. gallinae*; Fig. 4.7 (for figure see end of this chapter)] are important agents of disease affecting poultry worldwide, threatening the health and productivity of many farmed birds (Sparagano et al. 2009; De Luna et al. 2008; Marangi et al. 2009). Large amounts of such mites may cause enormous blood loss (anemia) and disease via transmitted viruses, bacteria, and/or parasites (Zeman et al. 1982; Hoglund et al. 1995; Vreecken-Buijis et al. 1998; Valiente-Moro et al. 2007; Abdel-Ghaffar et al. 2008a, b; Mehlhorn 2008). Large amounts of such mites may also stain the surface of eggs (by their bloody feces) in poultry farms thereby affecting sales (Chauve 1998).

Massive infestations occur especially in large chicken houses where inspections by the naked eye are only carried out in a few selected places, so that growing populations of mites may not be seen or only noted at a late stage. Furthermore, in the last years, control of *D. gallinae* mites has become more and more problematical due to increasing resistance to pyrethroids and other chemical products

(Marangi et al. 2009). The problem of control is also enlarged by the fact that the use of some chemical products requires the destruction of eggs laid during the spraying period or the request of a rather long waiting period before the chicken may be slaughtered (Hamscher et al. 2007). Thus, a beneficial control of the development of *D. gallinae* mites in chicken houses is greatly needed. Most antimite products act exclusively by contact, which is reduced by the many hiding places around cages or even in chicken houses with floor breeding.

One chemical product (ByeMite®, Fa. Bayer) and one biological product (MiteStop®, Fa. Alpha-Biocare) were proven to be active when used in chicken houses (Meyer-Kuehling et al. 2007; Abdel-Ghaffar et al. 2008a; Locher 2009). An in vitro study compared both products with respect to their required dosage and the duration of the required contact of mites to the applied product. The studied *Dermanyssus* mites [Fig. 4.7 (for figure see end of this chapter)] were collected at a commercial egg production farm with caged poultry close to Nancy (France) and at a scientific rearing farm close to Rommerskirchen (Germany), where poultry was kept free in and outside of stables. Collected mites were brought to the Institute of Parasitology in Düsseldorf and there exposed to the compounds in three repeated series on 3 days following the day of collection.

MiteStop® is a patented special formulation of an extract of the seeds of the neem tree (*A. indica*), which have been eaten by humans for centuries in India. The product was obtained from Fa. Alpha-Biocare (Düsseldorf, Germany) and was freshly prepared before each application by diluting the seed extract with tap water.

ByeMite® (the content is identical with the acaricide Sebacil® for pig scabies, containing the organophosphorous compound phoxim) was obtained from the veterinary trade in France and used according to the labeling on the bottle or was diluted as described below.

The mite stages [Fig. 4.7 (for figure see end of this chapter)] were collected in the farms, brought to the institute, kept at 25–27°C, and the experiments were started immediately afterwards. Under the dissection microscope, groups of 100 mites were formed (containing all stages of the life cycle – having sucked blood or not). The experiments were carried out by exposing the mites to the two products under the conditions described below.

The first two series of experiments aimed to evaluate the principal susceptibility of the mites to each of the two acaricide formulations in order to find out whether the mites (1) are killed upon permanent contact with the biocide or (2) die even after short contact with the biocide. This could occur, if a mite was hit by a sprayed droplet, but then could crawl to an untreated site eventually recovering. The third trial aimed to see whether the biocides also exert a toxic effect in the vapor phase, that is whether they work as fumigants and thus reach and kill mites hidden in crevices.

In experiments of type 1, the mites were kept sealed in 8.5-cm Petri dishes on a filter paper wetted with 400-μl biocide solution, giving a dosage of 7 μl/cm^2. In the experiments of type 2, the mites were immersed in the biocide solution on a small piece of filter paper (area of 2 cm^2) for 4 s and then this small piece of filter paper was transferred to a Petri dish lined with an untreated, dry filter paper. Any mite still mobile could leave the treated area. In the experiments of type 3, the mites were encased in small vials (1 cm diameter, 2 cm in length, one side covered with a fine nylon mesh to permit gas exchange). The vials were transferred into Petri dishes, dosed with 400 µl acaricide as describe above, and then the dishes were tightly sealed. All treated groups of mites were observed during the next hours after treatment. Results of mortality rate in percent were recorded 24 and 48 h after application.

Sebacil® (identical to ByeMite®) was used in the dilution recommended by Bayer AG for use in poultry, which is 100 ml concentrate diluted to 25 l water (2,000 ppm phoxim). MiteStop® (extract) was diluted 1:33 with water, the dilution recommended by Alpha-Biocare. Controls were carried out applying the same volumes of water instead of the test biocides. The *experiments of the first series* delivered the following results:

1. *Permanent exposure*: 100% of the mites originating from Rommerskirchen and being exposed to either ByeMite® (2,000 ppm) or Mite-Stop® 1:33) were killed within 24 h. The French mites showed different reactions. The exposition to MiteStop® killed 100% of the mites, while ByeMite® left survivors after 14 h of exposure, killing only 96.2%. This might be due to resistance.
2. *Brief contact*: After brief contact of only 4 s, MiteStop® (1:33) killed 84.5% of the Rommerskirchen mites and 100% of the Nancy mites. ByeMite® killed only 28.8% of the Rommerskirchen mites and 30% of the Nancy mites. This shows that ByeMite® needs longer contacts.

In the *second series of experiments*, only the half dose of the compounds was used. Permanent contact to 1,000 ppm ByeMite® led to a killing effect of 93.8% in the Rommerskirchen mites and 90.6% in the Nancy mites. On the other hand, MiteStop® killed 100% of the Rommerskirchen mites and only 98.2% of the Nancy mites. This shows that MiteStop® is highly efficient even in reduced dosage, which may occur at hidden crevices. After brief contact (4 s) with both diluted products, the following results were obtained ByeMite® killed only 18.8% of the Rommerskirchen mites and 28.3% of the Nancy mites, while MiteStop® killed 22.1% of the Rommerskirchen mites and 59.3% of the Nancy mites.

The *third series of experiments* tested the fumigant activity of both products. It was shown that ByeMite® (2,000 ppm) killed about 4% of the Rommerskirchen mites and about 5% of the Nancy mites (corresponding to the numbers of dead mites in the controls). The results with MiteStop® were different from those with ByeMite®. The Rommerskirchen mites were all killed, while the Nancy mites survived as those that had been exposed to ByeMite® at both farms.

These results show again that there are apparently different sensibilities in *D. gallinae* strains. While MiteStop® has apparently fumigant activities in the Rommerskirchen strain, both MiteStop® and ByeMite® do not affect the Nancy mites in hidden places by fumigant / gaseous transportation.

These in vitro studies in principle confirm the results of the in vivo treatments (Meyer-Kuehling et al. 2007; Abdel-Ghaffar et al. 2008a, b; Locher 2009; Table 4.6). However, it is also clearly shown that the biological product Mite-Stop® has some advantages in efficacy. Furthermore, Hamscher et al. (2007) showed that the organophosphorous phoxim (active compound in ByeMite®) when used to treat

4 The Neem Tree Story: Extracts that Really Work

Table 4.6 Trials on the efficacy of MiteStop®, respectively, Tre-san® (Fa. Alpha-Biocare) against two species of ticks (*Ixodes ricinus* and *Rhipicephalus sanguineus*), house dust mites (*Dermatophagoides pteronyssinus*, Tre-san®), bed bugs (*Cimex lectularius*), cockroach (*Gomphadorhina portentosa*) subadults, maggots of *Calliphora* fly larvae; two concentrations: water diluted 1:40, 1:66; experimental design A, sprayed directly on targets and experimental design B, mites or insects were placed on sprayed filter paper in plastic Petri dishes

Species/number of objects (n)	Dilution	Experimental design	Observation after ... hours				
			1 h	3 h	6 h	24 h	48 h
Ixodes (3)	1:66	A	3/3 motionless, legs are sensitive to contact	3/3 motionless, legs are sensitive to contact	All dead	–	–
Ixodes (4)	1:66	B	4/4 motionless, legs are sensitive to contact	4/4 motionless, legs are sensitive to contact	3/4 dead, 1/4 motionless, legs are sensitive to contact	All dead	–
Rhipicephalus (4)	1:66	A	4/4 alive, weak movement	4/4 alive, weak movement	All dead	–	–
Rhipicephalus (3)	1:66	B	3/3 alive, weak movement	3/3 alive, weak movement	3/3 alive, weak movement	3/3 alive, weak movement	2/3 alive, 1/3 dead
Dermatophagoides (ca. 300)	1:40	A	All dead	–	–	–	–
Dermatophagoides (ca. 300)	1:40	B	All dead	–	–	–	–
Dermatophagoides (ca. 300)	1:66	A	All dead	–	–	–	–
Dermatophagoides (ca. 300)	1:66	B	All dead	–	–	–	–
Cimex (20)	1:40	A	20/20 alive	20/20 alive	13/20 dead, 7/20 alive	All dead	–
Cimex (5)	1:40	B	5/5 alive	5/5 alive	2/5 dead, 3/5 alive	All dead	–
Cimex (4)	1:66	A	4/4 alive	4/4 alive	1/4 dead, 3/4 alive	3/4 dead, 1/4 alive	All dead
Cimex (7)	1:66	B	7/7 alive	7/7 alive	2/7 dead, 5/7 alive	3/7 dead, 4/7 alive	All dead
Gomphadorhina (2)	1:40	A	1/2 alive, 1/2 dead	1/2 alive, 1/2 dead	1/2 alive, 1/2 dead	1/2 dead, 1/2 alive	All dead

(*continued*)

Table 4.6 (continued)

Species/number of objects (n)	Dilution	Experimental design	Observation after ... hours				
			1 h	3 h	6 h	24 h	48 h
Gomphadorhina (1)	1:40	B	1/1 alive	1/1 alive	1/1 alive	All dead	–
Gomphadorhina (1)	1:66	A	1/1 alive	1/1 alive	1/1 alive	1/1 alive	1/1 alive
Gomphadorhina (1)	1:66	B	1/1 alive	1/1 alive	dead (4 h)	–	–
Maggots of flies (10)	1:40	A	10/10 alive	10/10 alive	1/10 dead, 9/10 alive	2/10 dead, 8/10 alive	2/10 dead, 8/10 alive
Maggots of flies (10)	1:40	B	10/10 alive	10/10 alive	10/10 alive	10/10 alive	10/10 alive, 3 of them pupated
Maggots of flies (10)	1:66	A	10/10 alive	10/10 alive	10/10 alive	10/10 alive	10/10 alive, 4 of them pupated
Maggots of flies (10)	1:66	B	10/10 alive	10/10 alive	10/10 alive	10/10 alive	10/10 alive, 3 of them pupated
Adult Calliphora flies (20)	1:40	A	20/20 alive	20/20 dead	–	–	–

chicken houses led to residues in eggs, so that they have to be removed after each spraying and chicken have then a withdrawal time of 25 days.

4.3.3.2 *Dermatophagoides pteronyssinus*

House dust mites (*D. pteronyssinus*), which cause severe allergic reactions in sensitive persons, are killed within 1 h when coming into contact either directly or indirectly (on filter paper) with the 1:40 water-diluted product (Table 4.6). The product sold is named **Tresan®**, which however has to be freshly prepared just before use. The mites [Fig. 4.8 (for figure see end of this chapter)] are killed by drying and shrinking within a short time.

4.3.3.3 *Neotrombicula autumnalis*

After spraying of the 1:66 diluted MiteStop® extract three times onto a mite-containing soil or grass area at intervals of 3–4 days, the numbers of mites were drastically reduced. A repetition of this treatment after 4 weeks made the mites disappear in the next year or afforded then only a single treatment series at the beginning of the summer season. This repetition was required due to the fact that the attacking larvae of the mites do not leave the soil daily and thus some might not come into contact with the extract during the first year of treatment. These stages will produce progeny that come out in the next year (only the larvae attack hosts). However, sprinkling of tap water is required about 2 h after each treatment so that the compound is washed into the soil, where the larvae hide.

4.3.4 Antitick Products (MiteStop®)

The results of the several times repeated experiments were documented as examples in Tables 4.1–4.8. It can be seen that the 1:40 aqueous dilution of MiteStop®) kills ticks of the genera *Ixodes* and *Rhipicephalus* within 5 h when sprayed on the surface or when the ticks come into contact with the compound just with their feet (Table 4.6). Even a dilution of 1:66 kills *Ixodes* ticks, while using this dose, *Rhipicephalus* stages die only after direct spraying onto their backside [Table 4.6; Figs. 4.9 and 4.10 (for figures see end of this chapter)].

4.3.5 Insecticidal Effects (MiteStop®)

4.3.5.1 **Mallophages**

Horses ($n = 100$) being heavily infested with Mallophaga (biting lice) of the rather common species *Werneckiella equi equi* [Fig. 4.11 (for figure see end of this chapter)] were treated with a 1:20 dilution of MiteStop®. Starting 1–2 h after the product had been applied to the hair with the help of a brush, hundreds of thousands of dead mallophages became visible at the tips of the hair. The complete surface of the hair seemed to be covered with "fine wool" representing a layer of dead mallophages. The horse owners also reported that blood-sucking and/or other molesting insects were repelled for often as much as 4 days, when treated horses grazed on the meadow.

4.3.5.2 Fleas

Larvae of Cat fleas (*Ctenocephalides felis*) died within 1 h after contact with the 1:33 dilution, while it took longer for adult fleas. Some of them died only after 24 h (Table 4.7), but had limited motility up to that point [Figs. 4.12 and 4.13 (for figure see end of this chapter)], so that they stopped blood sucking. However, finally there were no survivors among the treated fleas.

Table 4.7 Trials on the efficacy of MiteStop® (Fa. Alpha-Biocare) against larvae and adult stages of Cat fleas (*Ctenocephalides felis*); dilution of MiteStop® 1:33 and 1:40, being sprayed four times from a distance of 20 cm; the test objects were sprayed directly from above

Species/number of objects (n)	Dilution	Observation after ... hours					
		1 h	2 h	3 h	5 h	8 h	24 h
Ctenocephalides larvae (10)	1:33	All dead					
Ctenocephalides larvae (10)	1:40	All dead					
Ctenocephalides adult (30)	1:33	30/30 alive	22/30 dead, 8/30 alive	All dead			
Ctenocephalides adult (30)	1:40	30/30 alive	30/30 alive	11/30 dead, 19/30 alive	21/30 dead, 9/30 alive	28/30 dead, 2/30 alive	All dead

Table 4.8 Trials on the efficacy of MiteStop® (Fa. Alpha-Biocare) against two insect species (*Triatoma infestans* and *Tenebrio molitor*); Dilution of MiteStop® 1:33 and 1:40, being sprayed four times from a distance of 20 cm; the test objects were sprayed directly from above, on filter paper in plastic Petri dishes

Species/number of objects (n)	Dilution	Observation after ... hours				
		1 h	2 h	3 h	6 h	24 h
Triatoma subadult (6)	1:20	6/6 alive	6/6 alive	6/6 alive	1/6 dead, 5/6 alive	All dead
Triatoma subadult (10)	1:33	10/10 alive	2/10 dead, 8/10 alive	4/10 dead, 6/10 alive	6/10 dead, 4/10 alive	8/10 dead, 2/10 alive
Triatoma subadult (10)	1:40	10/10 alive	10/10 alive	10/10 alive	2/10 dead, 8/10 alive	3/10 dead, 7/10 alive
Tenebrio adult (10)	1:20	10/10 alive	10/10 alive	6/10 dead, 4/10 alive	All dead (5 h),	–
Tenebrio adult (10)	1:33	10/10 alive	10/10 alive	10/10 alive	10/10 alive	3/10 dead, 7/10 alive
Tenebrio adult (10)	1:40	10/10 alive	10/10 alive	10/10 alive	10/10 alive	4/10 dead, 6/10 alive

4.3.5.3 Raptor Bugs

The raptor bugs (*Triatoma infestans*) were very resistant even against the 1:20 dilution (Table 4.8), since several specimens [Fig. 4.14 (for figure see end of this chapter)] survived even after 24 h. The knock-out ratio after 24 h was rather similar after contacts to 1:20, 1:33, or 1:40 dilutions (Table 4.8).

4.3.5.4 Flour Beetles

The flour beetle (*Tenebrio molitor*), a common food pest, turned out to be rather resistant to the neem extract. Even at dilutions of 1:20, a larger number of survivors occurred in the same range as after contacts to dilutions of 1:33 or 1:40 [Table 4.8; Fig. 4.15 (for figures see end of this chapter)].

4.3.5.5 Cockroaches

Cockroaches turned out to be very sensitive to the extract. However, the killing rate varied according to the species (Tables 4.6 and 4.9). *Blattella germanica* (a small-sized

Table 4.9 Trials on the efficacy of MiteStop® (Fa. Alpha-Biocare) against juvenile and adult stages of cockroaches (*Blattella germanica* and *Blatta orientalis*), dilution of MiteStop® 1:20, 1:33, and 1:40, being sprayed four times from a distance of 20 cm; the test objects were sprayed directly from above, on filter paper in plastic Petri dishes

Species/number of objects (*n*)	Dilution	Observation after ... hours				
		1 h	3 h	7 h	18 h	24 h
B. germanica juv. (4)	1:33	4/4 alive	4/4 alive	3/4 dead, 1/4 alive	All dead	–
B. germanica juv. (4)	1:40	4/4 alive	4/4 alive	2/4 dead, 2/4 alive	All dead	–
B. germanica adult (7)	1:33	7/7 alive	7/7 alive	1/7 dead, 6/7 alive	All dead	–
B. germanica adult (6)	1:40	6/6 alive	6/6 alive	2/6 dead, 4/6 alive	All dead	–
B. orientalis juv. (8)	1:20	8/8 alive	8/8 alive	All dead (6 h)	–	–
B. orientalis juv. (4)	1:20	4/4 alive	4/4 alive	2/4 dead, 2/4 alive	3/4 dead, 1/4 alive	All dead
B. orientalis juv. (4)	1:33	4/4 alive	4/4 alive	1/4 dead, 3/4 alive	2/4 dead, 2/4 alive	3/4 dead, 1/4 alive
B. orientalis juv. (4)	1:40	4/4 alive	4/4 alive	1/4 dead, 3/4 alive	1/4 dead, 3/4 alive	2/4 dead, 2/4 alive
B. orientalis adult (6)	1:20	3/6 dead, 3/6 alive	All dead (2 h)	–	–	–
B. orientalis adult (4)	1:20	4/4 alive	4/4 alive	2/4 dead, 2/4 alive	All dead	–
B. orientalis adult (4)	1:33	4/4 alive	4/4 alive	4/4 alive	4/4 alive	1/4 dead, 3/4 alive
B. orientalis adult (4)	1:40	4/4 alive	4/4 alive	4/4 alive	4/4 alive	2/4 dead, 2/4 alive

species) finally died when coming into contact with the 1:40 dilution. This concentration did not work very well with the larger species (*Gomphadorhina* sp. or *Blatta orientalis*), which needed a treatment using the 1:20 dilution (Tables 4.6 and 4.9; Figs. 4.16 and 4.17).

4.3.5.6 Flies

While adult flies (*Musca, Lucilia, Calliphora*) [Fig. 4.18 (for figure see end of this chapter)] die if they are sprayed with a 1:40 dilution of MiteStop®, maggots survive (Fig. 4.19 (for figure see end of this chapter); Table 4.6).

4.3.5.7 Bed Bugs

The effects on bed bugs (*Cimex lectularius*) were considerable and very promising for an own product [Table 4.10; Fig. 4.20 (for figure see end of this chapter)].

4.4 Evaluation of the Results of the Efficacy Tests with MiteStop®

Thousands of older and recent publications show that plants contain thousands of components that have either an insecticidal/acaricidal activity or possess the potential property of repellency. These compounds had been developed by the plants during their evolution in order to protect themselves against feeding attacks of various organisms. Some of those components have a very strong activity (e.g., pyrethrum) and consequently were already used by humans as insecticides against a broad spectrum of pests attacking plants, animals, and/or humans. Many other compounds are present in plants only in small amounts, so that they have to become enriched via extraction methods before use. However, many of the acaricidal and insecticidal compounds cannot be used, since they induce severe side effects or are harmful to human skin. Thus, it must be considered that natural compounds may also be dangerous for humans or may remain ineffective, in case the active compounds are only present in low concentrations in the plants. Therefore, only intense tests may show the capacity of different plant extracts. In addition, it must be considered that the potential target organisms have an often strongly varying sensitivity against plant extracts.

Such a different sensitivity was shown in the present review on the activity of an extract of seeds of the neem tree. While *Calliphora* fly maggots, stages of the raptor bug *T. infestans*, and adult flour beetles (*T. molitor*) showed practically no sensitivity against the extract, on the other hand ticks, mites, cockroaches, blood-sucking lice, mallophages, bed bugs, and fleas were killed by the extract (Tables 4.1–4.10).

4 The Neem Tree Story: Extracts that Really Work

Table 4.10 Test on the insecticidal activity of MiteStop® against adult bed bugs (*Cimex lectularius*): *Experiment A*, five bed bugs were placed onto filter paper which had been sprayed with MiteStop® (1:40 diluted) *twice* from a distance of 20 cm. *Experiment B*, bed bugs had been sprayed *once* directly (from a distance of 20 cm, dilution 1:40) with MiteStop®

Experimental design	0 h	1 h	3 h	5 h	6 h	8 h
A	Moving with stiff legs, trying to stay away from the filter paper with their ventral side	Motionless, legs are sensitive to contact	Motionless, legs are sensitive to contact	3/5 supine, no reaction; 2/5 weak reflexes of legs	5/5 supine, no reaction; 1 weak reflexes of legs	All dead
B	Moving intensely	Motionless, legs are sensitive to contact	1/4 supine, dead; 2/4 motionless; 1/4 weak reflexes	3/4 supine, dead; 1/4 weak reflexes	All dead	All dead

The efficacy, however, varied depending on the target species or on the concentration of the extract. Therefore, in order to develop a useful product in any case intensive tests have to be done to evaluate the range of efficacy against each target species. Further tests are needed to determine the lowest effective dose and the highest security level for humans, animals, and environment before the product is launched. The present review showed the high efficacy of the products MiteStop®, Tre-san®, Wash Away Louse®, and Picksan Louse Stop®.

For blood-sucking lice, it was clearly shown that the activity of the neem extract is based on the fact that the shampoo with the extract covers the terminal ends of the tracheoles, thus blocking mechanically the oxygen transfer through the fine water layer into the cell. This blocking has the effect that the cells of the body muscles, heart etc., are disrupted from oxygen uptake. Thus, a rather quick *knock-out effect* occurs within short periods. Apparently, the different needs of oxygen, the size of the internal systems of tracheoles, the size and structure of the openings of the tracheoles, the thickness of the body cuticula, and the size of the internal cuticle layers induce the variations in the speed of the knock-down effects seen in the experiments within different pests presented here.

4.5 Conclusions

As seen in the present review, it has been proven that natural extracts may have high insecticidal and acaricidal activities. These effects depend, however, on the dosage/concentration used and on the sensitivity of the target organisms. Since it is known that plant extracts may lose their activity when having had a too long contact with water, the products used in the present study were stored in a completely water-free version or were used in a composition with different fine shampoos. In the case of the fresh preparation of the solution prior to use an extreme high efficacy was reached against a broad spectrum of very important parasites and pests of humans and animals. Since tests of skin irritations remained negative (Pittermann et al. 2008) and such neem extracts may even be used as food, the described products offer a considerable support to human and animal health.

The practical use of neem extracts against several human diseases besides other efficacies than those of own interest of the tree (i.e., repel and kill insects and acari that attack the tree) needs more investigation. This is needed, although numerous possibilities of applications have been published in the past (see Schmutterer 2002). However, the number and complexity of many ingredients of leaves, bark, fruits, or roots is so high that side effects might harm the users. This statement is valid for all plant extracts which, however, will have much more importance in the future.

4 The Neem Tree Story: Extracts that Really Work

Fig. 4.1 *Azadirachta indica* – tree

Fig. 4.2 *Azadirachta indica* – seeds and leaflets

Fig. 4.3 *Azadirachta indica* – seeds without peel

Fig. 4.4 Head louse (*Pediculus humanus capitis*) – adult, egg-bearing female, light micrograph

Fig. 4.5 Head louse (*Pediculus humanus capitis*) – nit, scanning electron micrograph; note the cover (operculum) with openings (aeropyles)

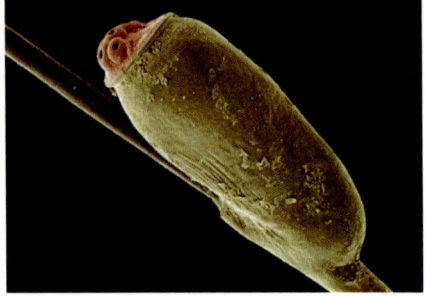

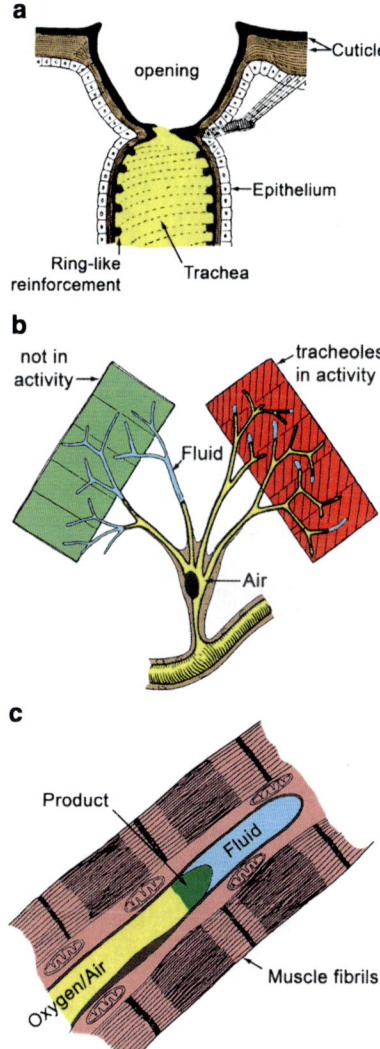

Fig. 4.6 Diagrammatic representation of the activity of Wash-Away Louse inside the tracheoles of the louse: (**a**) Opening of the tracheal system at the surface of the louse (stigma). (**b**) Position of the air (oxygen) in an inactive organ (*left*) and during activity (*right*). (**c**) The product Wash-Away louse covers the fine aqueous layer at the end of the tracheoles and thus blocks the uptake of the oxygen

Fig. 4.7 Amounts of different developmental stages of the red chicken mite *Dermanyssus gallinae* (fed and unfed stages) taken from the floor of a chicken house

Fig. 4.8 Scanning electron micrograph of a house dust mite *Dermatophagoides pteronyssinus* (untreated)

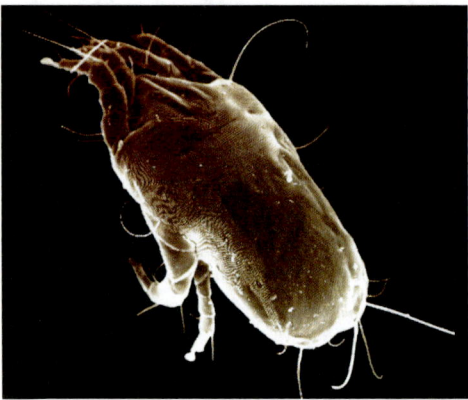

4 The Neem Tree Story: Extracts that Really Work

Fig. 4.9 Adult female *Ixodes ricinus*

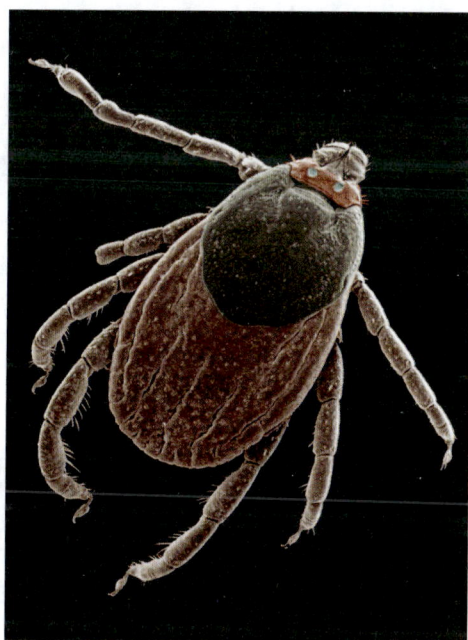

Fig. 4.10 Scanning electron micrograph of an adult tick of *Rhipicephalus sanguineus*

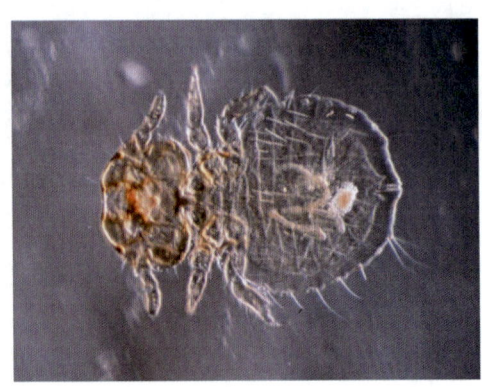

Fig. 4.11 Light micrograph of a mallophage

Fig. 4.12 Scanning electron micrographs of the cat flea (*Ctenocephalides felis*) – adult

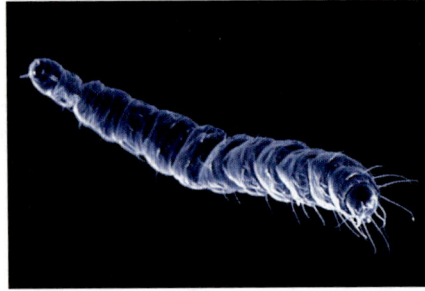

Fig. 4.13 Scanning electron micrographs of the cat flea (*Ctenocephalides felis*) – larva

Fig. 4.14 *Triatoma infestans*, a blood-sucking raptor bug from South America

Fig. 4.15 Flour beetle *Tenebrio molitor* (larvae and adult)

Fig. 4.16 Cockroaches – *Blattella germanica*

Fig. 4.17 Cockroaches – *Blatta orientalis*

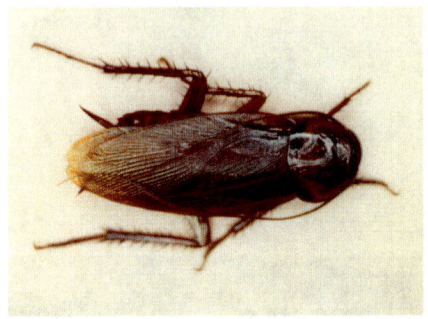

Fig. 4.18 Developmental stages of *Lucilia sericata* – adult

Fig. 4.19 Developmental stages of *Lucilia sericata* – larva

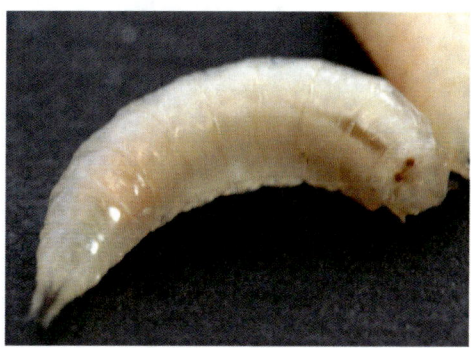

Fig. 4.20 Female bed bug *Cimex lectularius* and two freshly laid eggs

References

Abdel-Ghaffar F, Semmler M (2007) Efficacy of neem seed extract shampoo on head lice of naturally infected in Egypt. Parasitol Res 100:329–332

Abdel-Ghaffar F, Sobhy HM, Al-Quraishy S, Semmler M (2008a) Field study of an extract of neem seed (MiteStop®) against the red mite *Dermanyssus gallinae* naturally infecting poultry in Egypt. Parasitol Res 103:481–485

Abdel-Ghaffar F, Al-Quraishy S, Sobhy H, Semmler M (2008b) Neem seed extract shampoo, Wash Away Louse® is an effective plant agent against *Sarcoptes scabiei* mites infecting dogs in Egypt. Parasitol Res 104:145–148

Abdel-Ghaffar F, Semmler M, Al-Rashied K, Mehlhorn H (2009) In-vitro efficacy of ByeMite ® and MiteStop® on developmental stages of the red chicken mite *Dermanyssus gallinae*. Parasitol Res 105:1469–1471

Abdel-Ghaffar F, Semmler M, Al-Rasheid KAS, Klimpel S, Mehlhorn H (2010) Comparative in-vitro tests on the efficacy and safety of 13 anti- head lice products. Parasitol Res 106:423–429
Amer A, Mehlhorn H (2006) Repellency effect of forty-one essential oils against *Aedes*, *Anopheles*, and *Culex* mosquitoes. Parasitol Res 99:478–490
Athanasiadou S, Githori J, Kyriazakis I (2007) Medical plants for helminth parasite control: facts and fiction. Animal 1:1392–1400
Bäumler S (2007) Heilpflanzen, Praxis heute [Medicinal plants, praxis of today]. Urban and Fischer, München, 989
Biswas K, Chattopadhyay I, Banerjee RK, Bandyopadhyay U (2002) Biological activities and medical properties of neem (*Azadirachta indica*). Curr Sci 82:13336–13450
Brown D (1996) Encyclopedia of herbs and their uses. Dumont, Cologne
Chauve C (1998) The poultry red mite *Dermanyssus gallinae* (De Geer, 1778): current situation and future prospects for control. Vet Parasitol 79:239–245
Costa CTC, Bevilaqua CML, Maciel MV et al (2006) Anthelminthic activity of *Azadirachta indica* against sheep gastrointestinal nematodes. Vet Parasitol 137:306–310
De Luna CJ, Arkle S, Harrington D, George DR, Guy JH, Sparagano OA (2008) The poultry red mite *Dermanyssus gallinae* as a potential carrier of vector-borne diseases. Ann NY Acad Sci 1149:255–258
Ermel K (2002) Azadirachtin content of neem seed kernels from different regions of the world. In: Schmutterer H (ed) The neem tree. Neem Foundation, Mumbai
Fajimi AK, Taiwo AA (2005) Herbal remedies in animal parasitic diseases in Nigeria: a review. Afr J Biotechnol 4:303–307
Gemma EV, Beckmann E, Burke BJ, Boyer A, Maslen SL, Ley SV (2007) Synthesis of azadirachtin: a long but successful journey. Angew Chem Int Ed 46:76–79
Hamscher G, Prieb B, Nau H (2007) Determination of phoxim residues in eggs by using high-performance liquid chromatography diode array detection after treatment of stocked housing facilities for the poultry red mite (*Dermanyssus gallinae*). Anal Chim Acta 586:330–335
Heukelbach J, Oliveira FAS, Spear R (2006) A new shampoo based on neem (*Azadirachta indica*) is highly effective against head lice in vitro. Parasitol Res 99:353–356
Hoglund H, Nordenfors H, Uggla A (1995) Prevalence of the poultry red mite *Dermanyssus gallinae*, in different types of production system for egg layers in Sweden. Poultry Sci 74:1793–1798
Hussain A, Khan MN, Igbal Z, Sajiid MS (2008) An account of the botanical anthelmintics used in traditional veterinary practices in Satriwal district of Punjab, Pakistan. Int Ethnopharmacol 119:185–190
Kim SI, Na YE, Yi JH, Kim BS, Ahn YJ (2007) Contact and fumigant toxicity of oriental medical plant extracts against *Dermanyssus gallinae* (Acari: Dermanyssidae). Vet Parasitol 145:377–382
Koul O, Is S, Kapil RS (1996) The effect of neem allelochemicals on nutritional physiology of larval *Spodoptera litura*. Entomol Exp Appl 79:43–50
Koul O, Singh G, Singh R, Singh J, Daniewski WM, Berlozeckis K (2004) Bioefficacy and mode of action of some limonoids of salannin group from *Azadirachta indica* and their role in a multicomponent system against lepidopteran larvae. J Biosci 29:409–416
Kraus W (2002) Biologically active ingredients in neem tree (azadirachtin and other triterpenoids). In: Schmutterer H (ed) The neem tree. Neem Foundation, Mumbai
Leung AY (1985) Chinese medicinal plants. Diederichs, Munich
Locher N (2009) Life cycle and control of *Dermanyssus gallinae* mites. PhD. thesis, Free University of Berlin, Germany
Locher N, Al-Rasheid KAS, Abdel-Ghaffar F, Mehlhorn H (2010) In-vitro and field studies on the contact and fumigant toxicity of a neem-product (MiteStop®) against the developmental stages of the poultry red mite *Dermanyssus gallinae*. Parasitol Res 107:417–423

Lundh J, Wiktelius D, Chirico J (2005) Azadirachtin-impregnated traps for the control of *Dermanyssus gallinae*. Vet Parasitol 130:337–342

Marangi M, Cafiero MA, Capelli G, Camada A, Sparango OA, Gingaspero A (2009) Evaluation of the poultry red mite *Dermanyssus gallinae* (Acari: Dermanyssidae) susceptibility to some acaricides in field populations from Italy. Exp Appl Acarol 48:8

Mehlhorn H (2008) Encyclopedia of parasitology, vols 1 and 2, 3rd edn. Springer, New York

Mehlhorn H (2009) Not only in summer: mallophages. Dog Cat Horse J 1(09):22–25

Mehlhorn H, Schmahl G, Schmidt J (2005) Extract of seeds of the plant *Vitex agnus castus* proven to be highly efficacious as a repellent against ticks, fleas, mosquitoes and biting flies. Parasitol Res 95:363–365

Mehlhorn H, Schmahl G, Schmidt H (2006) Repellency of ticks by extracts of monk pepper seeds (*Vitex agnus castus*). Chemother J 15:175–178

Meyer-Kuehling B, Pfister K, Mueller J, Heine J (2007) Field efficacy of phoxim 50 % (ByeMite®) against the poultry red mite *Dermanyssus gallinae* in battery cages stocked with laying hens. Vet Parasitol 147:289–296

Michaelakis A, Papachristos P, Kimbaris A et al (2009) Citrus essential oils and four enantiomeric pinenes against *Culex pipiens*. Parasitol Res 105:769–774

Mitchell MJ, Smith SL, Johnson S, Morgan ED (1997) Effects of the neem tree compounds azadirachtin, salannin, nimbin and 6-desacetylnibin on ecdysone- 20-monooxygenase activity. Arch Insect Biochem Physiol 35:199–209

Mulla MS, Su T (1999) Activity and biological effects of neem products against arthropods of medical and veterinary importance. J Am Mosq Control Assoc 15(2):133–152

Mumcuoglu KY et al (1996) Repellency of essential oils and their components to the human body louse, *Pediculus humanus humanus*. Entomol Exp Appl 78:309–314

Niemann L, Stinchombe S, Hilbig V (2002) Toxicity to mammals including humans. In: Schmutterer H (ed) The neem tree, 2nd edn. Neem Foundation, Mumbai

Oladimeji FA, Orafidiya OO, Ogunniy TA, Adewunmi TA (2000) Pediculocidal and scabicidal properties of *Lippia multiflora* essential oil. J Ethnopharmacol 72:305–311

Oliveira LMB, Bevilaqua CML, Costa CTC et al (2009) Anthelmintic activity of *Cocos nucifera* against sheep gastrointestinal nematodes. Vet Parasitol 159:55–59

Pittermann W, Lehmacher W, Kietzmann M, Mehlhorn H (2008) Treatment against blood sucking insects without skin irritation. SOÉFW J 134:37–43

Rates SMK (2001) Plants as source of drugs. Toxicon 39:603–613

Schmahl G, Al-Rasheid KAS, Abdel-Ghaffar F, Klimpel S, Mehlhorn H (2010) The efficacy of neem seed extract (Tre-san®, MiteStop®) on a broad spectrum of pests and parasites. Parasitol Res 107:261–269

Schmutterer H (ed) (2002) The neem tree, 2nd edn. Neem Foundation, Mumbai, p 893

Semmler M, Abdel-Ghaffar F, Al-Rasheid KAS, Mehlhorn H (2009) Nature helps: from research to products against blood sucking arthropods. Parasitol Res 105:1483–1487

Senthil-Natan S, Kalaivani K, Murugan K, Chung G (2005a) The toxicity and physiological effects of neem limonoids on *Cnaphalocrocis medicinalis*: the rice leaffolder. Pesticide Biochem Physiol 81:112–118

Senthil-Natan S, Kalaivani K, Murugan K, Chung PG (2005b) Effects of neem limonoids on the malarial vector *Anopheles stephensi*. Acta Trop 96:74–85

Sparagano O, Pavlicevic A, Murano T, Lamarda A et al (2009) Prevalence and key figures for the poultry red mite *Dermanyssus gallinae* infections in poultry systems. Exp Appl Acarol 48:3–10

Tagboto S, Townson S (2001) Antiparasitic properties of medicinal plants and other naturally occurring products. Adv Parasitol 50:200–285

Valiente-Moro C, Chauve C, Zenner L (2007) Experimental infection of *Salmonella enteritidis* by the poultry red mite *Dermanyssus gallinae*. Vet Parasitol 146:329–336

von Bingen H (1974) Naturkunde/Reproduction of a medieval book of efficacies of plants, 2nd edn. Mueller, Wiss Buchgesellschaft, Salzburg

Vreecken-Buijis MJ, Hassink J, Brussard L (1998) Relationship of soil microarthropod biomass with organic matter and pore size distribution in soils under different land use. Soil Biol Biochem 30:97–106

Xiao PG, Fu SL (1986) Traditional antiparasite drugs in China. Parasitol Today 2:353–355

Zeman P, Stika V, Barik M, Dusbabek F, Lavickova M (1982) Potential role of *Dermanyssus gallinae* (De Geer, 1778) in the circulation of the agent of pullurosis-typhus in hens. Folia Parasitol 29:371–374

Chapter 5
The Efficacy of Extracts from Plants – Especially from Coconut and Onion – Against Tapeworms, Trematodes, and Nematodes

Heinz Mehlhorn, Gülendem Aksu, Katja Fischer, Bianca Strassen, Fathy Abdel Ghaffar, Khaled A.S. Al-Rasheid, and Sven Klimpel

Abstract Nematodes and cestodes are present in almost all animals and humans. These parasites spread to a considerable degree among animals that are continuously grazing because in the case of nematodes, the eggs or larvae excreted in the feces leas to and immediate subsequent infection when they are ingested orally with food. Infections also occur very simply in the case of tapeworms, even though in most cases intermediate hosts are involved, which are likewise infected by the egg- or larvae-containing piles of feces of infected grazing phase, particularly when the animals stay close together, for example in mass production farms. Infestations with nematodes and cestodes are therefore a major problem threatening the rearing of livestock animals, ruminants in particular, because of the reduction in weight related to the amount of worms. The present chapter gives an overviews on the efficiency of various plant extracts on trematodes, tapeworms and nematodes. In detail it is described, that extracts of onion bulbs and of coconut–given at the same time-are able to control infections due to platyhelminths and round worms in animals. These experiments are so convincing that products may become lounched soon onto the agricultural markets.

H. Mehlhorn (✉), G. Aksu, K. Fischer, and B. Strassen
Department of Zoology and Parasitology, Heinrich Heine University, D-40225 Düsseldorf, Germany
e-mail: mehlhorn@uni-duesseldorf.de

F.A. Ghaffar
Department of Zoology, Cairo University, Giza, Egypt

K.A.S. Al-Rasheid
Department of Zoology, College of Science, King Saud University, Riyadh, Saudi Arabia

S. Klimpel
Biodiversity and Climate Research Centre (BiK-F), Goethe-University, D-60 325 Frankfurt am Main, Germany

5.1 Introduction

Nematodes and cestodes are present in almost all animals and humans. These parasites spread to a considerable degree among animals that are continuously grazing because in the case of nematodes, the eggs or larvae excreted in the feces lead to an immediate subsequent infection when they are ingested orally with food. Infections also occur very simply in the case of tapeworms, even though in most cases intermediate hosts are involved, which are likewise infected by the egg- or larvae-containing piles of feces of infected grazing animals. The intestinal worm population thus increases dramatically as a true burden during the grazing phase, particularly when the animals stay close together, for example in mass production farms. Infestations with nematodes and cestodes are therefore a major problem threatening the rearing of livestock animals, ruminants in particular, because of the reduction in weight related to the amount of worms (Mehlhorn et al. 1993, 1995 Mehlhorn 2008; Eckert et al. 2008).

However, this problem was already known in the old cultures of the Chinese, Egyptians, etc., based on the observation that only a few or no worms were found when healthy animals were slaughtered, while a large number of worms were always seen in emaciated animals (Leung 1985; Tagboto and Townson 2001; Rates 2001; Hussain et al. 2008; Asase et al. 2005; Barrau et al. 2005). In times before the invention of chemical anthelminthics, worms were controlled by feeding specific plants which, based more on belief than on knowledge, were credited with having specific actions. Specific plants actually have an anthelminthic action, as has been shown also in recent tests. For example, the consumption of male fern parts, chili pods, or papaya kernels remove worms from the gut. However, many plant constituents are found to be extremely toxic. The so-called worm fern (*Dryopteris filix-mas*) for example causes – particularly in the case of overdose – poisoning with vomiting, dizziness with visual disturbances up to blindness, and even death (Schönfelder and Schönfelder 2001; Bäumler 2007).

Moreover, many of the plant products used in natural medicine have not been sufficiently tested in terms of either their action or their tolerability in accordance with the safety standards that are necessary today, so that they are today no longer used in practice and have been replaced by chemical products (Eckert et al. 2008; Mehlhorn et al. 1993; Rommel et al. 2006; Mehlhorn 2008; Besier 2007; Besier and Love 2003; Coles et al. 1992; Hosking et al. 2009, 2010; Kaminsky et al. 2008, 2011).

However, aside from the preference for a clearly demonstrated action, chemical substances have several disadvantages:

1. They are, in most cases, only active against one group of worms – that is to say, they act either against nematodes or against flatworms.
2. They lead to tissue depositions, and therefore, relatively long waiting times are needed before meat, milk, or eggs can be released for human consumption.

3. They are relatively expensive for the customer, because the search for new chemical compounds is extremely expensive.
4. Most of them lose their efficacy after about 20 years of use because resistance develops, particularly in the case of mass production of livestock animals. Thus, even today, mostly a plurality of different anthelminthics must be used in succession during the grazing season (Eckert et al. 2008) and other methods such as rotation of grazing fields (with all its logistic problems) have to be used to solve the problems of constantly needed chemotherapy for multiresistant parasites.
5. Chemical products cannot be used by ecofarms.

For these reasons, it is necessary to seek alternatives coming from the plant sector, especially since plants have successfully fought against worms and insects for millions of years and the consumption of useful plants is the daily practice of animals.

Thus, trials using plants as anthelminthic remedies go back to older methods of the prechemotherapeutic periods, however, they have become more and more important today. The evaluation of plant-derived products was officially recommended by an initiative of the World Health Organization (WHO) in the year 2000. Many trials have been started and their number increases constantly (Table 5.1).

Of course, in many regions of the world, representatives of "traditional medicine" use plants in the fight against parasites. However, although there are many reports of plant efficacy (e.g., Tagboto and Townson 2001; Soffar and Mokhtar 1991; von Bingen 1974; Zahir et al. 2009; Kamaraij et al. 2010; Bizimenyera et al. 2006; Tadesse et al. 2009; Table 5.1), only few of them are tested using significant scientific methods, while many of the effects are based more on belief than on sound background knowledge. This makes the use of plant extracts doubtful, since side effects may endanger the users. In general, the number of tests using plant extracts as remedies have increased in the last 10 years and will increase further in the future in many fields.

There are some reports from books dealing with natural medicine that parts of the garden onion (*Allium cepa*) and of garlic (*Allium sativum*) have anthelminthic activity (Giove 1996; Fajimi and Taiwo 2005). However, a distinction is not made in the reports between nematodes and flatworms, which both may occur together in the intestine but behave completely differently. For example, feeding green leaves from the garden onion to hens in Nigeria is said to help against "helminthiasis" (worm diseases) (Fajimi and Taiwo 2005). Abu-El-Ezz (2005) reports in Egypt more specifically that onion and garlic oil help against infestations with *Trichinella spiralis* in rats. However, it is not clear, whether this action is not based on an expelling process due to large amounts of included fatty acids. The oil-driven excretion of the adults from the intestine apparently decreases the number of muscle larvae. A killing effect of "*Allium*" oil inside muscles is, however, unlikely. Therefore, there is no definitive proof that onion extracts act against both worm

Table 5.1 Some plant-derived compounds known or claimed for their anthelminthic activity as cited in the literature

Plant species/used portion	Activity described against worm type	Name of known ingredient	Authors
Onion (*Allium cepa*), garlic (*Allium sativum*), bulbs	Worms in general	S-containing compounds, flavonoids	Giove (1996), Klimpel et al. (2011), Abdel-Ghaffer et al. (2011)
Onions and garlic (*Allium cepa, A. sativum*), oil	*Trichinella spiralis*	Not known	Abu-El-Ezz (2005), Rotscheidt (2008)
Garlic (*Allium sativum*)	*Trichinella spiralis*	Allicin, dialyl sulfate	Grundzinsky et al. (2001)
Allium schoenoprasum Liliaceae	Worms in general	Aliine, glykamylpeptide	Schönfelder and Schönfelder (2001)
Worm fern (*Dryopteris filix-mas*), roots	Nematodes	Filicin	Bown (1995)
Pumpkin (*Cucurbita*), seeds	Worms in general	Curcubin (amino acids)	Giove (1996), Rybaltovsky (1966), Xiao and Fu (1986)
Absinth (*Artemisia absinthium*) Abrotanum (*A. abrotanum*)	Worms in general	Absinthin (sesquirterpen lacton; leads to diarrhea)	Giove (1996)
Goose feet plant (*Chenopodium ambrosioides*)	Nematodes	Ascaridol (terpenoxyd)	Giove (1996), MacDonald et al. (2004)
White goosefoot (*Chenopodium album*)	Trichostrongylids	Not known	Jabbar et al. (2007)
Betel nut palm (*Areca catechu*)	Worms in general	Arecolin (alkaloid)	Roepke (1996)
Citrus fruits (*Citrus aromaticum, C. media*), juice	*Ascaris lumbricoides*	Not known	African communications
Ebony (*Diospyros mollis*)	Tape- and hookworms	Diospyrol	Sen et al. (1974)
Tobacco (*Nicotina tabacum*)	*Ascaris* worms	Not known, leads to diarrhea	Lorimer et al. (1996)
Papaya (*Carica papaya*), seed extracts	Nematodes	Cystein proteinases	Stepek et al. (2004)
Swamp trefoil (*Lotus pendunculatus*), leafs	Trichostrongylids	Tannin	Molan et al. (2000)
Spanish Esparsette (*Hedysarum coronarium*), leafs	Trichostrongylids	Tannin	Molan et al. (2000)
Chicoree (*Cichorium intybus*), roots	Trichostrongylids	Tannin	Molan et al. (2000)
Neem tree (*Azadirachta indica*), seeds	Trichostrongylids	Azadirachtin	Hördegen et al. (2003)
Pepper of Szechuan (*Zanthoxylum zanthoxyloides*), seeds	*Haemonchus contortus*	Not known	Hounzangbe-Adobe et al. (2004)

(*continued*)

Table 5.1 (continued)

Plant species/used portion	Activity described against worm type	Name of known ingredient	Authors
Efeu (*Hedera helix*), fruits	*Haemonchus contortus*	Not known	Eguale et al. (2007)
Legominosae (*Flemingia vestita*)	*Echinobothrium* sp. *Fasciolopsis buski*	Isoflavone, genistein	Tandon et al. (1997)
Compositae (*Vernonia amygdalina*)	Schistosomes	Vernodalin, sesquitepine lactone	Ohigashi et al. (1994)
Zingiberaceae (*Kaempferia galanga*) and other species	*Ascaris*, schistosomes	Alcoholic extract	Adewunmi et al. (1990)
Punicaceae (*Punia granatum*)	Tapeworms	Isopelletierin	Schönfelder and Schönfelder (2001)
Liliaceae (*Colchicium autumnale*)	Mediterranean coast fever	Colchicin	Schönfelder and Schönfelder (2001)

stages: adults and larvae. Thus, in many cases, activities are claimed, that are not verified in tests. This was also the case for *Azadirachta indica*, where Costa et al. (2006, 2008, 2009) showed that there is no anthelminthic efficacy against gastrointestinal worms.

Although studies on the action of garlic (*A. sativum*), which is systematically closely related to the garden onion (*A. cepa*), have shown that such extracts exhibit in vitro effects against nematodes, it was often not possible to confirm this in in vivo tests with infected animals (Burke et al. 2009; Igbal et al. 2001; Bastidas 1969), or the degree of action was very low unless very large amounts were used (Ayaz et al. 2008). These differences between in vitro and in vivo tests suggest that the reported positive effects in vitro might be attributable to possible killing effects of remnants of extracting agents such as chloroform, acetonitrile, alcohol, etc.

There are likewise cursory reports in the literature dealing with natural medicine relating to an "anthelminthic action" of coconut (*Cocos nucifera*) (Duke and Wain 1981; Blini and Lira 2005; Amorim and Borba 1994, 1995). The action of coconut milk was said to be directed against nematodes (genus *Syphacia*) and tapeworms (genus *Hymenolepis* syn. *Vampirolepis*) of naturally infected rodents. In experimental in vivo tests by Oliveira et al. (2009), however, coconut exhibited absolutely no effects on gastrointestinal nematodes in sheep. Studies by Costa et al. (2006) had already previously delivered similar results – positive effects in vitro but no effects in vivo. With respect to these above cited, often contradictory findings, in vivo studies should bring definite proofs on the usefulness of a combination of onion and coconut extracts in ruminants, since the first test objects with favorable results have been mice and/or rats (Abdel-Ghaffar et al. 2011; Klimpel et al. 2011).

5.2 Effects of Cocos and Onions on Trematodes, Cestodes, and Nematodes

5.2.1 The Plants

5.2.1.1 Coconut

The cocos palm (*C. nucifera*), Fam. Palmae and/or Arecaceae (Franke 1994) may reach a height of up to 30 m and is found today worldwide in all tropical and subtropical zones, although its origin is probably in Asia and it was transported (apparently before Columbus) to America. The large leaves [Fig. 5.1 (for figure see end of the chapter)] which form the top of the tree consist of 200–300 leaflets. In the axles of each newly formed leaf a new flower arises beginning from the sixth year of development. Along the upper portions of the lateral leaves about 200–300 small male flowers arise, while at the base of the lateral leaves 20–40 large female flowers are formed. In the case of the coconut palm fecundation always occurs with foreign sperm driven by wind or transported by insects. The growth of the fruits (nuts) with a size of up to 2.5 kg needs about 12–14 months [Fig. 5.2 (for figure see end of the chapter)]. The "nuts" possess a smooth exocarp, a thick fibrous mesocarp, and a "strong" endocarp with three germ grooves, one of which becomes later penetrated by the embryo [Fig. 5.3 (for figure see end of the chapter)]. The coconut contains inside a thick, oil-containing endosperm layer and a fluid endosperm (= cocos milk). Because of air chambers inside the mesocarp and its fibrous surface the coconut may swim at the surface of water and thus may be transported from one island to another. As soon as the thick white endosperm is dry (reducing the water content to only 5–6%) the final product is called khopra, which is the starting material for the production of cocos oil by pressure. Since all portions of the tree and of the fruit are used this tree is also called the "Tree of life" (Persley 1992; Schütt et al. 2004).

5.2.1.2 Onion

The onion [*A. cepa;* Fig. 5.4 (for figure see end of the chapter)] has been used for a long time by many people on Earth. In the Middle Ages, it was believed in Europe that the bulb (when placed by doors) prohibited the entrance of the pest. Romans and old Egyptians ate the bulb in large amounts and held it as a symbol of the universe. Medicinal effects of onions were detected rather early in human culture. Their sulfoxides (e.g., Allicin) and similar components make the juice of onions a good disinfectant. But also an efficacy in treating diseases of the respiratory and excretory system was claimed.

The onion (*A. cepa*) along with garlic (*A. sativum*), bear's garlic (*A. ursinum*), and chive (*A. schoenoprasum*) belongs to the family Alliaceae and to the Liliaceae sensu lato since they all possess (above fine roots) a bulbus-like swelling (with different layers of scales that surround the sprout which grows out after overwintering). This bulbus contains organic sulfide components. During dissection of an onion the enzyme allinase is set free thus introducing the formation of thiopropanol-*s*-oxide, which irritates the eyes and thus brings about the flood of tears. The onion has its homeland in Persia, but is now cultured worldwide in many variations and used as a spice in human food.

5.2.2 The Extracts

5.2.2.1 Preparation of Extracts

The following extracts from the coconut's hard endosperm and from onion bulbs were made at room temperature starting from minced and dried plant material at room temperature or at 37°C and under constant shaking for 24 h:

- Methanolic
- Ethanolic
- Aqueous
- Acetonitrile
- Chloroform
- 1:1 PEG/PC (polyethylene glycol/propylcarbonate)

In all cases, dried plant material (being produced at 40°C) was mixed with the extraction fluid in a relation of 1 g plant material within 10 ml extraction solution. In the case of the *Toxocara cati* test, the undiluted fruit water of a coconut was used.

5.2.2.2 Preparation of Extracts for Use in Sheep

1. Onions and coconuts for the different repeated experiments were bought on the market (Düsseldorf, Germany). The onions were of local origin, the coconuts came from the Dominican Republic being packed in the typical sacs containing about 20–25 specimens.
2. Grinding and/or optimal micronization of dried onion bulbs (without covers) or of coconut "fruit flesh" (dried endosperm) was done with laboratory grinders. The same results were obtained when conventionally micronized onion powder was used.
3. Moistening of the powder was done by spraying tap water.
4. Optimal ultrasound treatment was done for 5–10 min.
5. Drying of the powdered material was done in a conventional heating chamber at 40°C.

6. The dry powder was stored at −4°C until use.
7. For application, the dried mass was collected in tap water in a ratio of 1:10 plant preparation/tap water or in a 1:1 mixture of PEG (polyethylene glycol) and PC (propyl carbonate).
8. At three locations (Ratingen, Germany; Shandalat, Egypt; Riyadh, Saudi Arabia) sheep of 30–40 kg body weight were treated daily orally, with 60 g onion and 60 g coconut for 8 days. These sheep were naturally infected when treated with the food additive. Always two sheep were kept untreated as controls.
9. In order to ameliorate the taste of this plant combination for the sheep, 10 g of milk powder was added per treatment.
10. Similar worm-depriving results were obtained, when the same amount of dry onion, and coconut were fed to the sheep without milk powder. Thus, it does not make a difference to the anthelminthic effect whether milk powder is added or not.

5.2.3 The Test Parasites

5.2.3.1 Trematodes

1. *Echinostoma caproni* [Fig. 5.5 (for figure see end of the chapter)] lives as an adult fluke in the intestine of water birds and/or in mammals with contact to water, since the larval development occurs first in water snails (e.g., *Lymnaea, Biomphalaria*) and later (with the production of metacercariae) in other snails, tadpoles, or mussels before being eaten by rodents etc., as final hosts. Balb/C mice were experimentally infected by oral application of 20–25 metacercariae taken from experimentally infected snails.
2. *Fasciola hepatica* is the so-called large liver fluke of ruminants living in the bile ducts. It can, however, also be held in laboratory rats (as in our case) or in rabbits, which become infected if they swallow metacercariae, which in general, are attached to plants at the border of water reservoirs on meadows. Wistar rats were experimentally infected by oral application of 12 metacercariae [Fig. 5.6 (for figure see end of the chapter)].

5.2.3.2 Cestodes (tapeworms)

1. *Hymenolepis microstoma* [Fig. 5.7 (for figure see end of the chapter)] lives in its adult stage in the bile duct and with its posterior end in the intestine of mice, while the larva (cysticercoid) is found in beetles like *Tenebrio molitor*. Each of the mice (Balb/C) was experimentally infected by oral application of 15 cysticercoid larvae.

5 The Efficacy of Extracts from Plants – Especially from Coconut and Onion   117

2. *Hymenolepis diminuta* is found in its adult stage in the intestine of rodents (rats) and occasionally in humans, while their larvae are also found in beetles. Each of the rats was orally infected with three cysticercoids, humans might be infected by eggs [Fig. 5.8 (for figure see end of the chapter)].
3. *Taenia taeniaeformis* reaches a length of about 60 cm when found in the intestine of cats which become infected by uptake of a so-called strobilocercus-larva, being situated in the liver of rats, mice, or rabbits. In the present study, however, two naturally infected cats were used.

5.2.3.3 Nematodes

The intestinal worms of the species *Trichuris muris* [Fig. 5.9 (for figure see end of the chapter)] were kept in 8–9-week old female laboratory mice strains (C 57/BL 10 or NMRI) for years in the institute. The animals were orally infected with 250 larvae-containing eggs. This brings a mean of about 75–90 adults into untreated mice (depending on the infectiousness of the eggs). *T. muris* has a direct life cycle, where infection occurs by oral uptake of larvae-containing eggs. *Angiostrongylus cantonensis* worms [Figs. 5.10 and 5.11 (for figures see end of the chapter)] were cultivated in Wistar rats (*Rattus rattus*). These rats were orally infected with 40 larvae, which had been selected from the mucus or squeezed portions of infected snails (intermediate hosts = *Biomphalaria* species or *Archachatina* species). The adult worms live in the pulmonary arteries of their host, and their larvae 1 are excreted via saliva and/or feces [Fig. 5.12 (for figure see end of the chapter)].

5.2.4 The Test Methods

5.2.4.1 In Vitro Studies

The in vitro studies of trematodes and cestodes were done in 37°C warm Tyrode's medium, within which the parasites were incubated for 24 h containing one of the different extracts or not (controls).

The in vitro tests of the larvae 1 of *A. cantonensis* were done after they had been obtained with the help of a so-called Baermann funnel from feces of infected rats and placed into the culture medium at 37°C. Adult worms of *T. muris* were taken from the intestines of infected adult mice and incubated like the larvae of *A. cantonensis* in the test media. The vitality of the incubated worms was controlled at intervals of 1–2 h for 24 h.

In vitro dosages of test products: a broad variety of concentrations of the product was used at least in one-, two- and threefold steps of the initial dosage being added to the treated Tyrode's solution, wherein the worms became incubated. As controls, the same worm stages were incubated in untreated Tyrode's solution at 37°C.

5.2.4.2 In Vivo Studies

Rats, mice, and one *T. cati*-infected cat were orally treated either by dosage per kilogram of body weight or – in the case of the cat – by 6 ml of undiluted cocos fluid. Six mice were also fed for 12 days exclusively with dried coconut. The effects of the different treatments were measured either by fecal control for eggs/larvae or by intestinal/lung inspection of sacrificed animals.

The tests on nematodes using sheep in Ratingen, Egypt, and Saudi Arabia were done as described above (Sect. 5.2.2.2).

5.2.5 The In Vitro Results

In order to calculate the amount of the extracts obtained by different extraction methods and at different temperatures pilot studies were carried out (Table 5.2). These results clearly show that it is always required that the same method and temperature for the extraction is used in the test.

In order to determine whether the in vitro incubation tests of larvae or adult worms give reasonable results extracts of different plants were used as well as pure Tyrode's medium alone to check the motility and survival rate of tested worms (Table 5.3). It was seen that the worms may survive for a long time (up to 7 days) in 37°C warm Tyrode's solution, while killing effects occurred by adding different plant extracts.

5.2.5.1 Trematodes

Adult *F. hepatica* worms [Fig. 5.6 (for figure see end of the chapter)] from rats were placed into 1 ml Tyrode's medium containing 1 mg of the chloroform extract of coconut. The *Fasciola* worms that had been incubated in the coconut extract medium started immediately to exhibit strong contractions and died within 4 h of exposition, while the controls survived unchanged for at least 24 h.

Table 5.2 Amounts of extracted material obtained depending on the method of extraction and on the temperature (n.d. = not done)

Plant material	Fresh weight (g)	Dried weight (g)	Aqueous 22°C (g)	Aqueous 37°C (g)	Methanolic extract 22°C (g)	Acetonitrile 22°C (g)
Pineapple fruit	250	43.6	3	30	10.3	0.25
Pineapple leaves	65	17.0	0.5	n.d.	1.1	n.d.
Banana fruit	250	16.6	19.8	19.3	19.4	0.75
Chicory	250	227	14.1	5.4	2.2	0.2
Date tree fruit	250	224	21.6	54.4	62.2	0.25
Fig fruit	250	118	6.2	13.3	67.8	0.75
Pumpkin	250	92	1.5	2.5	n.d.	n.d.
Onion bulb	250	115	15.3	10.2	n.d.	n.d.

5 The Efficacy of Extracts from Plants – Especially from Coconut and Onion

Table 5.3 In vitro effects of aqueous extracts of different plants on the motility of adult *T. muris* worms ($n = 10$ per probe)

Plant extract	Doses (mg/ml)	Start 0 h	After 1 h	After 6 h	After 12 h	After 24 h
Pineapple fruit	27.5	3	3	2.5	1	0
Pineapple leaves	5.5	3	2.5	2.5	2.5	1.5
Banana fruit	44.5	3	3	2.5	2	0
Chicory	16.5	3	2.5	2.5	2.5	0
Date tree fruit	47.0	3	1	1.5	1.5	0
Fig fruit	60.0	3	1.5	1	2	0
Control Tyrode	-	3	3	3	3	2.5

Motility grading: 0 = no movements; 1 = low motility; 2 = continuous smooth movements; 3 = normal, intensive movements

Adult trematodes of the species *Echinostoma caproni* [Fig. 5.5 (for figure see end of the chapter)] showed the same reactions as *F. hepatica*. Those worms incubated in extract containing medium died within 4 h, while the controls survived for at least 24 h without change.

5.2.5.2 Cestodes

Hymenolepis microstoma [Fig. 5.7 (for figure see end of the chapter)]. One milligram of the chloroform extract was added to 1 ml Tyrode's medium at 37°C. In this medium five adult tapeworms were placed, while two tapeworms were kept in pure Tyrode's solution. About 15 min after the start of the experiment, all treated tapeworms showed strong contractions and started to dissolve their proglottids, while the two control worms remained unchanged and moved slightly. After 30 min, the treated tapeworms were dead, while the untreated ones lived, even after another 24 h.

5.2.5.3 Nematodes

When using in vitro aqueous, methanolic, or chloroform extracts on adult *T. muris* [Fig. 5.9 (for figure see end of the chapter)] worms similar results were obtained as depicted in the example given in Table 5.4, which also shows that the efficacy of an extract depends on the dosage used.

Similar results were obtained in larvae 1 of *A. cantonensis* [Fig. 5.12 (for figure see end of the chapter)], when PEG (polyethylen glycol) and PC (propyl carbonate) – acetonitrile extracts were used or when pure cocos milk was fed (Table 5.5).

The main conclusions that could be drawn from a very broad series of in vitro results were that there is clear evidence that plant extracts are able to kill intestinal worms. However, it became also clear that these effects depend not only on the mode of extraction but also on the doses used. Just the latter finding made it obviously difficult to transfer results of efficacy obtained in vitro to in vivo treatments. There remains always the crucial question, whether a certain compound in

Table 5.4 In vitro effects on the motility of a chloroform extract of coconut on adults of *Trichuris muris* during 24 h of incubation in Tyrode's medium at 37°C ($n = 10$ per probe)

Plant	Doses (mg/ml)	0 h	1 h	8 h	12 h	24 h
Coconut	0.005	3	3	2	0	0
Coconut	0.01	3	3	0	0	0
Coconut	0.1	3	3	0	0	0
Control Tyrode	–	3	3	3	3	3

Motility grading: 0 = no movements; 1 = low motility; 2 = continuous smooth movements; 3 = normal, intensive movements

Table 5.5 In vitro effects of PEG/PC acetonitrile extracts of neem seeds and cocos milk on the motility of larvae 1 of *A. cantonensis* ($n = 50$ per probe)

Plant extract	Doses (µl/ml)	0 h	1 h	8 h	12 h	24 h
Neem seeds	10	3	0	0	0	0
Cocos milk	500	3	2	1.5	0	0
Cocos milk (pure)	undiluted	3	2	1	0	0
PEG/PC alone	10	3	3	3	3	2
Control Tyrode solution	–	3	3	3	3	2.5

Motility grading: 0 = no movements; 1 = low motility; 2 = continuous smooth movements; 3 = normal, intensive movements

a plant extract may be taken up inside an intestine in such an amount, that it may attain the required concentrations in the normal fluid inside the intestine of animals or man hosting intestinal parasites (not even considering concentrations needed to kill parasites inside tissues or in blood). Therefore in vivo tests with experimentally infected hosts are absolutely required – otherwise everything remains speculative.

5.2.6 The In Vivo Results

5.2.6.1 Cestodes

A series of tests on the tapeworm *H. microstoma* showed that a treatment with a mixture of extracts of coconut plus other plant extracts has no effects, if these combinations are given for only 4 days (Table 5.6). The excreted number of worm eggs did not change considerably after 4 days when compared to egg numbers excreted before treatment.

The reason might be that this worm reaches only with its posterior end into the intestine, while the anterior portions enter the bile ducts, which the coconut plus other plant extracts might not have reached. This would explain why a 3-day treatment of rats with intestinal tapeworms of the species *H. diminuta* stops worm egg excretion after this 3-day treatment (Table 5.7).

The next experiment with *H. microstoma* – proceeding a 7-day feeding period with khopra – shows that coconut extracts are principally able to kill this tapeworm, where the dose is increased and the treating period is prolonged (Table 5.8).

5 The Efficacy of Extracts from Plants – Especially from Coconut and Onion

Table 5.6 Effects of some combinations of plant extracts on *Hymenolepis microstoma* given for 4 days. Egg count in 1 g feces (mean from five infected mice)

Plant/type of extract	Dosage (g/kg bw)	Egg number – day before treatment	Eggs 1 day after the 4-day treatment
Cocos (chloroform) + aqueous Pineapple	1; 20	9,451	9,498
Cocos (chloroform) + Chicory (aqueous)	1; 8	9,071	8,892
Cocos (chloroform) + Date tree (aqueous)	1; 15	10,372	12,940
Cocos (chloroform) + Fig (aqueous)	1; 20	11,253	10,331
Control (normal food)	–	13,084	12,950

bw body weight; *g* gram; *kg* kilogram

Table 5.7 Effects of a chloroform extract of coconut on *Hymenolepis diminuta* (in rats) after a 3-day treatment (at days 23, 25, and 27)

Rats no.	Dosage (g/kg/bw/day)	Eggs in feces (day 22)	Eggs in feces (day 24)	Eggs in feces (day 26)	Eggs in feces (day 28)	Cestodes in intestine on day 33
Control 1	Normal food	+++	+++	+++	+++	3
Control 1	Normal food	+++	+++	+++	+++	3
Treated 3, 4	0.1; 0.1	+++; +++	+++; +++	++; ++	–; –	2; 3
Treated 5, 6	0.4; 0.4	+++; +++	+++; +++	+; +	–; –	0; 0
Treated 7, 8	0.6; 0.6	++; ++	+++; +++	–; –	–; –	0; 0

bw body weight; *g* gram; *kg* kilogram; *d* day
+++ Very large number of eggs; ++ Many eggs; + Few eggs; – No eggs

Table 5.8 Effects of a 7-day pure khopra diet on *Hymenolepis microstoma* and the weight increase of the animals. Egg counts were done per gram feces as a mean of five mice

Plant	Mice number	Eggs per gram feces; 1 day before start of diet	Eggs after 1st day of diet	Eggs after 2nd day of diet	Eggs after 3rd day of diet	Eggs after 7th day of diet	Worms in mice after diet
Dried coconut (khopra)	5	26,782	15,913	6,887	0	0	0
Dried coconut (khopra)	5	21,210	11,398	5,117	0	0	0
Controls	5	15,456	13,072	3,298	0	0	9 + 1

Similar results were also obtained when different dosages of a combination of onion and coconut against *H. microstoma* were given for 8 days to infected mice (Table 5.9). It was clearly shown that onion and coconut act synergistically on this tapeworm, since 500 mg/kg/bw of both plant extracts gives the same result as 1,000 mg of each.

When feeding two naturally infected cats excreting proglottids of *T. taeniaeformis* daily for 6 days with 6 ml of undiluted cocos fluid (= fluid endosperm) the excretion of proglottids stopped after the first day of treatment. Then, for the next 2 days, a lot of single proglottids were excreted while at the beginning of the fifth day of treatment no further proglottids appeared in the feces (not even 1 month later).

Table 5.9 Effects of feeding a combination of onion and coconut for 8 days to mice

Mouse No.	Infected with	Administration of/ target dose	Weight of mouse (g)/ volume to be administered	Findings on day 9
1	H. microstoma	Infection control	–	10 Worms, normal
2	H. microstoma	Infection control	–	9 Worms, normal
3	H. microstoma	Infection control	–	10 Worms, normal
4	H. microstoma	Onion/500 mg/kg bw	36.8 g/0.184 ml	3 Worms with immature proglottids
5	H. microstoma	Onion/500 mg/kg bw	40.7 g/0.203 ml	3 Worms with immature proglottids
6	H. microstoma	Onion/500 mg/kg bw	37.8 g/0.189 ml	4 Worms without noticeable change under light microscope
7	H. microstoma	Onion/500 mg/kg bw	44.9 g/0.224 ml	2 Worms without noticeable change under light microscope
8	H. microstoma	Onion and coconut/ each 500 mg/kg bw	40.3 g/ 0.201 ml + 0.201 ml	0
9	H. microstoma	Onion and coconut/ each 500 mg/kg bw	40.1 g/ 0.201 ml + 0.201 ml	0
10	H. microstoma	41.5 g/415 µl	1,000 mg coconut + 1,000 mg onion + 100 mg milk powder/kg mouse bw	0
11	H. microstoma	37.8 g/378 µl	1,000 mg coconut + 1,000 mg onion + 100 mg milk powder/kg mouse bw	0
12	H. microstoma	42.8 g/428 µl	1,000 mg coconut + 1,000 mg onion + 100 mg milk powder/kg mouse bw	0

bw body weight; *mg* milligram; *g* gram; *kg* kilogram

5.2.6.2 Trematodes

The in vivo tests with trematodes were kept at the level of a pilot study. In addition to a 5-day application of 3.8 g of the chloroform extract of coconut to *F. hepatica*-infected rats as the oral feeding, application of the extract for 5 days to mice infected with *Echinostoma caproni* also did not produce convincing results. Treated and untreated animals showed, in both cases, no reduction of the excreted eggs and no reduction of the adult worms after scarifying the animals. This failure might be based on the possibility that the chloroform extract did not provide the really active

Table 5.10 Effects of a combination of onion and coconut on the intestinal fluke *Echinostoma caproni* after 8 days treatment

Mouse No.	Infected with	Administration	No worms in intestine
1	*Echinostoma* flukes in the intestine	– (control)	22/Normal
2	*Echinostoma* flukes in the intestine	– (control)	24/Normal
3	*Echinostoma* flukes in the intestine	1,000 mg onion/kg bw	12
4	*Echinostoma* flukes in the intestine	1,000 mg onion/kg bw	16
5	*Echinostoma* flukes in the intestine	1,000 mg coconut/kg bw	8
6	*Echinostoma* flukes in the intestine	1,000 mg coconut/kg bw	9
7	*Echinostoma* flukes in the intestine	500 mg onion/500 mg coconut/kg bw	0
8	*Echinostoma* flukes in the intestine	500 mg onion/500 mg coconut/kg bw	0

bw body weight; *mg* milligram; *kg* kilogram

compound. On the other hand, a trial using PEG/PC extracted/diluted components of coconut and onion in combination showed very convincing results in *E. caproni*, when each component was given consecutively for 8 days at a dosage of 500 mg or 1,000 mg/kg body weight (Table 5.10), while an application of either 1,000 mg of onion extract alone or 1,000 mg coconut alone left surviving worms in the intestine.

5.2.6.3 Nematodes

Trichuris muris

In the first experiments, the infected mice were treated only for 3 days with the different plant extracts. On these days and on days 1, 5, 12, and 22 after treatment, the excreted eggs were counted within 1 g of feces. The results, however, show that a 3-day treatment with the plant extracts is not sufficient to stop egg excretion. Furthermore, the egg excretion of the control mice was also varying and decreasing. Thus, in another series of experiments, a full coco diet was given to three infected mice for 12 days. These three mice were each fed daily a mean of 1.5 g. The mean excreted egg amount decreased from day 2 of the treatment from a mean of 16,089 per gram of feces to 196 on day 12 of the treatment, while the egg excretion of the controls remained rather stable. The controls started with a mean of 13,398 eggs per gram of feces and ended with 15,075 eggs on day 12 of treatment. Sacrificing the mice, it was seen that the four treated mice contained only two of four worms with signs of degeneration, while the two untreated control mice contained 68 and 72 fully motile worms without any degeneration, respectively. This result led us to the insight that a longer treatment should be made using, besides coco, other plant extracts – garlic, chives, and onion (Table 5.11).

However, as can be seen in Table 5.11, only the treatment with onion powder was able to eliminate fully the load of adult *T. muris* worms from the intestines of infected mice, while garlic or chives' powder was not very effective. Therefore, in a further experiment, a combination treatment was made with onion powder and coconut powder (Table 5.12). This experiment clearly shows that the combination

Table 5.11 In vivo tests of the efficacy of plant powders of three species of the genus *Allium* on *T. muris* when given for eight consecutive days. Infection of mice was done by application of 250 worm eggs. The food amount was prepared from 2 g of powder plus 8 ml of PEG/PC (1:1)

Mouse No.	Dose (mg/kg bw)	Findings on dissection after 8 days
1	Untreated infection control	89 Worms, vital
2	Untreated infection control	83 Worms, vital
3	500 mg wild garlic powder/kg bw	81 Worms, without noticeable damage
4	500 mg wild garlic powder/kg bw	72 Worms, without noticeable damage
5	1,000 mg wild garlic powder/kg bw	79 Worms, without noticeable damage
6	1,000 mg wild garlic powder/kg bw	84 Worms, without noticeable damage
7	500 mg wild onion powder/kg bw	48 Worms, sit loosely in the mucosa
8	500 mg wild onion powder/kg bw	12 Worms, sit loosely in the mucosa
9	1,000 mg wild onion powder/kg bw	0
10	1,000 mg wild onion powder/kg bw	0
11	500 mg wild chives powder/kg bw	68 Worms, without noticeable damage
12	500 mg wild chives powder/kg bw	77 Worms, without noticeable damage
13	1,000 mg wild chives powder/kg bw	58 Worms, without noticeable damage
14	1,000 mg wild chives powder/kg bw	66 Worms, without noticeable damage

bw body weight; *kg* kilogram

of 500 mg/kg body weight each of coconut and onion given for 8 days is fully successful, while 500 mg either of onion or coconut leaves residual worms in the intestine of *T. muris*-infected mice. The results were identical in both sexes when using female and male worms.

Pilot Study on *Toxocara cati*

The cat of one of the participants of the study excreted daily considerable numbers (equivalent to 360 per gram of feces) of eggs of the worm *T. cati* [Fig. 5.13 (for figure see end of the chapter)]. The owner fed this cat for 5 days daily with 6 ml cocos fluid (fluid endosperm) of a coconut. This resulted in the stopping of egg excretion after the fourth day of treatment. The egg excretion did not start again afterwards.

Test Using Naturally Infected Sheep

Test Series 1

In Shandalat (Egypt), naturally infected sheep were investigated coproscopically for intestinal worms. All sheep (weighing around 40 kg) were infected with specimens of the genera *Toxocara, Nematodirus, Trichostrongylus, Haemonchus, Trichuris,* and/or *Monieza*. Four of those animals obtained once daily for 8 days an amount of 60 g onion powder and 60 g coconut powder plus dry milk powder spread on their normal food. On the ninth day after the beginning of the experiment (first day after the end of the treatment), microscopic examination of the feces showed that there were no more eggs of the previously seen species. Feces controls

Table 5.12 In vivo experiments of the effects of coconut and onion alone and in combination against adult *T. muris* worms. The infection was done with 250 eggs

Mouse No.	Parasite	Treatment	Findings on dissection/ number of worms
1	*T. muris*	Untreated infection control	64 Living worms, normal
2	*T. muris*	Untreated infection control	74 Living worms, normal
3	*T. muris*	Untreated infection control	77 Living worms, normal
4	*T. muris*	Onion/500 mg per kg bw	13 Worms, no noticeable change under light microscope
5	*T. muris*	Onion/500 mg per kg bw	18 Worms, no noticeable change under light microscope
6	*T. muris*	Coconut/500 mg per kg bw	7 Worms, no noticeable change under light microscope
7	*T. muris*	Coconut/500 mg per kg bw	11 Worms, no noticeable change under light microscope
8	*T. muris*	Onion and coconut/each 500 mg per kg bw	1 Worm, male, with disrupted spermiogenesis; damaged
9	*T. muris*	Onion and coconut/each 500 mg per kg bw	0
10	*T. muris*	1,000 mg coconut + 1,000 mg onion + 100 mg milk powder/kg bw	0
11	*T. muris*	1,000 mg coconut + 1,000 mg onion + 100 mg milk powder/kg bw	1 Female, loose in the intestine, moribund
12	*T. muris*	1,000 mg coconut + 1,000 mg onion + 100 mg milk powder/kg bw	0
13	*T. muris*	1,000 mg coconut + 1,000 mg onion + 100 mg milk powder/kg bw	2 Females, loose in the intestine, moribund
14	*T. muris*	1,000 mg coconut + 1,000 mg onion + 100 mg milk powder + 100 mg cocoa/kg bw	0

were repeated on day 16 and 23 after the beginning of the experiments. Both tests showed also no parasitic stages in the feces which were also inspected in vain for proglottids.

Test Series 2

In Ratingen (Germany), 12 sheep of about 30 kg body weight had been selected for the test (their feces contained stages and/or eggs of trichostrongylid worms, specimens of eggs of the genera *Trichuris*, *Haemonchus*, *Nematodirus*, and proglottids of *Monieza*). Ten of 12 sheep obtained daily orally 60 g coconut powder and 60 g onion powder, both diluted in milk. On days 9 and 20, after the beginning of the experiment, the feces were controlled by coproscopical methods. It was seen that on both the 9th and on the 20th day, there were no longer worm stages in the feces.

Test Series 3

In Riyadh (Saudi Arabia), six naturally infected sheep were selected weighing 30–40 kg. Two were kept as untreated controls. Four sheep received daily for 8 days 60 g each of onion and coconut powder being previously diluted in 1:1 PEG (Polyethylene glycol) and PC (propyl-carbonate) in order to see, whether PEG alone has an effect on the intestinal worms, through feces control. As seen via microscopic fecal control techniques on day 9 and 20 after the beginning of the trial, the control animals still excreted worm stages, while this stopped in the treated animals. This proves that the addition of onion and coconut to the food was successful and that the addition of PEG/PC alone (without coconut and onion) has no influence on the worm burden.

Body Weight

All investigators observed 4 weeks after the end of the addition of onion and coconut powder to the food of the test sheep that the development of the latter was ameliorated in comparison to the control animals. In all cases, a visibly better increase of body weight of the treated animals was noted by the investigators.

5.3 Discussion

The broad testing of extracts obtained with different methods from about 21 different plants in our institute and comparison of our results with other publications (Aksu 2009; Strassen 2007; Fischer 2007; Abdel-Ghaffer et al. 2011; Klimpel et al. 2011) resulted in some not very surprising results, but also led to some insights, which gave hope of the production soon of a reliable and relatively cheap product. If this thus-developed plant product is added to food, deworming (e.g., of ruminants) might occur without the usual waiting time (Mehlhorn et al. 2011).

It was not surprising that different methods (aqueous, chloroform, alcoholic, acetonitrile, and PEG/PC) delivered different extracts with different grades of efficacy. Thus, the literature must be checked as to whether results reported there are really comparable. Not really surprising was that several of the plants – even when extracted with the favorable aqueous extraction mode – had no or only slight effects on the stages of our nematode models (*T. muris* and *A. cantonensis*). On the other hand, it became clear that in vitro experiments alone will not be sufficient to find a reliable product since – although the results in vitro might be good – in vivo the needed concentration could never be reached.

Surprising was that chive, onion, and garlic – although all belonging to the plant genus *Allium* – had completely different effects on nematodes, so that only the onion (*A. cepa*) was used further on in the trials.

When comparing the literature, many contradictory results are reported, for example Oliveira et al. (2009) obtained in their experiment an in vitro anthelminthic activity of an ethyl acetate extract of the green coconut husk fiber (which is a byproduct of the food industry) against the larvae of *Haemonchus contortus*. However, when feeding this extract daily over 3 days at a dose of 400 mg/kg body weight, no effects on the gastrointestinal nematodes were seen. This failure is – when looking at our results – apparently due to the fact that Oliveira et al. (2009) used another type of extract, which originated from another portion of the coconut, and that it was given only for 3 days, while in the present in vivo studies on mice and in our in vivo study with sheep, we fed another coconut extract for 8 days (Mehlhorn et al. 2011).

Our in vitro and in vivo studies on the effects of plant extracts on cestodes and trematodes confirm the findings of parallel in vitro and in vivo studies on nematodes (Mehlhorn et al. 2011; Klimpel et al. 2011). All these studies proved that plant extracts may have significant killing effects in vitro and in vivo on nematodes, cestodes, and trematodes. Similar effects had been described for many plants for example for pumpkin or *Agrimonia pilosa* (Rosaceae) extracts against *Taenia solium* or *Schistosoma japonicum* (Xiao and Fu 1986), *Punica granatum* (Hukkeri et al. 1993) against *T. solium*, *Thymus serrulatus* (Desta 1995) against *Taenia saginata* or *Flemingia vestita* (Tandon et al. 1997; Liu et al. 2011) against trematodes.

Many other reports on effective plants are listed in the survey of Tagboto and Townson (2001) (see Table 5.1). In most cases, only reports on accidental tests were given. In these papers selected by Tagboto and Townson (2001) or by Hussain et al. (2008) no hints were given of a synergistic action of onion and coconut as was found in the papers of Abdel-Ghaffar et al. (2011), Klimpel et al. (2011), and Mehlhorn et al. (2011). The total failure or partial failure of extracts of coconut and onion alone as reported in the papers of Abu-El-Ezz (2005) or Oliveira et al. (2009), depend apparently on the different extracts used and/or on the use of only one plant extract over rather short treatment periods. But all these tests and most in the literature had/have laboratory character, which provides us with good hints but they may not be convincing and successful in practice when used against large animals like sheep and/or other ruminants.

Helminthic infections pose enormous problems on the overcrowded Earth, since billions of animals and humans suffer from such infections (Mehlhorn 2008). Furthermore, nearly seven billion humans need food [Fig. 5.14 (for figure see end of the chapter)], which is partly obtained through rearing ruminants, pigs, rabbits, chickens, etc. These farm animals are often infected by helminthic parasites, especially at locations where these animals are kept in large monocultures. Such helminths were mostly transmitted via oral uptake of parasitic stages (e.g., worm eggs, worm larvae) included in fecally contaminated food (grass etc.). Such infections, however, not only occur as animals graze on meadows but also even in stables. Then livestock animals come into direct contact with potentially worm-infected feces or infectious stages might be mechanically transmitted via the mouthparts or feet of flies (Förster et al.

2007). Because of the often enormous worm burden, growth of farmed animals is reduced leading to enormous financial losses, since these animals have to be fed considerably longer until they reach the normal slaughter weight. Own observations during a trial with an anthelminthic product (doramectin) left no doubt that the treated animals increased their body weight by a mean of 25–30 kg within a period of 6 weeks compared to untreated ones (if the latter had a considerable worm burden; Mehlhorn unpublished).

Therefore nowadays, the use of anthelminthic drugs is obligatory in animal farming. However, most of the medicaments are rather expensive, act only against a limited range of worms which they often don't fully eliminate and/or these medicaments may introduce resistances in a broad spectrum of worms. Furthermore, all of them require so-called waiting times until meat or milk can be used again as human food. This waiting time may stretch from 3 days, for example with fenbendazole, up to 35–60 days for the avermectins and milbemycines (Eckert et al. 2008). Furthermore there are growing dring resistanes (Blake and Coles 2007).

Plants or plant extracts may help to control parasites of many kinds (Semmler et al. 2009; Schmahl et al. 2010) – and are of use especially against worms (Table 5.1). However, although they have been used since ancient times, their use never translated into real success in mass production systems of farming. The reason was that there is no plant which really acts anthelminthically against the various species of nematodes and cestodes living at the same time in a broad spectrum of varying niches inside humans and animals. Of course there was no doubt at all, that plants can be used, since they had to develop anthelminthic abilities (e.g., against soil nematodes) in order to survive the struggle for life during evolution. This phenomenon becomes clear when considering that around 40% of the compounds listed in the "Dictionary of Natural Products" are not synthetically produced and that 40% of the 520 medicaments registered between 1983 and 1994 are natural remedies (Chiej 1984; Henkel 1999: Cragg et al. 1997; Costa et al. 2009; Athanasiadou et al. 2007). This is even underlined by the fact that nine of the 20 most often sold medicaments in the year 1999 were derivates of plants (Harvey 2000). Therefore, more research must be done on the efficacious use of plant extracts against the background that only mixtures of different plant derivates will have a broader, significant effect.

The tests in sheep presented by Mehlhorn et al. (2011) were based on the results of a broad laboratory screening of herbal products against nematodes, cestodes, and trematodes (Abdel-Ghaffar et al. 2011; Klimpel et al. 2011; and on several unpublished in vitro and in vivo assays by our group). In general, it was found that it depends on the extraction fluid (methanolic, ethanolic, aqueous, acetonitrile, chloroform, etc.), whether an extract reaches a considerable efficacy against a given worm load. Furthermore, it was seen that always a longer exposition period is needed when using plant extracts compared to chemical compounds. And in addition, the experiments showed that in case of an anthelminthic efficacy, the most common compound that a plant produces must not necessarily be responsible for the effect. This, for example, was shown when pure allicin (the leading substance in onion or garlic) remained ineffective against worms, while minced powder of these plants had clear but species-specific anthelminthic effects (Mehlhorn unpublished results).

These observations explain why in the literature contradictory results are published on the efficacy of extracts of coconut, onions, or garlic either on cestodes or on nematodes or on both together. In some cases coconut extracts were described to be effective against nematodes, in others not. For example, Oliveira et al. (2009) found an activity of coconut against the hatching and development of larvae of *Haemonchus*, while adult gastrointestinal worms were not significantly affected. This was different in *T. muris* experiments (Abdel-Ghaffar et al. 2011), where 2–6 mg of a chloroform extract per kilogram body weight reached reduction rates up to 98% of intestinal adult worms.

The tests of Mehlhorn et al. (2011) using many plant extracts, including garlic, onion, and/or coconut, showed that a combination of both against *A. cepa* and *C. nucifera* could be successful, since both alone reached a relatively high efficacy against either nematodes (*A. cepa*) or cestodes, and trematodes (*C. nucifera*) (Klimpel et al. 2011; Abdel-Ghaffar et al. 2011).

As a *first conclusion* for a practical application in farm animals a combination of powder of coconut and onion turned out to be most effective against both cestodes and nematodes. The synergistic effect of both plants had been shown in our previous laboratory tests (Klimpel et al. 2011; Abdel-Ghaffar et al. 2011). The only condition was that the sheep were fed daily for at least 8 days with 60 g coconut and 60 g onion powder, which of course, were more effective than using true medicaments and which have the advantage that there is no waiting time after application, since they are of plant origin without chemical additions.

5.4 Conclusions

1. The present review clearly showed that plants have many properties that had been used already in the past, but many of them had been forgotten due to the development of a splendid chemotherapy.
2. The "rediscovery" of the various plant derived efficacies, however, needs intensive research since the determination of definitive efficacies depends on the mode of extraction of certain compounds, while the feeding of the whole plant or its fruit alone will never have the full effect.
3. In order to consistently obtain the same effects the same extraction mode has to be used constantly after its determination as the best one.
4. Even if a substance obtained from plants shows high efficacy in in vitro tests, that is no guarantee that the needed effective dose can be reached in vivo inside a parasitized host.
5. Therefore, in vivo studies always need to be carried out– even before publication, since otherwise the paper is filled with useless findings.
6. The present review showed that additions of the properties of different plants may have synergistic effects, which may increase the chance of developing

a useful product, since often the efficacies of a single plant do not surmount the needed dose or the high dose that has to be applied might be too toxic or too expensive for practical use.
7. The present review based on the results of three papers in Parasitology Research prove that it is possible to develop a useful, relatively cheap product that helps against intestinal parasites, which can be reached much more easily than parasites hidden inside organs or inside the blood.
8. From all these studies it can be concluded that plant-derived extracts should be
   (a) Mixtures of several plants
   (b) Produced by the same methods
   (c) Effective at low dosages
   (d) Tested urgently in vivo
   (e) Mainly (and most easily) developed against parasites of inner cavities (mouth, intestine, sexual system, bladder) and against blood suckers on the body surface, since for these localities low doses will mostly be sufficient to attain the goal of killing the parasites.

5 The Efficacy of Extracts from Plants – Especially from Coconut and Onion

Fig. 5.1 Crown of a coconut tree showing the green fruits

Fig. 5.2 *Open and closed coconuts* showing the fibrous mesocarp and the white, thick inner stiff layer of the endosperm

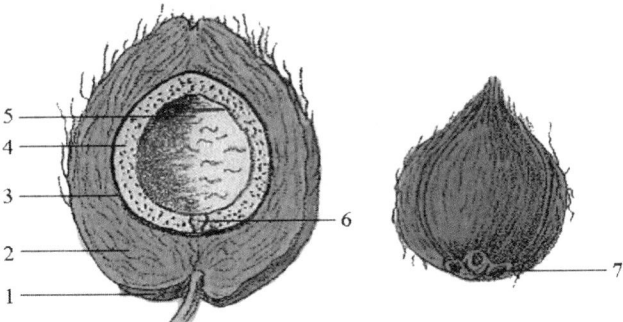

Fig. 5.3 Diagrammatic representation of the layers of a coconut. (1) Thin outer layer (*green to yellow-brown* outer layer = exocarp). (2) Thick fibrous middle layer (mesocarp). (3) Thin, very hard inner layer (endocarp). (4) White, 1–2-cm thick inner layer (khopra, oil-containing stiff endosperm). (5) The fluid = cocos water fills the center of the nut. (6) Sprout. (7) Three pores ("eyes") as possible sprout exits

Fig. 5.4 Onion bulbs showing the ventral degenerating roots

Fig. 5.5 Light micrograph of the anterior end of an adult *Echinostoma* worm. Note the tegumental hooks

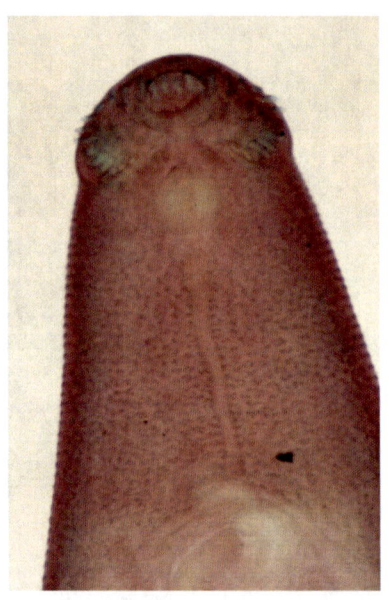

Fig. 5.6 Scanning electron micrograph of the anterior pole of *Fasciola hepatica* showing both suckers (oral and ventral sucker), the two openings of the uterus and the male system (adjacent to each other) and the numerous stiff tegumental hooks

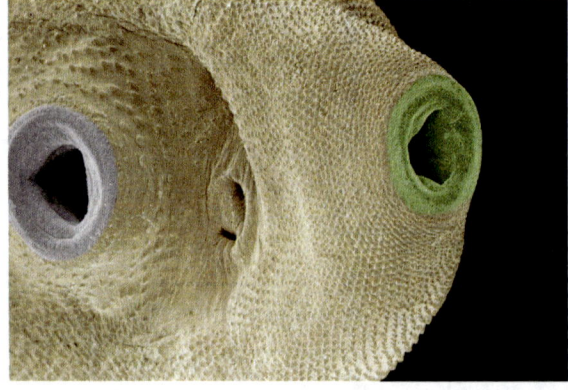

5 The Efficacy of Extracts from Plants – Especially from Coconut and Onion

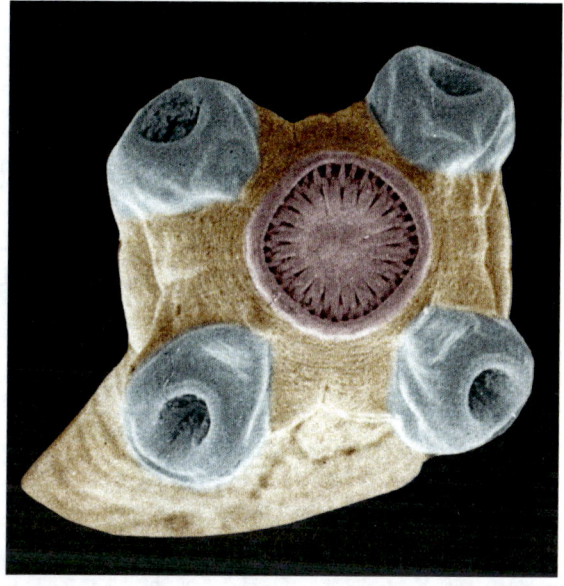

Fig. 5.7 Anterior end of the tapeworm *Hymenolepis microstoma*. Note the four suckers and the retracted hooked rostellum

Fig. 5.8 Light micrograph of the egg of the tapeworm *Hymenolepis diminuta*. Note (in focus) four of the six hooks of the centrally situated oncosphaera larva

Fig. 5.9 Scanning electron micrograph of an adult *Trichuris* worm. Note the whip-like (*thin*) anterior portion of the worm

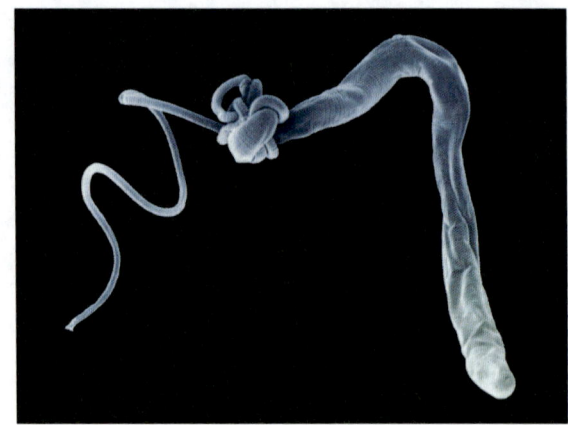

Fig. 5.10 Light micrograph of adult *Angiostrongylus* worms

Fig. 5.11 Scanning electron micrograph of the anterior pole of an *Angiostrongylus* worm

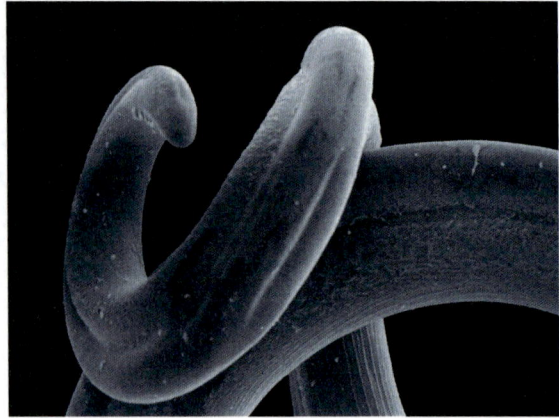

5 The Efficacy of Extracts from Plants – Especially from Coconut and Onion

Fig. 5.12 Light micrograph of larva 1 of *Angiostrongylus cantonensis* from rat feces

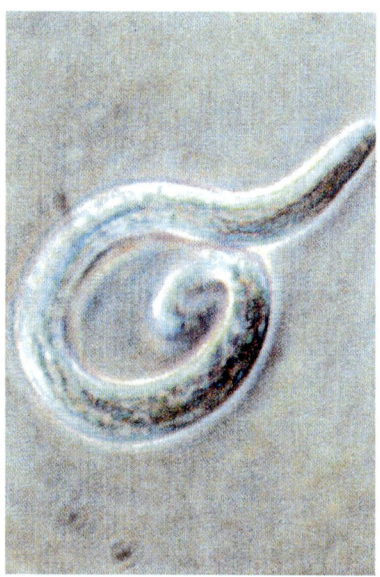

Fig. 5.13 Scanning electron micrograph of the anterior end of a *Toxocara* worm

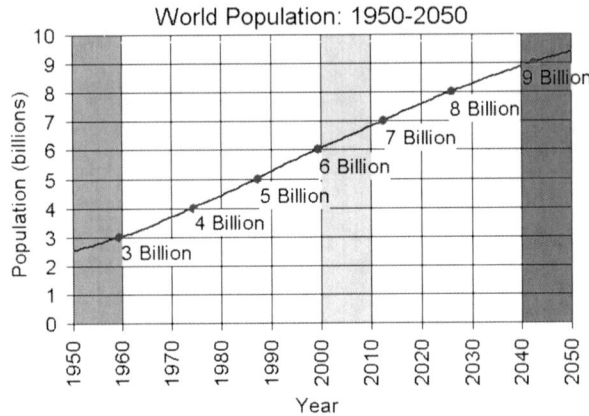

Fig. 5.14 Scheme and projection of the increase of the human world population

References

Abdel-Ghaffar F, Semmler M, Al-Rasheid SB, Akus G, Fischer K, Klimpel S, Mehlhorn H (2011) The effects of different plant extracts on intestinal nematodes and trematodes. Parasitol Res. 108:979–984

Abu-El-Ezz NM (2005) Effects of Nigella sativa and *Allium cepa* oils on *Trichinella spiralis* in experimentally infected rats. J Egypt Soc Parasitol 35:511–523

Adewunmi CO, Babajide OO, Furo P (1990) Molluscicidal and antischistosomal activities of *Zingiber officinale*. Planta Med 56:374–376

Aksu G (2009) Plant extracts in the fight against worms. Diploma thesis. Duesseldorf, Germany

Amorim A, Borba HR (1994) Acav anthelminthica de plantas X. Testes in-vivo com extractros brutos de Cocos nucifera. Rev Bras Farm 75:91–92

Amorim A, Borba HR (1995) Acav anthelminthica de plantas XI. Influencia de extractros brutos de Cocos nucifera. Rev Bras Farm 76:99–113

Asase A, Oteng-Yeboah AA, Odamtten GT, Simmonds MSJH (2005) Ethnobotanical study of some Ghanaian anti-malarial plants. J Ethnopharmacol 99:273–279

Athanasiadou S, Eithiori J, Kryiazakis I (2007) Medical plants for helminth parasitic control: facts and fictions. Animal 1:1392–1400

Ayaz E, Türel GA, Silmaz O (2008) Evaluation of the anthelminthic activity of garlic (*Allium sativum*) in mice naturally infected with *Aspiculuris traptera*. Recent Pat Antiinfect Drug Disc 3:149–152

Barrau E, Fabre N, Fouraste I, Hoste H (2005) Effect of bioactive compounds from sainfoin (*Onobrychis viciifolia* Scop.) on the in-vitro larval migration of *Haemonchus contortus*: role of tannins and flavonol glycosides. Parasitology 131:531–538

Bastidas G (1969) Effects of ingested garlic on *Necator americanus* and *Ancylostoma caninum*. Am J Trop Med Hyg 18:9209–9923

Bäumler S (2007) Heilpflanzen. Praxis heute. Urban und Fischer, Munich

Besier RB (2007) New anthelmintics for livestock: the time is right. Trends Parasitol 23:21–24

Besier RB, Love SCJ (2003) Anthelminthic resistance in sheep nematodes in Australia: the need for new approaches. Aust J Exp Agric 43:1383–1391

Bizimenyera ES, Githioris JB, Eloff JN, Swan GE (2006) In-vitro activity of *Peltophourum africanum* Sond. (Fabaceae) extracts on the egg hatching and larval development of the parasitic nematode *Trichostrongylus colubriformis*. Vet Parasitol 142:336–342

Blake B, Coles GC (2007) Flock cull due to anthelminthic-resistant nematodes. Vet Rec 161:36

Blini W, Lira CM (2005) Salvado vidas com a medicina natural. Unier, Sao Paulo, p 407

Bown D (1995) Encyclopedia of herbs and their uses. Dorling Kindersley, London

Burke JM, Wells IJ, Wsey P, Miller JE (2009) Garlic and papaya lack control over gastrointestinal nematodes in goats and lambs. Vet Parasitol 159:171–174

Chiej R (1984) Encyclopedia of medicinal plants. MacDonald, London

Coles GC, Bauer C, Borgsteede F, Geerts S, Klei TR, Taylor MA, Waller PJ (1992) World Association for Advancement of Veterinary Parasitology (WAAVP) methods for the detection of anthelmintic resistance in nematodes of veterinary importance. Vet Parasitol 44:35–43

Costa CTC, Bevilaqua CML, Maciel BV et al (2006) Anthelmintic activity or *Azadirachta indica* A. Juss against sheep gastrointestinal nematodes. Vet Parasitol 137:306–310

Costa CTC, Bevilaqua CML, Camurca-vasconcelos ALF, Maciel MV, Morais SM, Castor CMS, Braga RR, Oliveira LMB (2008) In-vitro ovicidal and larvacidal activity of *Azadirachta indica* ectracts on *Haemonchus contortus*. Small Rum Res 74:284–287

Costa CTC, Oliveira LMB, Camuirga Vasconalos ALF et al (2009) Atividade anthelminthica de *Cocos nucifera*. Vet Parqasitol 157:211–213

Cragg GM, Newman DJ, Snader KM (1997) Natural products in drug discovery and development. J Nat Prod 60:52–60

Desta B (1995) Ethiopian traditional herbal drugs. Part I: studies on the toxicity and therapeutic activity of local taenicidal medications. J Ethnopharmacol 45:27–33

Duke HA, Wain KK (1981) Medicinal plants of the world: a computer index with more than 85,000 entries, 3 vols. Longman, UK

Eckert J, Friedhoff KT, Zahner H, Deplazes P (2008) Lehrbuch der Parasitologie für die Tiermedizin, 2nd edn. Enke, Stuttgart, Germany

Eguale T, Tilahun G, Debella A, Feleke A, Makonnen E (2007) *Haemonchus contortus:* in vitro and in-vivo anthelmintic activity of aqueous and hydro-alcoholic extracts of *Hedera helix*. Exp Parasitol 116:340–345

Fajimi AK, Taiwo AA (2005) Herbal remedies min animal parasitic diseases in Nigeria: a review. Afr J Biotechnol 4:303–307

Fischer K (2007) The effects of plant extracts on worm stages. Diploma thesis, University of Duesseldorf, Germany

Förster M, Klimpel S, Mehlhorn H, Sievert K, Messler S, Pfeffer K (2007) Pilot study on synanthropic flies (e.g. Musca, Sarcophagas, Calliphora, Fannia, Lucilia, Stomoxys) as vectors of pathogenic microorganisms. Parasitol Res 101:243–246

Franke W (1994) Nutzpflanzenkunde (knowledge on cultured plants) Thieme, Stuttgart, 4th edition

Giove N (1996) Traditional medicine in the treatment of enteroparasites. Rev Gastroenterol Peru 16:197–202

Grundzinsky IP, Frankiewitz-Kosko A, Bany J (2001) Dialyl sulfide – a flavour component from garlic (*Allium sativum*) attenuates lipid peroxidation in mice infected with *Trichinella spiralis*. Phytomedicine 8:174–177

Harvey A (2000) Strategies for discovering drugs from previously unexplored natural products. Drug Discov Today 5:294–300

Henkel T (1999) Drugs. Angew Chemie Int Ed Engl 38:643–653

Hördegen P, Hertzbverg H, Heilmann J, Langhans W, Maurer V (2003) The efficacy of five plant products against gastrointestinal trichostrongylids in artificially infected lambs. Vet Parasitol 117:51–60

Hosking BC, Griffiths TM, Woodgate RG, Besier RB, Leeuvre AS, Nilon P, Trengove C, Vanhoff KJ, Kaye-Smith BG, Seewald W (2009) A clinical field study to evaluate the efficacy of monepantel against gastro-intestinal nematodes of sheep, in comparison with registered anthelmintics, in Australia. Aust Vet J 87:455–462

Hosking BC, Kaminsky R, Sager H, Rolfe PF, Seewald W (2010) A pooled analysis of the efficacy of monepantel, an amino-acetonitrile derivative against gastrointestinal nematodes of sheep. Parasitol Res 106:529–532

Hounzangbe-Adobe MS, Paolini V, Fouraste I, Moutairon K, Hoste H (2004) In-vitro effects of four tropical plants on three life-cycle stages of the parasitic nematode *Haemonchus contortus*. Res Vet Sci 78:155–160

Hukkeri VI, Kalyani GA, Hatpaki BCV, Manvi FV (1993) In-vitro anthelmintic activity of aqueous extract of fruit rind of *Punica granatum*. Fitotherapia 64:69–70

Hussain A, Khan MN, Igbal Z, Sajid MS (2008) An account of the botanical anthelmintics used in traditionally veterinary practices in Sahiwal district of Punjab (Pakistan). J Ethnopharmacol 119:185–190

Igbal Z, Khadid-Nadeem Q, Khan MN, Akthar MSS, Waraich FN (2001) In vitro anthelmintic activity of *Allium sativum*, *Zingiber officinale* and *Ficus religiosus*. Int J Agric Biol 3:454–457

Jabbar A, Zaman MA, Igbal Z, Yassen M, Shamin A (2007) Anthelmintic activity of *Chenopodium album* and *Caesalpinia crista* against trichostrogylid nematodes on sheep. J Ethnopharmacol 114(1):86–91

Kamaraij C, Rahuman AA, Bagavan A, Mohamed MJ, Elango G, Rajakumar G, Zahir AA, Santhoshkumar T, Marimuthu S (2010) Ovicidal and larvicidal activity of crude extracts of *Melia azedarach* against *Haemonchus contortus* (Strongylida). Parasitol Res 106:1071–1077

Kaminsky R, Ducray P, Jung M, Clover R, Rufener L, Bouvier J, Schorderet Weber S, Wenger A, Wieland-Berghausen S, Goebel T, Gauvry N, Pautrat F, Skripsky T, Froelich O, Komoin-Oka C, Westlund B, Sluder A, Mäser P (2008) A new class of anthelminthics effective against drug-resistant nematodes. Nature 452:176–180

Kaminsky R, Bapst B, Stein P, Strehlan G, Brooke A, Hosking B, Rolfe B, Sager H (2011) Differences in efficacy of monopantel, derquantel and abamectin against multiresistant nematodes of sheep. Parasitol Res

Klimpel S, Abdel-Ghaffar F, AL-Rasheid KAS, Aksu G, Fischer K, Strassen B, Mehlhorn H (2011) The effects of different plant extracts on nematodes. Parasitol Res 108:1047–1054

Leung AY (1985) Chinese medical plants. Diederichs, Munich

Liu YT, Wan F, Wang GX, Han J, Wang Y, Wang YH (2010) In vivo anthelmintic activity of crude extracts of *Radix angtelicae pubescentis*, *Fructus bruceae*, *Caulis spathgolobi*, *Semen aesculi*, and *Semen pharbitidis* against *Dactylogyrus intermedius* (Monogenea) in goldfish (*Carassius auratus*). Parasitol Res 106(5):1233–1239

Lorimer SD, Perry NB, Foster LM, Burgess EJ (1996) A nematode larval motility inhibition assay for screening plant extracts and natural products. J Agric Food Chem 44:2842–2845

MacDonald D, Van Crey K, Harrison P, Rangachari PK, Rosenfeld J, Warren C, Sorger G (2004) Acaridolless infusion *Chenopodium ambrosicoides* contain a nematocide that is not toxic to mammalian smooth muscle. J Ethnopharmacol 92:215–221

Mehlhorn H (2008) Encyclopedia of parasitology, 3rd edn. Springer, New York

Mehlhorn H, Düwel D, Raether W (1993) Diagnosis and therapy of parasitosis of house and farm animals, 2nd edn. G Fischer, Stuttgart, Germany

Mehlhorn H, Eichenlaub D, Löscher T, Peters W (1995) Diagnosis and therapy of the parasites of men, 2nd edn. G. Fischer, Stuttgart, Germany

Mehlhorn H, Al-Quaraishy S, Al-Rasheid KAS, Jatzlau A, Abdel-Ghaffar F (2011) Addition of a combination of onion (*Allium cepa*) and coconut (*Cocos nucifera*) to food of sheep stops gastrointestinal helminthic infections. Parasitol Res 108:1041–1046

Molan AL, Waghorn GTC, Min BR, McNabb WC (2000) The effects of condensed tannins from seven herbals on *Trichostrongylus colubriformis* larval migration in vitro. Folia Parasitol 47:39–44

Ohigashi O, Yoshikawa T, Osaki E (1994) Effects of *Vernonia* extracts on schistosomes. J Ethnopharmacol 82:115–123

Oliveira LMB, Bevilagne CML, Cosha CZC et al (2009) Anthelminthic activity of *Cocos nucifera* against sheep gastrointestinal nematodes. Vet Parasitol 159:55–59

Persley GJ (1992) Replantin the tree of life. CAB Intrernational, Wallingford, UK

Rates SMK (2001) Plants as source of drugs. Toxicon 39:603–613

Roepke DA (1996) Traditional and reapplied veterinary medicine in East Africa. In: McCokle et al (eds) Ethnoveterinary research and development. Intermediate Technol, London

Rommel M, Eckert J, Kutzer E, Boch J, Supperer R (2006) Veterinärmedizinische Parasitologie, 5th edn. Blackwell, Berlin

Rotscheidt W (2008) Kann man Schafe indirekt entwurmen? Schafzucht-Tiergesundheit 17:36–37

Rybaltovsky OV (1966) On the discovery of curcurbitin – a component of pumpkin seed with anthelminthic action. Med Parazitol (Mosk) 57:299–304

Schmahl G, Al-Rasheid KAS, Abdel-Ghaffar F, Klimpel S, Mehlhorn H (2010) The efficacy of neem seed extracts (Tre-san®, MiteStop®) and a broad spectrum of pests and parasites. Parasitol Res 107:261–269

Schönfelder A, Schönfelder P (2001) Heilpflanzenführer. Franckh-Kosmos, Stuttgart

Schütt P, Weisgerber H, Schuck HJ, Lang U, Stimm B, Roloff A (2004) Bäume der Tropen. Nikol Verlagsgesellschaft, Hamburg

Semmler M, Abdel-Ghaffar F, Al-Rasheid KAS, Mehlhorn H (2009) Nature helps: from research to products against blood sucking arthropods. Parasitol Res 105:1483–1487

Sen HG, Joshi BS, Parthasarathy P, Kamast VN (1974) Anthelminthic efficacy of diospyrol and its derivates. Arzneimittelforschung 24:2000–2003

Soffar SA, Mokhtar GM (1991) Evaluation of the antiparasitic effect of aqueous garlic (*Allium sativum*) extract in *Hymenolepis nana* and giardiasis. J Egypt Soc Parasitol 21:497–502

Stepek G, Behnke JM, Buttle DJ, Duce IR (2004) Natural plant cysteine proteinases as anthelmintics. Trends Parasitol 20:322–327

Strassen B (2007) *Cocos nucifera* investigations on the efficacy of extracts on parasites. PhD thesis, University of Duesseldorf, Germany

Tadesse D, Eguale T, Giday M, Mussa A (2009) Ovicidal and larvicidal activity of crude extracts of *Maesa lanceolata* and *Plectranthus punctatus* against *Haemonchus contortus*. J Ethnopharmacol 122:240–244

Tagboto S, Townson S (2001) Antiparasitic properties of medical plants and other naturally occurring products. Adv Parasitol 50:199–295

Tandon V, Pal PO, Rtoy B, Rao HPS, Reddy LES (1997) In vitro anthelmintic activity of root-tuber extract of *Flemingia vestita*, an indigenous plant in Shillong, India. Parasitol Res 83:492–498

von Bingen H (1974) Naturkunde – reproduction of a medieval book, 2nd edn. Mueller-Wiss, Buchgesellschaft, Salzburg

Xiao P, Fu SC (1986) Traditional antiparasitic drugs in China. Parasitol Today 2:353–355

Zahir AA, Rahuman AA, Kamaraj C, Bagavan A, Elango G, Sangaran A, Senthil Kumar B (2009) Laboratory determination of efficacy of indigenous plant extracts for parasites control. Parasitol Res 105(2):453–461

Chapter 6
Curcumin: A Natural Herb Extract with Antiparasitic Properties

Md. Shahiduzzaman and Arwid Daugschies

Abstract This short review addresses the knowledge on curcumin use against parasite infections from traditional to modern medicine. Curcumin is the active ingredient of turmeric (*Curcuma longa*). The extract of the rhizome of turmeric has been traditionally used against various diseases including parasitic infections. Recently, the crude extract of turmeric and its active ingredient curcumin have been explored with respect to the biological and molecular activity against many pathogens. The antioxidant, antitumor, and anti-inflammatory properties of curcumin make it a promising natural drug to be used against bacterial, fungal, and viral agents. Antiparasitic effects of curcumin have attracted considerable attention over the last decades. Curcumin has been found to display activity against various parasites both in vitro and in vivo. However, the effects of curcumin become obvious in most cases at relatively high dose levels. The bio-molecular and cellular processes involved in curcumin effects on parasites are not sufficiently understood at present and more research in this area is inevitable to define the actual applicability of curcumin in parasite control measures and therapy. We here review the available information on the therapeutic potential of curcumin against parasites obtained from in vitro studies, animal models, and clinical trials.

6.1 Introduction

The plant turmeric, *Curcuma longa* L. (Zingiberaceae family), has a long history of therapeutic use in Indian and Chinese medicines for the treatment of flatulence, dyspepsia, liver disorder, jaundice, urinary tract diseases, wound, inflammation, and

---

Md. Shahiduzzaman
Department of Parasitology, Bangladesh Agricultural University, Mymensingh 2202, Bangladesh
e-mail: szamanpara@yahoo.com

A. Daugschies (✉)
Institute of Parasitology, University of Leipzig, An den Tierkliniken 35, 04103 Leipzig, Germany
e-mail: daugschies@vetmed.uni-leipzig.de

other diseases. It has also been used in the treatment of intestinal parasites and of parasitic skin infection. The scientific backgrounds for the curcumin effects have been discovered gradually. A crude extract of turmeric contains curcumin and curcuminoids (Rasmussen et al. 2000) where curcumin is the major constituent. Curcumin is a polyphenolic orange–yellow compound and the main coloring principle of curcumin is 1,7-bis-(4-hydroxy-3-methoxy-phenyl)-hepta-1,6-diene-3,5-dione (diferuloylmethane). Curcumin has attracted considerable attention in recent years due to its remarkable pharmacological activities, including antioxidant, anti-tumor, and anti-inflammatory activities (Surh 2002; Aggarwal et al. 2003; Choi et al. 2006; Shen and Ji 2007; Goel et al. 2008). It exerts influence on a variety of biological and cellular processes. Curcumin has been suggested to be useful in antidiabetic (Srinivasan and Menon 2003), anti-HIV (Jordan and Drew 1996; Barthelemy et al. 1998), antibacterial (Negi et al. 1999; Mahady et al. 2002), and antifungal treatment (Kim et al. 2003). In addition to inhibiting the growth of a variety of pathogens, curcumin has been shown to have anthelmintic and antiprotozoal activities (Nose et al. 1998; Araujo et al. 1998, 1999; Koide et al. 2002). The current experience regarding curcumin effects in various scenarios rewards more research into the drug targets that might be suitable for therapeutic invention. The purpose of this review is to provide a brief summary of the current published knowledge of curcumin as a potential antiparasitic drug.

6.2 Curcumin Effects on Parasites

6.2.1 Helminths

Juice of turmeric traditionally has long been used to cure worm infections in south and south-east Asia (Nadkarni 1976). Recently, it has been reported that the extract of *C. longa* is active against *Schistosoma mansoni* (El-Ansary et al. 2007; El-Banhawey et al. 2007) and *Caenorhabditis elegans* (Atjanasuppat et al. 2009). Curcumin treatment modulates cellular and humoral immune responses of infected mice and leads to a significant reduction of parasite burden and liver pathology in acute murine schistosomiasis. It also possesses in vitro activity against adult *S. mansoni* worms (Magalhaes et al. 2009) and is suspected to have therapeutic potential in the treatment and prevention of schistosomiasis (Allam 2009). Antifibrogenic and anti-inflammatory properties of curcumin may reduce *Opisthorchis viverrini*-induced fibrosis and prevent cholangiocarcinoma development that may be associated with opisthorchiasis (Boonjaraspinyo et al. 2009; Pinlaor et al. 2010). Thus, curcumin may be useful as a chemopreventive drug by reducing the severity of *O. viverrini*-associated disease and the risk of cholangiocarcinoma. It has been suggested that this property of curcumin can be explained by reduction of the oxidative and nitrogenic DNA damage that may be due to suppression of oxidant-generating genes and enhancement of antioxidant genes in the nucleus of bile duct epithelial and inflammatory cells (Pinlaor et al. 2009). Although particularly trematode infections have been studied, curcumin seems also to have a positive effect in nematode infections such as *Toxocara canis* in dogs (Kiuchi et al. 1993). *T. canis* is a zoonotic parasite and, like other ascarid infections, is very prevalent worldwide and particularly in many developing countries. It would be of considerable benefit if curcumin effects would also be seen in human infections with *Ascaris lumbricoides* or *A. suum* in swine, however, no scientific data has been published in this respect so far.

6.2.2 Protozoa

Antiprotozoal activities of curcumin have been reported extensively over the last decade. The spice rhizome of turmeric (1% crude extract), as well as its main medicinal component, curcumin (0.05%), appear effective in reducing upper- and mid-small-intestinal infections caused by *Eimeria acervulina* and *Eimeria maxima* but they do not act beneficially in *Eimeria tenella* infections (Allen et al. 1998). However, in vitro incubation of *E. tenella* sporozoites with curcumin showed considerable effects on sporozoite morphology and viability and resulted in decreased invasion of MDBK cells (Khalafalla et al. 2010). Curcumin, as an alcoholic extract, was found to have antiprotozoal activity against *Entamoeba histolytica* (Dhar et al. 1968). Antiprotozoal activities of curcumin were also described for *Plasmodium* (Reddy et al. 2005; Cui et al. 2007), *Leishmania* (Araujo et al. 1998; Rasmussen et al. 2000; Koide et al. 2002; Saleheen et al. 2002; Das et al. 2008), *Trypanosoma* (Nose et al. 1998), and *Giardia lamblia* (Pérez-Arriaga et al. 2006), both in vitro and in vivo. Curcumin was able to reduce parasitemia by 80–90% in *Plasmodium berghei*-infected mice (Reddy et al. 2005). Recently, curcumin was found to be effective against *Cryptosporidium parvum* in cell culture. *C. parvum* appears to be more sensitive to curcumin than *Plasmodium, Giardia*, and *Leishmania* (Shahiduzzaman et al. 2009). Synergistic antiprotozoal effects were shown when curcumin was applied in combination with other drugs. For instance, the combination of artemisinin and curcumin shows additive activity in killing *Plasmodium falciparum* in culture and enabled experimentally *P. berghei*-infected mice to survive (Nandakumar et al. 2006). Drug resistance of *Plasmodium* strains is a major threat to malaria control. However, chloroquine-resistant *P. falciparum* (Reddy et al. 2005) and artemisinin-resistant *Plasmodium chabaudi* (Martinelli et al. 2008) were found to be sensitive to curcumin in culture and in mice, respectively. These are promising data that may open alternative options for malaria control, particularly where drug resistance has become a relevant issue.

6.3 Mode of Action and Perspectives

It appears that curcumin acts against parasites through unique biomolecular mechanisms which would explain its activity on both drug-sensitive and drug-resistant parasite strains. Various studies have shown that curcumin has antioxidant and anti-inflammatory properties and that it modulates numerous targets and cell signaling pathways. These include growth factors, growth factor receptors, transcription factors, cytokines, enzymes, and genes regulating apoptosis.

The discovery of the antioxidant properties of curcumin explains many of its wide-ranging pharmacological activities. Curcumin is an effective antioxidant and scavenges superoxide radicals, hydrogen peroxide, and nitric oxide (NO) from activated macrophages (Joe and Lokesh 1994). Curcumin is associated with the maintenance of reactive oxygen species (ROS) and activity of NO in a dose-depending manner. High doses (20–50 μM) of curcumin induce formation of ROS (Balasubramanyam et al. 2003) by elevation of cytosolic calcium through the release of calcium ions from intracellular stores. Moreover, influx of extracellular calcium leads to depolarization of mitochondrial membrane potential, release of cytochrome c into the cytosol and concomitant nuclear alterations. For instance, deoxynucleotidyltransferase-mediated dUTP end labeling and DNA fragmentation in *P. falciparum* (Cui et al. 2007) and *Giardia* (Pérez-Arriaga et al. 2006) were observed and promising antileishmanial activity (Das et al. 2008) was supposed to be related to this molecular mechanism. In leishmaniasis inducible nitric oxide synthase (iNOS) is correlated with increasing doses of curcumin leading to intracellular killing of *Leishmania major* (Liew et al. 1990, 1991; Green et al. 1990). Leitch and Qing (1999) reported that both reactive nitrogen species (RNS) and ROS play protective roles in experimental cryptosporidiosis in mice. *Cryptosporidium* has a poor capacity to scavenge ROS (Entrala et al. 1997), making it potentially more susceptible to killing by such oxygenic compounds. Certain ROS, especially hydroxyl radicals and hydrogen peroxide, produced as a result of parasite exposure to ultraviolet irradiation, resulted in inactivation (photo-toxicity) of *C. parvum* oocysts (Gerrity et al. 2008; Ryu et al. 2008). Conversely, low doses (1–15 μM) of curcumin activate peroxisome proliferator-activated receptor gamma, deactivate type 1 response, inhibit iNOS, and interferes with adaptive immunity thus exacerbating the pathogenic effects of *Leishmania donovani* infection (Adapala and Chan 2008). Curcumin in low doses also is capable of blocking the action of both NO and NO congeners on intracellular *Leishmania* or scavenges ROS produced as a result of activation of macrophages in leishmaniasis infection. This contributes to protection of promastigotes and amastigotes of the visceral species, *L. donovani*, and promastigotes of the cutaneous species, *L. major* (Chan et al. 2005) from host attack after phagocytosis by for example macrophages. Despite the wide evidence that NO can be regarded as a natural antiprotozoal weapon, little efforts have been made to develop and test NO-based drugs. This is mainly due to the difficulty in designing suitable chemical carriers that are able to release the right amount of NO, in the right place and at the right time, to avoid toxic effects against nontarget host cells. The curcumin diphasic effect against parasites that depends on applied dose and exposure time may be advantageous for development of such an antiparasitic drug.

Antiparasitic activities of curcumin are achieved through effects on transcription of genes. Recent studies find that histone acetylation plays an important role in eukaryotic gene transcription, carcinogenesis, and the therapy of cancer. Generally, histone acetylation contributes to the formation of a transcriptionally competent environment by "opening" chromatin and permits access of transcription factors to DNA (Fry and Peterson 2002; Lehrmann et al. 2002) whereas histone deacetylation contributes to a "closed" chromatin state and transcriptional repression. The histone acetylation–deacetylation balance is accurately maintained through a balance of histone acetyltransferase (HAT) and histone deacetylase. Curcumin induces histone hypoacetylation in vivo mainly through inhibition of HAT and concomitant generation of ROS by curcumin effects also contribute to inhibition of HAT activity (Kang et al. 2005). Histone deacetylase is a novel therapeutic target for fungal-derived antiprotozoal agents (like acipidin). Such drugs may alter proliferation of apicomplexan parasites (Darkin-Rattray et al. 1996). In vitro, curcumin effects on recombinant *P. falciparum* have been attributed to inhibition of histone deacetylation (Cui et al. 2007) and inhibition of HAT activity (Balasubramanyam et al. 2004). A new member of the apicomlexan histone deacetylase family has been recently described in *C. parvum* (Rider and Zhu 2009), and it seems to be possible that a respective mechanism is involved in inhibition of growth of *Cryptosporidium* by curcumin. Curcumin induces hypoacetylation of histone H3 at K9 and K14, but not of H4 at K5, K8, K12, and K16. The specific inhibition of the PfGCN5 HAT and generation of ROS have been supposed to be responsible for curcumin-related cytotoxicity for malaria parasites (Cui et al. 2007). Curcumin inhibits the intracellular adhesion molecules that contribute to sequestration and establishment of *Toxoplasma* (Barragan et al. 2005) and *Plasmodium* (Chakravorty and Craig 2005). The orthologue of mammalian sarcoplasmic–endoplasmic reticulum Ca^{2+}–ATPase in *P. falciparum*, PfATP6, is the molecular target of artemisinins, which are the most potent antimalarials available (Eckstein-Ludwig et al. 2003). It has been demonstrated by docking simulation that curcumin can efficiently inhibit PfATP6, which provides some deeper insights into the antimalarial mechanism of curcumin (Ji and Shen 2009).

Curcumin is found to significantly increase adhesion but remarkably reduce viability of *Giardia* trophozoites (Pérez-Arriaga et al. 2006). It has been found that biomolecular discharge from apical organelles of *C. parvum* and *Toxoplasma gondii* is essential for host cell invasion and this depends on parasite intracellular calcium levels (Chen et al. 2004; Lovett et al. 2002; Lovett and Sibley 2003). Reduced intracellular calcium levels in free sporozoites decrease secretion from the apical complex, thus reducing invasion and infection by the zoites. Curcumin mechanistically interferes with protein kinase C (PKC) and calcium regulation through increased ROS generation (Balasubramanyam et al. 2003). PKC-like enzymes play a critical role in attachment and in internalization of *Leishmania mexicana* (Varez-Rueda et al. 2009). Hypocalcimic action and inhibition of PKC by curcumin therefore should be taken into account to develop suitable and novel drugs for control of infection of parasite into host cells.

A significant inhibition of *C. parvum* sporozoite invasion of HCT cells by curcumin was reported in vitro (Shahiduzzaman et al. 2009). Infectivity of sporozoites is mediated by interaction of molecules secreted from sporozoites with matching receptors present on both parasite and host cell (Nesterenko et al. 1999). Phospholipase A2 (PLA2), a secretory mammalian host cell enzyme involved in arachidonic acid metabolism, has been supposed to be associated with infectivity of *Toxoplasma* (Saffer et al. 1989; Saffer and Schwartzman 1991) and *Cryptosporidium* (Pollok et al. 2003). Curcumin was found to inhibit mammalian phospholipase (Huang et al. 1991; Rao et al. 1995) and it appears possible that similar effects on parasite PLA2 may reduce infectivity of sporozoites. However, other enzymes are also thought to play a pivotal role in apicomplexan host cell invasion. Inhibition of *C. parvum* serine protease (Forne et al. 1996) and arginine aminopeptidase (Okhuysen et al. 1994, 1996) was reported to reduce the ability of sporozoites to infect host cells. Blocking of *T. gondii* serine protease (Conseil et al. 1999) significantly reduced the level of infection. Curcumin is able to suppress both serine protease and aminopetidase activity in a variety of tissues and cells (Shim et al. 2003; Ukil et al. 2003) and thus it may well be that curcumin acts as an inhibitor of host cell invasion by impairing the function of respectively relevant parasite enzymes. Curcumin has been proposed as a HIV-1 or HIV-2 protease inhibitor (Sui et al. 1993). Parasite infections like cryptosporidiosis and toxoplasmosis are known to be opportunistic pathogens that cause severe disease in immunocompromised individuals. Thus, a protease inhibitor such as curcumin could be of double benefit by directly acting on HIV and on opportunistic protozoa.

A glutathione transferase (PfGST) isolated from *P. falciparum* has been associated with chloroquine resistance. Curcumin is a potent inhibitor of PfGST which may open alternative perspectives for management of drug resistance in malaria (Mangoyi et al. 2010).

Curcumin inhibits metalloproteinase activity including classical matrix metalloproteinase inhibitors such as EDTA, EGTA, phenantroline, and also tetracycline in *Trypanosoma brucei* infection (de Sousa et al. 2010). It is suspected to inhibit the matrix metalloproteinase-9-like molecules in *Trypanosoma cruzi* (Nogueira de Melo et al. 2010) and *Theileria annulata*-infected bovine leukocytes (Baylis et al. 1995). Metalloproteinases mediate the metastatic phenotype of *T. annulata*-transformed cells (Adamson and Hall 1996). This property of curcumin has not been extensively studied in terms of antiparasitic potential but deserves further research.

The apoptotic response of infected intestinal epithelial cells is actively suppressed by *C. parvum* via up regulation of survivin, favoring parasite replication (Liu et al. 2008). Curcumin-mediated down regulation of survivin induces apoptosis in tumor cells and similarly may enhance apoptosis of *C. parvum*-infected cells. Caspase-dependent apoptosis during infection with *C. parvum* raises the possibility that therapeutic interference with host cell death could alter the course of the pathology in vivo (Ojcius et al. 1999). In *T. gondii*-infected cells, the termination of $NF_{-kappa}B$ ($_kB$) signaling is associated with reduced phosphorylation of p65/RelA, an event involved in the ability of $NF_{-k}B$ to translocate to the nucleus and to bind to DNA. The phosphorylation of p65/RelA represents an event downstream of $I\alpha B$ degradation that may be targeted by pathogens to subvert $NF_{-k}B$ signaling (Shapira et al. 2005). Curcumin can effectively down-regulate $NF_{-k}B$, thus inhibiting IkappaBalpha kinase and reducing IkappaBalpha phosphorylation, leading to cell cycle arrest, apoptosis, and suppression of proliferation (Shishodia et al. 2005) of parasite-infected cells. Thioredoxin reductases (TrxRs) are essential for cell growth and survival and they appear good targets for antitumor therapy. The parasitic nematode

Haemonchus contortus contains two TrxRs, a cytoplasmic enzyme HcTrxR1 with a selenocysteine in the active site (Gly–Cys–SeCys–Gly), similar to the mammalian TrxR, and a mitochondrial enzyme HcTrxR2 with a Gly–Cyc–Cys–Gly active site which is unique to nematodes. Curcumin inhibition of TrxRs (Hudson et al. 2010) may reduce parasite proliferation which is attractive for control strategies. It has been shown that curcuminoids strongly bind to *P. falciparum* thioredoxin (PfTrxR) and glutathione (PfGR) reductases which has allowed the development of automated high-throughput screening to rapidly determine the binding affinity of respective enzyme ligands (Mulabagal and Calderon 2010).

6.4 Curcumin Analogs

Synthetic curcumin analogs exhibit more potent antiparasitic effects than curcumin extracted from turmeric. The natural curcuminoids (curcumin, demethoxycurcumin, bisdemethoxycurcumin) exhibit low antitrypanosomal and antileishmanial activity. In contrast, curcumin derivatives (methylcurcumin) with des-*O*-methylcentrolobine are very active against the extracellular form (promastigotes) and intracellular form (amastigotes) of *Leishmania amazonensis* (Araujo et al. 1999). Curcuminoids synthesized by the condensation of 2,4-pentanedione with differently substituted benzaldehydes and the compound 1,7-bis-(2-hydroxy-4-methoxyphenyl)-1,6-heptadiene-3,5-dione are highly effective in vitro against *L. amazonensis* promastigotes (Gomes et al. 2002b). Chemically modified curcumin for example 1,7-bis-(4-propargyl-3-methoxyphenyl)-1,6-heptadiene-3,5-dione, is about ten times more efficient against *L. amazonensis* promastigotes than the original curcumin (Gomes et al. 2002a). The highly active analog 1,7-bis(4-hydroxy-3-methoxyphenyl) hept-4-en-3-one (40) is particularly active against kinetoplastid parasites. The diminazene-resistant strain of *Trypanosoma brucei brucei* (TbAT1-KO) B48 is susceptible to curcuminoids carrying a conjugated keto (enone) motif. The enone motif 40 was found to exert particularly high trypanocidal activity against all *Trypanosoma* species and strains tested (Changtam et al. 2010).

6.5 Conclusions

In recent years, the molecular basis for curcumin efficacy has been extensively investigated. Understanding of curcumin effects on the biomolecular or cellular level will hopefully help to identify and develop new therapeutic options. Curcumin is nontoxic to mammals at even high doses which makes it attractive as a potential antiparasitic drug, however, synthetic analogs of curcumin are superior in efficiency as compared to natural curcumin. In any case, curcumin extracted from turmeric represents an accessible and low-cost alternative for control of parasites in populations living in risk areas and thus further research into the potential of this natural plant product appears very rewarding.

References

Adamson RE, Hall FR (1996) Matrix metalloproteinases mediate the metastatic phenotype of *Theileria annulata*-transformed cells. Parasitology 113:449–455

Adapala N, Chan MM (2008) Long-term use of an antiinflammatory, curcumin, suppressed type 1 immunity and exacerbated visceral leishmaniasis in a chronic experimental model. Lab Invest 88:1329–1339

Aggarwal BB, Kumar A, Bharti AC (2003) Anticancer potential of curcumin: preclinical and clinical studies. Anticancer Re 23:363–398

Allam G (2009) Immunomodulatory effects of curcumin treatment on murine schistosomiasis mansoni. Immunobiology 214:712–727

Allen PC, Danforth HD, Augustine PC (1998) Dietary modulation of avian coccidiosis. Int J Parasitol 28:1131–1140

Araujo CAC, Alegrio LV, Castro D, Lima MEF, Leon LL (1998) *Leishmania amazonensis*: in vivo experiments with diarylhetanoids from Leguminosae and *Zingiberaceae* plants. Mem Inst Oswaldo Cruz 93:306–310

Araujo CA, Alegrio LV, Gomes DC, Lima ME, Gomes-Cardoso L, Leon LL (1999) Studies on the effectiveness of diarylheptanoids derivatives against *Leishmania amazonensis*. Mem Inst Oswaldo Cruz 94:791–794

Atjanasuppat K, Wongkham W, Meepowpan P, Kittakoop P, Sobhon P, Bartlett A, Whitfield PJ (2009) In vitro screening for anthelmintic and antitumour activity of ethnomedicinal plants from Thailand. J Ethnopharmacol 123:475–482

Balasubramanyam M, Koteswari AA, Kumar RS, Monickaraj SF, Maheswari JU, Mohan V (2003) Curcumin-induced inhibition of cellular reactive oxygen species generation: novel therapeutic implications. J Biosci 28:715–721

Balasubramanyam K, Varier RA, Altaf M, Swaminathan V, Siddappa NB, Ranga U, Kundu TK (2004) Curcumin, a novel p300/CREB-binding protein-specific inhibitor of acetyltransferase, represses the acetylation of histone/nonhistone proteins and histone acetyltransferase-dependent chromatin transcription. J Biol Chem 279:51163–51171

Barragan A, Brossier F, Sibley LD (2005) Transepithelial migration of *Toxoplasma gondii* involves an interaction of intercellular adhesion molecule 1 (ICAM-1) with the parasite adhesin MIC2. Cell Microbiol 7:561–568

Barthelemy S, Vergnes L, Moynier M, Guyot D, Labidalle S, Bahraoui E (1998) Curcumin and curcumin derivatives inhibit Tat-mediated transactivation of type 1 human immunodeficiency virus long terminal repeat. Res Virol 149:43–52

Baylis HA, Megson A, Hall R (1995) Infection with *Theileria annulata* induces expression of matrix metalloproteinase 9 and transcription factor AP-1 in bovine leucocytes. Mol Biochem Parasitol 69:211–222

Boonjaraspinyo S, Boonmars T, Aromdee C, Srisawangwong T, Kaewsamut B, Pinlaor S, Yongvanit P, Puapairoj A (2009) Turmeric reduces inflammatory cells in hamster opisthorchiasis. Parasitol Res 105:1459–1463

Chakravorty SJ, Craig A (2005) The role of ICAM-1 in *Plasmodium falciparum* cytoadherence. Eur J Cell Biol 84:15–27

Chan MM, Adapala NS, Fong D (2005) Curcumin overcomes the inhibitory effect of nitric oxide on *Leishmania*. Parasitol Res 96:49–56

Changtam C, de Koning HP, Ibrahim H, Sajid MS, Gould MK, Suksamrarn A (2010) Curcuminoid analogs with potent activity against *Trypanosoma* and *Leishmania* species. Eur J Med Chem 45:941–956

Chen XM, O'Hara SP, Huang BQ, Nelson JB, Lin JJ, Zhu G, Ward HD, LaRusso NF (2004) Apical organelle discharge by *Cryptosporidium parvum* is temperature, cytoskeleton, and intracellular calcium dependent and required for host cell invasion. Infect Immun 72:6806–6816

Choi H, Chun YS, Kim SW, Kim MS, Park JW (2006) Curcumin inhibits hypoxia-inducible factor-1 by degrading aryl hydrocarbon receptor nuclear translocator: a mechanism of tumor growth inhibition. Mol Pharmacol 70:1664–1671. doi: 10.1124/mol.106.025817

Conseil V, Soete M, Dubremetz JF (1999) Serine protease inhibitors block invasion of host cells by *Toxoplasma gondii*. Antimicrob Agents Chemother 43:1358–1361

Cui L, Miao J, Cui L (2007) Cytotoxic effect of curcumin on malaria parasite *Plasmodium falciparum*: inhibition of histone acetylation and generation of reactive oxygen species. Antimicrob Agents Chemother 51:488–494

Darkin-Rattray SJ, Gurnett AM, Myers RW, Dulski PM, Crumley TM, Allocco JJ, Cannova C, Meinke PT, Colletti SL, Bednarek MA, Singh SB, Goetz MA, Dombrowski AW, Polishook JD, Schmatz DM (1996) Apicidin: a novel antiprotozoal agent that inhibits parasite histone deacetylase. Proc Natl Acad Sci USA 93:13143–13147

Das R, Roy A, Dutta N, Majumder HK (2008) Reactive oxygen species and imbalance of calcium homeostasis contributes to curcumin induced programmed cell death in *Leishmania donovani*. Apoptosis 13:867–882

de Sousa KP, Atouguia J, Silva MS (2010) Partial biochemical characterization of a metalloproteinase from the bloodstream forms of *Trypanosoma brucei brucei* parasites. Protein J 29:283–289

Dhar ML, Dhar MM, Dhawan BN, Mehrotra BN, Ray C (1968) Screening of Indian plants for biological activity: I. Indian J Exp Biol 6:232–247

Eckstein-Ludwig U, Webb RJ, Van G, East JM, Lee AG, Kimura M, O'Neill PM, Bray PG, Ward SA, Krishna S (2003) Artemisinins target the SERCA of *Plasmodium falciparum*. Nature 424:957–961

El-Ansary AK, Ahmed SA, Aly SA (2007) Antischistosomal and liver protective effects of Curcuma longa extract in *Schistosoma mansoni* infected mice. Indian J Exp Biol 45 (9):791–801

El-Banhawey MA, Ashry MA, El-Ansary AK, Aly SA (2007) Effect of *Curcuma longa* or praziquantel on Schistosoma mansoni infected mice liver: histological and histochemical study. Indian J Exp Biol 45(10):877–889

Entrala E, Mascaro C, Barrett J (1997) Anti-oxidant enzymes in Cryptosporidium parvum oocysts. Parasitology 114(Pt 1):13–17

Forne JR, Yang S, Du C, Healey MC (1996) Efficacy of serine protease inhibitors against *Cryptosporidium parvum* infection in a bovine fallopian tube epithelial cell culture system. J Parasitol 82:638–640

Fry CJ, Peterson CL (2002) Unlocking the gates to gene expression. Science 295:1847–1848

Gerrity D, Ryu H, Crittenden J, Abbaszadegan M (2008) Photocatalytic inactivation of viruses using titanium dioxide nanoparticles and low-pressure UV light. J Environ Sci Health A Toxicol Hazard Subst Environ Eng 43:1261–1270

Goel A, Kunnumakkara AB, Aggarwal BB (2008) Curcumin as "Curecumin": from kitchen to clinic. Biochemical Pharmacology 75:787–809

Gomes DC, Alegrio LV, de Lima ME, Leon LL, Araujo CA (2002a) Synthetic derivatives of curcumin and their activity against *Leishmania amazonensis*. Arzneimittelforschung 52:120–124

Gomes DC, Alegrio LV, Leon LL, de Lima ME (2002b) Total synthesis and anti-leishmanial activity of some curcumin analogues. Arzneimittelforschung 52:695–698

Green SJ, Crawford RM, Hockmeyer JT, Meltzer MS, Nacy CA (1990) *Leishmania major* amastigotes initiate the L-arginine dependent killing mechanism in IFN-c-stimulated macrophages by induction of tumor necrosis factor-a. J Immunol 145:4290–4297

Huang MT, Lysz T, Ferraro T, Abidi TF, Laskin JD, Conney AH (1991) Inhibitory effects of curcumin on in vitro lipoxygenase and cyclooxygenase activities in mouse epidermis. Cancer Res 51:813–819

Hudson AL, Sotirchos IM, Davey MW (2010) Substrate specificity of the mitochondrial thioredoxin reductase of the parasitic nematode Haemonchus contortus. Parasitol Res 107:487–493

Ji HF, Shen L (2009) Interactions of curcumin with the PfATP6 model and the implications for its antimalarial mechanism. Bioorg Med Chem Lett 19:2453–2455

Joe B, Lokesh BR (1994) Role of capsaicin, curcumin and dietary n-3 fatty acids in lowering the generation of reactive oxygen species in rat peritoneal macrophages. Biochim Biophys Acta 1224:255–263

Jordan WC, Drew CR (1996) Curcumin – a natural herb with anti-HIV activity. J Natl Med Assoc 88:333

Kang J, Chen J, Shi Y, Jia J, Zhang Y (2005) Curcumin-induced histone hypoacetylation: the role of reactive oxygen species. Biochem Pharmacol 69:1205–1213

Khalafalla RE, Müller U, Shahiduzzaman M, Dyachenko V, Desouky AY, Alber G, Daugschies A (2010) Effects of curcumin (diferuloylmethane) on *Eimeria tenella* sporozoites in vitro. Parasitol Res. doi:10.1007/s00436-010-2129-y

Kim GM, Choi KJ, Lee HS (2003) Fungicidal property of *Curcuma longa* L. rhizome-derived curcumin against phytopathogenic fungi in a greenhouse. J Agric Food Chem 51:1578–1581

Kiuchi F, Goto Y, Sugimoto N, Akao N, Kondo K, Tsuda Y (1993) Nematocidal activity of turmeric: synergistic action of curcuminoids. Chem Pharm Bull 41:1640–1643

Koide T, Nose M, Ogihara Y, Yabu Y, Ohta N (2002) Leishmanicidal effect of curcumin *in vitro*. Biol Pharm Bull 25:131–133

Lehrmann H, Pritchard LL, Harel-Bellan A (2002) Histone acetyltransferasesand deacetylases in the control of cell proliferation and differentiation. Adv Cancer Res 86:41–65

Leitch GJ, Qing HE (1999) Reactive nitrogen and oxygen species ameliorate experimental cryptosporidiosis in the neonatal BALB/ c mouse model. Infect Immun 67:5885–5891

Liew FY, Millott S, Parkinson C, Palmer RM, Moncada S (1990) Macrophage killing of *Leishmania* parasite in vivo is mediated by nitric oxide from L-arginine. J Immunol 144:4794–4797

Liew FY, Li Y, Moss D, Parkinson C, Rogers MV, Moncada S (1991) Resistance to *Leishmania major* infection correlates with the induction of nitric oxide synthase in murine macrophages. Eur J Immunol 21:3009–3014

Liu J, Enomoto S, Lancto CA, Abrahamsen MS, Rutherford MS (2008) Inhibition of apoptosis in *Cryptosporidium parvum*-infected intestinal epithelial cells is dependent on survivin. Infect Immun 76(8):3784–3792

Lovett JL, Sibley LD (2003) Intracellular calcium stores in *Toxoplasma gondii* govern invasion of host cells. J Cell Sci 116:3009–3016

Lovett JL, Marchesini N, Moreno SN, Sibley LD (2002) Toxoplasma gondii microneme secretion involves intracellular Ca(2+) release from inositol 1, 4, 5-triphosphate (IP(3))/ryanodine-sensitive stores. J Biol Chem 277:25870–25876

Magalhaes LG, Machado CB, Morais ER, Moreira EB, Soares CS, da Silva SH, Da Silva Filho AA, Rodrigues V (2009) In vitro schistosomicidal activity of curcumin against *Schistosoma mansoni* adult worms. Parasitol Res 104:1197–1201

Mahady GB, Pendland SL, Yun G, Lu ZZ (2002) Turmeric (*Curcuma longa*) and curcumin inhibit the growth of *Helicobacter pylori*, a group 1 carcinogen. Anticancer Res 22:4179–4181

Mangoyi R, Hayeshi R, Ngadjui B, Ngandeu F, Bezabih M, Abegaz B, Razafimahefa S, Rasoanaivo P, Mukanganyama S (2010) Glutathione transferase from *Plasmodium falciparum* – Interaction with malagashanine and selected plant natural products. J Enzyme Inhib Med Chem 25:854–862

Martinelli A, Rodrigues LA, Cravo P (2008) *Plasmodium chabaudi*: efficacy of artemisinin + curcumin combination treatment on a clone selected for artemisinin resistance in mice. Exp Parasitol 119:304–307

Mulabagal V, Calderon AI (2010) Development of binding assays to screen ligands for *Plasmodium falciparum* thioredoxin and glutathione reductases by ultrafiltration and liquid chromatography/mass spectrometry. J Chromatogr B Analyt Technol Biomed Life Sci 878:987–993

Nadkarni KM (1976) Indian Materia Medica. Popular Prakashan, Bombay, p 1074

Nandakumar DN, Nagaraj VA, Vathsala PG, Rangarajan P, Padmanaban G (2006) Curcumin-artemisinin combination therapy for malaria. Antimicrob Agents Chemother 50:1859–1860

Negi PS, Jayaprakasha GK, Jagan L, Rao M, Sakariah KK (1999) Antibacterial activity of turmeric oil: a byproduct from curcumin manufacture. J Agric Food Chem 47:4297–4300

Nesterenko MV, Woods K, Upton SJ (1999) Receptor/ligand interactions between *Cryptosporidium parvum* and the surface of the host cell. Biochim Biophys Acta 1454:165–173

Nogueira de Melo AC, de Souza EP, Elias CG, dos Santos AL, Branquinha MH, vila-Levy CM, dos Reis FC, Costa TF, Lima AP, de Souza Pereira MC, Meirelles MN, Vermelho AB (2010) Detection of matrix metallopeptidase-9-like proteins in *Trypanosoma cruzi*. Exp Parasitol 125:256–263

Nose M, Koide T, Ogihara Y, Yabu Y, Ohta N (1998) Trypanocidal effects of curcumin in vitro. Biol Pharm Bull 21:643–645

Ojcius DM, Perfettini JL, Bonnin A, Laurent F (1999) Caspase-dependent apoptosis during infection with *Cryptosporidium parvum*. Microbes Infect 1(14):1163–1168

Okhuysen PC, DuPont HL, Sterling CR, Chappell CL (1994) Arginine aminopeptidase, an integral membrane protein of the *Cryptosporidium parvum* sporozoite. Infect Immun 62:4667–4670

Okhuysen PC, Chappell CL, Kettner C, Sterling CR (1996) *Cryptosporidium parvum* metalloaminopeptidase inhibitors prevent in vitro excystation. Antimicrob Agents Chemother 40:2781–2784

Pérez-Arriaga L, Mendoza-Magaña ML, Cortés-Zárate R, Corona-Rivera A, Bobadilla-Morales L, Troyo-Sanromán R, Ramírez-Herrera MA (2006) Cytotoxic effect of curcumin on *Giardia lamblia* trophozoites. Acta Trop 98(2):152–161

Pinlaor S, Yongvanit P, Prakobwong S, Kaewsamut B, Khoontawad J, Pinlaor P, Hiraku Y (2009) Curcumin reduces oxidative and nitrative DNA damage through balancing of oxidant–antioxidant status in hamsters infected with *Opisthorchis viverrini*. Mol Nutr Food Res 53:1316–1328

Pinlaor S, Prakobwong S, Hiraku Y, Pinlaor P, Laothong U, Yongvanit P (2010) Reduction of periductal fibrosis in liver fluke-infected hamsters after long-term curcumin treatment. Eur J Pharmacol 638:134–141

Pollok RC, McDonald V, Kelly P, Farthing MJ (2003) The role of *Cryptosporidium parvum*-derived phospholipase in intestinal epithelial cell invasion. Parasitol Res 90:181–186

Rao CV, Rivenson A, Simi B, Reddy BS (1995) Chemoprevention of colon carcinogenesis by dietary curcumin, a naturally occurring plant phenolic compound. Cancer Res 55:259–266

Rasmussen HB, Christensen SB, Kuist LP, Karazmi AA (2000) Simple and effective separation of the curcumins, the antiprotozoal constituents of *Curcuma longa*. Planta Med 66:396–398

Reddy RC, Vatsala PG, Keshamouni VG, Padmanaban G, Rangarajan PN (2005) Curcumin for malaria therapy. Biochem Biophys Res Commun 326:472–474

Rider SD Jr, Zhu G (2009) An apicomplexan ankyrin-repeat histone deacetylase with relatives in photosynthetic eukaryotes. Int J Parasitol 39:747–754

Ryu H, Gerrity D, Crittenden JC, Abbaszadegan M (2008) Photocatalytic inactivation of *Cryptosporidium parvum* with TiO(2) and low-pressure ultraviolet irradiation. Water Res 42:1523–1530

Saffer LD, Schwartzman JD (1991) A soluble phospholipase of *Toxoplasma gondii* associated with host cell penetration. J Protozool 38:454–460

Saffer LD, Long Krug SA, Schwartzman JD (1989) The role of phospholipase in host cell penetration by *Toxoplasma gondii*. Am J Trop Med Hyg 40:145–149

Saleheen D, Ali SA, Ashfaq K, Siddiqui AA, Agha A, Yasinzai MM (2002) Latent activity of curcumin against leishmaniasis in vitro. Biol Pharm Bull 25:386–389

Shahiduzzaman M, Dyachenko V, Khalafalla RE, Desouky AY, Daugschies A (2009) Effects of curcumin on *Cryptosporidium parvum* in vitro. Parasitol Res 105:1155–1161

Shapira S, Harb OS, Margarit J, Matrajt M, Han J, Hoffmann A, Freedman B, May MJ, Roos DS, Hunter CA (2005) Initiation and termination of NF-kappaB signaling by the intracellular protozoan parasite *Toxoplasma gondii*. J Cell Sci 118:3501–3508

Shen L, Ji HF (2007) Theoretical study on physicochemical properties of curcumin. Spectrochim Acta A Mol Biomol Spec- trosc 67: 619–623

Shim JS, Kim JH, Cho HY, Yum YN, Kim SH, Park HJ, Shim BS, Choi SH, Kwon HJ (2003) Irreversible inhibition of CD13/aminopeptidase N by the antiangiogenic agent curcumin. Chem Biol 10:695–704

Shishodia S, Amin HM, Lai R et al (2005) Curcumin (diferuloylmethane) inhibits constitutive NF-kappaB activation, induces G1/S arrest, suppresses proliferation, and induces apoptosis in mantle cell lymphoma. Biochem Pharmacol 70:700–713

Srinivasan A, Menon VP (2003) Protection of pancreatic β-cell by the potential antioxidant bis-o-hydroxycinnamoyl methane, analogue of natural curcuminoid in experimental diabetes. J Pharm Pharm Sci 6:327–333

Sui Z, Salto R, Li J, Craik C, Ortiz de Montellano PR (1993) Inhibition of the HIV-1 and HIV-2 proteases by curcumin and curcumin boron complexes. Bioorg Med Chem 1:415–422

Surh YJ (2002) Anti-tumor promoting potential of selected spice ingredients with antioxidative and anti-inflammatory activities: a short review. Food Chem Toxicol 40:1091–1097

Ukil A, Maity S, Karmakar S, Datta N, Vedasiromoni JR, Das PK (2003) Curcumin, the major component of food flavour turmeric, reduces mucosal injury in trinitrobenzene sulphonic acid-induced colitis. Br J Pharmacol 139:209–218

Varez-Rueda N, Biron M, Le PP (2009) Infectivity of *Leishmania mexicana* is associated with differential expression of protein kinase C-like triggered during a cell-cell contact. PLoS One 4:7581

Chapter 7
Marine Organisms and Their Prospective Use in Therapy of Human Diseases

Sherif S. Ebada and Peter Proksch

Abstract Marine ecosystems show remarkable chemical, biological, and ecological diversities which have inspired marine natural product chemists for many years to identify novel chemical entities possessing potent pharmacological activities that could be eventually developed into therapeutics for human diseases such as cancer, analgesia, microbial infections, allergy, and immune diseases. By 1974, cytarabine (Ara-C, Cytosar-U®) and vidarabine (Ara-A, Vira-A®) were the first two marine-derived pharmaceuticals in the US pharmacopeia to treat human diseases. Since then, it took more than three further decades to introduce the next marine-derived drugs namely ω-conotoxin MVIIA (ziconotide, Prialt®), which was approved in December 2004 as an analgesic against severe chronic pain, and ecteinascidin (ET-743, Yondelis®) which was approved in October 2007 for the treatment of advanced soft tissue sarcomas, to the drug market. In addition many other marine-derived natural products are presently either in clinical or preclinical trial phases.

In this chapter, we survey the history of marine-derived therapeutics and the current status of the marine pharmaceutical pipeline with particular attention regarding the obstacles that may hinder transferring the marine-derived natural products into clinical trials and suggest possible solutions for introducing more drugs from the sea to the pharmaceutical market.

S.S. Ebada
Institute of Pharmaceutical Biology and Biotechnology, Heinrich-Heine University, Universitaetsstrasse 1, 40225 Duesseldorf, Germany
and
Department of Pharmacognosy and Phytochemistry, Faculty of Pharmacy, Ain-Shams University, Organisation of African Unity 1, 11566 Cairo, Egypt
e-mail: sherif.elsayed@uni-duesseldorf.de

P. Proksch (✉)
Institute of Pharmaceutical Biology and Biotechnology, Heinrich-Heine University, Universitaetsstrasse 1, 40225 Duesseldorf, Germany
e-mail: proksch@uni-duesseldorf.de

7.1 Introduction

Natural products continue to be an important source of the active constituents for the treatment of various human diseases. This was widely accepted when applied to drug discovery before the advent of high-throughput screening and the postgenomic era (Butler 2004; Koehn and Carter 2005). More than 60% of active drug ingredients were either natural products or derived thereof (Harvey 2008). However, it is arguably still true today: according to surveys performed by David J. Newman and Gordon M. Cragg of the National Cancer Institute (Newman and Cragg 2004, 2007), a total of 1,010 small molecules have been introduced as drugs all over the world from January 1981 to June 2006 and 65% thereof were either natural products or inspired by natural products (Newman and Cragg 2007). These data were further interpreted and it was observed that it includes 4% natural products, 23% natural product derivatives, 5% synthetic compounds with natural product-derived chromophores, and 31% totally synthetic products, the latter often found by random screening or modification of an existing agent (Newman and Cragg 2007). The percentage of therapeutics based on natural products is even higher when only antibacterial (69%) and anticancer (68%) compounds were considered (Newman and Cragg 2007).

Some examples of natural product-derived anticancer drugs include paclitaxel (**1**) (Taxol®, Fig. 7.1) – a highly functionalized diterpenoid approved by the Food and Drug Administration (FDA) in 1993 for the treatment of breast, lung, and ovarian cancers – which has evolved and become a blockbuster drug with commercial sales of well over US $3 billion in 2004 (Staniek et al. 2009). Paclitaxel was primarily obtained from the inner bark of the Pacific yew *Taxus brevifolia*, a relatively rare and slow-growing tree (Wani et al. 1971). For commercial purposes, other sources of paclitaxel were explored. Fortunately, it was possible to obtain paclitaxel synthetically from baccatin III, a secondary metabolite obtained from *Taxus baccata* and it was also disclosed to be biosynthesized by *Taxomyces andreana* – an endophytic fungus first reported from the genus *Taxus* in 1993 (Stierle et al. 1993). Recently, paclitaxel was also found to be produced by *Phyllosticta citricarpa*, a leaf spot fungus of the Angiosperm *Citrus medica* (Kumaran et al.

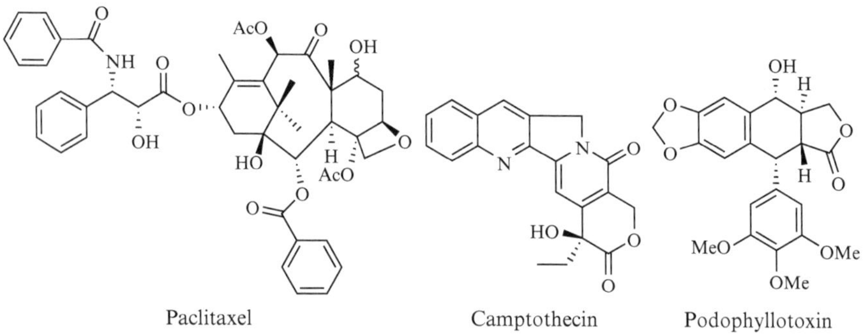

Fig. 7.1 Chemical structures of paclitaxel, camptothecin, and podophyllotoxin

2008). Irinotecan (Campto®) was obtained based on the lead structure of camptothecin (Fig. 7.1), first reported in 1966 from *Camptotheca acuminata* (Wall et al. 1966), and it received FDA approval in 1994 (Newman and Cragg 2007). In 2006, an endophytic fungus *Entrophospora infrequens*, isolated from the inner bark of *Nothapodytes foetida* growing in the Jammu region of India, was reported to biosynthesize camptothecin and may be a new source to fulfill market demands (Amna et al. 2006). Etoposide (Etopophos®) was derived by partial synthesis from the lignan podophyllotoxin (Fig. 7.1) – first isolated from the rhizomes of *Podophyllum peltatum* (Sultan et al. 2010) – and was approved by FDA in 1996 (Newman and Cragg 2007). Other traditional examples of drug molecules based on natural products from terrestrial plants and microbes include morphine from poppies, cardiotonic digitalis glycosides from foxgloves, and penicillins from fungi.

Most natural product-derived pharmaceuticals have so far been isolated or based on lead compounds obtained from terrestrial plants or their endophytic symbionts rather than from the marine ecosystem which covers more than two thirds of the planet's surface. As human pathogens and cancers develop resistance to existing therapeutics, the search for novel leads has attracted considerable scientific interest.

Oceans represent a treasure of useful products awaiting discovery for the treatment of human diseases. Marine plants and animals have adapted to polar, temperate, and tropical regions of the oceans which may hold 1,000 species per m^2 in some areas especially in the Indo-Pacific Ocean region (Pomponi 1999). Ecological pressures, including competition for space, surface fouling, predation, and successful reproduction have resulted in the evolution of unprecedented marine natural products with potent biological activities (Donia and Hamann 2003).

Oceans started to attract the interests of the pharmaceutical companies and research institutions around 60 years ago since the discovery of the sponge-derived nucleosides spongothymidine, spongosine, and spongouridine from the marine sponge *Cryptotethia crypta* by Bergmann and Feeney (1951). These sponge-derived nucleosides were the basis for the first two marine-derived pharmaceuticals, cytarabine (Ara-C) and vidarabine (Ara-A), which were introduced to the US pharmacopeia by 1974.

Since then, over 17,000 different natural products have been reported from marine resources (Singh et al. 2008). Discovery of these marine natural products has been documented in more than 8,200 research articles (Blunt et al. 2010). In addition over 12,000 publications dealing with the syntheses, reviews, biological and ecological studies and over 300 patents have been issued on bioactive marine natural products in the last 50 years since marine natural products emerged as a discipline (Singh et al. 2008; Blunt et al. 2010).

Marine-derived natural products are diverse in chemical structure ranging from linear peptides to complex macrocyclic polyethers and exhibit a vast array of bioactivities such as antifungal, antibacterial, antimalarial, antiviral, anthelmintic, anticancer activity etc. (Singh et al. 2008; Blunt et al. 2010).

Despite the medicinal potential and scientific efforts that have been invested in marine natural product chemistry, it has taken over 30 years for another marine-derived natural product to gain approval and become part of the pharmacopeia.

In December 2004, ziconotide (Prialt®) was approved for the treatment of moderate to severe pain, whereas trabectedin (ET-743, Yondelis®) received European approval in October 2007 for the treatment of soft tissue sarcoma (STS), and in 2009 for ovarian carcinoma (Mayer et al. 2010). Simultaneously, more than 1,000 marine natural products or derivatives thereof are in different phases of clinical and preclinical trials (Mayer et al. 2010).

In this chapter, we survey marine-derived natural products undergoing Phases I–III clinical trials providing details of the marine sources of these compounds, mode of action, and clinical relevance. In addition, this chapter will discuss the main obstacles facing the emergence of further marine-derived therapeutics and propose some possible solutions. This will help in turn to improve the diversity of the pharmacopeia by introducing more marine-derived pharmaceuticals to treat a wide array of human diseases.

7.2 Marine Pharmaceuticals: FDA-Approved Drugs

In the US pharmacopeia, there are currently three marine-derived natural products approved by the FDA, namely cytarabine (Ara-C, Cytpsar-U®, and Depocyt®), vidarabine (Ara-A, Vira-A®), and ω-conotoxin MVIIA (ziconotide, Prialt®). In addition, a fourth member, trabectedin (ET-743, Yondelis®) has been approved by the European Agency for the Evaluation of Medicinal Products (EMEA), and now is completing key Phase III clinical trials for FDA approval. This section will discuss in detail the history of these four marine-derived natural products which have been successfully developed into pharmaceuticals.

7.2.1 Cytarabine (Cytosar-U® and Depocyt®)

Cytarabine (cytosine arabinoside or arabinofuranosyl cytosine, Ara-C) (Fig. 7.2) is a synthetic pyrimidine nucleoside whose synthesis was inspired by spongothymidine (Fig. 7.2), a nucleoside originally isolated from the Caribbean sponge *C. crypta*, which was first reported in 1950 (Bergmann and Feeney 1950) and one

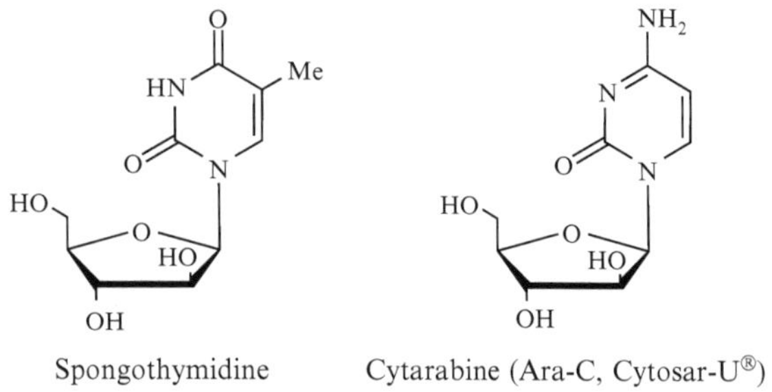

Fig. 7.2 Chemical structures of spongothymidine and cytarabine

Spongothymidine

Cytarabine (Ara-C, Cytosar-U®)

7 Marine Organisms and Their Prospective Use in Therapy of Human Diseases   157

year later its chemical structure was depicted together with other sponge-derived nucleosides, namely spongosine and spongouridine (Bergmann and Feeney 1951).

The cell cycle is functionally divided into four phases: G_1, S, G_2, and M. Cells synthesize DNA and replicate themselves during the DNA synthesis (S) phase, while DNA chromatids become separated and cells divide into two cells in the mitosis (M) phase. M and S phases are separated by gap phases 1 (G_1) and 2 (G_2). Following mitosis, cells can divide again by entering G_1 phase or cells can exit the cell cycle to the inactive (G_0) phase (Dermatakis 2002; Nguyen and Tepe 2009). Cytarabine (Ara-C) is an S-phase-specific antimetabolite cytotoxic agent, converted intracellularly to cytosine arabinoside triphosphate, which competes with the physiological substrate deoxycitidine triphosphate resulting in inhibition of both DNA polymerase and DNA synthesis (Mayer et al. 2010).

Cytarabine was approved by FDA in 1969 and it is currently available as either conventional cytarabine (Cytosar-U®) or a liposomal formulation (Depocyt®). FDA approved Cytosar® to be used for treatment of acute lymphocytic leukemia, acute myelocytic leukemia, and the blast crisis phase of chronic myelogenous leukemia and meningeal leukemia (Absalon and Smith 2009; Thomas 2009). Depocyt®, liposomal cytarabine, is indicated for intrathecal (IT) treatment of lymphomatous meningitis (Mayer et al. 2010). Cytosar® and Depocyt® are marketed by Bedford Laboratories and Enzon Pharmaceuticals, respectively.

7.2.2 Vidarabine (Vira-A®)

Vidarabine (adenine arabinoside or arabinofuranosyl adenine, Ara-A) (Fig. 7.3) is a synthetic purine nucleoside whose synthesis (Lee et al. 1960) was based on the model of a sponge-derived nucleoside, spongosine (Fig. 7.3), which was reported together with spongothymidine and spongouridine from the Caribbean marine sponge *C. crypta* (Bergmann and Feeney 1951).

Vidarabine (arabinofuranosyl adenine, Ara-A) and its 3′-*O*-acetyl derivative were later isolated from the gorgonian *Eunicella cavolini* collected off the bay of Naples (Cimino et al. 1984). Currently, vidarabine (Ara-A) is produced by *Streptomyces antibioticus* (Suhadolnik et al. 1989).

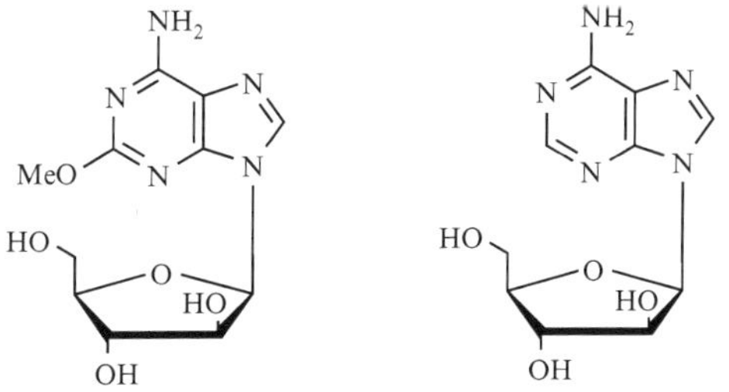

Fig. 7.3 Chemical structures of spongosine and vidarabine

Spongosine

Vidarabine (Ara-A, Vira-A®)

Vidarabine (adenine arabinoside, Ara-A) is readily converted intracellularly into its triphosphate derivative, which inhibits viral DNA polymerase and DNA synthesis of herpes, vaccinia and Varicella zoster viruses. In 1976, vidarabine received FDA approval, Vira-A® ophthalmic ointment 3%, is indicated for treatment of acute keratoconjunctivitis, recurrent epithelial keratitis caused by herpes simplex virus types 1 and 2, and superficial keratitis caused by herpes simplex virus that has not responded to topical idoxuridine (Herplex®) (Mayer et al. 2010).

Vidarabine (Ara-A, Vira-A®) was previously marketed by King Pharmaceuticals, however, its marketing was discontinued in June 2001 in the US market by an executive decision from the FDA, probably due to the diminished therapeutic window of Vira-A® compared to other antiviral compounds recently launched into the pharmaceutical market (Mayer et al. 2010).

7.2.3 Zidovudine (AZT, Retrovir®)

From the starfish *Acanthaster planci*, 2′-deoxythymidine and 2′-deoxyuridine (Fig. 7.4) were isolated in 1980 (Komori et al. 1980). One of the thymidine analogs that has received particular worldwide attention from both chemists and biologists alike is 3′-azido-2′,3′-dideoxythymidine (AZT, zidovudine) (Fig. 7.4) due to its potent inhibitory activity against human immunodeficiency virus (HIV) (De Clercq 1986; Varmus 1988; Parang et al. 2000; Mavromoustakos et al. 2001).

Zidovudine was the first reverse transcriptase (RT) inhibitor approved by FDA for HIV treatment in 1987 (Mavromoustakos et al. 2001). Structurally, zidovudine is a synthetic pyrimidine analog that differs from 2′-deoxythymidine in having azido functionality instead of a hydroxyl group at the 3′ position of the deoxyribofuranose moiety. Zidovudine inhibits the action of RT, the enzyme that HIV utilizes to produce a double-stranded DNA copy of its single-stranded RNA. Only the viral double-stranded DNA can integrate into the genetic material of the infected host cell, where it is then known as a provirus (Yarchoan et al. 1986; Mitsuya et al. 1990).

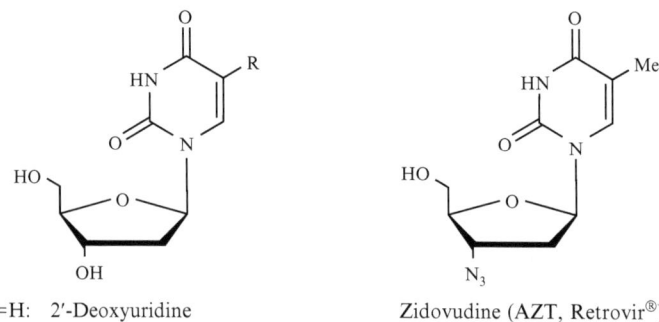

R=H: 2′-Deoxyuridine
R=Me: 2′-Deoxythymidine

Zidovudine (AZT, Retrovir®)

Fig. 7.4 Chemical structures of 2′-deoxyuridine, 2′-deoxythymidine, and zidovudine

Zidovudine (AZT) does not eradicate the HIV infection, but only delays the progression of the disease and the viral replication, even at very high doses. With prolonged AZT treatment, HIV proved able to gain an increased resistance to AZT by mutation of its RT. To slow down the development of resistance, physicians recommend a combination therapy of AZT together with another RT inhibitor and an antiretroviral from another group, such as a protease inhibitor or a nonnucleoside RT inhibitor. This therapy is known as HAART (Highly Active Anti Retroviral Therapy). Zidovudine is currently marketed by GlaxoSmithKline under the brand name, Retrovir®. Interestingly, the time relapsed between the first report of AZT activity against HIV and its FDA approval was 25 months which is one of the shortest periods of drug development in recent history.

Other thymidine nucleoside analogs, which have been approved by FDA for HIV treatment include 2′,3′-dideoxy-3′-thiacytidine (3TC, lamivudine) and 2′,3′-didehydro-3′-deoxythymidine (D4T, stavudine) (Fig. 7.5).

Lamivudine (3TC) (Fig. 7.5) is a synthetic cytidine analog that inhibits both types 1 and 2 of HIV reverse transcriptase and also RT of hepatitis B virus (HBV). Intracellularly, its phosphorylated metabolite competes for incorporation into viral DNA where the lack of a 3′-OH group in lamivudine prevents the formation of the 5′ to 3′ phosphodiester linkage essential for DNA chain elongation, and hence terminates the viral DNA synthesis (Schinazi et al. 1992). Lamivudine received FDA approval in 1995 for use with zidovudine (AZT, Retrovir®) and in 2002 as a once-a-day dosed medication for treatment of HIV. It is currently marketed as Epivir® 150 mg or 300 mg tablets. For treatment of HBV, Epivir-HBV® is also marketed by GlaxoSmithKline which contains lower doses of lamivudine, 100 mg tablets.

Stavudine (D4T) (Fig. 7.5) is an analog of thymidine which is phosphorylated by cellular kinases into an active triphosphate derivative. Stavudine triphosphate inhibits the HIV reverse transcriptase by competing with natural substrate, thymidine triphosphate and terminates viral DNA synthesis. Concomitant use of stavudine (D4T) and zidovudine (AZT) is not recommended, since the latter can competitively inhibit the intracellular phosphorylation of stavudine (Balzarini et al. 1987).

Stavudine received FDA approval in June, 1994 for adults and in September, 1996 for pediatric use and again as a prolonged-release formulation for once-a-day dosing, stavudine was approved in 2001. Stavudine is marketed by Bristol-Myers

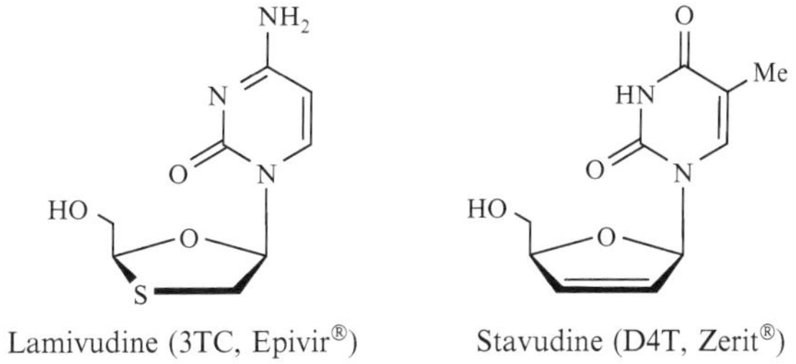

Fig. 7.5 Chemical structures of other reverse transcriptase (RT) inhibitors

Lamivudine (3TC, Epivir®)

Stavudine (D4T, Zerit®)

Squibb with the brand name, Zerit, and is available in 20, 30, or 40 mg hard-gelatin capsule dosage form.

7.2.4 Ziconotide (Prialt®)

Ziconotide is the synthetic form of the naturally occurring peptide, ω-conotoxin MVIIA (Fig. 7.6), originally isolated from the Pacific piscivorous marine cone snail *Conus magus* by Olivera and coworkers in 1979 (Olivera et al. 1985). It took ziconotide more than two decades of research and development to achieve FDA approval and to become the first marine-derived pharmaceutical for the specific indication of severe chronic pain.

Marine snails of the genus *Conus* feed by hunting small fish or invertebrates and injecting venom into their prey. This venom contains a plethora of conotoxins, neurotoxic peptides acting synergistically to paralyze prey by targeting its neuromuscular system. ω-Conotoxin MVIIA (Fig. 7.6) has been an exceptional lead for drug development in the management of severe and chronic pain. It elicited a

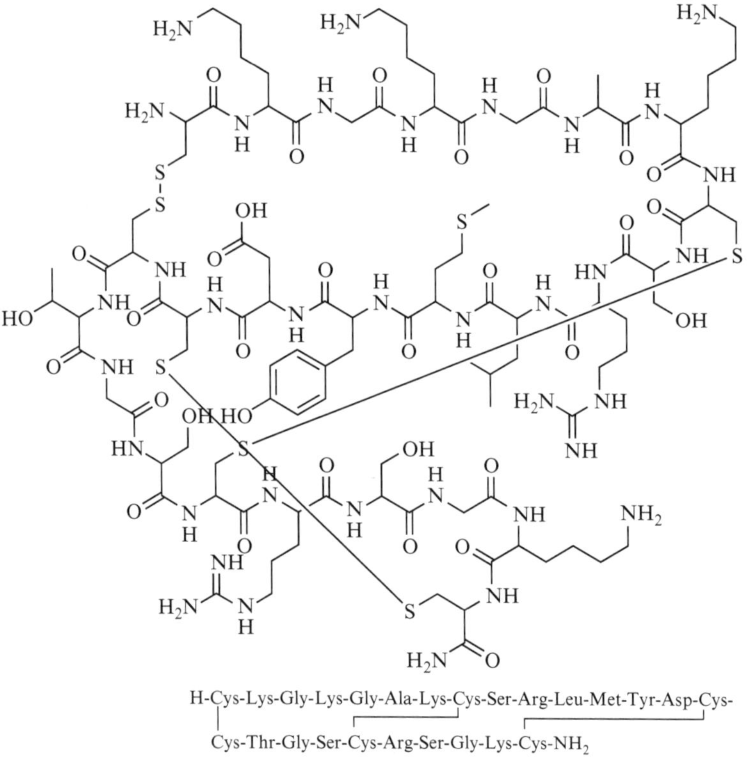

Fig. 7.6 Chemical structure of ω-conotoxin MVIIA (Ziconotide, Prialt®)

characteristic shaking behavior in mice after an intrathecal injection (Olivera et al. 1985). Structurally, ω-conotoxin MVIIA is a linear polycationic 25-amino acid peptide containing six cysteine residues linked by three disulfide bridges that stabilize its well-defined characteristic 3D structure (Olivera et al. 1994; Price-Carter et al. 1998).

A complete chemical synthesis of ziconotide was successfully achieved in 1987 (Molinski et al. 2009). Ziconotide revealed potent analgesic activity with a completely novel mechanism of action in which N-type voltage-sensitive calcium channels (NVSCCs) were disclosed to be its target site (Kerr and Yoshikami 1984; Olivera et al. 1987; Bingham et al. 2010). Various subtypes of voltage-gated calcium channels have been recognized in the nervous system (Olivera et al. 1987). Ziconotide reversibly blocks NVSCCs which are found exclusively in presynaptic neurons. NVSCCs regulate membrane depolarization induced by calcium influx, which subsequently controls a variety of calcium-dependent cellular processes. NVSCCs are abundantly located on primary nociceptive afferent neurons in the superficial lamina of the dorsal horn of the spinal cord where they play an important role in the spinal processing of nociceptive afferent (pain signaling) activity (Gohil et al. 1994; Bowersox et al. 1996). Ziconotide potently inhibits the conduction of nerve signals (K_i value of 0.5 μM) (Yeager et al. 1987) by specifically blocking the NVSCCs. In the complex with NVSCC, ziconotide forms a compact folded structure with a binding loop between Cys8 and Cys15 that also contains Tyr13, a crucial amino acid residue at the binding site (Kim et al. 1995; Atkinson et al. 2000).

ω-Conotoxin MVIIA (Ziconotide) revealed significant affinity towards NVSCC (K_d value of 9 pM) (Kristipati et al. 1994) which attracted interest to develop this peptide into an antinociceptive agent. To meet the supply needs for clinical trials, Neurex Corporation of Elan pharmaceuticals successfully synthesized ω-conotoxin MVIIA which was then named ziconotide (Jones et al. 2001). Preliminary studies disclosed that ziconotide has outstanding antinociceptive properties in animal models with acute, persistent, and neuropathic pain after intrathecal administration (Bowersox and Luther 1998), whereas for postoperative pain in a rat incisional model, ziconotide proved more potent and displayed longer activity than intrathecal morphine (ED50s of 0.049 and 2.1 nM, respectively) (Wang et al. 2000a).

The promising analgesic activity exhibited in animal studies supported ziconotide to enter clinical trials in USA and in Europe for the treatment of severe chronic pain. Because of the poor blood brain barrier penetration and the hypotensive effect of ziconotide when systemically administered, the compound is delivered intrathecally via an implantable pump or temporarily by an external microinfusion device (Gaur et al. 1994; Staats et al. 2004; McGivern 2006). In contrast to opiate-based therapies, in which tolerance is a major limiting factor, tolerance to ziconotide was not a problem over time (Wang et al. 2000b).

In December 2004, Elan pharmaceuticals achieved FDA approval for ziconotide as an intrathecal infusion formulation under the trade name Prialt®. The compound is currently labeled for the management of severe chronic pain in patients with cancer or AIDS (Staats et al. 2004; Rauck et al. 2009) for whom intrathecal treatment is warranted and who are intolerant of or refractory to other treatments, such as systemic analgesics, adjunctive therapies, or IT morphine. The FDA approval of ziconotide supported other conotoxins that are currently under investigation for therapeutic potential and probable clinical applications (Molinski et al. 2009).

7.2.5 Trabectedin (Yondelis®, EU registered)

Trabectedin (ecteinascidin-743, ET-743) (Fig. 7.7) is a marine natural product isolated from extracts of the marine tunicate *Ecteinascidia turbinata* found in the Caribbean (Rinehart et al. 1990, 1991) and the Mediterranean Sea (Wright et al. 1990). Trabectedin (ET-743) is a renaissance for marine-derived therapeutics; however, it took almost 17 years after the first report of its structure for it to be launched into the pharmaceutical market in the European Union as the first marine-derived anticancer drug for patients with soft tissue sarcoma (Verweij 2009) and patients with relapsed platinum-sensitive ovarian cancer (Yap et al. 2009).

Trabectedin (ecteinascidin, ET-743) (Fig. 7.7) and its *N*-demethyl analog (ET-729) are the most abundant active components of the extract of *E. turbinata* whose chemical structures are formed by three fused tetrahydroisoquinoline rings through a 10-membered thioether lactone bridge – a distinctive structural feature of ecteinascidins. Initially, ecteinascidins were found to possess in vitro cytotoxic activity against L1210 leukemia cells with an IC_{50} value in the low nanomolar range (Rinehart et al. 1990) and were later shown to possess potential in vivo antitumor effects in various mouse models with different murine and human cancer xenografts (Rinehart et al. 1990; Sakai et al. 1992). Further studies that compared ET-743 and ET-729 revealed that they have comparable potency. Subsequently,

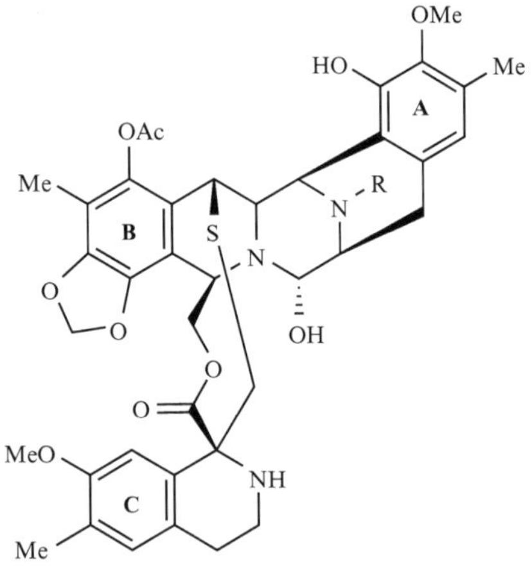

R = Me: Trabectedin (ET-743, Yondelis®)
R = H: ET-729

Fig. 7.7 Chemical structures of trabectedin (ET-743) and ET-729

ET-743 was selected for further development because of its higher abundance in *E. turbinata*.

To increase the supply of ET-743, different strategies have been explored such as large-scale aquafarming of the tunicate and chemical synthesis. Unfortunately, neither of these two strategies proved promising since the yields were either variable or very low. To fulfill the supply needs, a breakthrough was achieved by PharmaMar, the licensee of natural trabectedin (ET-743), who succeeded in developing a large-scale semisynthetic protocol for ET-743 starting with cyanosafracin B, an antibiotic that can be produced in multikilogram scale by fermentation of *Pseudomonas fluorescens* (Cuevas et al. 2000). Today, trabectedin (ET-743) is licensed by PharmaMar to Johnson & Johnson/OrthoBiotech for drug development in the United States (Aune et al. 2002).

The mechanism of action of trabectedin (ET-743) is attributed to a covalent reversible modification of DNA through guanine-specific alkylation at the minor-groove N2 position inducing a bend in the DNA helix directed towards the major groove (Pommier et al. 1996; Zewail-Foote and Hurley 1999). Furthermore, ET-743 was found to interact with different binding proteins of the Nucleotide Excision Repair (NER) system (Takebayashi et al. 2001; Soares et al. 2007; Herrero et al. 2006). Whereas, other known DNA-interacting agents require a deficient NER for their activity, trabectedin (ET-743) requires an intact NER system to exert its cytotoxic activity. Cell cycle studies on tumor cells revealed that ET-743 arrests cells at G2/M phase and induces apoptosis independent of p53 (Erba et al. 2001).

Trabectedin (ET-743) was approved by the European Agency for the Evaluation of Medicinal Products (EMEA) to PharmaMar in July 2007, under the trade name Yondelis®, for indications of refractory soft tissue sarcoma and ovarian cancer.

Currently, ET-743 is undergoing Phase II trials in breast, lung, prostate and pediatric cancers, and Phase III trials for first-time therapy in STS in the United States (Molinski et al. 2009). The safety profile of Yondelis® includes the most frequent adverse effect as neutropenia, which is reversible and elevated transaminases which were also transient (Sessa et al. 2009). Interestingly, no mucositis, alopecia, neurotoxicity, cardiotoxicity, or cumulative toxicities have been observed.

7.3 Marine Pharmaceuticals in Phase III Clinical Trials

Marine-derived pharmaceuticals in Phase III clinical trials include, in addition to trabectedin (ET-743, Yondelis®) for US approval, soblidotin (TZT-1027) and eribulin mesylate (E7389) which will be discussed in this section.

7.3.1 Soblidotin (TZT-1027)

Soblidotin (TZT-1027) is a synthetic derivative of dolastatin 10 (Fig. 7.8) which was first isolated, in addition to a series of cytotoxic peptides, from the sea hare

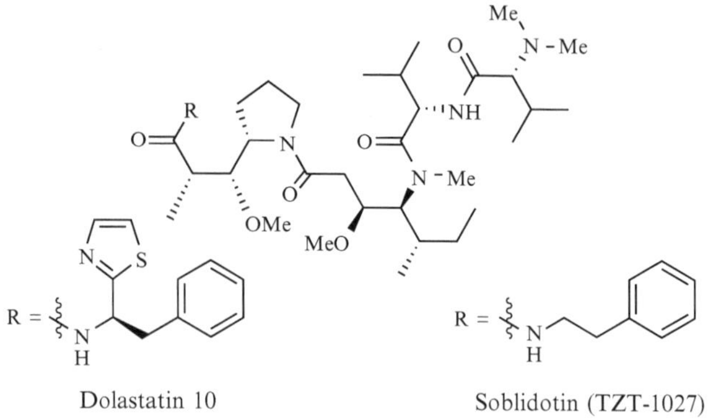

Fig. 7.8 Chemical structures of dolastatin 10 and soblidotin (TZT-1027)

Dolabella auricularia collected during an explorative expedition off the island state of Mauritius in the Western Indian Ocean in 1972 (Pettit 1997). Soblidotin (TZT-1027) (Fig. 7.8) differs from dolastatin 10 in the replacement of the terminal dolaphenine amino acid residue with a simple phenethylamine group.

Both dolastatin 10 and soblidotin (TZT-1027) proved to have a potent antimitotic activity by inhibiting microtubule assembly and tubulin polymerization by binding to tubulin at a distinct site close to the vinca peptide binding site with equal potency (IC_{50} values of 2.3 and 2.2 µM, respectively) (Bai et al. 1990a, b). Consequently, they arrest cells at the mitosis (M) phase and depleted cellular microtubules (Bai et al. 1992). In addition to its tubulin inhibitory activity, TZT-1027 was also found to be a vascular disrupting agent (VDA), collapsing the vasculature in the tumor (Watanabe et al. 2007; Lippert 2007).

Soblidotin (TZT-1027) entered Phase I clinical trials in Europe, Japan, and the USA under the auspices of either Teikoku Hormone, the originator or the licensee, Daiichi Pharmaceuticals. It exhibited an interesting development path, as after Phase I and Phase II clinical trials, the licensing agreement with Daiichi was terminated (Schöffski et al. 2004; De Jonge et al. 2005). Currently, TZT-1027 is under the auspices of Asaka Pharmaceuticals which was formed through a merger between Teikoku Hormone and Grellan Pharmaceuticals that has licensed the compound to Yakult for world-wide development.

7.3.2 Eribulin Mesylate (E7389)

Eribulin mesylate (E7389) is a synthetic analog of halichondrin B (HB) (Fig. 7.9), a polyether macrolide natural product that was first isolated by Uemura and co-workers in 1986 together with norhalichondrin B and homohalichondrin B from the marine sponge *Halichondria okadai*, that also sequesters the phosphatase inhibitor okadaic acid (Hirata and Uemura 1986; Uemura et al. 1985). Halichondrins

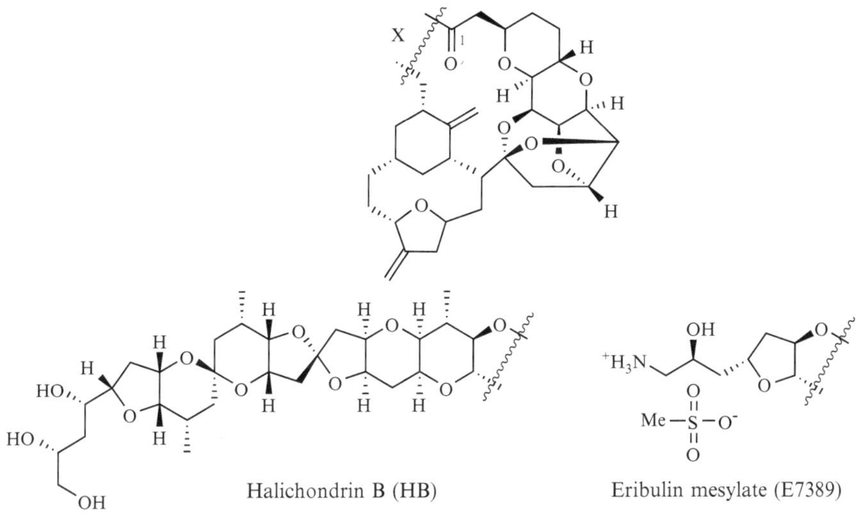

Fig. 7.9 Chemical structures of halichondrin B (HB) and eribulin mesylate (E7389)

revealed a potent antiproliferative activity with IC_{50} values in the nanomolar range (Pettit et al. 1991) by binding to tubulin at the vinca binding site (Bai et al. 1991; Dabydeen et al. 2006). To accomplish the supply needs of halichondrin B, total synthesis (Aicher et al. 1992) and aquaculture of the sponge source (Munro et al. 1999) have been performed, however, none of them can afford a sufficient supply of halichondrins.

A crucial breakthrough was achieved by academia together with Eisai company in Japan, its US subsidiary, and the Eisai Institute, who successfully synthesized HB analogs with almost 70% of the molecular weight of HB and with equipotent activity to the natural product against tumor cells (Stamos et al. 1997; Wang et al. 2000c). In particular, the halichondrin B analog, eribulin mesylate (E7389) (Fig. 7.9) was the best candidate, retaining the promising biological properties of HB as well as the favorable pharmaceutical attributes including water solubility and chemical stability (Towle et al. 2001).

Whereas eribulin mesylate (E7389) and halichondrin B (HB) (Fig. 7.9) are tubulin-targeting agents resembling the taxanes and vinca chemotherapeutics, they have a unique mechanism of action different from those of vinca alkaloids and taxanes primarily through suppression of microtubule growth (Jordan et al. 2005; Okouneva et al. 2008). Eribulin mesylate (E7389) was found to exert potent irreversible antimitotic effects against cancer cells leading to apoptotic cell death (Kuznetsov et al. 2004). In Phase I clinical trials, E7389 was intravenously administered with a maximal tolerated dose of 1–2 mg/m^2, whereas dose-limiting adverse effects including neutropenia, leukopenia, and fatigue were reported (Goel et al. 2009; Tan et al. 2009). In addition, pharmacokinetics of E7389 was dose proportional with a half-life elimination of 1.5–2 days (Goel et al. 2009; Tan et al. 2009).

Phase II clinical trials of eribulin mesylate (E7389) in patients with multiple cancer types have been carried out. In those with breast cancer, the response rate was 9.3–11.5% in intensively treated patients, with responses occurring in patients refractory to taxanes or other agents (Vahdat et al. 2009).

Recently, two Phase III clinical studies evaluating E7389 against capecitabine (NCT00337103) and treatment of physician's choice (NCT00388726) have been started. Preliminary results of the latter disclosed a statistically significant improvement in overall survival rate (Eisai Inc., 2009). Eribulin mesylate (E7389) is currently developed and marketed by Eisai Incorporation.

7.4 Marine Pharmaceuticals in Phase II Clinical Trials

Eight Marine-derived pharmaceuticals in Phase II clinical trials, namely DMXBA (GTS-21), plinabulin (NPI-2358), plitidepsin (Aplidin®), elisidepsin (Irvalec®), PM1004 (Zalypsis®), tasidotin (synthadotin, ILX-651), cryptophycins, and pseudopterosins, will be described in this section with regards to their discovery, course of development, and clinical effects.

7.4.1 DMXBA (GTS-21)

DMXBA (GTS-21) (Fig. 7.10) is 3-(2,4-dimethoxybenzylidene)-anabaseine, a synthetic derivative of the marine alkaloid anabaseine which was first isolated from the intertidal Pacific nemertine *Paranemertes peregrina* (Kem et al. 1971) and also disclosed to be present in several species of marine worms belonging to the Phylum Nemertea (Kem et al. 2006).

DMXBA (GTS-21) (Fig. 7.10) was proven to selectively stimulate α7 nicotinic acetylcholine receptors (Kem et al. 2004), which are present on CNS neurons, CNS astrocytes, and on peripheral macrophages. DMXBA was also found to improve cognition (Buccafusco et al. 2005) and deficient sensory gating (Olincy and Stevens 2007) in animal models. Moreover, DMXBA (GTS-21) and related

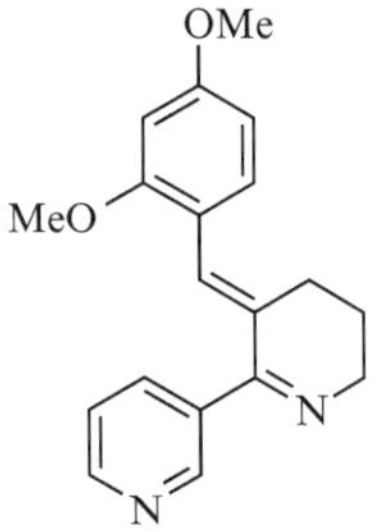

Fig. 7.10 Chemical structure of DMXBA (GTS-21)

arylidene–anabaseines have also been disclosed to possess in vitro and in vivo neuroprotection properties (Buckingham et al. 2009; Thinschmidt et al. 2005).

DMXBA was also able to ameliorate the deleterious effects of β-amyloid in cell cultures of cerebral cortex neurons (Shimohama 2009). In addition, DMXBA (GTS-21) revealed anti-inflammatory activity in animal models by its agonistic effect on macrophage α7 receptors (Pavlov et al. 2007; Rosas-Ballina et al. 2009). Recently, GTS-21 was found to improve survival of rats undergoing experimental hemorrhage (Yeboah et al. 2008; Cai et al. 2009).

Phase I clinical studies of DMXBA (GTS-21) revealed significant cognitive improvement in healthy young males (Kitagawa et al. 2003) and schizophrenics (Olincy et al. 2006). A recent Phase II clinical trial using DMXBA (GTS-21) for schizophrenics has been undertaken and it revealed improved cognitive functions compared to placebo but only with higher DMXBA doses (Freedman et al. 2008).

7.4.2 Plinabulin (NPI-2358)

Plinabulin (NPI-2358) (Fig. 7.11) is a fully synthetic analog of the natural diketopiperazine halimide (Fig. 7.11) from the marine fungus *Aspergillus* sp. CNC-139 isolated from the alga *Halimeda lacrimosa* collected in the Bahamas (Mayer et al. 2010) and (−)-phenylahistin from *Aspergillus ustus* (Kanoh et al. 1997).

Mechanistically, plinabulin (NPI-2358) binds to a region between α- and β-tubulin near the colchicine binding site and inhibits tubulin polymerization (Nicholson et al. 2006; Yamazaki et al. 2008), resulting in a destabilization of tumor endothelial vasculature. Thus, plinabulin is considered as a VDA inducing selective collapse of established tumor vasculature, in addition to its direct apoptotic effect on tumor cells (Nicholson et al. 2006).

In 2006, a Phase I clinical trial on patients with solid tumors or lymphomas was initiated by Nereus Pharmaceuticals. Results revealed that plinabulin (NPI-2358) had a measurable effect on tumor vasculature at doses ≥ 13.5 mg/m^2 and was tolerated up to 30 mg/m^2 (Mayer et al. 2010). These findings, together with the arguments that VDAs can synergize with chemotherapeutics and antiangiogenesis

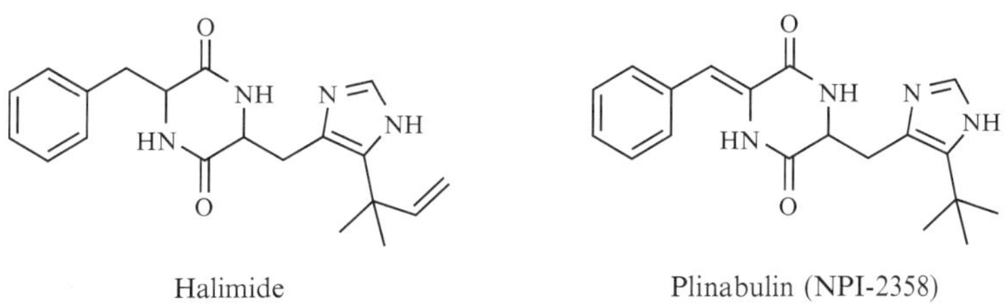

Halimide

Plinabulin (NPI-2358)

Fig. 7.11 Chemical structures of halimide and plinabulin (NPI-2358)

agents, supported the initiation of the ADVANCE (Assessment of Docetaxel and Vascular disruption in Nonsmall Cell Lung Cancer) Phase I/II clinical trials in 2009 (Mayer et al. 2010). Currently, plinablulin (NPI-2358) is developed by Nereus Pharmaceuticals Inc. for cancer.

7.4.3 Plitidepsin (Aplidine, Aplidin®)

Didemnin B and plitidepsin (aplidine, Aplidin®) (Fig. 7.12) are two closely related marine-derived peptides from different organisms which supports the argument that their actual producer is possibly a common symbiont. Didemnin B (Fig. 7.12) was first reported by Rinehart and coworkers in 1981 from the Caribbean tunicate *Trididemnum solidum* (Rinehart et al. 1981a) and displayed in vitro antiviral and cytotoxic activities at nanomolar concentrations (Rinehart et al. 1981a, b). Aplidine (plitidepsin, Aplidin®) (Fig. 7.12), a dehydrodidemnin B in which the *N*-lactyl side chain is replaced by a pyruvate moiety, was first isolated also by the Rinehart group from the Mediterranean tunicate *Aplidium albicans* in 1991 (Rinehart and Lithgow-Bertelloni 1991).

In spite of the activity exhibited by didemnin B (Fig. 7.12) during Phase I and II clinical trials in patients with advanced pretreated cancers, these trials are currently suspended due to serious adverse effects including severe fatigue (Kucuk et al. 2000), and anaphylaxis (Nuijen et al. 2000). Plitidepsin (aplidine, Aplidin®) (Fig. 7.12), which is currently developed by total synthesis by PharmaMar, showed similar levels of antitumor activities compared to didemnin B in cultured tumor cells (Nuijen et al. 2000) and proved to induce apoptosis by depleting glutathione (GSH) and to increase the cellular oxidative stress with IC_{50} values in the low nanomolar range (Rinehart and Lithgow-Bertelloni 1991).

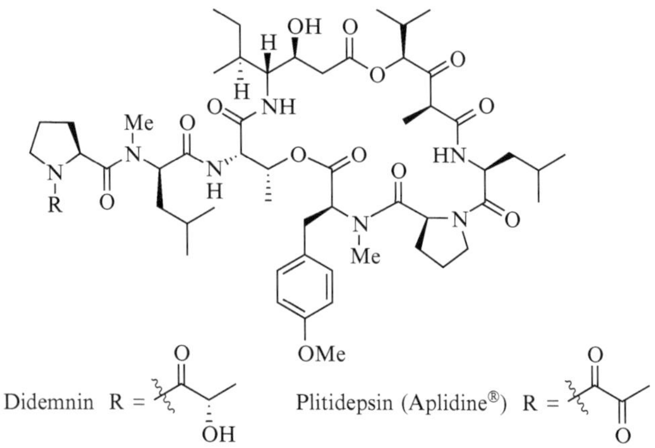

Fig. 7.12 Chemical structures of didemnin and plitidepsin

Preclinical in vitro and in vivo studies of plitidepsin against different tumor types were the basis for the selection and design of Phase I and Phase II clinical trials. Clinically, plitidepsin (aplidine, Aplidin®) revealed preliminary efficacy in two different Phase II trials against relapsing and refractory multiple myeloma and T cell lymphomas (Mitsiades et al. 2008). The results of these clinical trials encouraged further clinical research especially as a cotreatment with other anticancer agents. The most common toxicity accompanied with plitidepsin therapeutic schedules included muscular toxicity, transient increase in transaminases, fatigue, diarrhea, and cutaneous rash (Le Tourneau et al. 2007), whereas no severe bone marrow toxicity has been recognized.

7.4.4 Elisidepsin (PM02734, Irvalec®)

Elisidepsin (PM02734, Irvalec®) (Fig. 7.13) is a novel marine-derived cyclic peptide belonging to the kahalalide family of peptides especially the novel antitumor depsipeptide kahalalide F, first isolated by Mark T. Hamann and Paul J. Scheuer from seasonal collections of the marine sacoglossan *Elysia rufescens* and the green alga *Bryopsis* sp. in 1993 (Scheuer and Hamann 1993). The absolute stereochemistry of the compound was determined in 1999 (Goetz et al. 1999) and then in 2001 kahalalide F was chemically synthesized (López-Macià et al. 2001).

Elisidepsin (PM02734, Irvalec®) (Fig. 7.13) revealed a potent in vitro antiproliferative activity against a vast array of human cancer cell lines and it is currently undergoing Phase II clinical studies with preliminary evidence of antitumor activity and a favorable therapeutic index (Ling et al. 2009). However, the mechanism of cytotoxicity of elisidepsin has not yet been clarified, and it has been shown to induce oncolytic rather than apoptotic cell death (Suárez et al. 2003). Elisidepsin (PM02734, Irvalec®) is currently developed by PharmaMar.

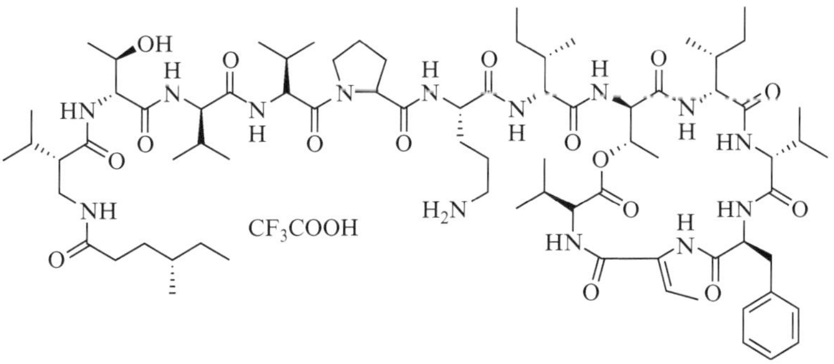

Fig. 7.13 Chemical structure of elisidepsin (PM02734, Irvalec®)

7.4.5 PM1004 (Zalypsis®)

PM1004 (Zalypsis®) (Fig. 7.14) is a new synthetic DNA-binding alkaloid related to the tetrahydroisoquinolines renieramycin, isolated from the Mexican marine sponge *Reniera* sp. (Frincke and Faulkner 1982; He and Faulkner 1989) and the Fijian sponge *Xestospongia caycedoi* (Davidson 1992), and jorumycin (Fig. 7.14), obtained from the skin and mucus of the Pacific marine nudibranch *Jorunna funebris* (Fontana et al. 2000). Renieramycins, jorumycin, and PM1004 (Zalypsis®) were successfully synthesized (Scott and Williams 2002).

PM1004 binds to guanines in selected DNA triplets, and eventually breaks the DNA helix, arrests the cancer cells at the DNA synthesis (S) phase and induces apoptosis (Leal et al. 2009). Cellular studies of Zalypsis® disclosed that cell lines with mutant p53 or lacking p53 are more prone to the treatment than cell lines with wild type p53 (Leal et al. 2009).

Preclinical in vivo studies of PM1004 (Zalypsis®) revealed a significant antitumor activity against breast, prostate, and renal cancer, while it displayed a moderate profile in colon cancer. The main adverse event recorded with PM1004 during Phase I clinical trials was hematological disorders and/or elevated transaminases which were transient and reversible (Leal et al. 2009). PM1004 (Zalypsis®) is currently licensed and developed by PharmaMar and is now under Phase II clinical studies.

7.4.6 Tasidotin (Synthadotin, ILX-651)

Tasidotin (synthadotin, ILX-651) (Fig. 7.15) is a synthetic derivative of dolastatin 15, first reported from the sea hare *Dolabella auricularia* in 1992 (Bai et al. 1992) from which dolastatin 10 (Fig. 7.8) was also reported in 1972 (Pettit 1997).

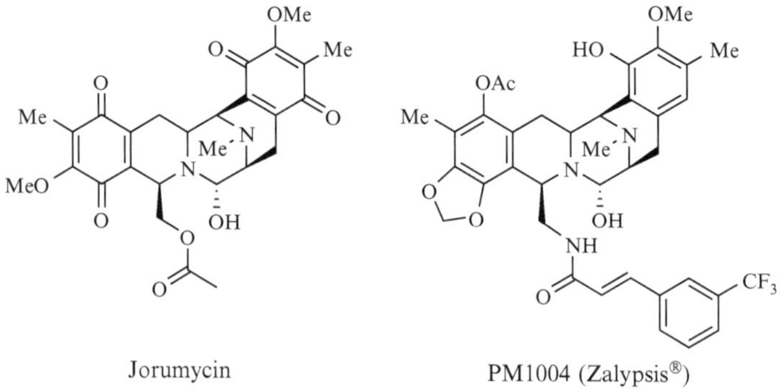

Fig. 7.14 Chemical structures of jorumycin and PM1004 (Zalypsis®)

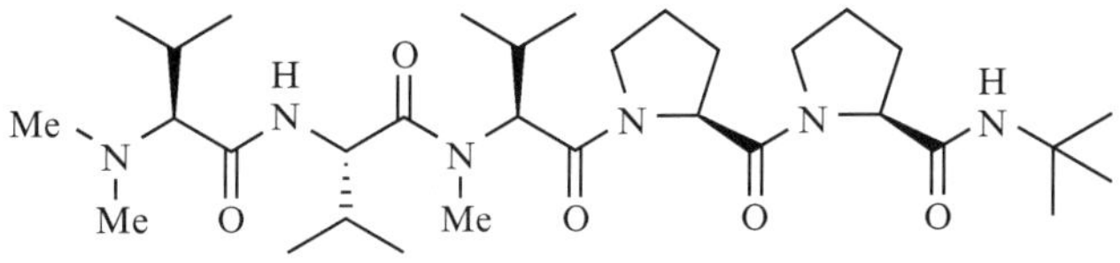

Fig. 7.15 Chemical structure of tasidotin (synthadotin, ILX-651)

Fig. 7.16 Chemical structures of cryptophycin derivatives

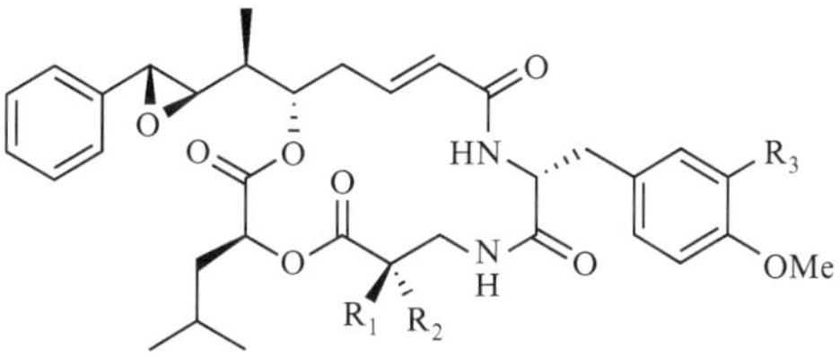

Cryptophycin-1 R_1 = Me, R_2 = H, R_3 = Cl
Arenastatin A R_1 = R_2 = R_3 = H
Cryotphycin-52 R_1 = R_2 = Me, R_3 = Cl

Dolastatin 15 differs from dolastatin 10 (Fig. 7.8) in lacking the dolaisoleucine residue and in having an ester linkage in its unit bonding (Bai et al. 1992).

However, tasidotin (synthadotin, ILX-651) is proven to be a potent inhibitor of tubulin assembly, and further investigation has recently been performed on its mechanism of action (Ray et al. 2007; Bai et al. 2009) which disclosed that the active metabolite is probably the pentapeptide resulting from hydrolysis of the C-terminal amide bond. ILX-651 is orally active and it has advanced to Phase II clinical trials against a wide range of cancer initially under Ilex Pharmaceuticals, and then under the auspices of Genzyme Corporation following the purchase of Ilex. Results of Phase II clinical trials have recently been reported and they revealed that ILX-651 was well tolerated but the efficacy was not favorable enough to proceed as a single agent therapy (Mayer et al. 2010). Therefore, tasidotin (synthadotin, ILX-651) has returned to preclinical studies to determine better routes and targets including advanced refractory neoplasms (Mayer et al. 2010).

7.4.7 Cryptophycins

Cryptophycin-1 (Fig. 7.16), the prototype of a group of 27 marine cyanobacteria-derived tubulin-binding natural products, was first obtained from the marine cyanobacteria *Nostoc* sp. strain ATCC 53789 (Schwartz et al. 1990). Cryptophycin-1 exhibited extremely potent in vitro and in vivo cytotoxic activities with IC_{50} values in the low picomolar range (Trimurtulu et al. 1994). Arenastatin A (Fig. 7.16), a related cytotoxic compound, was isolated from the marine sponge *Dysidea arenaria*

(Kobayashi et al. 1994). Again, this may be a proof of the production of these metabolites by a symbiotic microorganism. Cryptophycin-1 was successfully prepared by chemical synthesis in 1995 (Barrow et al. 1995). Currently, the natural and synthetic cryptophycins are licensed from the University of Hawaii and Wayne State University, USA to Eli Lilly for drug development (Molinski et al. 2009).

Cryptophycins are potent cytotoxic and antimitotic agents whose mechanism of action is ascribed to the destabilization of microtubule dynamics and the induction of hyperphosphorylation of the antiapoptotic protein B-cell leukemia/lymphoma 2 (BCL-2), thus inducing apoptosis (Lu et al. 2001), whereas the antimitotic property is due to its tubulin binding at the vinca peptide binding site – similar to hemiasterlin (see below) – thereby inducing microtubule depolymerization and arresting of the cell cycle at the G2/M phase (Lu et al. 2001; Mooberry et al. 1996; Kerksiek et al. 1995; Smith and Zhang 1996; Panda et al. 1997).

Cryptophycin-52 (LY355703) (Fig. 7.16), a synthetic analog of cryptophycin-1, was developed by Eli Lilly to improve hydrolytic stability and formulation. Interestingly, cryptophycin-52 revealed 40- to 400-fold higher potency than either vinca alkaloids or paclitaxel in the in vitro cytotoxicity assays (Wagner et al. 1999). In the late 1990s, cryptophycin-52 was selected for Phase I clinical studies in patients with refractory solid tumors (Sessa et al. 2002; Stevenson et al. 2002) where dose-related toxicities included neuropathy, myalgia, long-lasting neuroconstipation, cardiac dysrhythmia, and mild alopecia. In a Phase II clinical study involving patients with Stage IIIb or IV nonsmall cell lung cancer (NSCLC) previously treated with platinum-containing chemotherapeutics, the results were not favorable due to the nonsignificant response produced by cryptophycin-52 at a dose of 1.5 mg/m^2, high enough to induce neurological toxicities (Edelman et al. 2003).

Currently, cryptophycin-52 has been excluded from clinical trials, however, new analogs, such as cryptophycins-309 and -249, have been preclinically investigated, and they were found to be promising enough to advance into the clinical trial phase (Liang et al. 2005).

7.4.8 Pseudopterosins

Pseudopterosin A (Fig. 7.17) is the parent compound of a family of diterpene glycosides comprising to date a further 25 congeners which were isolated from

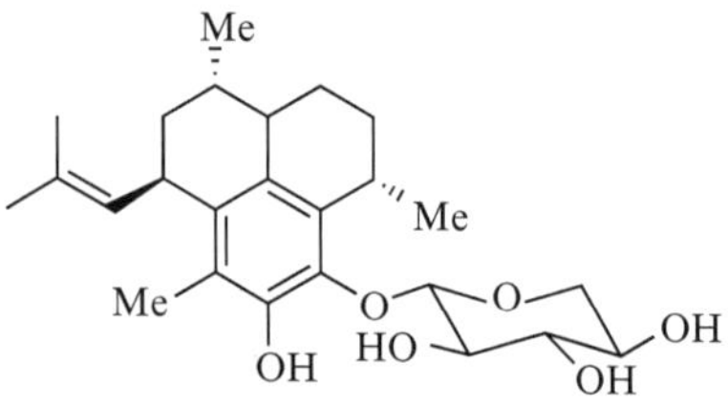

Fig. 7.17 Chemical structure of pseudopterosin A

the marine octocoral *Pseudopterogorgia elisabethae* (Look et al. 1986a; Roussis et al. 1990; Duque et al. 2004). The basic chemical skeleton of pseudopterosins features a tricarbocyclic core possessing four chiral carbons and a sugar moiety attached at either C-9 or C-10 of a catechol subunit that includes one of the three core rings.

Pseudopterosin A inhibits phorbol myristate acetate-induced topical inflammation in mice (Look et al. 1986b) and stabilizes cell membranes (Ettouati and Jacobs 1987). In addition it was found to prevent the release of prostaglandins and leukotrienes from zymosan-stimulated murine macrophages (Mayer et al. 1998) and to inhibit degranulation of human polymorphonuclear leukocytes and phagosome formation in *Tetrahymena* cells (Moya and Jacobs 2006). A pretreatment with pertussis toxin prior to pseudopterosin A administration blocked its antiphagocytic activity, supporting an investigation of the role of pseudopterosins to act upon G-protein-coupled receptors of the adenosine variety (Zhong et al. 2008). Moreover, pseudopteroin A-10-*O*-methyl ether revealed potent anti-inflammatory and wound-healing properties (Montesinos et al. 1997). Several preclinical studies and Phase I/II clinical trials disclosed potent reepithelialization and wound-healing activities attributed to pseudopterosin A and its 10-*O*-methyl ether derivative.

7.5 Marine Pharmaceuticals in Phase I Clinical Trials

Marine pharmaceuticals undergoing Phase I clinical trials will be discussed in this section including bryostatin 1 (NSC 339555), hemiasterlin (E7974), and salinosporamide A (marizomib, NPI-0052).

7.5.1 Bryostatin 1 (NSC 339555)

Bryostatin 1 (NSC 339555) (Fig. 7.18), a 26-membered macrocyclic lactone with 11 stereocenters and an unprecedented polyacetate carbon backbone, was first

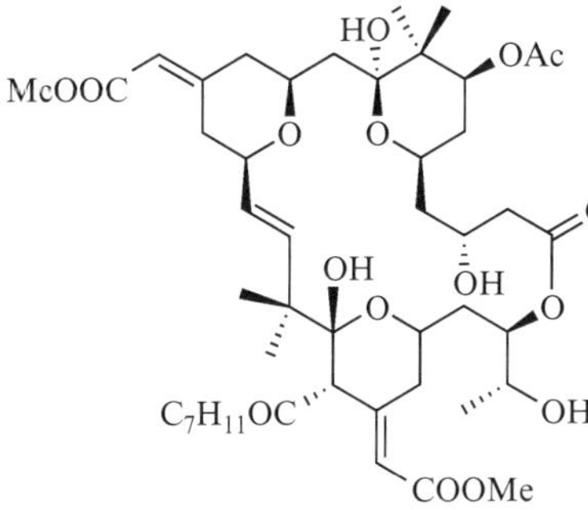

Fig. 7.18 Chemical structure of bryostatin-1 (NSC 339555)

isolated from the bryozoan *Bugula neritina* collected from the Gulf of Mexico in 1982 (Pettit et al. 1982). Subsequently, 19 additional bryostatin congeners were isolated from *B. neritina* which differ mainly in substitution at C_7 and C_{20} by different acyloxy functionalities (Pettit 1996). Bryostatin 1 revealed a potent in vitro cytotoxicity against the P388 murine lymphocytic leukemia cell line with an IC_{50} value of 1.0 μM (Pettit et al. 1982). The low and variable abundance of bryostatin 1 from the natural bryozoan collections hindered the sufficient supply of the compound for preclinical and clinical studies thereafter. However, in 1991, 18 g of cGMP quality bryostatin 1 were obtained from a large-scale collection of *B. neritina* from Californian waters (Schaufelberger et al. 1991).

In the meantime, the Southern Californian company CalBioMarine Technologies was interested in developing the aquacultural techniques and obtaining bryostatin 1 from its natural source (Rouhi 1995), as the total synthesis was not promising due to the unique stereochemistry, however, the first total synthesis of bryostatin 1 and its C9-deoxy derivative has recently been reported (Keck et al. 2010).

Bryostatin 1 (Fig. 7.18) was proven to modulate protein kinase C (PKC) but no details have been defined (Hennings et al. 1987). However, bryostatin 1 was disclosed to bind to the regulatory domain of PKC at the same binding site of phorbol esters at low nanomolar concentrations and it exhibited physiological effects other than tumor-promoting activity (Wender et al. 1988). Only upon prolonged exposure did bryostatin 1 deplete PKC from the cells probably through proteolysis by the proteasome (Isakov et al. 1993).

In addition bryostatin 1 revealed other biological activities such as immunomodulation (Berkow et al. 1993), induction of cell differentiation (Hu et al. 1993), radioprotection (Grant et al. 1992), and synergistic interactions with other anticancer agents (Mutter and Wills 2000). Bryostatin 1 has been investigated either alone or as a cotreatment with other chemotherapeutics in Phase I and II clinical trials against a vast array of cancers (Clamp and Jayson 2002). The most common toxicity accompanying bryostatin 1 administration as a single agent therapy was myalgia probably due to muscular vasoconstriction resulting from impaired oxidative metabolism and proton efflux from the cells (Hickman et al. 1995). Unfortunately, the outcomes from Phase II clinical trials of bryostatin 1 alone against squamous cell carcinoma, melanoma, and ovarian, renal and colorectal cancers were disappointing with no significant clinical improvement being recognized (Clamp and Jayson 2002). However, the results of a Phase I clinical study using bryostatin 1 as a combination therapy with cytarabine (Ara-C, Cytosar-U®) against acute leukemia (Cragg et al. 2002) and vincristine in B-cell malignancies (Dowlati et al. 2003) were promising. Currently, five Phase II clinical trials, using bryostatin 1 as a cotherapy with other chemotherapeutic agents against different cancers, are being carried out under the auspices of the US National Cancer Institute. In combination with paclitaxel (Taxol®), bryostatin 1 revealed significant response rates in patients with untreated, advanced gastric or gastro-esophageal junction adenocarcinoma compared to paclitaxel single treatment (Ajani et al. 2006).

7.5.2 Hemiasterlin (E7974)

Hemiasterlin (Fig. 7.19), an antimitotic tripeptide first reported by Kashman and coworkers from the marine sponge *Hemiasterella minor* in 1994, exhibited potent cytotoxicity against the P388 leukemia cell line ($IC_{50} = 0.02$ μM) (Talpir et al. 1994). Two additional congeners, hemiasterlins A and B (Fig. 7.19) were isolated from sponges of the genus *Auletta* and *Cymbastella* in 1995 (Coleman et al. 1995) whereas a fourth derivative, hemiasterlin C (Fig. 7.19), was obtained from the marine sponge *Siphonochalina* sp. collected off the coast of Papua New Guinea in 1999 (Gamble et al. 1999). In 1996, the linear structure and the presence of nonproteinogenic amino acids in hemiasterlin were proven by an X-ray crystallography study of its methyl ester derivative (Coleman et al. 1996).

Hemiasterlins revealed significant in vitro cytotoxicity against a vast array of human and murine cell lines with IC_{50} values in the nanomolar range (Coleman et al. 1995) which was attributed to the inhibition of cell cycle in the mitosis (M) phase resembling other tubulin-binding chemotherapeutics, paclitaxel or vinblastine, at ED_{50} values between 0.5 nM (hemiasterlin) and 28 nM (hemiasterlin B) (Anderson et al. 1997).

Comprehensive structure–activity relationship (SAR) studies disclosed that HTI-286 (Fig. 7.19), a simpler synthetic derivative of hemiasterlin featuring a phenyl moiety replacing the abrine (*N*-methyltryptophan) residue, is more potent than hemiasterlin (Coleman et al. 1996) whereas the analog of HTI-286 with a *para*-methoxyl functionality on the benzene ring was even more potent (Gamble et al. 1999). In addition the geminal β,β-dimethyl and the *N*-methyl groups on the first amino acid residue (N terminus), the isopropyl and an olefin in the homologated γ-amino acid (C terminus) and a terminal carboxylic acid or methyl ester were found to be essential for activity. Moreover, replacement of the aryl side chain on the N terminus with alkyl groups (e.g., *tert*-butyl) revealed no objective effect on activity (Nieman et al. 2003; Zask et al. 2004a, b; Yamashita et al. 2004).

Preclinical studies on mice with human xenografts revealed that HTI-286 (Fig. 7.19) results in tumor regression and growth inhibition (Loganzo et al. 2003). Furthermore, an open-label Phase I clinical trial of HTI-286 in patients

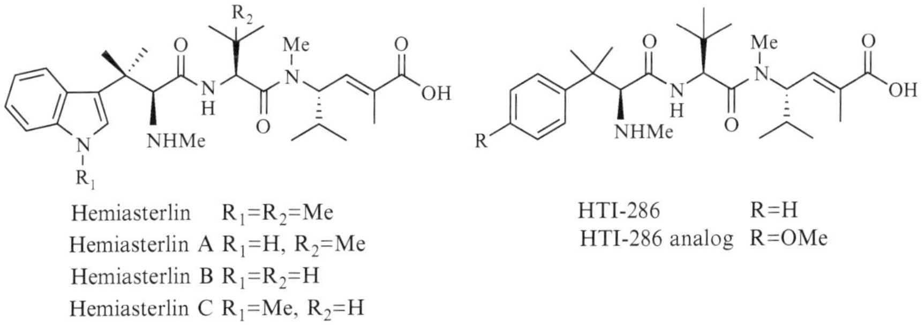

Hemiasterlin $R_1=R_2=Me$
Hemiasterlin A $R_1=H, R_2=Me$
Hemiasterlin B $R_1=R_2=H$
Hemiasterlin C $R_1=Me, R_2=H$

HTI-286 R=H
HTI-286 analog R=OMe

Fig. 7.19 Chemical structures of hemiasterlins and related derivatives

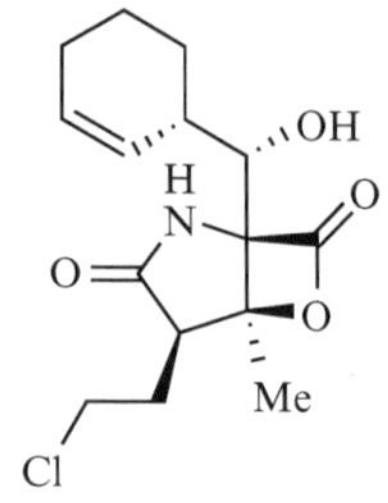

Fig. 7.20 Chemical structure of salinosporamide A (marizomib, NPI-0052)

with advanced solid tumors was completed; however, the results showed no objective responses and common toxicities observed included neutropenia, nausea, alopecia, and pain (Ratain et al. 2003). Therefore, Phase II trials of HTI-286 have been reconsidered. Nevertheless, HTI-286 is still an interesting subject due to the recent results illustrating its high antitumor activity in androgen-dependent and androgen-independent mouse models of refractory prostate cancer, and in a newly established in vitro taxane-resistant prostate PC-3 cell line (Hadaschik et al. 2008).

7.5.3 Salinosporamide A (Marizomib, NPI-0052)

Salinosporamide A (marizomib, NPI-0052) (Fig. 7.20) is an unusual β-lactone produced by the marine actinomycete *Salinospora tropica* recovered from deep-sea sediments in 2003 (Feling et al. 2003; Fenical et al. 2009). Salinosporamide A revealed a potent selective inhibition of the p26 proteasome (Feling et al. 2003; Fenical et al. 2009; Groll et al. 2006; Chauhan et al. 2006), a multicatalytic enzyme complex responsible for nonlysosomal cellular protein degradation and an interesting target for cancer therapy. Mechanistically, salinosporamide A inhibits proteasome activity by acylation of the N-terminal Thr1O$^\gamma$ residue followed by chloride displacement (Groll et al. 2006) similar to that of the known proteasome inhibitor bortezomib (Velcade®; Millennium/Janssen-Cilag), which is approved for indications of multiple myeloma. However, salinosporamide A was able to overcome resistance to bortezomib in patients with refractory myeloma (Chauhan et al. 2005).

Salinosporamide A is currently being developed by Nereus Pharmaceuticals (La Jolla, California) which has initiated and completed Phase I clinical trials of salinosporamide A in patients with multiple myeloma, lymphoma, leukemia, and solid tumors (Fenical et al. 2009; Chauhan et al. 2008).

7.6 Marine Pharmaceuticals in Preclinical Phase

The number of marine pharmaceuticals in preclinical phase is vast comprising more than 1,200 marine natural products from 1998 to 2006 (Mayer 1999; Mayer and

Lehmann 2001; Mayer and Gustafson 2003, 2004, 2006, 2008) that can be further divided according to pharmacological significance into 592 marine compounds revealing antitumor and cytotoxic properties (Mayer and Gustafson 2008) and 666 exhibiting a vast array of pharmaceutical activities including antimicrobial, anti-inflammatory, anticoagulant, antiplasmodial, and also some affecting the cardiovascular, immune and nervous systems (Mayer et al. 2009). The status of marine pharmaceuticals in preclinical phases is promising and their contribution to enlarging the pharmacopea by introducing more therapeutics to treat a wide variety of human ailments seems to be on the horizon.

Some of the marine natural products exhibited promising activities during the in vitro and in vivo preclinical assessment. Hence, they have been selected as interesting candidates for drug leads and for extended preclinical and clinical trials thereafter. Examples of these interesting candidates include manzamine A (antimalarial, antituberculosis, anti-HIV), lasonolides (antifungal), jaspamides and geodiamolides (antiproliferative), and azumamides and psammaplin A (histone deacetylaste inhibitory activity) (Laport et al. 2009).

7.7 Troubleshooting to Develop Marine Pharmaceuticals

On the basis of the everlasting need to find new therapeutics to treat human diseases, especially those prone to becoming resistant to the currently used pharmaceuticals, collaborative research efforts between academia and pharmaceutical companies have been drawn toward marine-derived pharmaceuticals. This interest in marine natural products as possible pharmaceuticals has apparently developed as terrestrial drug lead resources proved to be unreliable in supplying either new entities or more derivatives even with extensive agricultural studies and exhaustive harvesting.

In spite of the increasing interest in marine natural product chemistry, there are numerous obstacles to the development of marine pharmaceuticals starting from the initial collection of the marine samples from their natural habitat, then the low yield of marine natural product which may be less than $10^{-6}\%$, and thereafter the large-scale production of marine compounds for clinical use. Some of these obstacles have been overcome while others remain unresolved. In this section, the main obstacles and their "troubleshooting" are discussed.

The first obstacle is collecting a marine specimen from its natural habitat and this has been overcome and improved since the advent of SCUBA techniques almost 65 years ago followed by the development of remote-controlled submersibles which can be directed toward previously unattainable depths allowing untouched marine macro- and microorganisms to be brought to the surface.

In addition to the collection problem, the vanishingly low yield of marine compounds from their natural sources has been a problem in terms of unambiguous identification of their chemical structures. To overcome this obstacle, new advances in analytical methodology including sophisticated and hyphenated techniques, such as NMR spectroscopy (up to 1,000 MHz), and mass spectroscopic methods coupled to chromatographic methods (HPLC-MS) to isolate and elucidate the chemical components from the marine organisms have been developed. These new technologies in analytical spectroscopy have pushed the limits of observation so that the discovery and structural elucidation of new entities require only a few micrograms – a fraction of the material that was required even 10 years ago.

The third challenge with regard to the development of marine pharmaceuticals is large-scale production to fulfill the supply needs for clinical use. To achieve a reasonable and steady supply of marine natural products, environmentally sound and economically feasible strategies are required.

Chemical synthesis is the first alternative that has been investigated. Unfortunately, there are not so many success stories due to the structural complexity of marine metabolites with novel mechanisms of action and high selectivity which has resulted in only a few successful examples such as ziconotide (Prialt®; Elan Pharmaceuticals) (Olivera 2000). A second alternative is to define the potential pharmacophores of marine natural products and then attempt to develop pharmaceuticals based on marine-derived pharmacophores via chemical synthesis, degradation, modification, or a combination of these. Many of the marine pharmaceuticals mentioned above under different clinical phases represent good examples of the chemical synthesis of potential pharmacophores of the natural marine products.

A third strategy seems to be interesting – if successfully performed – which is aquaculture of the source organisms, including sponges, tunicates, and bryozoans, with the aim of securing a sustainable supply of the active constituent(s). This method has progressed notably in the field of compounds for cancer therapy. However, there are many environmental factors that may affect the yield afforded through aquaculture which lead in many cases to either insufficient biomass or low yield of the active marine compound(s) (Mendola 2000). Furthermore, the cultivation of invertebrates in their natural environment is subject to several hazards and threats, such as destruction by storms or diseases.

An intriguing strategy has been to identify the real producers of bioactive compounds and to explore whether or not they are of microbial origin including bacteria, cyanobacteria, or fungi that live within the tissues of marine invertebrates. If bacterial or other associated microorganisms prove to produce the compounds of interest, the next step will be to carefully design special culture media for large-scale fermentation. A success story of this strategy is trabectedin (ET-743, Yondelis®; PharmaMar) production which is currently produced by a large-scale semisynthetic protocol starting with cyanosafracin B that can be produced in multikilogram scale by fermentation of *P. fluorescens*. Only 5% or less of the symbiotic bacteria present in marine specimens can be cultivated to date under standard conditions (Fenical 1993). Consequently, molecular approaches provide particularly promising alternatives through the transfer of biosynthetic gene clusters to a vector suitable for large-scale fermentation, thereby avoiding the obstacles in culturing symbiotic bacteria.

In the future, Oceans will play a potential role in treating, controlling and/or relieving the global disease burden. To achieve this, more research needs to be carried out by marine natural product chemists and pharmacologists alike to identify novel drug leads from marine resources with unique mechanisms of action.

Acknowledgement Continuous support by BMBF to P.P. is gratefully acknowledged. A predoctoral fellowship granted and financed by the Egyptian government (Ministry of High Education) to S.S.E. is acknowledged.

References

Absalon MJ, Smith FO (2009) Treatment strategies for pediatric acute myeloid leukemia. Exp Opin Pharmacother 10:57–79
Aicher TD, Buszek KR, Fang FG, Forsyth CJ, Jung SH, Kishi Y, Matelich MC, Scola PM, Spero DM, Yoon SK (1992) Total synthesis of halichondrin B and norhalichondrin B. J Am Chem Soc 114:3162–3164
Ajani JA, Jiang Y, Faust J, Chang BB, Ho L, Yao JC, Rousey S, Dakhil S, Cherny RC, Craig C, Bleyer A (2006) A multi-center phase II study of sequential paclitaxel and bryostatin 1 (NSC 339555) in patients with untreated, advanced gastric or gastroesophageal junction adenocarcinoma. Investig New Drugs 24:353–357
Amna T, Puri SC, Verma V, Sharma JP, Khajuria RK, Musarrat J, Spiteller M, Qazi GN (2006) Bioreactor studies on the endophytic fungus *Entrophospora infrequens* for the production of an anticancer alkaloid camptothecin. Can J Microbiol 52:189–196
Anderson HJ, Coleman JE, Andersen RJ, Roberge M (1997) Cytotoxic peptides hemiasterlin, hemiasterlin A and hemiasterlin B induce mitotic arrest and abnormal spindle formation. Cancer Chemother Pharmacol 39:223–226
Atkinson RA, Kieffer B, Dejaegere A, Sirockin F, Lefèvre JF (2000) Structural and dynamic characterization of ω-conotoxin MVIIA: the binding loop exhibits slow conformational exchange. Biochemistry 39:3908–3919
Aune GA, Furuta T, Pommier Y (2002) Ecteinascidin 743: a novel anticancer drug with a unique mechanism of action. Anticancer Drugs 13:545–555
Bai R, Pettit GR, Hamel E (1990a) Dolastatin 10, a powerful cytostatic peptide derived from a marine animal – inhibition of tubulin polymerization mediated through the vinca alkaloid binding domain. Biochem Pharmacol 39:1941–1949
Bai R, Pettit GR, Hamel E (1990b) Binding of dolastatin 10 to tubulin at a distinct site for peptide antimitotic agents near the exchangeable nucleotide and vinca alkaloid sites. J Biol Chem 265:17141–17149
Bai R, Paull KD, Herald CL, Malspeis L, Pettit GR, Hamel E (1991) Halichondrin B and homohalichondrin B, marine natural products binding in the vinca domain of tubulin. J Biol Chem 266:15882–15889
Bai R, Friedman SJ, Pettit GR, Hamel E (1992) Dolastatin 15, a potent antimitotic depsipeptide derived from *Dolabella auricularia* – interaction with tubulin and effects on cellular microtubules. Biochem Pharmacol 43:2637–2645
Bai R, Edler MC, Bonate PL, Copeland TD, Pettit GR, Luduena RF, Hamel E (2009) Intracellular activation and deactivation of tasidotin, an analog of dolastatin 15: correlation with cytotoxicity. Mol Pharmacol 75:218–226
Balzarini J, Kang GJ, Dalal M, Herdewijn P, De Clercq E, Broder S, Johns DG (1987) The anti-HTLV-III (anti-HIV) and cytotoxic activity of 2′, 3′-didehydro-2′, 3′-dideoxyribonucleosides: a comparison with their parental 2′, 3′-dideoxyribonucleosides. Mol Pharmacol 32:162–167

Barrow RA, Hemscheidt T, Liang J, Paik S, Moore RE, Tius MA (1995) Total synthesis of cryptophycins. Revision of the structures of cryptophycins A and C. J Am Chem Soc 117:2479–2490

Bergmann W, Feeney RJ (1950) The isolation of a new thymine pentoside from sponges. J Am Chem Soc 72:2809–2810

Bergmann W, Feeney RJ (1951) Contributions to the study of marine products. XXXII. The nucleosides of sponges. I. J Org Chem 16:981–987

Berkow RL, Schlabach L, Dodson R, Benjamin WH, Pettit GR, Rustagi P, Kraft AS (1993) In vivo administration of the anticancer agent bryostatin 1 activates platelets and neutrophils and modulates protein kinase C activity. Cancer Res 53:2810–2815

Bingham JP, Mitsunaga E, Bergeron ZL (2010) Drugs from slugs–past, present and future perspectives of ω-conotoxin research. Chem Biol Interact 183:1–18

Blunt JW, Copp BR, Munro MHG, Northcote PT, Prinsep MR (2010) Marine natural products. Nat Prod Rep 27:165–237

Bowersox SS, Luther R (1998) Pharmacotherapeutic potential of omega-conotoxin MVIIA (SNX-111), an N-type neuronal calcium channel blocker found in the venom of *Conus magus*. Toxicon 36:1651–1658

Bowersox SS, Gadbois T, Singh T, Pettus M, Wang YX, Luther RR (1996) Selective N-type neuronal voltage-sensitive calcium channel blocker, SNX-111, produces spinal antinociception in rat models of acute persistent and neuropathic pain. J Pharmacol Exp Ther 279:1243–1249

Buccafusco JJ, Letchworth SR, Bencherif M, Lippiello PM (2005) Long-lasting cognitive improvement with nicotinic receptor agonists: mechanisms of pharmacokinetic-pharmacodynamic discordance. Trends Pharmacol Sci 26:352–360

Buckingham SD, Jones AK, Brown LA, Sattelle DB (2009) Nicotinic acetylcholine receptor signalling: roles in Alzheimer's disease and amyloid neuroprotection. Pharmacol Rev 61:39–61

Butler MS (2004) The role of natural product chemistry in drug discovery. J Nat Prod 67:2141–2153

Cai B, Chen F, Ji Y, Kiss L, De Jonge WJ, Conejero-Goldberg C, Szabo C, Deitch EA, Ulloa L (2009) Alpha7 cholinergic-agonist prevents systemic inflammation and improves survival during resuscitation. J Cell Mol Med 13:3774–3785

Chauhan D, Catley L, Li G, Podar K, Hideshima T, Velankar M, Mitsiades C, Mitsiades N, Yasui H, Letai A, Ovaa H, Berkers C, Nicholson B, Chao TH, Neuteboom STC, Richardson P, Palladino MA, Anderson KC (2005) A novel orally active proteasome inhibitor induces apoptosis in multiple myeloma cells with mechanisms distinct from bortezomib. Cancer Cell 8:407–419

Chauhan C, Hideshima T, Anderson KC (2006) A novel proteasome inhibitor NPI-0052 as an anticancer therapy. Br J Cancer 95:961–965

Chauhan D, Singh A, Brahmandam M, Podar K, Hideshima T, Richardson P, Munshi N, Palladino MA, Anderson KC (2008) Combination of proteasome inhibitors bortezomib and NPI-0052 trigger in vivo synergistic cytotoxicity in multiple myeloma. Blood 111:1654–1664

Cimino G, De Rosa S, De Stefano S (1984) Antiviral agents from a gorgonian, *Eunicella cavolini*. Experientia 40:339–340

Clamp A, Jayson GC (2002) The clinical development of the bryostatins. Anticancer Drugs 13:673–683

Coleman JE, de Silva ED, Kong F, Andersen RJ (1995) Cytotoxic peptides from the marine sponge *Cymbastela* sp. Tetrahedron 51:10653–10662

Coleman JE, Patrick BO, Andersen RJ, Rettig SJ (1996) Hemiasterlin methyl ester. Acta Crystallogr Sect C C52:1525–1527

Cragg LH, Andreeff M, Feldman E, Roberts J, Murgo A, Winning M, Tombes MB, Roboz G, Kramer L, Grant S (2002) Phase I trial and correlative laboratory studies of bryostatin 1 (NSC 339555) and high-dose 1-β-D-arabinofuranosylcytosine in patients with refractory acute leukemia. Clin Cancer Res 8:2123–2133

Cuevas C, Pérez M, Martín MJ, Chicharro JL, Fernández-Rivas C, Flores M, Francesch A, Gallego P, Zarzuelo M, de la Calle F, García J, Polanco C, Rodríguez I, Manzanares I (2000) Synthesis of ecteinascidin ET-743 and phthalascidin Pt-650 from cyanosafracin B. Org Lett 2:2545–2548

Dabydeen DA, Burnett JC, Bai R, Verdier-Pinard P, Hickford SJH, Pettit GR, Blunt JW, Munro MHG, Gussio R, Hamel E (2006) Comparison of the activities of the truncated halichondrin B analog NSC 707389 (E7389) with those of the parent compound and a proposed binding site on tubulin. Mol Pharmacol 70:1866–1875

Davidson BS (1992) Renieramycin G, a new alkaloid from the sponge *Xestospongia caycedoi*. Tetrahedron Lett 33:3721–3724

De Clercq E (1986) Chemotherapeutic approaches to the treatment of the acquired immune deficiency syndrome (AIDS). J Med Chem 29:1561–1569

De Jonge MJA, van der Gaast A, Planting AST, van Doorn L, Lems A, Boot I, Wanders J, Satomi M, Verweij J (2005) Phase I and pharmacokinetic study of the dolastatin 10 analogue TZT-1027, given on days 1 and 8 of a 3-week cycle in patients with advanced solid tumors. Clin Cancer Res 11:3806–3813

Dermatakis A (2002) ATP-competitive inhibitors of cyclin-dependent kinases. Front Biotech Pharm 3:125–156

Donia M, Hamann MT (2003) Marine natural products and their potential applications as anti-infective agents. Lancet Infect Dis 3:338–348

Dowlati A, Lazarus HM, Hartman P, Jacobberger JW, Whitacre C, Gerson SL, Ksenich P, Cooper BW, Frisa PS, Gottlieb M, Murgo AJ, Remick SC (2003) Phase I and correlative study of combination bryostatin 1 and vincristine in relapsed B-cell malignancies. Clin Cancer Res 9:5929–5935

Duque C, Puyana M, Narváez G, Osorno O, Hara N, Fujimoto Y (2004) Pseudopterosins P–V, new compounds from the gorgonian octocoral *Pseudopterogorgia elisabethae* from Providencia island, Colombian Caribbean. Tetrahedron 60:10627–10635

Edelman MJ, Gandara DR, Hausner P, Israel V, Thornton D, DeSanto J, Doyle LA (2003) Phase 2 study of cryptophycin 52 (LY355703) in patients previously treated with platinum based chemotherapy for advanced non-small cell lung cancer. Lung Cancer 39:197–199

Erba E, Bergamaschi D, Bassano L, Damia G, Ronzoni S, Faircloth GT, D'Incalci M (2001) Ecteinascidin-743 (ET-743), a natural marine compound, with a unique mechanism of action. Eur J Cancer 37:97–105

Ettouati WS, Jacobs RS (1987) Effect of pseudopterosin A on cell division, cell cycle progression, DNA, and protein synthesis in cultured sea urchin embryos. Mol Pharmacol 31:500–505

Feling RH, Buchanan GO, Mincer TJ, Kauffman CA, Jensen PR, Fenical W (2003) Salinosporamide A: a highly cytotoxic proteasome inhibitor from a novel microbial source, a marine bacterium of the new genus *Salinospora*. Angew Chem Int Ed 42:355–357

Fenical W (1993) Chemical studies of marine bacteria: developing a new resource. Chem Rev 93:1673–1683

Fenical W, Jensen PR, Palladino MA, Lam KS, Lloyd GK, Potts BC (2009) Discovery and development of the anticancer agent salinosporamide A (NPI-0052). Bioorg Med Chem 17:2175–2180

Fontana A, Cavaliere P, Wahidulla S, Naik CG, Cimino G (2000) A new antitumor isoquinoline alkaloid from the marine nudibranch *Jorunna funebris*. Tetrahedron 56:7305–7308

Freedman R, Olincy A, Buchanan RW, Harris JG, Gold JM, Johnson L, Allensworth D, Guzman-Bonilla A, Clement B, Ball MP, Kutnick J, Pender V, Martin LF, Stevens KE, Wagner BD, Zerbe GO, Soti F, Kem WR (2008) Initial Phase 2 trial of a nicotinic agonist in schizophrenia. Am J Psychiatry 165:1040–1047

Frincke JM, Faulkner DJ (1982) Antimicrobial metabolites of the sponge *Reniera* sp. J Am Chem Soc 104:265–269

Gamble WR, Durso NA, Fuller RW, Westergaard CK, Johnson TR, Sackett DL, Hamel E, Cardellina JH II, Boyd MR (1999) Cytotoxic and tubulin-interactive hemiasterlins from *Auletta* sp. and *Siphonochalina* sp. sponges. Bioorg Med Chem 7:1611–1615

Gaur S, Newcomb R, Rivnay B, Bell JR, Yamashiro D, Ramachandran J, Miljanich GP (1994) Calcium channel antagonist peptides define several components of transmitter release in the hippocampus. Neuropharmacology 33:1211–1219

Goel S, Mita AC, Mita M, Rowinsky EK, Chu QS, Wong N, Desjardins C, Fang F, Jansen M, Shuster DE, Mani S, Takimoto CH (2009) A Phase I study of eribulin mesylate (E7389), a mechanistically novel inhibitor of microtubule dynamics, in patients with advanced solid malignancies. Clin Cancer Res 15:4207–4212

Goetz G, Yoshida WY, Scheuer PJ (1999) The absolute stereochemistry of kahalalide F. Tetrahedron 55:7739–7746

Gohil K, Bell JR, Ramachandran J, Miljanich GP (1994) Neuroanatomical distribution of receptors for a novel voltage-sensitive calcium-channel antagonist, SNX-230 (ω-conopeptide MVIIC). Brain Res 653:258–266

Grant S, Pettit GR, McCrady C (1992) Effect of bryostatin 1 on the *in vitro* radioprotective capacity of recombinant granulocyte macrophage colony-stimulating factor (Rgm-Csf) toward committed human myeloid progenitor cells (Cfu-Gm). Exp Hematol 20:34–42

Groll M, Huber R, Potts BCM (2006) Crystal structures of salinosporamide A (NPI-0052) and B (NPI-0047) in complex with the 20S proteasome reveal important consequences of β-lactone ring opening and a mechanism for irreversible binding. J Am Chem Soc 128:5136–5141

Hadaschik BA, Ettinger S, Sowery RD, Zoubeidi A, Andersen RJ, Roberge M, Gleave ME (2008) Targeting prostate cancer with HTI-286, a synthetic analogue of the marine sponge product hemiasterlin. Int J Cancer 122:2368–2376

Harvey AL (2008) Natural products in drug discovery. Drug Discov Today 13:894–901

He HY, Faulkner DJ (1989) Renieramycins E and F from the sponge *Reniera* sp.: reassignment of stereochemistry of the renieramycins. J Org Chem 54:5822–5824

Hennings H, Blumberg PM, Pettit GR, Herald CL, Shores R, Yuspa SH (1987) Bryostatin 1, an activator of protein kinase C, inhibits tumor promotion by phorbol esters in SENCAR mouse skin. Carcinogenesis 8:1343–1346

Herrero AB, Martin-Castellanos C, Marco E, Gago F, Moreno S (2006) Cross-talk between nucleotide excision and homologous recombination DNA repair pathways in the mechanism of action of antitumor trabectedin. Cancer Res 66:8155–8162

Hickman PF, Kemp GJ, Thompson CH, Salisbury AJ, Wade K, Harris AL, Radda GK (1995) Bryostatin 1, a novel antineoplastic agent and protein kinase C activator, induces human myalgia and muscle metabolic defects: a ^{31}P magnetic resonance spectroscopic study. Br J Cancer 72:998–1003

Hirata Y, Uemura D (1986) Halichondrins–antitumor polyether macrolides from a marine sponge. Pure Appl Chem 58:701–710

Hu ZB, Ma WL, Uphoff CC, Lanotte M, Drexler HG (1993) Modulation of gene-expression in the acute promyelocytic leukemia-cell line Nb4. Leukemia 7:1817–1823

Isakov N, Galron D, Mustelin T, Pettit GR, Altman A (1993) Inhibition of phorbol ester-induced T cell proliferation by bryostatin is associated with rapid degradation of protein kinase C. J Immunol 150:1195–1204

Jones RM, Cartier GE, McIntosh JM, Bulaj G, Farrar VE, Olivera BM (2001) Composition and therapeutic utility of conotoxins from genus *Conus*. Patent status 1996–2000. Exp Opin Ther Patents 11:603–623

Jordan MA, Kamath K, Manna T, Okouneva T, Miller HP, Davis C, Littlefield BA, Wilson L (2005) The primary antimitotic mechanism of action of the synthetic halichondrin E7389 is suppression of microtubule growth. Mol Cancer Ther 4:1086–1095

Kanoh K, Kohno S, Asari T, Harada T, Katada J, Muramatsu M, Kawashima H, Sekiya H, Uno I (1997) (−)Phenylahistin: A new mammalian cell cycle inhibitor produced by *Aspergillus ustus*. Bioorg Med Chem Lett 7:2847–2852

Keck GE, Poudel YB, Rudra A, Stephens JC, Kedei N, Lewin NE, Peach ML, Blumberg PM (2010) Molecular modeling, total synthesis, and biological evaluations of C-9 deoxy bryostatin 1. Angew Chem Int Ed 49:4580–4584

Kem WR, Abbott BC, Coates RM (1971) Isolation and structure of a hoplonemertine toxin. Toxicon 9:15–22

Kem WR, Mahnir VM, Prokai L, Papke RL, Cao X, LeFrancois S, Wildeboer K, Prokai-Tatrai K, Porter-Papke J, Soti F (2004) Hydroxy metabolites of the Alzheimer's drug candidate 3-[(2, 4-dimethoxy)benzylidene]-anabaseine dihydrochloride (GTS-21): Their molecular properties, interactions with brain nicotinic receptors, and brain penetration. Mol Pharmacol 65:56–67

Kem W, Soti F, Wildeboer K, LeFrancois S, MacDougall K, Wei DQ, Chou KC, Arias HR (2006) The nemertine toxin anabaseine and its deriative DMXBA (GTS-21): chemical and pharmacological properties. Mar Drugs 4:255–273

Kerksiek K, Mejillano MR, Schwartz RE, Georg GI, Himes RH (1995) Interaction of cryptophycin 1 with tubulin and microtubules. FEBS Lett 377:59–61

Kerr LM, Yoshikami D (1984) A venom peptide with a novel presynaptic blocking action. Nature 308:282–284

Kim JI, Takahashi M, Ohtake A, Wakamiya A, Sato K (1995) Tyr13 is essential for the activity of ω-conotoxin MVIIA and GVIA, specific N-type calcium channel blockers. Biochem Biophys Res Commun 206:449–454

Kitagawa H, Takenouchi T, Azuma R, Wesnes KA, Kramer WG, Clody DE, Burnett AL (2003) Safety, pharmacokinetics, and effects on cognitive function of multiple doses of GTS-21 in healthy, male volunteers. Neuropsychopharmacology 28:542–551

Kobayashi M, Aoki S, Ohyabu N, Kurosu M, Wang W, Kitagawa I (1994) Arenastatin A, a potent cytotoxic depsipeptide from the Okinawan marine sponge *Dysidea arenaria*. Tetrahedron Lett 35:7969–7972

Koehn FE, Carter GT (2005) The evolving role of natural products in drug discovery. Nat Rev Drug Discov 4:206–220

Komori T, Sanechika Y, Ito Y, Matsuo J, Nohara T, Kawasaki T, Schulten HR (1980) Biologically active glycosides from Asteroidea. I. Structures of a new cerebroside mixture and of two nucleosides from the starfish Acanthaster planci. Leibigs Ann Chem 1980:653–668

Kristipati R, Nádasdi L, Tarczy-Hornoch K, Lau K, Miljanich GP, Ramachandran J, Bell JR (1994) Characterization of the binding of omega-conopeptides to different classes of non-L-type neuronal calcium channels. Mol Cell Neurosci 5:219–228

Kucuk O, Young ML, Habermann TM, Wolf BC, Jimeno J, Cassileth PA (2000) Phase II trial of didemnin B in previously treated non-Hodgkin's lymphoma: an Eastern cooperative oncology group (ECOG) study. Am J Clin Oncol 23:273–277

Kumaran RS, Muthumary J, Hur BK (2008) Taxol from *Phyllosticta citricarpa*, a leaf spot fungus of the angiosperm *Citrus medica*. J Biosci Bioeng 106:103–106

Kuznetsov G, Towle MJ, Cheng H, Kawamura T, TenDyke K, Liu D, Kishi Y, Yu MJ, Littlefield BA (2004) Induction of morphological and biochemical apoptosis following prolonged mitotic blockage by halichondrin B macrocyclic ketone analog E7389. Cancer Res 64:5760–5766

Laport MS, Santos OCS, Muricy G (2009) Marine sponges: potential sources of new antimicrobial drugs. Curr Pharm Biotechnol 10:86–105

Le Tourneau C, Raymond E, Faivre S (2007) Aplidine: a paradigm of how to handle the activity and toxicity of a novel marine anticancer poison. Curr Pharm Des 13:3427–3439

Leal JFM, García-Hernández V, Moneo V, Domingo A, Bueren-Calabuig JA, Negri A, Gago F, Guillén-Navarro MJ, Avilés P, Cuevas C, García-Fernández LF, Galmarini CM (2009) Molecular pharmacology and antitumor activity of *Zalypsis*® in several human cancer cell lines. Biochem Pharmacol 78:162–170

Lee WW, Benitez A, Goodman L, Baker BR (1960) Potential anticancer agents. XL. Synthesis of the β-anomer of 9-(D-arabinofuranosyl)-adenine. J Am Chem Soc 82:2648–2649

Liang J, Moore RE, Moher ED, Munroe JE, Al-awar RS, Hay DA, Varie DL, Zhang TY, Aikins JA, Martinelli MJ, Shih C, Ray JE, Gibson LL, Vasudevan V, Polin L, White K, Kushner J, Simpson C, Pugh S, Corbett TH (2005) Cryptophycins-309, 249 and other cryptophycin analogs: preclinical efficacy studies with mouse and human tumors. Investig New Drugs 23:213–224

Ling YH, Aracil M, Jimeno J, Perez-Soler R, Zou Y (2009) Molecular pharmacodynamics of PM02734 (elisidepsin) as single agent and in combination with erlotinib; synergisitic activity in human non-small cell lung cancer cell lines and xenograft models. Eur J Cancer 45:1855–1864

Lippert JW III (2007) Vascular disrupting agents. Bioorg Med Chem 15:605–615

Loganzo F, Discafani CM, Annable T, Beyer C, Musto S, Hari M, Tan X, Hardy C, Hernandez R, Baxter M, Singanallore T, Khafizova G, Poruchynsky MS, Fojo T, Nieman JA, Ayral-Kaloustian S, Zask A, Andersen RJ, Greenberger LM (2003) Antimirotubule agent that circumvents P-glycoprotein-mediated resistance in vitro and in vivo. Cancer Res 63: 1838–1845

Look SA, Fenical W, Matsumoto GK, Clardy J (1986a) The pseudopterosins: a new class of anti-inflammatory and analgesic diterpene pentosides from the marine sea whip *Pseudopterogorgia elisabethae* (Octocorallia). J Org Chem 51:5140–5145

Look SA, Fenical W, Jacobs RS, Clardy J (1986b) The pseudopterosins: anti-inflammatory and analgesic natural products from the sea whip *Pseudopterogorgia elisabethae*. Proc Natl Acad Sci USA 83:6238–6240

López-Macià À, Jiménez JC, Royo M, Giralt E, Albericio F (2001) Synthesis and structure determination of kahalalide F. J Am Chem Soc 123:11398–11401

Lu K, Dempsey J, Schultz RM, Shih C, Teicher BA (2001) Cryptophycin-induced hyperphosphorylation of Bcl-2, cell cycle arrest and growth inhibition in human H460 NSCLC cells. Cancer Chemother Pharmacol 47:170–178

Mavromoustakos T, Calogeropoulou T, Koufaki M, Kolocouris A, Daliani I, Demetzos C, Meng Z, Makriyannis A, Balzarini J, De Clercq E (2001) Ether phospholipid-AZT conjugates possessing anti-HIV and antitumor cell activity. Synthesis, conformational analysis, and study of their thermal effects on membrane bilayers. J Med Chem 44:1702–1709

Mayer AMS (1999) Marine pharmacology in 1998: antitumor and cytotoxic compounds. The Pharmacologist 41:159–164

Mayer AMS, Gustafson KR (2003) Marine pharmacology in 2000: antitumor and cytotoxic compounds. Int J Cancer 105:291–299

Mayer AMS, Gustafson KR (2004) Marine pharmacology in 2001–2: antitumor and cytotoxic compounds. Eur J Cancer 40:2676–2704

Mayer AMS, Gustafson KR (2006) Marine pharmacology in 2003–2004: anti-tumor and cytotoxic compounds. Eur J Cancer 42:2241–2270

Mayer AMS, Gustafson KR (2008) Marine pharmacology in 2005–2006: antitumor and cytotoxic compounds. Eur J Cancer 44:2357–2387

Mayer AMS, Lehmann VKB (2001) Marine pharmacology in 1999: antitumor and cytotoxic compounds. Anticancer Res 21:2489–2500

Mayer AMS, Jacobson PB, Fenical W, Jacobs RS, Glaser KB (1998) Phamracological characterization of the pseudopterosins: novel anti-inflammatory natural products isolated from the Caribbean soft coral, *Pseudopterogorgia elisabethae*. Life Sci 62:PL401–PL407

Mayer AMS, Rodríguez AD, Berlinck RGS, Hamann MT (2009) Marine pharmacology in 2005–6: Marine compounds with anthelmintic, antibacterial, anticoagulant, antifungal, anti-inflammatory, antimalarial, antiprotozoal, antituberculosis, and antiviral activities; affecting the cardiovascular, immune and nervous systems, and other miscellaneous mechanisms of action. Biochim Biophys Acta 1790:283–402

Mayer AMS, Glaser KB, Cuevas C, Jacobs RS, Kem W, Little RD, McIntosh JM, Newman DJ, Potts BC, Shuster DE (2010) The odyssey of marine pharmaceuticals: a current pipeline perspective. Trends Pharmacol Sci 31:255–265

McGivern JG (2006) Targeting N-type and T-type calcium channels for the treatment of pain. Drug Discov Today 11:245–253

Mendola D (2000) Aquacultural production of bryostatin 1 and ecteinascidin 743. In: Fusetani N (ed) Drugs from the Sea. Basel, Switzerland

Mitsiades CS, Ocio EM, Pandiella A, Maiso P, Gajate C, Garayoa M, Vilanova D, Montero JC, Mitsiades N, McMullan CJ, Munshi NC, Hideshima T, Chauhan D, Aviles P, Otero G, Faircloth G, Mateos MV, Richardson PG, Mollinedo F, San-Miguel JF, Anderson KC (2008) Aplidin, a marine organism-derived compound with potent antimyeloma activity in vitro and in vivo. Cancer Res 68:5216–5225

Mitsuya H, Yarchoan R, Broder S (1990) Molecular targets for AIDS therapy. Science 249:1533–1544

Molinski TF, Dalisay DS, Lievens SL, Saludes JP (2009) Drug development from marine natural products. Nat Rev Drug Discov 8:69–86

Montesinos MC, Gadangi P, Longaker M, Sung J, Levine J, Nilsen D, Reibman J, Li M, Jiang CK, Hirschhorn R, Recht PA, Ostad E, Levin RI, Cronstein BN (1997) Wound healing is accelerated by agonists of Adenosine A_2 ($G_{\alpha s}$-linked) receptors. J Exp Med 186:1615–1620

Mooberry SL, Taoka CR, Busquets L (1996) Cryptophycin 1 binds to tubulin at a site distinct from the colchicine binding site and at a site that may overlap the vinca binding site. Cancer Lett 107:53–57

Moya CE, Jacobs RS (2006) Pseudopterosin A inhibits phagocytosis and alters intracellular calcium turnover in a pertussis toxin sensitive site in *Tetrahymena thermophila*. Comp Biochem Physiol B Toxicol Pharmacol 143:436–443

Munro MHG, Blunt JW, Dumdei EJ, Hickford SJH, Lill RE, Li S, Battershill CN, Duckworth AR (1999) The discovery and development of marine compounds with pharmaceutical potential. J Biotechnol 70:15–25

Mutter R, Wills M (2000) Chemistry and clinical biology of the bryostatins. Bioorg Med Chem 8:1841–1860

Newman DJ, Cragg GM (2004) Advanced preclinical and clinical trials of natural products and related compounds from marine sources. Curr Med Chem 11:1693–1713

Newman DJ, Cragg GM (2007) Natural products as sources of new drugs over the last 25 years. J Nat Prod 70:461–477

Nguyen TNT, Tepe JJ (2009) Preparation of hymenaldisine, analogues and their evaluation as kinase inhibitors. Curr Med Chem 16:3122–3143

Nicholson B, Lloyd GK, Miller BR, Palladino MA, Kiso Y, Hayashi Y, Neuteboom STC (2006) NPI-2358 is a tubulin-depolymerizing agent: in-vitro evidence for activity as a tumor vascular-disrupting agent. Anticancer Drugs 17:25–31

Nieman JA, Coleman JE, Wallace DJ, Piers E, Lim LY, Roberge M, Andersen RJ (2003) Synthesis and antimitotic/cytotoxic activity of hemiasterlin analogues. J Nat Prod 66:183–199

Nuijen B, Bouma M, Manada C, Jimeno JM, Schellens JHM, Bult A, Beijnen JH (2000) Pharmaceutical development of anticancer drugs derived from marine sources. Anticancer Drugs 11:793–811

Okouneva T, Azarenko O, Wilson L, Littlefield BA, Jordan MA (2008) Inhibition of centromere dynamics by eribulin (E7389) during mitotic metaphase. Mol Cancer Ther 7:2003–2011

Olincy A, Stevens KE (2007) Treating schizophrenia symptoms with an α7 nicotinic agonist, from mice to men. Biochem Pharmacol 74:1192–1201

Olincy A, Harris JG, Johnson LL, Pender V, Kongs S, Allensworth D, Ellis J, Zerbe GO, Leonard S, Stevens KE, Stevens JO, Martin L, Adler LE, Soti F, Kem WR, Freedman R (2006) Proof-of-concept trial of an α7 nicotinic agonist in schizophrenia. Arch Gen Psychiatry 63:630–638

Olivera BM (2000) ω-Conotoxin MVIIA: from marine snail venom to analgesic drug. In: Fusetani N (ed) Drugs from the Sea. Basel, Switzerland

Olivera BM, Gray WR, Zeikus R, McIntosh JM, Varga J, Rivier J, de Santos V, Cruz LJ (1985) Peptide neurotoxins from fish-hunting cone snails. Science 230:1338–1343

Olivera BM, Cruz LJ, de Santos V, LeCheminant GW, Griffin D, Zeikus R, McIntosh JM, Galyean R, Varga J, Gray WR, Rivier J (1987) Neuronal calcium channel antagonists. Discrimination between calcium channel subtypes using ω-conotoxin from Conus magus venom. Biochemistry 26:2086–2090

Olivera BM, Miljanich GP, Ramachandran J, Adams ME (1994) Diversity and neurotransmitter release: the ω-conotoxins and ω-agatoxins. Annu Rev Biochem 63:823–867

Panda D, Himes RH, Moore RE, Wilson L, Jordan MA (1997) Mechanism of action of the unusally potent microtubule inhibitor cryptophycin 1. Biochemistry 36:12948–12953

Parang K, Wiebe LI, Knaus EE (2000) Novel approaches for designing 5′-O-ester prodrugs of 3′-azido-2′, 3′-dideoxythymidine (AZT). Curr Med Chem 7:995–1039

Pavlov VA, Ochani M, Yang LH, Gallowitsch-Puerta M, Ochani K, Lin X, Levi J, Parrish WR, Rosas-Ballina M, Czura CJ, LaRosa GJ, Miller EJ, Tracey KJ, Yousef Al-Abed (2007) Selective [alpha]7-nicotinic acetylcholine receptor agonist GTS-21 improves survival in murine endotoxemia and severe sepsis. Crit Care Med 35:1139–1144

Pettit GR (1996) Progress in the discovery of biosynthetic anticancer drugs. J Nat Prod 59:812–821

Pettit GR (1997) The dolastatins. Prog Chem Org Nat prod 70:1–79

Pettit GR, Herald CL, Doubek DL, Herald DL, Arnold E, Clardy J (1982) Isolation and structure of bryostatin 1. J Am Chem Soc 104:6846–6848

Pettit GR, Herald CL, Boyd MR, Leet JE, Dufresne C, Doubek DL, Schmidt JM, Cerny RL, Hooper JNA, Rutzler KC (1991) Antineoplastic agents. 219. Isolation and structure of the cell growth inhibitory constituents from the Western Pacific marine sponge *Axinella* sp. J Med Chem 34:3339–3340

Pommier Y, Kohlhagen G, Bailly C, Waring M, Mazumder A, Kohn KW (1996) DNA sequence- and structure-selective alkylation of guanine N2 in the DNA minor groove by ecteinascidin 743, a potent antitumor compound from the Caribbean tunicate *Ecteinascidia turbinata*. Biochemistry 35:13303–13309

Pomponi SA (1999) The bioprocess-technological potential of the sea. J Biotechnol 70:5–13

Price-Carter M, Hull MS, Goldenberg DP (1998) Roles of individual disulfide bonds in the stability and folding of a ω-conotoxin. Biochemistry 37:9851–9861

Ratain MJ, Undevia S, Janisch L, Roman S, Mayer P, Buckwalter M, Foss D, Hamilton BL, Fischer J, Bukowski RM (2003) Phase 1 and pharmacological study of HTI-286, a novel antimicrotubule agent: correlation of neutropenia with time above a threshold serum concentration. Proc Am Soc Clin Oncol 22:516

Rauck RL, Wallace MS, Burton AW, Kapural L, North JM (2009) Intrathecal ziconotide for neuropathic pain: a review. Pain Pract 9:327–337

Ray A, Okouneva T, Manna T, Miller HP, Schmid S, Arthaud L, Luduena R, Jordan MA, Wilson L (2007) Mechanism of action of the microtubule-targeted antimitotic depsipeptide tasidotin (formerly ILX651) and its major metabolite tasidotin C-carboxylate. Cancer Res 67:3767–3776

Rinehart KL, Lithgow-Bertelloni AM (1991) Dehydrodidemnin B. WO9104985

Rinehart KL, Gloer JB, Cook JC, Mizsak SA, Scahill TA (1981a) Structures of the didemnins, antiviral and cytotoxic depsipeptides from a Caribbean tunicate. J Am Chem Soc 103:1857–1859

Rinehart KL, Gloer JB, Hughes RG, Renis HE, McGovren JP, Swynenberg EB, Stringfellow DA, Kuentzel SL, Li LH (1981b) Didemnins: antiviral and antitumor depsipeptides from a Caribbean tunicate. Science 212:933–935

Rinehart KL, Holt TG, Fregeau NL, Stroh JG, Keifer PA, Sun F, Li LH, Martin DG (1990) Ecteinascidins 729, 743, 745, 759A, 759B, and 770: Potent antitumor agents from the Caribbean tunicate *Ecteinascidia turbinata*. J Org Chem 55:4512–4515

Rinehart KL, Holt TG, Fregeau NL, Stroh JG, Keifer PA, Sun F, Li LH, Martin DG (1991) Ecteinascidins 729, 743, 745, 759A, 759B, and 770: potent antitumor agents from the Caribbean tunicate *Ecteinascidia turbinata*. J Org Chem 56:1676 (Erratum)

Rosas-Ballina M, Goldstein RS, Gallowitsch-Puerta M, Yang L, Valdés-Ferrer SI, Patel NB, Chavan S, Yousef Al-Abed, Yang H, Tracey KJ (2009) The selective α7 agonist GTS-21 attenuates cytokine production in human whole blood and human monocytes activated by ligands for TLR2, TLR3, TLR4, TLR9, and RAGE. Mol Med 15:195–202

Rouhi MA (1995) Supply issues complicate track of chemicals from sea to market. Chem Eng News 73:42–44
Roussis V, Wu Z, Fenical W, Strobel SA, van Duyne GD, Clardy J (1990) New anti-inflammatory pseudopterosins from the marine octocoral *Pseudopterogorgia elisabethae*. J Org Chem 55:4916–4922
Sakai R, Rinehart KL, Guan Y, Wang AHJ (1992) Additional antitumor ecteinascidins from a Caribbean tunicate: crystal structures and activities *in vivo*. Proc Natl Acad USA 89:11456–11460
Schaufelberger DE, Koleck MP, Beutler JA, Vatakis AM, Alvarado AB, Andrews P, Marzo LV, Muschik GM, Roach J, Ross JT, Lebherz WB, Reeves MP, Eberwein RM, Rodgers LL, Testerman RP, Snader KM, Forenza S (1991) The large-scale isolation of bryostatin 1 from *Bugula neritina* following current good manufacturing practices. J Nat Prod 54:1265–1270
Scheuer PJ, Hamann MT (1993) Kahalalide F: a bioactive depsipeptide from the sacoglossan mollusk *Elysia rufescens* and the green alga *Bryopsis* sp. J Am Chem Soc 115:5825–5826
Schinazi RF, Chu CK, Peck A, McMillan A, Mathis R, Cannon D, Jeong LS, Beach JW, Choi WB, Yeola S, Liotta DC (1992) Activities of the four optical isomers of 2′, 3′-dideoxy-3′-thiacytidine (BCH-189) against human immunodeficiency virus type 1 in human lymphocytes. Antimicrob Agents Chemother 36:672–676
Schöffski P, Thate B, Beutel G, Bolte O, Otto D, Hofmann M, Ganser A, Jenner A, Cheverton P, Wanders J, Oguma T, Atsumi R, Satomi M (2004) Phase I and pharmacokinetic study of TZT-1027, a novel synthetic dolastatin 10 derivative, administered as a 1-hour intravenous infusion every 3 weeks in patients with advanced refractory cancer. Ann Oncol 15:671–679
Schwartz RE, Hirsch CF, Sesin DF, Flor JE, Chartrain M, Fromtling RE, Harris GH, Salvatore MJ, Liesch JM, Yudin K (1990) Pharmaceuticals from cultured algae. J Ind Microbiol 5:113–124
Scott JD, Williams RM (2002) Chemistry and biology of the tetrahydroisoquinoline antitumor antibiotics. Chem Rev 102:1669–1730
Sessa C, Weigang-Köhler K, Pagani O, Greim G, Mor O, De Pas T, Burgess M, Weimer I, Johnson R (2002) Phase I and pharmacological studies of the cryptophycin analogue LY355703 administered on a single intermittent or weekly schedule. Eur J Cancer 38:2388–2396
Sessa C, Perotti A, Noberasco C, De Braud F, Gallerani E, Cresta S, Zucchetti M, Viganò L, Locatelli A, Jimeno J, Feilchenfeldt JW, D'Incalci M, Capri G, Ielmini N, Gianni L (2009) Phase I clinical and pharmacokinetic study of trabectedin and doxorubicin in advanced soft tissue sarcoma and breast cancer. Eur J Caner 45:1153–1161
Shimohama S (2009) Nicotinic receptor-mediated neuroprotection in neurodegenerative disease models. Biol Pharm Bull 32:332–336
Singh R, Sharma M, Joshi P, Rawat DS (2008) Clinical status of anti-cancer agents derived from marine sources. Anti-Cancer Agents Med Chem 8:603–617
Smith CD, Zhang X (1996) Mechanism of action of cryptophycin – interaction with the *vinca* alkaloid domain of tubulin. J Biol Chem 271:6192–6198
Soares DG, Escargueil AE, Poindessous V, Sarasin A, de Gramont A, Bonatto D, Henriques JAP, Larsen AK (2007) Replication and homologous recombination repair regulate DNA double-strand break formation by the antitumor alkylator ecteinascidin 743. Proc Natl Acad Sci USA 10:13062–13067
Staats PS, Yearwood T, Charapata SG, Presley RW, Wallace MS, Byas-Smith M, Fisher R, Bryce DA, Mangieri EA, Luther RR, Mayo M, McGuire D, Ellis D (2004) Intrathecal ziconotide in the treatment of refractory pain in patients with cancer or AIDS. A randomized controlled trial. J Am Med Assoc 291:63–70
Stamos DP, Chen SS, Kishi Y (1997) New synthetic route to the C.14–C.38 segment of halichondrins. J Org Chem 62:7552–7553
Staniek A, Woerdenbag HJ, Kayser O (2009) *Taxomyces andreane*: a presumed paclitaxel producer demystified? Planta Med 75:1561–1566
Stevenson JP, Sun W, Gallagher M, Johnson R, Vaughn D, Schuchter L, Algazy K, Hahn S, Enas N, Ellis D, Thornton D, O'Dwyer PJ (2002) Phase I trial of the cryptophycin analogue

LY355703 administered as an intravenous infusion on a day 1 and 8 schedule every 21 days. Clin Cancer Res 8:2524–2529

Stierle A, Strobel GA, Stierle D (1993) Taxol and taxane production by *Taxomyces andreane*, an endophytic fungus of Pacific yew. Science 260:214–216

Suárez Y, González L, Cuadrado A, Berciano M, Lafarga M, Muñoz A (2003) Kahalalide F, a new marine-derived compound, induces oncosis in human prostate and breast cancer cells. Mol Cancer Ther 2:863–872

Suhadolnik RJ, Pornbanlualap S, Wu JM, Baker DC, Hebbler AK (1989) Biosynthesis of 9-β-D-arabinofuranosyladenine: hydrogen exchange at C-2$'$ and oxygen exchange at C-3$'$ of adenosine. Arch Biochem Biophys 270:363–373

Sultan P, Shawl AS, Abdellah AA, Ramteke PW (2010) Isolation, characterization and comparative study on podophyllotoxin and related glycosides of *Podophyllum hexandrum*. Curr Res J Biol Sci 2:345–351

Takebayashi Y, Pourquier P, Zimonjic DB, Nakayama K, Emmert S, Ueda T, Urasaki Y, Kanzaki A, Akiyama SI, Popescu N, Kraemer KH, Pommier Y (2001) Antiproliferative activity of ecteinascidin 743 is dependent upon transcription-coupled nucleotide-excision repair. Nat Med 7:961–966

Talpir R, Benayahu Y, Kashman Y, Pannell L, Schleyer M (1994) Hemiasterlin and geodiamolide TA; two new cytotoxic peptides from the marine sponge *Hemiasterella minor* (Kirkpatrick). Tetrahedron Lett 35:4453–4456

Tan AR, Rubin EH, Walton DC, Shuster DE, Wong YN, Fang F, Ashworth S, Rosen LS (2009) Phase I study of eribulin mesylate administered once every 21 days in patients with advanced solid tumors. Clin Cancer Res 15:4213–4219

Thinschmidt JS, Frazier CJ, King MA, Meyer EM, Papke RL (2005) Septal innervation regulates the function of α7 nicotinic receptors in CA1 hippocampal interneurons. Exp Neurol 195:342–352

Thomas X (2009) Chemotherapy of acute leukemia in adults. Exp Opin Pharmacother 10:221–237

Towle MJ, Salvato KA, Budrow J, Wels BF, Kuznetsov G, Aalfs KK, Welsh S, Zheng W, Seletsky BM, Palme MH, Habgood GJ, Singer LA, DiPietro LV, Wang Y, Chen JJ, Quincy DA, Davis A, Yoshimatsu K, Kishi Y, Yu MJ, Littlefield BA (2001) In vitro and in vivo anticancer activities of synthetic macrocyclic ketone analogues of halichondrin B. Cancer Res 61:1013–1021

Trimurtulu G, Ohtani I, Patterson GML, Moore RE, Corbett TH, Valeriote FA, Demchik L (1994) Total sources of cryptophycins, potent antitumor depsipeptides from the blue-green alga *Nostoc* sp. strain GSV 224. J Am Chem Soc 116:4729–4737

Uemura D, Takahashi K, Yamamoto T (1985) Norhalichondrin A: an antitumor polyether macrolide from a marine sponge. J Am Chem Soc 107:4796–4798

Vahdat LT, Pruitt B, Fabian CJ, Rivera RR, Smith DA, Tan-Chiu E, Wright J, Tan AR, DaCosta NA, Chuang E, Smith J, O'Shaughnessy J, Shuster DE, Meneses NL, Chandrawansa K, Fang F, Cole PE, Ashworth S, Blum JL (2009) Phase II study of eribulin mesylate, a halichondrin B analog, in patients with metastatic breast cancer previously treated with an anthracycline and a taxane. J Clin Oncol 27:2954–2961

Varmus H (1988) Retroviruses. Science 240:1427–1435

Verweij J (2009) Soft tissue sarcoma trials: one size no longer fits all. J Clin Oncol 27:3085–3087

Wagner MM, Paul DC, Shih C, Jordan MA, Wilson L, Williams DC (1999) *In vitro* pharmacology of cryptophycin 52 (LY355703) in human tumor cell lines. Cancer Chemother Pharmacol 43:115–125

Wall ME, Wani MC, Cook CE, Palmer KH, McPhill AT, Sim GA (1966) Plant antitumor agents. I. The isolation and structure of camptothecin, a novel alkaloid leukemia and tumor inhibitor from *Camptotheca acuminata*. J Am Chem Soc 88:3888–3890

Wang YX, Pettus M, Gao D, Phillips C, Bowersox SS (2000a) Effects of intrathecal administration of ziconotide, a selective neuronal N-type calcium channel blocker, on mechanical allodynia and heat hyperalgesia in a rat model of postoperative pain. Pain 84:151–158

Wang YX, Gao D, Pettus M, Phillips C, Bowersox SS (2000b) Interactions of intrathecally administered ziconotide, a selective blocker of neuronal N-type voltage-sensitive calcium channels, with morphine on nociception in rats. Pain 84:271–281

Wang Y, Habgood GJ, Christ WJ, Kishi Y, Littlefield BA, Yu MJ (2000c) Structure–activity relationships of halichondrin B analogues: modifications at C.30–C.38. Bioorg Med Chem Lett 10:1029–1032

Wani MC, Taylor HL, Wall ME, Coggon P, McPhail AT (1971) Plant antitumor agents. IV. The isolation and structure of taxol, a novel antileukemic and antitumor agent from *Taxus brevifolia*. J Am Chem Soc 93:2325–2327

Watanabe J, Natsume T, Kobayashi M (2007) Comparison of the antivascular and cytotoxic activities of TZT-1027 (soblidotin) with those of other anticancer agents. Anticancer Drugs 18:905–911

Wender PA, Cribbs CM, Koehler KF, Sharkey NA, Herald CL, Kamano Y, Pettit GR, Blumberg PM (1988) Modeling of the bryostatins to the phorbol ester pharmacophore on protein kinase C. Proc Natl Acad Sci USA 85:7197–7201

Wright AE, Forleo DA, Gunawardana GP, Gunasekera SP, Koehn FE, McConnell OJ (1990) Antitumor tetrahydroisoquinoline alkaloids from the colonial ascidian *Ecteinascidia turbinata*. J Org Chem 55:4508–4512

Yamashita A, Norton EB, Kaplan JA, Niu C, Loganzo F, Hernandez R, Beyer CF, Annable T, Musto S, Discafani C, Zask A, Ayral-Kaloustian S (2004) Synthesis and activity of novel analogues of hemiasterlin as inhibitors of tubulin polymerization modification of the A segment. Bioorg Med Chem Lett 14:5317–5322

Yamazaki Y, Kohno K, Yasui H, Kiso Y, Akamatsu M, Nicholson B, Deyanat-Yazdi G, Neuteboom S, Potts B, Lloyd GK, Hayashi Y (2008) Tubulin photoaffinity labeling with biotin-tagged derivatives of potent diketopiperazine antimicrotubule agents. ChemBioChem 9:3074–3081

Yap TA, Carden CP, Kaye SB (2009) Beyond chemotherapy: targeted therapies in ovarian cancer. Nat Rev Cancer 9:167–181

Yarchoan R, Klecker R, Weinhold K, Markham P, Lyerly H, Durack D, Gelmann E, Lehrman S, Blum R, Barry D (1986) Administration of 3′-azido-3′-deoxythymidine, an inhibitor of HTLV-III/LAV replication, to patients with AIDS or AIDS-related complex. Lancet 1:575–580

Yeager RE, Yoshikami D, Rivier J, Cruz LJ, Miljanich GP (1987) Transmitter release from presynaptic terminals of electric organs: inhibition by the calcium channel antagonist omega *Conus* toxin. J Neurosci 7:2390–2396

Yeboah MM, Xue X, Duan B, Ochani M, Tracey KJ, Susin M, Metz CN (2008) Cholinergic agonists attenuate renal ischemia-reperfusion injury in rats. Kidney Int 74:62–69

Zask A, Birnberg G, Cheung K, Kaplan J, Niu C, Norton E, Suayan R, Yamashita A, Cole D, Tang Z, Krishnamurthy G, Williamson R, Khafizova G, Musto S, Hernandez R, Annable T, Yang X, Discafani C, Beyer C, Greenberger LM, Loganzo F, Ayral-Kaloustian S (2004a) Synthesis and biological activity of analogues of the antimicrotubule agent N, β, β-trimethyl-L-phenylalanyl-N1-[(1S, 2E)-3-carboxy-1-isopropylbut-2-enyl]-N1, 3-dimethyl-L-valinamide (HTI-286). J Med Chem 47:4774–4786

Zask A, Birnberg G, Cheung K, Kaplan J, Niu C, Norton E, Yamashita A, Beyer C, Krishnamurthy G, Greenberger LM, Loganzo F, Ayral-Kaloustian S (2004b) D-piece modifications of the hemiasterlin analogue HTI-286 produce potent tubulin inhibitors. Bioorg Med Chem Lett 14:4353–4358

Zewail-Foote M, Hurley LH (1999) Ecteinascidin 743: a minor groove alkylator that bends DNA toward the major groove. J Med Chem 42:2493–2497

Zhong W, Moya C, Jacobs RS, Little RD (2008) Synthesis and an evaluation of the bioactivity of the *C*-glycoside of pseudopterosin A methyl ether. J Org Chem 73:7011–7016

Chapter 8
Benefits and Failure of Imported Animals in the Fight Against Pests

Volker Walldorf

Abstract The employment of living organisms in the fight against pests is called *biological control*. The target organisms are completely eradicated, but should be suppressed at a tolerable level. As agents for biological control predators, parasitoids and pathogens are used. Different vertebrates (fox, toads etc.) and a number of free-living insects (ladybeetles, lacewings etc.) serve as predators. Parasitoids used were found within the insect orders Hymenoptera (*Trichogramma, Apantheles* etc.) and Diptera (*Compsilura, Cryptochaetum* etc.). Pathogens employed belong to bacteria (*Bacillus thuringiensis* strains) (Miller and Wansbrough 2002) and to the group of viruses (myxomatosis virus, rabbit calicivirus). Agents such as the entomopathogenic nematodes *Steinernema* and *Heterorhabditis*, which are assembled for the control of arthropod pests – for instance against weevils like *Otiorhynchus sulcatus* – act as parasites (the nematodes) and pathogens (symbiontic bacteria such as *Xenorhabdus* and *Photorhabdus*). The present chapter describes the tactic of the invaders and the problems that they pose inside their new environment for other domestic species but also for humans. To understand how important such invaders may be can be calculated from the fact, that in Germany about 1,500 new species arrived during the last 20 years (800 plant species, 700 animal species) or by the fact that the import of only 60,000 little Aga toads into Australia led to a population of 200 million specimens – today.

8.1 Who Are the Actors?

The employment of living organisms in the fight against pests is called *biological control*. The target organisms are not completely eradicated, but should be suppressed at a tolerable level. As agents for biological control predators, parasitoids, and pathogens are used. Different vertebrates (fox, toads etc.) (Hoddle 2002) and

---

V. Walldorf
Institute of Zoomorphology, Cell Biology and Parasitology, Heinrich-Heine-University, Düsseldorf, Germany
e-mail: walldorf@uni-duesseldorf.de

a number of free-living insects (ladybeetles, lacewings etc.) (Pappas et al. 2011) serve as predators. Parasitoids used were found within the insect orders Hymenoptera (*Trichogramma, Apantheles* etc.) and Diptera (*Compsilura, Cryptochaetum* etc.) (Hoddle 2002). Pathogens employed belong to bacteria (*Bacillus thuringiensis* strains) (Miller and Wansbrough 2002) and to the group of viruses (myxomatosis virus, rabbit calicivirus) (Hoddle 2002). Agents such as the entomopathogenic nematodes (Kaya and Gaugler 1993) *Steinernema* and *Heterorhabditis*, which are assembled for the control of arthropod pests – for instance against weevils like *Otiorhynchus sulcatus* –act as parasites (the nematodes) and pathogens (symbiontic bacteria such as *Xenorhabdus* and *Photorhabdus*) (Burnell and Stock 2000).

8.2 What Are the Strategies and Problems?

The most common strategy is the support of native organisms by giving them shelter or raising their effectiveness by breeding and releasing high numbers of cultured organisms. However, in combination with a well-balanced agriculture this strategy would only be successful, if local (native) pests are the targets of control.

As soon as men started to import different crops into new regions all over the world, where they were established, increasingly unwelcome passengers that in part became hazardous pests, were imported, too. The same occurs every day, since numerous stowaways are carried together with people and goods *on the various pathways of globalization*. This leads to frequent introduction of weeds, invertebrates, and vertebrates into new areas. Some of them turn out to be a pest for established imported crops and native plants and animals as well.

To combat those invading exotic organisms it was and is necessary to search for powerful antagonists in order to get rid off or at least to become able to control those intruders. Hopefully useful antagonists were mostly searched for in the invaders' country of origin. If assumed enemies were found, they were used to combat the new pest. However, this had in the past sometimes fatal consequences for the flora and fauna of their new environment.

It is not the aim of this chapter to give a complete, detailed representation of the biological control of pests, but this short abstract should direct attention to the difficulties and risks that might accompany biological control, if it is not done in a decisive manner. Therefore, only a short summary of its history and some examples – positive and negative – are given herein.

8.3 Examples

At first glance one tends to think above all positively of biological control and that it is a practice of dealing with pests in a very elegant way, because natural-borne agents are used. But as had to be learnt by many failures, account has to be taken of some basic principles that are summarized later on in this chapter.

About 1,700 years ago biological control was already practiced with great effort by citrus growers in southern China. They used the weaver ant, *Oecophylla smaragdina* [Figs. 8.1–8.4 (for figures see end of the chapter)], as a predator to control a number of arthropods affecting their citrus plantations. In those times weaver ant nests were gathered, sold and placed in citrus trees to reduce the numbers of insect pests. To support this arboreal ant in climbing from tree to tree and spreading within the plantations, bamboo rods as well as ropes were put up as bridges between trees (McCook 1882; DeBach 1964; Hölldobler and Wilson 1990). This technique is still used today in Southeast Asia and Australia as an alternative to chemical control of insect pests in a variety of fruit plantations (mango, cashew, citrus). Furthermore, in Asian countries the larvae of the weaver ants are harvested and serve as food for local populations. In Africa a second weaver ant species, *Oecophylla longinoda*, is used in the same way, for example in Ghanaian cashew plantations (Offenberg and Wiwatwitaya 2009).

However, it is important to put on record, that the weaver ants are organisms originating from the region of their deployment as precautionary pest controllers and remain embedded into their native community.

On the other hand for newly introduced pests exotic antagonists had and have to be found and imported. This was not always successful. Herein are provided examples for different cases where the following occurred:

1. The successful control of the target
2. Complete failure of the control with no or minimal negative influence on nontarget organisms
3. Where the indigenous flora and/or fauna of some regions has to fight with the newly introduced antagonists even today (sometimes an obviously forlorn struggle)

Sugar cane (*Saccharum officinarum*) was cultivated first in India from where it spread to the West and the East. It reached the West via the Arabs and was then carried by the Portuguese and Spanish to the West Indies Islands and to Middle- and South America and is cultivated in Australia, too. After the sugar cane had been established in these countries, it was only a question of time until a number of pests appeared in the monocultured fields. The cane was attacked by several insects – often at the same time. One very harmful species is a scarabaeid beetle, *Dermolepida albohirtum*, the larvae of which bore inside the stems of the plant.

The story of the cane toad, *Bufo marinus* (Easteal 1981; Freeland 1985) is closely connected to that of sugar cane. This big toad [Fig. 8.5 (for figure see end of the chapter)] is found in regions of southern Texas and its territory reaches down to the Amazon basin. This giant toad may grow up to a body length of 24 cm. It has large, toxin-producing parotid glands and is characterized by its very large mouth and its high reproduction rates. The cane toad engorges nearly every other animal that can pass down its throat. This toad is an example of an *inconsiderate introduction* of a predaceous animal into a new environment to reduce or eliminate insects that are pests to sugar cane.

Before 1840 the first toads were imported to the Caribbean islands of Martinique and Barbados. In 1844 they were carried to Jamaica to control not only insect pests but also the rats and mice on this island, too. But the struggle against rats failed, because adult rats are too large to be a prey for this toad.

Nevertheless in 1920 the "cane" or "Aga" toad was brought to Puerto Rico, in order to reduce the damage done by the larvae of the already mentioned sugar cane borer, *D. albohirtum*. By 1932 the toad species *B. marinus* had spread throughout the island, while simultaneously there was observed a striking reduction of sugar cane pests. It was thought that this effect might be due to the introduced predator and this belief resulted in the introduction of the "Aga" toad to other countries: Hawaii 1932, Australia 1935, and several islands in the Pacific region. Even in Egypt attempts were made to naturalize the "Aga" toads, however, this also failed – luckily, when looking at it from today's point of view and knowledge on this toad.

Years later it became obvious that the supposed successful pest control in Puerto Rico was not caused by the cane toad, but by extreme weather conditions during the years 1931 to 1936. In these years very high rainfalls for weeks were followed by extremely dry periods, which led to the high reduction of the cane borers. In some of the following years (1943, 1948) the harvest of sugar cane was again very low, despite the fact that meanwhile the toad population had reached a high density. This proves that there was no direct influence due to the feeding activity of the toad.

Also in Australia the cane toad again did not fulfill expectations. After the release of more than 60,000 little toads in Queensland it started to spread from there to the North, West and South of Australia. Today more than 200 million toads are approximated to live in the western regions of Australia, where this invader has become an extreme threat to the native fauna, not only because it feeds on many Australian animals covering insects, amphibians, reptiles, and mammals, but especially due to its poisonousness. The toxin of the cane toad is extremely dangerous for many potential predators of this toad, the death of which often is the result of feeding on this highly toxic prey. Also humans (children) are highly endangered by contact with this toad (The Cane Toad 2003).

To date the cane toad is still spreading and there is still no answer in sight. In some regions toad populations have reached such high population densities, that a politician proposed a cane toad killing day (Squires 2008), where people catch the toads, which – after brief examination by specialists to be sure that they are cane toads – are killed.

The basic mistake made when introducing the cane toad to Australia was neglection of the fact that it is a *generalist predator* that feeds on a broad spectrum of prey. This is amplified by the fact that there are nearly no predators that feed on this toad because of its toxicity. There is little hope that in the future a solution will be found.

Over the last 40 years releases of imported predators have taken place in many countries, which turned out to be a threat to the native flora and fauna, because of bad consideration of the pros and cons of their application, in part owing to the lack of knowledge of the biology and ecology of all participants in the "party," i.e., predator, target, and nontarget organisms. Another reason appears to be ignorance of the risks and neglecting them by using only economics as the guideline. The latter at least in part could be overcome by the countries setting limits and installing control mechanisms, although it is clear that uncontrolled introduction of predators, parasitoids, and pathogens is hardly controllable, if done unofficially by people believing their governments are not doing what has to be done. There are some countries regulating the introduction of biological control agents by law, whereas others do not.

Comparing case reports of unsuccessful application of biological control, which turned out to be a threat to nontarget organisms, it is noticeable that in these trials *generalist predators* were deployed as control agents. Some examples are given in Table 8.1.

Numerous efforts have been made to control the rabbit menace in Australia as well as in New Zealand. The rabbit, *Oryctolagus cuniculus*, was introduced several times to Australia, but first became established in 1879, as wild European rabbits were imported. It spread all over Australia except in the tropics, becoming a major pest. The same had taken place earlier in New Zealand, where they were introduced in the 1860s (Godfrey 1974). From 1871 to 1884 various generalists were introduced including for instance the European fox to Australia, the stoat, ferret, and weasel to New Zealand. The rabbits are still there, but native marsupials, birds, lizards, and insects are eaten, too, and thus suffer from the newly introduced enemies. The reason for this failure is that rabbits are quicker and have a higher reproductive rate than lizards, birds, or marsupials. The foxes also courted the resentment of the Australian stockman, because they kill lambs (Saunders et al. 1995; King 1990a, b; Lavers and Clapperton 1990).

In 1896 a viral disease, myxomatosis, was first discovered in European rabbits that had been imported to Uruguay. Later it turned out that the common Brazilian rabbit, *Sylvilagus brasiliensis*, is its natural host. Whereas the European rabbits became critically ill and died, the Brazilian rabbits suffered only from comparatively mild illness. The vectors for the virus were mosquitoes and fleas. Although already in 1918 Aragaõ suggested that this virus could be a suitable agent to reduce the rabbit menace in Australia, it took more than 30 years, until after a lot of research and some unsuccessful trials that in 1950 a new effort was started, which again failed at first. But at the end of 1950 and in the beginning of 1951 there was a sudden outbreak of the disease that spread with remarkable speed over Southeast Australia and killed millions of rabbits (Fenner 1959, 2010). Before the field trials were completed it had been shown that the virus was rabbit specific and no threat to Australian native fauna. This was tested on a wide range of Australian animals and birds, both native and domestic (Bull and Dickinson 1937).

Initially 90% of the infected rabbits were killed, however mortality declined to about 50% within the next 4–5 years, due to the appearance of virus strains with differing virulence and development of increasing resistance of the rabbits against the *Myxoma* virus (Fenner 1959). After the release of the European rabbit flea,

Table 8.1 In this table some examples of generalist biological control agents, target pests, country and year of first introduction, and the unintended impact on nontarget organisms are listed (from Hoddle 2002, where the complete table with references is found)

Biological control agent and origin	Target pest	Country and year of first introduction	Nontarget impacts
European red fox, *Vulpes vulpes* L. Palearctic regions	Rabbits, *Oryctolagus cuniculus* L.	Australia, 1871	Native marsupials and birds, lambs
Stoat, *Mustela erminea* L. Eurasia and North America	Rabbits	New Zealand, 1884	Native birds, insects, and lizards
Weasel, *Mustela nivalis vulgaris* Erxleben Eurasia and North America			
Ferret, *Mustela furo* L. Central Europe and the Mediterranean	Rabbits	New Zealand, 1879	Native birds
Small Indian mongoose, *Herpestes javanicus* (Saint Hilaire) Iraq to the Malay Peninsula	Rats, *Rattus* spp.	Trinidad, 1870; Jamaica, 1872; Puerto Rico, 1877; Barbados, 1877; Hawaii, 1883; St. Croix, 1884; Cuba, 1886; Hispaniola, 1895; Surinam, 1900	Native birds and reptiles
Cane toad, *Bufo marinus* L. Northwestern Mexico through southern Brazil	White grubs, *Phyllophaga* sp.; potato hawk moth, *Agrius convolvuli* L.; Gray-backed cane beetle, *Dermolepida albohirtum* (Waterhouse)	Jamaica, 1844; Bermuda 1855; Puerto Rico, 1920; Hawaii, 1932; Phillipines, 1934; Australia, 1935; Fiji, 1936; Guam, 1937; New Guinea, 1937	Native insects, amphibians, and reptiles. Toads are toxic to native wildlife that consume it. Outcompetes native amphibians for shelter and breeding sites.
Mosquito fish, *Gambusia affinis* (Baird and Girard) Eastern USA and Mexico	Worldwide dissemination for control of mosquito larvae	Releases began around 1900 Established in about 70 countries including: Afghanistan, Australia, Canada, China, Ethiopia, Grand Cayman Islands, Greece, Hawaii, Iran, Korea, New Zealand, Somalia, Turkey, Ukraine	Substantial nontarget attacks on native aquatic invertebrates and vertebrates outside its native range

(*continued*)

Table 8.1 (continued)

Biological control agent and origin	Target pest	Country and year of first introduction	Nontarget impacts
Tachinid Fly, *Compsilura concinata* Meigen Europe	Gypsy Moth, *Lymantria dispar* L., and other lepidopteran pests	USA, 1906	Regional declines of native saturniid moths because of heavy parasitized larvae
Cactus moth, *Cactoblastis cactorum* (Bergroth) Northern Argentina, Uruguay, Paraguay, Southern Brazil	*Opuntia* spp. of cacti	Australia 1925; South Africa 1933; Hawaii, 1950; Mauritius, 1950; Caribbean, 1957; *accidental* into USA. 1989	Invaded mainland USA from the Caribbean in 1989, attacks on *Opuntia* spp., thus threatening native endangered as well as economically important species

Spilopsyllus cuniculi, in Australia in the 1960s it spread widely and was able to transmit the *Myxoma* virus throughout the year (Fenner 2010). Efforts to establish myxomatosis in New Zealand as well failed because of the lack of suitable vectors (O'Hara 2006).

In 1952 another strain of the *Myxoma* virus was used by Dr. Delille infecting two rabbits on his private property in France. From there the disease spread all over France and to the bordering countries and Great Britain. Myxomatosis became established in Europe and it became a threat to domestic rabbits that were kept as pets or for meat production. Therefore, a protective vaccination was developed. This vaccine is forbidden in Australia, where the view on wild rabbits and their control is considerably different. Whereas in Australia the rabbits are considered as a pest that successfully could be controlled – but not eradicated – by the Myxoma virus, in Europe myxomatosis is an unwanted disease rather than an agent to control rabbit populations (Fenner 1959).

Meanwhile in Australia another rabbit disease had been introduced. The Rabbit Hemorrhagic Disease (RHD) was first recognized in 1984 in China. It is caused by a Calici virus that is a pathogen specific for rabbits. Later it was tested on an Australian island, to determine whether it was a suitable agent for rabbit control. In 1995 the RHD virus escaped from the field test area and infected rabbits on the Australian mainland. After the escape it turned out that various blowflies, for instance *Calliphora dubia*, could serve as vectors, which had been unknown before, and laboratory tests showed that mosquitoes may transmit it as well. Nevertheless, RHD became a better control agent than myxomatosis in the arid regions of Australia, probably because flies are more common there than mosquitoes. In 1997 the RHD virus was first illegally introduced to New Zealand by farmers after legal import had been rejected. Now the virus has become endemic in New Zealand (Hoddle 2002; O'Hara 2006).

To combat rats being a pest in several Caribbean islands, Hawaii, and Surinam the small Indian mangoose was imported. The rat populations survived, but birds and reptiles are afflicted by the new predator (Hinton and Dunn 1967; Loope et al. 1988).

Up to 1982 the World Health Organization sponsored the dissemination of the fish *Gambusia affinis* to control mosquito larvae. However, extensive use led to significant impacts on aquatic invertebrates and vertebrates outside of the native range (Diamond 1996; Gamradt and Kats 1996; Legner 1996; Meisch 1985; Rupp 1996; Walton and Mulla 1991).

After the gypsy moth [Figs. 8.6–8.9 (for figures see end of the chapter)], *Lymantria dispar*, had been transmitted accidently from Europe to North America, a parasitoid from its place of origin, the tachinid *Compsilura concinata*, whose larvae parasitize the caterpillars, was used as a generalist agent to control the moths. Nearly 100 years later a regional decline of native saturniid moths was observed, due to heavy infestation of their larvae by this tachinid fly (Boettner et al. 2000).

Coccinellid beetles or so-called ladybeetles are known as effective predators of various aphids and their relatives and are mostly beloved by people for their efforts in controlling those pests.

An example of very effective biological control is the following story. From 1877 to 1879 the Australian ladybird, *Rodolia cardinalis*, was released in California. This so-called vedalia beetle or cardinal ladybird in combination with the parasitic Australian fly (*Crypotchaetum iceryae*) successfully reduced the cottony cushion scale, *Icerya purchasi*, in the citrus plantations to neglectable levels. Through this effective intervention the just-developing citrus industry was protected from severe damage (Sawyer 1996).

There are a number of ladybeetle species that are used or have been used in biological control of insect pests. In recent years attention has been focused on the Asian lovebug (Figs. 8.10–8.13 (for figures see end of the chapter)] *Harmonia axyridis,* originating from China, Japan, and Siberia. This coccinellid beetle had been transferred to the USA first in 1916 for biological control. In 1988 it was found to be established there and free living in nature (Chapin and Brou 1991). In the 1980s and 1990s it was introduced to several European countries. About 10 years later it was found to be free living and widespread in Europe. *H. axyridis* appears to be a polyphagous predator, that not only feeds on aphids, but also on a variety of other insects including the eggs, larvae, and pupae of native coccinellids, thus not only competing for the aphid prey, but also seeing their native competitors as prey, too. Besides that the Asian ladybeetle injures the North American wine- and fruit industry by consuming soft fruits. It disturbs humans not only by building up large aggregations in buildings for overwintering, but mainly by secreting its bad smelling hemolymph, when it is distressed. Therefore, *Harmonia* has changed sides at least in the minds of many people. In the USA this former biological controller meanwhile is classified as a pest species itself (Mannix 2001; Koch 2003; Roy and Wajnberg 2008; Michaud et al. 2002).

Similar negative impacts on native coccinellids arose and still arise from the European ladybeetle [Fig. 8.14 (for figure see end of the chapter)], *Coccinella septempunctata*, which was released in the USA in 1957 as a control agent for aphid pests. Competition with indigenous coccinellids leads to local displacement of native species. Moreover, the seven-spotted ladybeetle is reported to feed on the eggs of certain North American Lepidoptera (Elliot et al. 1996; Obrycki et al. 2000).

Within the parasitoids the common species *Trichogramma brassicae* shall serve as another example for the effectivity of biological control. This mass-reared egg parasite has been used in Germany as an effective control agent for the European corn borer, *Ostrinia nubilalis*, for more than 25 years. To ensure the efficiency of the parasite a continuous quality control is carried out according to international standards (IOBC, International Organization for Biological Control) (Zimmermann 2007).

Meanwhile there exist a number of insect pests belonging to the Lepidoptera, Coleoptera, Diptera and other insect orders that are targets of entomopathogenic nematodes. These nematodes, that have been mentioned above represent an especially important option for the protection of greenhouse and nursery plants, where chemical pest control should be avoided or used to a minimum. For instance the black vine weevil, *O. sulcatus*, an important pest of cranberries and strawberries can be controlled by *Heterorhabditis* species. The same applies for *Steinernema feltiae* controlling fungus gnats (Sciaridae), the larvae of which can damage various ornamental plants (Georgis et al. 2006). But to be successful the deployed nematode must match the target host regarding virulence, host-finding strategy, and ecological factors (Gaugler 1999).

So far only attempts – successful as well as unsuccessful – to control animal pests have been presented. The following example deals with the control of a plant that became a weedy pest, namely the prickly pear. *Opuntia stricta* was carried to Australia as an ornamental plant in the early 1800s. After escaping the cultivation area it spread and had infested about 25,000,000 ha in Queensland and New South Wales by 1925. The enormous populations made it impossible to utilize the areas occupied by the opuntias. An effective control agent for the prickly pear was found in the Argentine Cactus moth, *Cactoblastis cactorum*. This pyralid moth naturally occurs in Northern Argentina, Uruguay, Paraguay, and Southern Brazil, where its larvae feed on nearly all *Opuntia* (*Platyopuntia*) spp. (Zimmerman et al. 2004).

In 1926 it was imported into Australia and successfully destroyed the opuntia stands within about 4 years. With diminishing opuntias the moth itself declined, too. From 1933 on *Cactoblastis* was brought to a number of other countries including South Africa, New Caledonia, and Hawaii to perform its work on the prickly pear pests there. But in 1957 the moth was introduced to the Caribbean island of Nevis to control the abnormally dense growth of native cacti. In retrospect it is likely to have been the wrong decision to release *Cactoblastis* on this island with indigenous *Opuntia* species. This mistake may become more serious as Nevis is situated in the center of a region that is the homeland of many other *Opuntia* species. After successful control of *Opuntia* on Nevis the moth was carried to several other Caribbean islands in the years from 1960 to 1973. In 1989 *C. cactorum* was detected for the first time in the Florida Keys before making its way to the adjacent states (Georgia, South Carolina) in the following years (Zimmerman et al. 2004).

Once within the USA the control agent – so successful in several countries, especially Australia – has become a threat to Texas and Northern Mexico, if it spreads along the Gulf Coast States. *Opuntia* species that are native in that region play an important role within the biological community, serving for instance as food, shelter, or nesting places for numerous animals.

Aside from this *Opuntia* species are important from an economic point of view as an agricultural resource in a number of other countries in South America, Europe, Africa, and Asia as well [Fig. 8.15 (for figure see end of the chapter)]. Their fruits [Fig. 8.16 (for figure see end of the chapter)] and young pads are harvested and eaten, and they are the basis for juice, spirituous beverages etc. Certain opuntias serve as host plants for the cochineal, *Dactylopius coccus* [Fig. 8.17 (for figure see end of the chapter)], a scale insect that produces the carmine dye. That means there are thousands of people living on the cultivation of these cacti (Zimmerman et al. 2004).

In South Africa control of *Cactoblastis* has been practiced since 1950 as it became obvious that the cultivation of plants in cactus pear orchards was impossible without protection against the cactus moths. As the moth is needed in South Africa and Australia as a control agent for unwanted *Opuntia* stands, biological control by a newly introduced predator is not suitable. One possibility in countries where both native Opuntias and commercial plantations of the cacti need to be protected is the search for possible predators. Approaches using the sterile insect technique (SIT) have been made, although in the order Lepidoptera it is not as easy as in Diptera to produce sterile males. Dependent on the case chemical control could be an alternative strategy, although it cannot be used in all areas, especially, when nursery products are produced (Zimmerman et al. 2004). Therefore, it is absolutely necessary to keep an eye on *Cactoblastis* in those countries, where *Opuntia* species are indigenous or are of commercial interest.

The *Cactoblastis* story clearly demonstrates how important it is to define specific criteria before deciding how and where an introduced control agent should be employed. In some way this control agent shows both sides of a coin (brings to mind the story of Dr. Jekyl and Mr. Hide, containing a good and a bad side in one personality) being a story of success in Australia, where no indigenous or commercially used prickly pears were present, the *oligophagous moth* in contrast turns out to be a threat in countries with native and/or commercially used *Opuntia* species.

8.4 Conclusions

The conclusion from the selected cases, especially in the light of the failures with introduced nonindigenous predators, is the requirement for basic principles that should be followed before any release of exotic control agents. The main claim is that there is a need for a *high natural enemy host speciality* (Hoddle 2002). When straying from this need, nontarget impacts and lack of control presumably occur. The introduction of generalist natural enemies lacking high host and habitat specificity includes a high risk for native species through competition, becoming prey, or being intoxicated. Each loss of a native species leaves not only an empty niche, but also may influence other species that are connected with this species (in the food chain). Since all organisms in an environment are cross-linked, the loss of members could have an impact on various numbers of other organisms connected to this community. Shortly, further amplification of numerous unintended effects will take place, if generalist natural enemies are released, as has been demonstrated by the examples given above.

These dangers exist also in the case of deployment of control agents within closed cavities such as greenhouses and stockrooms. Commonly those areas are not sealed completely for different reasons and organisms released within these structures tend to look for freedom and find their way out. There they may spread, if they can withstand the new environment, and may become a threat to nontarget organisms.

A few words should be allowed concerning the use of pathogenic agents. The introduction of the RHD virus into New Zealand by private individuals highlights the threat of release of pathogens such as viruses illegally and uncontrollably, which is by some authors referred to as bioterrorism. Certainly the farmer's motives were not that of a terrorist; but if the farmers were able to import via a still unknown route and distribute the Calici virus, then there is reasonable fear that it is nearly impossible to control the introduction of other – possibly more hazardous – pathogens (O'Hara 2006).

Today it is realized that potential agents for biological control that are considered for introduction into a new environment, should fulfill at least the criteria enumerated above.

The first step must be the identification of the pest, to determine its origin and search there for an enemy that complies with basic prerequisites to maintain risks as low as possible. There is a need to increase knowledge on the biology, ecology, and variability of organisms, and if lacking it has first to be acquired. There are numerous newly described organisms that require a scientific name, need to be measured and their origin has to be identified. These details all need to be compared with native species. This extends to eggs, instars, pupae, life cycles, and feeding habits and this must be carried out first.

This means that there is much work to be done. Equipped with such knowledge there may be no or only low risk on the introduction of enemies that may serve as control agents of previously invaded pests. Examining organisms as to whether they are suitable for biological control is like looking for a medicine within the pharmacy of nature. It needs knowledge and the power of deduction to detect an adequate solution. Depending on its origin, properties, collection and form of application, it can be very helpful but also very harmful to health within a whole biotope.

Fig. 8.1 Big nests of the weaver ant, *Oecophylla smaragdina*, in a tree

Fig. 8.2 Weaver ant nest made out of several leaves

Fig. 8.3 Weaver ant subsidiary made of only one leaf

8 Benefits and Failure of Imported Animals in the Fight Against Pests

Fig. 8.4 Weaver ant workers employing their silk-producing instars for "sewing" leaves together to build a new shelter

Fig. 8.5 The cane toad, *Bufo marinus*

Fig. 8.6 Female of the gypsy moth, *Lymantria dispar*

Fig. 8.7 The hairy caterpillars of *Lymantria dispar* with *flashy red-* and *blue*-colored warts, are capable – especially when appearing in great numbers – of affecting forests

Fig. 8.8 Gypsy moth larva hiding and resting in a cavity of the oak tree bark

Fig. 8.9 Gathering of *Lymantria dispar* larvae on the trunk of an oaktree

Fig. 8.10 Larva and pupa of the Asian Ladybeetle, *Harmonia axyridis*

Fig. 8.11 Mating *Harmonia axyridis*

Fig. 8.12 Other phenotypes of *Harmonia axyridis* mating within an aphid colony

Fig. 8.13 Different phenotypes of *Harmonia axyridis*

Fig. 8.14 European ladybeetle, *Coccinella septempunctata*

Fig. 8.15 *Opuntia* orchard, Lanzarote, Canary Islands, Spain

Fig. 8.16 *Opuntia* cladodes with fruits

Fig. 8.17 *Opuntia* sp. with cochineal, *Dactylopius coccus*

References

Boettner GH, Elkinton JS, Boetnner CJ (2000) Effects of biological control introduction on three nontarget native species of saturniid moths. Conserv Biol 14:1798–1806

Bull LB, Dickinson CGC (1937) The Specificity of the virus of rabbit myxomatosis. Journal council Scientific Industrial Res (Australia) 10:291–294

Burnell AM, Stock SP (2000) *Heterorhabditis, Steinernema* and their bacterial symbionts – lethal pathogens of insects. Nematology 2:31–42

Chapin JB, Brou VA (1991) *Harmonia axyridis* (Pallas), the third species of the genus to be found in the United States (Coleoptera:Coccinellidae). Proc Entomol Soc Washington 93:630–635

DeBach P (1964) Biological control of insect pests and weeds. Chapman and Hall, London

Diamond JM (1996) A-bombs against amphibians Nature 383:386–387

Easteal S (1981) The history of introductions of *Bufo marinus* (Amphibia:Anura); a natural experiment in evolution. Biol J Linn Soc 16:93–113

Elliot NR, Kieckhefer R, Kauffman W (1996) Effects of an invading coccinellid on native coccinellids in an agricultural landscape. Oecologia 105:537–544

Fenner F (1959) Myxomatosis. Br Med Bull 15:240–245

Fenner F (2010) Deliberate introduction of the European rabbit, *Oryctolagus cuniculus*, into Australia. Rev Sci Tech Off Int Epiz 29:103–111

Freeland WJ (1985) The need to control cane toads. Search 16:211–215

Gamradt SC, Kats LB (1996) Effects of introduced crayfish and mosquitofish on California newts. Conserv Biol 10:1155–1162

Gaugler R (1999) Matching nematode and insect to achieve optimal field performance. In: Polavarapu S (ed) Optimal use of insecticidal nematodes in pest management. Rutgers University Press, New Brunswick, NJ, pp 9–14

Georgis R et al (2006) Successes and failures in the use of parasitic nematodes for pest control. Biol Control 38:103–123

Godfrey MER (1974) The European rabbit problem in New Zealand. In: Proceedings of the 6th vertebrate pest conference. http://digitalcommons.unl.edu/vpc6/16

Hinton HE, Dunn AMS (1967) Mongooses: their natural history and behaviour. University of California Press, Berkeley, CA, 144 pp

Hoddle MS (2002), Classical biological control of arthropods in the 21st century. In: Keynote Presentation held on the 1. International Symposium on Biological Control of Arthropods, 14–18 Jan 2002, Honululu, Hawaii, USA

Hölldobler B, Wilson EO (1990) The ants. Belknap, Harvard University Press, Cambridge, MA

Kaya HK, Gaugler R (1993) Entomopathogenic nematodes. Annu Rev Entomol 38:181–206

King CM (1990a) Stoat. In: King CM (ed) The handbook of New Zealand mammals. Oxford University Press, Auckland, New Zealand, pp 288–312

King CM (1990b) Weasel. In: King CM (ed) The handbook of New Zealand mammals. Oxford University Press, Auckland, New Zealand, pp 313–320

Koch RL (2003) The multicolored Asian lady beetle, *Harmonia axyridis*: a review of its biology, uses in biological control, and non-target impacts. J Insect Sci 3:32

Lavers RB, Clapperton BK (1990) Ferret. In: King CM (ed) The handbook of New Zealand mammals. Oxford University Press, Auckland, New Zealand, pp 320–330

Legner EF (1996) Comments on adverse assessments of *Gambusia affinis*. J Am Mosq Control Assoc 12:161

Loope L, Hamann O, Stone CP (1988) Comparative conservation biology of oceanic archipelagos. BioScience 38:272–282

Mannix L (2001) *Harmonia axyridis*, a new biological control... or new insect pest? http://www.colostate.edu/Depts/Entomology/courses/en507/papers_2001/mannix.htm

McCook HC (1882) cited by Debach (1964)

Meisch MV (1985) *Gambusia affinis affinis*. In Chapmann HC (ed) Biological control of mosquitoes. American Mosquito Control Association Bulletin No.6, Washington DC, pp 3–17

Michaud JP, McCoy CW, Futch SH (2002) Ladybeetles as biological control agents in Citrus. HS-873 edis.ifas.ufl.edu/hs138

Miller C, Wansbrough D (2002) Towards a strategy for using Bt toxins in New Zealand. MAF Technical Paper No: 2002/20. http://www.maf.govt.nz/publications

Obrycki JJ, Elliot NC, Giles KL (2000) Coccinellid introductions: potential for and evaluation of nontarget effects. In: Follet PA, Duan JJ (eds) Nontarget effects of biological control. Kluwer Academic, Boston, MA, pp 127–144

Offenberg J, Wiwatwitaya D (2009) Weaver ants convert pest insects into food – prospects for the rural poor. Tropentag 2009, University of Hamburg, Oct. 6–8, 2009, Conference on International Research on Food Security, Natural Resource Management and Rural Development

O'Hara P (2006) The illegal introduction of rabbit haemorrhagic disease virus in New Zealand. Rev Sci Tech Off Int Epiz 25:119–123

Pappas ML et al (2011) Chrysopid predators and their role in biological control. J Entomol 8:301–326

Roy H, Wajnberg E (2008) From biological control to invasion: the ladybird, Harmonia axyridis as a model species. Biol Control. doi:DOI 10.1007/s10526-007-9127-8

Rupp HR (1996) Adverse assessments of Gambusia affinis: an alternative view for mosquito control practitioners. J Am Mosq Control Assoc 12:155–166

Saunders G, Coman B, Kinnear J, Braysher M (1995) Managing vertebrate pests: foxes. Australian Government Printing Service, Canberra, p 141

Sawyer RC (1996) To make a spotless orange: biological control in California. Iowa State University Press, Ames, Iowa, p 290

Squires N (2008) Australian MP proposes can toad – killing day. The Telegraph 02.04.2008, Telegraph Media Group, UK

The Cane Toad (2003) http://www.csiro.au/files/files/p7rh.pdf

Walton WE, Mulla MS (1991) Integrated control of Culex tarsalis larvae using Bacillus sphaericus and *Gambusia affinis*: effects on mosquitoes and nontarget organisms in field mesocosms. Bull Soc Vector Ecol 16:203–211

Zimmerman H et al (2004) Biology, history, threat, surveillance and control of the cactus moth, *Cactoblastis cactorum*. IAEA, Vienna, IAEA/FAO-BSC/CM

Zimmermann O (2007) Biologische Bekämpfung des maiszünslers: Qualitätsstandards bei *Trichogramma*-Anwendungen. Nachrichtenbl Deut Pflanzenschutzd 59:289–292

Chapter 9
Helminth Therapy to Treat Crohn's and Other Autoimmune Diseases

Jeff Bolstridge, Bernard Fried, and Aditya Reddy

Abstract This review updates a previous one (Reddy and Fried, Parasitol Res 104:217–221, 2009) on Crohn's and other autoimmune diseases and helminth therapy. Our review considers the use of various helminths or their worm products to treat the following autoimmune diseases: Inflammatory Bowel Disease, Multiple Sclerosis, Rheumatoid Arthritis, Type-1 Diabetes, and Asthma and allergic disorders. The use of the following helminths or their worm products are considered in this review: the nematodes *Trichuris suis*, *Necator americanus*, *Heligomosomoides polygyrus*, *Trichinella spiralis*, *Ancanthocheilonema viteae*, *Ascaris suum*, *Nippostrongylus brasiliensis*, *Syphacia obvelata*, *Dirofilaria immitis*, *Litomosoides sigmodontis*; the cestodes *Hymenolepis diminuta* and *Taenia crassiceps*; and the trematodes *Schistosoma mansoni*, *Schistosoma japonicum*, and *Fasciola hepatica*. Our review discusses the potential role of worm products rather than live worms in future studies using helminths to treat autoimmune diseases.

9.1 Introduction

This review updates the previous ones by Reddy and Fried (2007, 2009) on the use of helminths to treat Crohn's and other autoimmune diseases. Helminth (or worm) therapy can be defined as the use of larval or adult helminths or products derived from these helminths to treat a disease. The current review considers some of the major findings in the earlier reviews, but mainly focuses on newer studies not mentioned previously. A current area of study uses worm products to treat autoimmune diseases as opposed to infection with live worms; the body of literature concerned with this new area of research is considered here. This review also

---

J. Bolstridge
Department of Chemistry, Lafayette College, Easton, PA 18042, USA

B. Fried (✉) and A. Reddy
Department of Biology, Lafayette College, Easton, PA 18042, USA
e-mail: friedb@lafayette.edu

covers helminths and autoimmune diseases not previously discussed in Reddy and Fried (2007, 2009).

An autoimmune disease is an illness caused by a self-destructive immune system damaging its own body tissues. The autoimmune diseases covered in this review are as follows: Inflammatory Bowel Disease (IBD), Multiple Sclerosis, Rheumatoid Arthritis, Type-1 Diabetes, and Asthma and allergic disorders. The diseases mentioned in this review are the most important for which helminth therapy has been shown to be a possible treatment.

9.2 Inflammatory Bowel Disease

IBD is characterized by chronic inflammation of the gastrointestinal tract and includes ulcerative colitis (UC) and Crohn's disease (CD) (Braus and Elliott 2009). While both UC and CD have the common symptom of periodic bouts of inflammation, UC is usually limited to the rectum, while CD affects all areas of the gastrointestinal tract (Reddy and Fried 2007). The inflammation presented in CD extends deep into the mucosal lining, commonly leading to abdominal pain and diarrhea (Reddy and Fried 2007). Genetic and environmental factors play a role in the development of these severe inflammatory responses (Ruyssers et al. 2009). Countries with low incidence of helminth infection, typically industrialized nations, have relatively high rates of IBD (Alic 2000). The well-known hygiene hypothesis suggests that infection with helminths from an early age reduces the occurrence of IBD, among other disorders (Reddy and Fried 2007).

The key to treating IBD is to induce long-term remission in patients in order to relieve them of their painful symptoms. Corticosteroids and mesalamine are the drugs of choice for the treatment of IBD, and can induce remission in patients (Ruyssers et al. 2009). Other drug therapies have been developed, but with many of these treatment options the patient may still require surgery because of the increasing dependency of the patient on the drugs (Ruyssers et al. 2009). Novel therapies that reduce the requirement of surgery in patients are needed for people suffering from IBD because of the risks inherent in gastrointestinal surgery. Helminth therapy involves the use of whole larval or adult worms or viable eggs or worm products (such as various proteins) to treat a disease. Table 9.1 details the various diseases where helminth therapy has undergone experimental trials in animal or human models. This table also lists the helminths reported to have beneficial effects on these diseases. Many studies have been conducted using helminth therapy to treat both UC and CD, both in animal and human models.

The cytokine profile of CD is characterized by a T helper type 1 (Th1) inflammatory response (Braus and Elliott 2009). Key developments in CD include intense local B cell stimulation, high IgG production, and an uncharacteristically low secretory IgA and IgM response (Reddy and Fried 2007). This characteristic response is different than in UC, which is thought to be a Th2-type inflammation (Kuijk and Van Die 2010). The complete immune responses to helminth infections

9 Helminth Therapy to Treat Crohn's and Other Autoimmune Diseases

Table 9.1 The use of various helminths or helminth products to treat autoimmune diseases

Helminth used	Worm stage or product used	Host used	References
Inflammatory Bowel Disease			
Trichuris suis (N)	Egg	Humans	Elliott et al. (2005), Summers et al. (2005a, b, c, 2006)
Hymenolepis diminuta (C)	Larva	Rats, Mice	Hunter et al. (2007), Johnston et al. (2010)
Schistosoma mansoni (T)	Egg	Mice	Elliott et al. (2003)
	Larva	Rats, Mice	Moreels et al. (2004), Smith et al. (2007)
	Worm products	Mice	Ruyssers et al. (2010)
Necator americanus (N)	Larva	Human	Croese et al. (2006)
Heligomosomoides polygyrus (N)	Larva	Mice	Elliott et al. (2004)
Trichinella spiralis (N)	Larva	Mice	Khan et al. (2002), Motomura et al. (2009)
S. japonicum (T)	Egg	Mice	Mo et al. (2007)
Multiple sclerosis			
S. mansoni (T)	Larva	Mice, Humans	Correale and Farez (2007), Correale et al. (2008, 2009), La Flamme et al. (2003)
	Egg antigens	Mice	Sewell et al. (2003)
S. japonicum (T)	Egg antigens	Mice	Zheng et al. (2008)
Fasciola hepatica (T)	Larva	Mice	Walsh et al. (2009)
T. spiralis (N)	Larva	Mice	Gruden-Movsesijan et al. (2008)
Rheumatoid Arthritis			
Ancanthocheilonema viteae (N)	Worm antigen	Mice	McInnes et al. (2003)
S. mansoni (T)	Larva	Mice	Osada et al. (2009)
Ascaris suum (N)	Worm antigen	Rats, Mice	Rocha et al. (2008)
Nippostrongylus brasiliensis (N)	Larva	Mice	Salinas-Carmona et al. (2009)
Syphacia obvelata (N)	Larva	Rat	Pearson and Taylor (1975)
Type-1 Diabetes			
S. mansoni (T)	Larva	Mice	Cooke et al. (1999)
	Egg antigens	Mice	Zaccone et al. (2003, 2009)
Dirofilaria immitis (N)	Worm antigen	Mice	Imai et al. (2001)
Litomosoides sigmodontis (N)	Larva	Mice	Hübner et al. (2009)
	Worm antigen	Mice	Hubner et al. (2009)
H. polygyrus (N)	Larva	Mice	Liu et al. (2009), Saunders et al. (2007)
T. spiralis (N)	Larva	Mice	Saunders et al. (2007)
Taenia crassiceps (C)	Larva	Mice	Espinoza-Jimenez et al. (2010)
Allergic Rhinitis			
S. mansoni (T)	Larva	Mice	Mangan et al. (2004, 2006)
S. japonicum (T)	Egg antigen	Mice	Yang et al. (2006)
	Viable eggs	Mice	Yang et al. (2006)
H. polygyrus (N)	Larva	Mice	Wilson et al. (2005)
A. suum (N)	Worm antigen	Mice	Itami et al. (2005)
L. sigmodontis (N)	Larva	Mice	Dittrich et al. (2008)

C Cestoda; *N* Nematoda; *T* Trematoda

still remain largely unknown (Ehrhardt 1996; Sartor 2006). It is known that helminths lead to a characteristic Th2-type phenotype with Th2-associated cytokines interleukin-4 (IL-4), IL-5, and IL-13 (Fukumoto et al. 2009). IL-33 is produced by intestinal epithelial cells (IEC) and it is seen in the earliest stages of infections (Humphreys et al. 2008). IL-33 is known to bind ST2 receptors, which are thought to play a role in the Th2 response (Yoshimoto and Nakanishi 2006).

Trichuris suis, the pig whipworm, has undergone multiple clinical studies for patients with IBD, and the results of these studies showed a reduction in disease activity, as measured by a CD activity index, following treatment with *T. suis* eggs (TSE) (Elliott et al. 2005; Summers et al. 2005a, b, c, 2006). It is thought that the domination of the Th2 phenotype due to the helminth overrides the Th1 response of the CD inflammation (Weinstock et al. 2005). Patients were exposed to repeated doses of TSE, and these repetitious doses led to concerns about the potential spread of the larvae in patients (Van Kruiningen and West 2005). Because the patients went into remission after ingestion of TSE future studies on the use of TSE as a realistic and viable treatment option for patients with IBD are warranted. Figure 9.1 shows the life cycle of *T. suis*. In many of the referenced studies on pig whipworm, the authors use the term *T. suis* ova (TSO) rather than

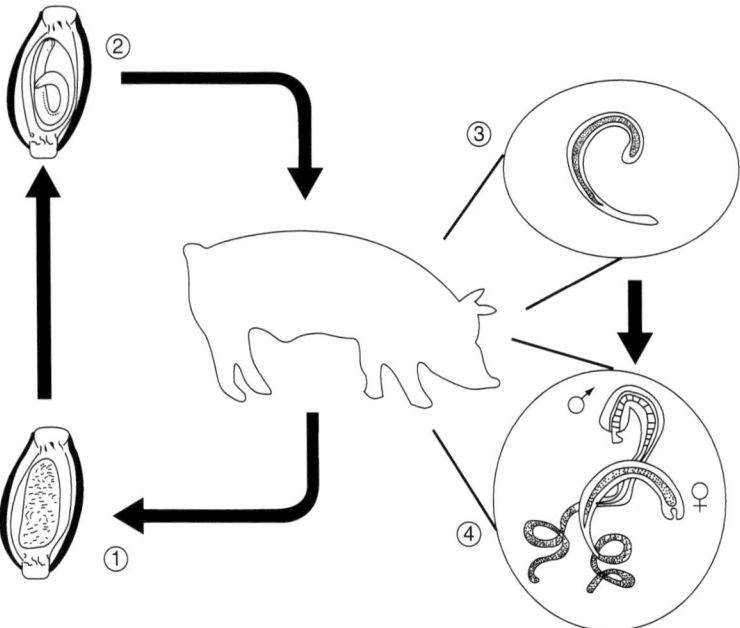

Fig. 9.1 Life cycle of the nematode *Trichuris suis*. (1) Unembryonated eggs are passed in the feces of the vertebrate host. (2) Unembryonated eggs develop into embryonated eggs. (3) The eggs are ingested by the vertebrate host and then hatch into larvae. (4) The larvae develop into male and female adult worms

TSE. It should be noted that TSE is the correct term, because the egg has a shell, residual yolk, and an ovum.

Hymenolepis diminuta (the rat tapeworm) has been examined for its effect on oxazolone-induced colitis in the rat (Hunter et al. 2007). Hunter et al. (2007) found that *H. diminuta* infection severely worsened the chemically induced Th2 model colitis in rats, as measured by macroscopic clinical scores, histologic damage scores, myeloperoxidase and eosinophil peroxidase activity, and cytokine synthesis. Independent of colitis, the *H. diminuta* infection increased the Th2 markers IL-4, IL-5, IL-10, and IL-13 (Hunter et al. 2007). The most interesting result from this study is that Hunter et al. (2007) found that *H. diminuta* can be helpful in treating other chemically induced colitis models.

H. diminuta has also been studied in mice with dinitrobenzenesulfonic acid (DNBS) induced colitis (Hunter et al. 2010; Johnston et al. 2010). These studies found that infection with *H. diminuta* alleviated the pathology of the DNBS treatment (Hunter et al. 2010; Johnston et al. 2010). Johnston et al. (2010) found that the mice cotreated with DNBS and three intraperitoneal injections of the parasite had decreased levels of tumor necrosis factor-alpha (TNF-α), and increased levels of IL-10 and IL-6. Furthermore, cotreated mice that were also administered anti-IL-10 antibodies showed a slightly blunted effect of *H. diminuta* on the colitis, suggesting that IL-10 is critical in the use of *H. diminuta* as a novel therapy for colitis. The specific factors that allow *H. diminuta* to alleviate colitis pathology would be preferable as a treatment rather than the use of the adult worm because there would be no risk of worm infection. Studies that identify specific immunosuppressive factors are needed. Hunter et al. (2010) transferred alternatively activated macrophages (AAMs) from mice infected with *H. diminuta* to mice with DNBS-induce colitis, and found that AAMs were able to alleviate the pathology of the colitis. They did this based on the knowledge that the DNBS/*H. diminuta* cotreatment shows a high expression of arginase-1, a marker for AAMs (Hunter et al. 2010). The promising results from this study suggest that AAM transfer could be developed into a viable therapy for patients suffering from IBD. Figure 9.2 shows the major stages of the *H. diminuta* life cycle.

Schistosoma mansoni (Fig. 9.3) has been used extensively in studies treating trinitrobenzenesulfonic acid (TNBS) colitis in animal models, as a live infection (Moreels et al. 2004; Smith et al. 2007), as a treatment using egg exposure (Elliott et al. 2003), and using soluble worm products (Ruyssers et al. 2010). Elliott et al. (2003) demonstrated a shift from Th1 to Th2 response upon exposure to *S. mansoni* eggs in TNBS-treated mice, and this resulted in prevention of lethal inflammation. This general shift was also found in studies that used live *S. mansoni* adult worms (Moreels et al. 2004; Smith et al. 2007). Ruyssers et al. (2010) found that cotreatment with TNBS and *S. mansoni* proteins led to the amelioration (measured at day 5 post-cotreatment) of colitis-induced peristaltic activity and gross disease scores as measured by validated inflammation parameters such as gastric emptying and cytokine profiles. Interestingly, Ruyssers et al. (2010) found that *S. mansoni* had no effect on the cytokine profiles of the TNBS treatment 5 days after cotreatment, which contradicted earlier studies that

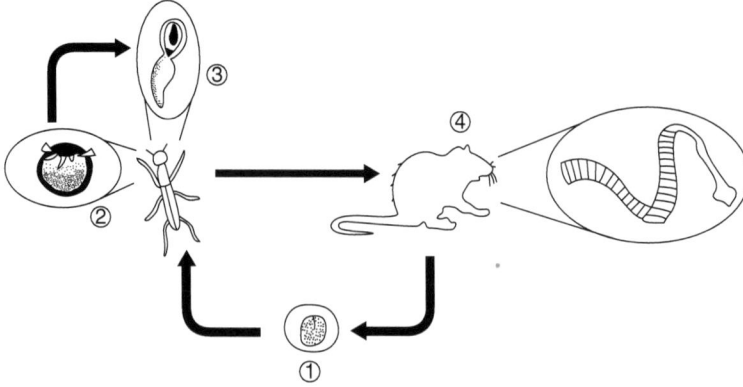

Fig. 9.2 Life cycle of the cestode *Hymenolepis diminuta*. (1) Eggs are passed in the feces of the vertebrate host. (2) Eggs are ingested by an insect intermediate host and oncospheres hatch upon ingestion. (3) Oncospheres develop into cysticerci within the insect. (4) The vertebrate host ingests the insect (which contains cystercerci), and adult worms develop within the vertebrate host

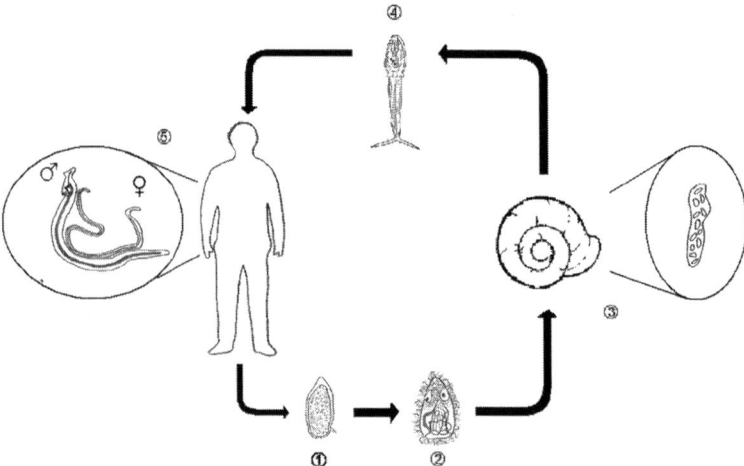

Fig. 9.3 Life cycle of the trematode *Schistosoma mansoni*. (1) Eggs are passed in the feces of the vertebrate host. (2) Upon contact with fresh water, eggs hatch into miracidia. (3) Miracidia penetrate into snail tissue and develop into sporocysts. (4) Cercariae are released by the snail host. (5) Cercariae penetrate through vertebrate skin. They then shed their tails, circulate, and develop into male and female adult worms

reported *S. mansoni* affected cytokine profiles 3 days after cotreatment. This result implies that repeated treatments with *S. mansoni* proteins may be required for the inducement of long-term remission in colitis patients.

Other parasites as noted in Table 9.1 have been used in the treatment of IBD. These include *Necator americanus* (Croese et al. 2006), *Heliogomosomoides polygyrus* (Elliott et al. 2004), *Trichinella spiralis* (Khan et al. 2002; Motomura et al.

2009), and *Schistosoma japonicum* (Mo et al. 2007). Most of these studies involved treating subjects with experimental or clinical colitis with infection of adult worms or treatment with live eggs. However, we believe that new studies that use worm products rather than whole worms are very important and crucial to the advancement of helminth therapy; by not infecting humans with viable eggs or adult worms there is minimal risk of worm infection developing in patients. Helminth therapy, particularly with the use of worm-derived products, appears to be promising as a future treatment option for the millions of people affected by IBD.

9.3 Multiple Sclerosis

Multiple sclerosis (MS) is a neurological disorder that is caused by an inflammatory response against the myelin sheath of a patient's central nervous system (Correale and Mauricio 2009). MS is commonly thought to be autoimmune in origin, with self-reactive Th1 and Th17 cells recognizing the myelin sheath (McFarland and Martin 2007). Studies have shown that MS has high rates of incidence in developed, industrialized countries (Kurtzke 2000). This disease is thought to be caused by a combination of genetic and environmental factors, but these factors are unknown at present (Correale and Mauricio 2009).

Studies on the potential use of helminth therapy as a treatment for MS often involve the use of mice with experimental autoimmune encephalitis (EAE), which is a mouse model for this disease (La Flamme et al. 2003; Sewell et al. 2003; Zheng et al. 2008). La Flamme et al. (2003) studied mice with EAE that were infected and uninfected with *S. mansoni* and found that the infected mice had a blunted inflammatory response as measured by the cellular composition of mouse spinal cords and brains. In conjunction with these results, they found that the infected mice had lower levels of IL-12p40, a Th1-associated cytokine (La Flamme et al. 2003). Similarly, Sewell et al. (2003) found decreased interferon-gamma (IFN-γ), a Th1-associated cytokine, and an increased presence of Th2-related cytokines IL-4, IL-10, and transforming growth factor-beta (TGF-β). Furthermore, infected mice which were deficient in the signal transducer and activator of transcription 6, a gene that plays a critical role in IL-4-mediated responses, lacked the decreased pathology of EAE that was seen in the infected control mice (Sewell et al. 2003). Similar results were noted in mice treated with soluble egg antigen (SEA) from *S. japonicum* (Zheng et al. 2008) and in mice infected with *Fasciola hepatica* (Walsh et al. 2009) and *Trichinella spiralis* (Gruden-Movsesijan et al. 2008).

Studies with MS patients have also been conducted, and a recent one found that helminth infection, as measured by many indicators of MS disease, was able to alleviate many symptoms of MS in patients; these positive results were associated with increased production of Th2-type cytokines IL-10 and TFG-β (Correale and Farez 2007). Additionally, it was found that MS patients with helminth infections had induced regulatory B cells, and through the increased production of IL-10 these B cells blunted the characteristic immune response of MS (Correale et al. 2008).

On the basis of these findings, Correale and Farez (2007) studied the effects of toll-like receptors (TLRs) on dendritic cell (DC) and B cell regulation in helminth-infected MS patients. TLR2 expression on DC and B cells was required for the SEA modulation of these cells to an anti-inflammatory profile (Correale and Mauricio 2009). These clinical studies, as well as the experimental mice studies, showed that the ability of helminths to shift towards a Th2-type immune response could be used as a future treatment for patients with MS.

9.4 Rheumatoid Arthritis

Rheumatoid Arthritis (RA) is a common autoimmune disease that mainly affects the bones, cartilage, and the synovium of patients (Kuijk and Van Die 2010). It is unknown what initiates this disease, but it is thought that autoreactive T cells and autoantibodies play an important role in the sustained inflammatory response (Kuijk and Van Die 2010). There are multiple drug therapies available, but for many patients the drugs never fully alleviate the disease.

Animal models have been used in the majority of studies on the effects of helminth therapy on RA with the majority of these studies using collagen-induced arthritis (CIA) as a model for RA. Early studies found that the incidence of arthritis is less severe in rats infected with the pinworm *Syphacia oblevata* (Pearson and Taylor 1975). *Nippostrongylus brasiliensis* (a rat nematode) has also been shown to ameliorate the severity of CIA in mice (Salinas-Carmona et al. 2009). In a study which compared the cytokine production in CIA-administered uninfected and *S. mansoni*-infected mice, Th1-type (IFN-γ) and inflammatory-type (TNF-α and IL-17A) cytokines were significantly lower in infected mice, while Th2-type (IL-4, IL-10) cytokines were upregulated in infected mice (Osada et al. 2009). Furthermore, *S. mansoni* infection prevented the augmentation of IL-1-β, IL-6, and receptor activation of NFκB (Osada et al. 2009). These findings show that *S. mansoni* infection has anti-inflammatory effects which prevent bone destruction (Osada and Kanazawa 2010). In addition to helminth infection, worm products have been used in various experiments on animals afflicted with CIA. Rocha et al. (2008) used an extract from *Ascaris suum* (a pig roundworm) to treat mice and rats with zymosan-induced arthritis (ZMA) and CIA. By all measures of clinical severity, *A. suum* extract functioned species-independently to reduce the severity of the arthritis pathology.

One of the more promising areas of research of helminth therapy as a treatment for arthritis concerns a protein termed ES-62 (Kuijk and Van Die 2010). ES-62 is a protein originally derived from *Ancanthocheilonema viteae* (a rodent filarial nematode), which contains phosphorylcholine (PC) in its natural form, and is known to induce general anti-inflammatory Th2-type properties (Harnett et al. 2003). McInnes et al. (2003) studied mice afflicted with CIA, and observed that mice that were subcutaneously administered PC-containing ES-62 had an inhibited

Th1-type inflammatory cytokine (TNF-α, IL-6, and IFN-γ) response. In addition to this, cells derived from human RA donors had suppressed TNF-α and IL-6 production (McInnes et al. 2003). Harnett et al. (2008) found that these immunosuppressive properties of ES-62 are dependent on the PC moiety. The research into helminth therapy as a treatment for RA, particularly using ES-62, may eventually lead to an effective treatment for patients suffering from RA.

9.5 Type-1 Diabetes

Type-1 Diabetes (T1D) is a disease caused by the autoimmune destruction of pancreatic β cells, which produce the hormone insulin. The full mechanism behind this destruction is unknown, but $CD4^+$ and $CD8^+$ T cells are known to play important roles in this autoimmune reaction (Tisch and McDevitt 1996). The disease is characterized as a Th1-type response, but studies have been found which implicate Th17 in T1D (Osada and Kanazawa 2010). Though T1D is manageable, the lifelong symptoms and therapy are difficult for patients, and this make preventative and regenerative therapies important to the millions of people afflicted with this disease (Kuijk and Van Die 2010).

The second half of the twentieth century has seen a rise of T1D (Gale 2002). Onkamo et al. (1999) found a 3% per year increase in T1D from 1960 to 1996. Diabetes has a particularly high rate in Western countries, and environmental factors, including the decrease in chronic helminth infections, are likely a major cause of this increase in disease incidence (Gale 2002).

Infections with a variety of parasites have been found to have a beneficial effect on induced diabetes in nonobese diabetic (NOD) mice (Table 9.1). Cooke et al. (1999) infected mice with *S. mansoni* and found a 40% decrease in the incidence of diabetes in infected NOD mice. Thirty percent of *S. mansoni*-infected mice still developed diabetes, and this could be explained by the length of time it takes for the infection to produce eggs (Cooke et al. 1999). This study also found that injection of eggs into the mice completely blocked the incidence of diabetes (Cooke et al. 1999). Zaccone et al. (2003) observed that the effects of *S. mansoni* SEA on NOD mice were characterized by a switch from diabetes-associated Th1 to Th2 response, as indicated by IL-4, IL-5, IL-10, and IL-13 production. Zaccone et al. (2009) investigated the protective mechanism of SEA further, and observed that NOD mice that were administered SEA had a significant number of pancreatic Treg cells positive for expression of forkhead box P3 (FoxP3). This protection by $FoxP3^+$ Tregs was dependent on TGF-β and DC cells (Zaccone et al. 2009). Studies conducted with *Litomosoides sigmodontis* (a mouse filarial worm) have also found a shift to a Th2-type response in conjunction with increased incidence of $FoxP3^+$ Treg cells (Hübner et al. 2009). A possible therapy that targets these specific $FoxP3^+$ Tregs would be preferable, because it would allow for the specific suppression of immune response while allowing response to other antigens

(Zaccone and Cooke 2010). Multiple molecules have been implicated as possible triggers for a Treg response that would be beneficial for patients with diabetes (Zaccone and Cooke 2010).

Many other helminths have been used in the treatment of NOD mice, including the nematodes *Dirofilaria immitus* (Imai et al. 2001), *Heligomosomoides polygyrus* (Liu et al. 2009; Saunders et al. 2007), *Trichinella spiralis* (Saunders et al. 2007), and the cestode *Taenia crassiceps* (Espinoza-Jimenez et al. 2010). Saunders et al. (2007) studied both *H. polygyrus* and *T. spiralis*, and found that these two nematodes had different mechanisms in response to NOD mice; this is probably due to differences in the life cycles of the two parasites. This is an important observation, because the diversity of life stages, and the molecules present in these life stages, could make a difference in the efficacy of helminth therapy on patients with T1D. Figures 9.1–9.3 show the life cycles of three different helminths, and they illustrate the diversity of the life cycles of different helminths. Helminth therapy could solve the problem of monotonous, life-spanning therapies that current patients with T1D must endure, and may be used as a regenerative or preventative treatment.

9.6 Asthma and Allergic Responses

The immune system responds to a seemingly infinite amount of harmful foreign antigens while ignoring harmless antigens (Smits et al. 2010). As many of the current world health problems are being treated, many nations, particularly Western nations, have seen an epidemic rise in childhood allergies, including rhinitis, atopic dermatitis, and allergic and nonallergic asthma (Schaub et al. 2006). The hygiene hypothesis has been proposed as a possible environmental factor behind this rise in allergic diseases.

The typical allergic response is a result of Th2-promoted immunity (Folkerts et al. 2000). This makes it difficult to understand how helminths, which are thought to promote a Th2-type environment, can blunt the self-reactive response seen in typical childhood allergy. What has been suggested to date is that helminths work to create a regulatory network of Tregs, B cells, DCs, macrophages, and local stromal cells to create an anti-inflammatory environment rich in IL-10 and TGF-β (Smits et al. 2010).

Some helminths have been found to use a Treg mechanism to reduce inflammation of allergic responses in mouse models (Table 9.1), including *S. japonicum* (Yang et al. 2006), *H. polygyrus* (Wilson et al. 2005), and *L. sigmodontis* (Dittrich et al. 2008). In summary, the aforementioned studies found a decrease in IL-5 and other Th2 cytokines, as well as a rise in Tregs and IL-10, when mice were administered helminths for suppression of sensitization and inflammation in a murine asthma model. It should be noted that Wilson et al. (2005) found that mice that were deficient in IL-10 still had suppression of asthma by helminth infection, illustrating that IL-10 is not necessary for this effect.

Other mechanisms have been proposed for helminth protection against allergies. Mangan et al. (2004) studied the effects of *S. mansoni* infection on anaphylaxis in mice, and found that IL-10-producing B cells had a critical role in the protection that this helminth provided. Studies have also looked into the effects of various worm antigens and life-cycle stages on allergic response. Mangan et al. (2006) found that infection with male-only *S. mansoni* worms blunted allergen-induced airway hyperresponsiveness in mice, while traditional infection with male and female worms exacerbated this harmful immune response. Figure 9.3 shows the life cycle of *S. mansoni*. Itami et al. (2005) studied the effects of the allergenic protein of *A. suum* (APAS-3) and the suppressive protein of *A. suum* (PAS-1) on murine experimental asthma. APAS-3 induced a Th2-type immune response while exacerbating IgE antibody production, eosinophilic airway inflammation, and hyperresponsiveness (Itami et al. 2005). Mice that were administered PAS-1 had reduced values comparable to control mice (Itami et al. 2005). Various life cycle stages and worm products can induce different immune responses in experimental models of asthma.

T. suis live eggs were used in a study on human patients with grass pollen-induced allergic rhinitis (Bager et al. 2010). No significant differences were found in mean symptom scores, total histamine, or grass-specific IgE (Bager et al. 2010). It could be that treatment with *T. suis* does not create an immune-suppressive environment which quells the symptoms of a patient with allergies. Other worm antigens or other helminths should be investigated for their efficacy as a treatment for patients with asthma and other allergies.

9.7 Concluding Remarks

There has been a global increase in autoimmune diseases, particularly the ones mentioned in this review. With the rise in the incidence of such diseases comes an increased need for novel therapies to treat or ameliorate the pathology of these illnesses. Helminth therapy is a promising area of research on these diseases particularly because of the many successful studies that have been done using animal models. The efficacy of helminth therapy to treat autoimmune disorders has been demonstrated unequivocally. What must now be done is to develop actual treatments for patients suffering from these diseases. Of most importance in this regard is the development of treatments which do not involve the use of live worms or viable eggs, thus lowering the risk of infection and further justifying use of the treatment. If advances in this field continue, helminth therapy will someday become a viable treatment option for patients suffering from autoimmune disease.

References

Alic M (2000) Inflammatory bowel diseases are diseases of higher socioeconomic status: dogma or reality? Am J Gastroenterol 95:3332–3333

Bager P, Arnved J, Ronborg S, Wohlfahrt J, Poulsen LK, Westergaard T, Petersen HW, Kristensen B, Thamsborg S, Roepstorff A, Kapel C, Melbye M (2010) *Trichuris suis* ova therapy for allergic rhinitis: a randomized, double-blind, placebo-controlled clinical trial. J Allergy Clin Immunol 125:123–130

Braus NA, Elliott DE (2009) Advances in the pathogenesis and treatment of IBD. Clin Immunol 132:1–9

Cooke A, Tonks P, Jones FM, O'Shea H, Hutchings P, Fulford AJC, Dunne DW (1999) Infection with *Schistosoma mansoni* prevents insulin dependent diabetes mellitus in non-obese diabetic mice. Parasite Immunol 21:169–176

Correale J, Farez M (2007) Association between parasite infection and immune responses in multiple sclerosis. Ann Neurol 61:97–108

Correale J, Mauricio F (2009) Helminth antigens modulate immune responses in cells from multiple sclerosis patients through TLR2-dependent mechanisms. J Immunol 183:5999–6012

Correale J, Farez M, Razzite G (2008) Helminth infections associated with multiple sclerosis induce regulatory B cells. Ann Neurol 64:187–199

Croese J, O'Neil J, Masson J, Cooke S, Melrose W, Pritchard D, Speare R (2006) A proof of concept study establishing *Necator americanus* in Crohn's patients and reservoir donors. Gut 55:136–137

Dittrich AM, Erbacher A, Specht S, Diesner F, Krokowski M, Avagyan A, Stock P, Ahrens B, Hoffmann WH, Hoerauf A, Hamelmann E (2008) Helminth infection with Litomosoides sigmodontis induces regulatory T cells and inhibits allergic sensitization, airway inflammation, and hyperreactivity in a murine asthma model. J Immunol 180:1792–1799

Ehrhardt RO (1996) New insights into the immunopathology of chronic inflammatory bowel disease. Semin Gastrointest Dis 7:144–150

Elliott DE, Li J, Blum A, Metwali A, Qadir K, Urban JF Jr, Weinstock JV (2003) Exposure to schistosome eggs protects mice from TNBS-induced colitis. Am J Physiol 284:385–391

Elliott DE, Setiawan T, Metwali A, Blum A, Urban JF Jr, Weinstock JV (2004) *Heligmosomoides polygyrus* inhibits established colitis in IL-10-deficient mice. Eur J Immunol 34:2690–2698

Elliott DE, Summers RW, Weinstock JV (2005) Helminths and the modulation of mucosal inflammation. Curr Opin Gastroenterol 21:51–58

Espinoza-Jimenez A, Rivera-Montoya I, Cardenas Arreola R, Moran L, Terrazas LI (2010) *Taenia crassiceps* infection attenuates multiple low-dose streptozotocin-induced diabetes. J Biomed Biotechnol. doi:10.1155/2010/850541

Folkerts G, Walzl G, Openshaw PJM (2000) Do common childhood infections 'teach' the immune system not to be allergic? Immunol Today 21:118–120

Fukumoto S, Iriko H, Otsuki H (2009) Helminth infections prevent autoimmune diseases through Th2-type immune response. Yonago Acta Med 52:85–104

Gale EA (2002) The rise of childhood type 1 diabetes in the 20th century. Diabetes 51:3353–3361

Gruden-Movsesijan A, Ilic N, Mostarica Stojkovic M, Stosic-Grujicic S, Milic M, Sofronic-Milosavljevic L (2008) *Trichinella spiralis*: modulation of experimental autoimmune encephalomyelitis in DA rats. Exp Parasitol 118:641–647

Harnett W, Harnett MM, Byron O (2003) Structural/functional aspects of ES-62: a secreted immunomodulatory phosphorylcholine-containing filarial nematode glycoprotein. Curr Prot Peptide Sci 4:59

Harnett MM, Kean DE, Boitelle A, McGuiness S, Thalhamer T, Steiger CN, Egan C, Al-Riyami L, Alcocer MJ, Houston KM, Gracie JA, McInnes IB, Harnett W (2008) The phosphorycholine moiety of the filarial nematode immunomodulator ES-62 is responsible for its anti-inflammatory action in arthritis. Ann Rheum Dis 67:518–523

Hübner MP, Stocker JT, Mitre E (2009) Inhibition of type 1 diabetes in filaria-infected non-obese diabetic mice is associated with a T helper type 2 shift and induction of FoxP3+ regulatory T cells. Immunology 127:512–522

Humphreys NE, Xu D, Hepworth MR, Liew FY, Grencis RK (2008) IL-33, a potent inducer of adaptive immunity to intestinal nematodes. J Immunol 180:2443–2449

Hunter MM, Wang A, McKay D (2007) Helminth infection enhances disease in a murine TH2 model of colitis. Gastroenterology 132:1320–1330

Hunter MM, Wang A, Parhar KS, Johnston MJG, Van Rooijen N, Beck PL, McKay DM (2010) In vitro-derived alternatively activated macrophages reduce colonic inflammation in mice. Gastroenterology 138:1395–1405

Imai S, Tezuka H, Fujita K (2001) A factor of inducing IgE from a filarial parasite prevents insulin-dependent diabetes mellitus in nonobese diabetic mice. Biochem Biophys Res Commun 286:1051–1058

Itami DM, Oshiro TM, Araujo CA, Perini A, Martins MA, Macedo MS, Macedo-Saores MF (2005) Modulation of murine experimental asthma by *Ascaris suum* components. Clin Exp Allergy 35:873–879

Johnston MJG, Wang A, Catarino MED, Ball L, Phan VC, MacDonald JA, McKay DM (2010) Extracts of the rat tapeworm, *Hymenolepis diminuta*, suppress macrophage activation *in vitro* and alleviate chemically induced colitis in mice. Infect Immunol 78:1364–1375

Khan WI, Blennerhasset PA, Varghese AK, Chowdhury SK, Omsted P, Deng Y, Collins SM (2002) Intestinal nematode infection ameliorates experimental colitis in mice. Infect Immunol 70:5931–5937

Kuijk LM, Van Die I (2010) Worms to the rescue: can worm glycans protect from autoimmune diseases? Life 62:303–312

Kurtzke JF (2000) Multiple sclerosis in time and space – geographic clues to cause. J Neurovirol 6: S134–S140

La Flamme AC, Ruddenklau K, Backstrom BT (2003) Schistosomiasis decreases central nervous system inflammation and alters the progression of experimental autoimmune encephalomyelitis. Infect Immunol 71:4996–5004

Liu Q, Sundar K, Mishra PK, Mousavi G, Liu ZG, Gaydo A, Alem F, Lagunoff D, Bleich D, Gause WC (2009) Helminth infection can reduce insulitis and type 1 diabetes through CD25- and IL-10-independent mechanisms. Infect Immunol 77:5347–5358

Mangan NE, Fallon RE, Smith P, Van Rooijen N, McKenzie AN, Fallon PG (2004) Helminth infection protects mice from anaphylaxis via IL-10-producing B cells. J Immunol 173:6346–6356

Mangan NE, Van Rooijen N, McKenzie ANJ, Fallon PG (2006) Helminth-modified pulmonary immune response protects mice from allergen-induced airway hyperresponsiveness. J Immunol 176:138–147

McFarland HF, Martin R (2007) Multiple sclerosis: a complicated picture of autoimmunity. Nat Immunol 8:913–919

McInnes IB, Leung BP, Harnett M, Gracie JA, Liew FY, Harnett W (2003) A novel therapeutic approach targeting articular inflammation using the filarial nematode-derived phosphorylcholine-containing glycoprotein ES-62. J Immunol 171:2127–2133

Mo HM, Liu WQ, Lei JH, Cheng YL, Wang CZ, Li YL (2007) *Schistosoma japonicum* eggs modulate the activity of $CD4^+$ $CD25^+$ Tregs and prevent development of colitis in mice. Exp Parasitol 116:385–389

Moreels TG, Nieuwendijk RJ, De Man JG, De Winter BY, Herman AG, Van Marck EA, Pelckmans PA (2004) Concurrent infection with *Schistosoma mansoni* attenuates inflammation induced changes in colonic morphology, cytokine levels, and smooth muscle contractility of trinitrobenzene sulphonic acid induced colitis in rats. Gut 53:99–107

Motomura Y, Wang H, Deng Y, El-Sharkawy RT, Verdu EF, Khan WI (2009) Helminth antigen-based strategy to ameliorate inflammation in an experimental model of colitis. Clin Exp Immunol 155:88–95

Onkamo P, Vaananen S, Karvonen M, Tuomilehto J (1999) Worldwide increase in incidence of type 1 diabetes – the analysis of the data on published incidence trends. Diabetologia 42:1395–1403

Osada Y, Kanazawa T (2010) Parasitic helminths: new weapons against immunological disorders. J Biomed Biotechnol. doi:10.1155/2010/743758

Osada Y, Shimizu S, Kumagai T, Yamada S, Kanazawa T (2009) *Schisosoma mansoni* infection reduces severity of collagen-induced arthritis via down-regulation of proinflammatory mediators. Int J Parasitol 39:457–464

Pearson DJ, Taylor G (1975) The influence of the nematode *Syphacia oblevata* on adjuvant arthritis in the rat. Immunology 29:391–396

Reddy A, Fried B (2007) The use of *Trichuris suis* and other helminth therapies to treat Crohn's disease. Parasitol Res 100:921–927

Reddy A, Fried B (2009) An update on the use of helminths to treat Crohn's and other autoimmune diseases. Parasitol Res 104:217–221

Rocha FAC, Leite AKRM, Pompeu MML, Cunha TM, Verri WA, Soares FM, Castro RR, Cunha FQ (2008) Protective effect of an extract from *Ascaris suum* in experimental arthritis models. Infect Immunol 76:2736–2745

Ruyssers NE, De Winter BY, De Man JG, Loukas A, Pearson MS, Weinstock JV, Van den Bossche RM, Martinet W, Pelckmans PA, Moreels TG (2009) Therapeutic potential of helminth soluble proteins in TNBS-induced colitis in mice. Inflamm Bowel Dis 15:491–500

Ruyssers NE, De Winter BY, De Man JG, Ruyssers ND, Van Gils AJ, Loukas A, Pearson MS, Weinstock JV, Pelckmans PA, Moreels TG (2010) *Schistosoma mansoni* proteins attenuate gastrointestinal motility disturbances during experimental colitis in mice. World J Gastroentrol 16:703–712

Salinas-Carmona MC, De la Cruz-Galacia G, Perez-Rivera I, Solis-Soto JM, Segoviano-Ramirez JC, Vazquez AV, Garza MA (2009) Spontaneous arthritis in MRL/lpr mice is aggravated by *Staphylococcus aureus* and ameliorated by *Nippostrongylus brasiliensis* infections. Autoimmunity 42:25–32

Sartor RB (2006) Mechanisms of disease: pathogenesis of Crohn's disease and ulcerative colitis. Nat Clin Prac Gastroenterol Hepatol 3:390–407

Saunders KA, Raine T, Cooke A, Lawrence CE (2007) Inhibition of autoimmune type 1 diabetes by gastrointestinal helminth infection. Infect Immunol 75:397–407

Schaub B, Lauener R, von Mutius E (2006) The many faces of the hygiene hypothesis. J Allergy Clin Immunol 117:969–977

Sewell D, Qing Z, Reinke E, Elliot D, Weinstock J, Sandor M, Fabry Z (2003) Immunomodulation of experimental autoimmune encephalomyelitis by helminth ova immunization. Int Immunol 15:59–69

Smith P, Mangan NE, Walsh CM, Fallon RE, McKenzie AN, Van Rooijen N, Fallon PG (2007) Infection with a helminth parasite prevents experimental colitis via a macrophage-mediated mechanism. J Immunol 178:4557–4566

Smits HH, Everts B, Hartgers FC, Yazdanbakhsh M (2010) Chronic Helminth infections protect against allergic diseases by active regulatory processes. Curr Allergy Asthma Rep 10:3–12

Summers RW, Elliott DE, Urban JF Jr, Thompson R, Weinstock JV (2005a) *Trichuris suis* therapy in Crohn's disease. Gut 54:87–90

Summers RW, Elliott DE, Weinstock JV (2005b) Is there a role for helminths in the therapy of inflammatory bowel disease? Nat Clin Prac Gastroenterol Hepatol 2:62–63

Summers RW, Elliott DE, Weinstock JB (2005c) Why *Trichuris suis* should prove safe for use in inflammatory bowel diseases. Inflamm Bowel Dis 11:783–784

Summers RW, Elliott DE, Urban JF Jr, Thompson RA, Weinstock JV (2006) *Trichuris suis* therapy for active ulcerative colitis: a randomized controlled trial. Gastroenterology 128:825–832

Tisch R, McDevitt H (1996) Insulin-dependent diabetes mellitus. Cell 85:291–297

Van Kruiningen HJ, West AB (2005) Potential danger in the medical use of *Trichuris suis* for the treatment of inflammatory disease. Inflamm Bowel Dis 11:515

Walsh KP, Brady MT, Finlay CM, Boon L, Mills KH (2009) Infection with a helminth parasite attenuates autoimmunity through TGF-β-mediated suppression of Th17 and Th1 responses. J Immunol 183:1577–1586

Weinstock JV, Summers RW, Elliott DE (2005) Role of helminths in regulating mucosal inflammation. Springer Semin Immunopathol 27:249–271

Wilson MS, Taylor MD, Balic A, Finney CAM, Lamb JR, Maizels RM (2005) Suppression of allergic airway inflammation by helminth-induced regulatory T cells. J Exp Med 202:1199–1212

Yang JH, Zhao JQ, Yang YF, Zhang L, Yang X, Zhu X, Ji MJ, Sun NX, Su C (2006) *Schistosoma japonicum* egg antigens stimulate $CD4^+CD25^+$ T cells and modulate airway inflammation in a murine model of asthma. Immunology 120:8–18

Yoshimoto T, Nakanishi K (2006) Roles of IL-18 in basophils and mast cells. Allergol Int 55:105–113

Zaccone P, Cooke A (2010) Harnessing $CD8^+$ regulatory T cells: therapy for type-1 diabetes? Immunity 32:504–506

Zaccone P, Fehervari Z, Jones FM, Sidobre S, Kronenberg M, Dunne DW, Cooke A (2003) *Schistosoma mansoni* antigens modulate the activity of the innate immune response and prevent onset of type 1 diabetes. Eur J Immunol 33:1439–1449

Zaccone P, Burton O, Miller N, Jones FW, Dunne DW, Cooke A (2009) *Schistosoma mansoni* egg antigens induce Treg that participate in diabetes prevention in NOD mice. Eur J Immunol 39:1098–1107

Zheng X, Hu X, Zhou G, Lu Z, Qiu W, Bao J, Dai Y (2008) Soluble egg antigen from *Schistosoma japonicum* modulates the progression of chronic progressive experimental autoimmune encephalomyelitis via Th2-shift response. J Neuroimmunol 194:107–114

Chapter 10
Insects Help to Solve Crimes

Jens Amendt

Abstract Insects are found in almost any ecosystem. Even a corpse is a habitat with a typical succession of more or less specialized insect species, providing a resource for about 300–400 insect species. In addition to their ecological importance in decomposition of organic matter, those insects may represent important tools in criminal investigations. Forensic entomology, defined as the use of insects and other arthropods in medicocriminal investigations, is best known as a means of providing evidence of the time since death of a body based on the fauna associated with it, although the discipline covers other topics too, all of which are described in the following chapter (Greenberg and Kunich, Entomology and the Law. Cambridge University Press, Cambridge, 2002; Amendt et al. Int J Legal Med 121:90–104, 2007).

10.1 Insects and Death: A Short History

Insects were first used in the context of a legal matter in the thirteenth century in China, when a farmer had been killed in a rice field. All suspects were called together and had to place their sickles on the ground. No obvious evidence was recognizable, but one sickle attracted numerous blowflies, apparently by invisible traces of blood on the blade. The owner of the sickle, when confronted with this entomological evidence, confessed to the killing. In Europe, despite many realistic and detailed illustrations during Medieval times (Fig. 10.1), which demonstrated the relation between death, decomposition and insects, the correlation between maggots on a cadaver and the oviposition of adult flies was not recognized. Just in the seventeenth century, the metamorphosis of insects became common knowledge and finally at the beginning of the eighteenth century it was noted that flies are

---

J. Amendt
Forensic Biology/Entomology, Institute of Forensic Medicine, Kennedyallee 104, 60 596 Frankfurt am Main, Germany
e-mail: amendt@em.uni-frankfurt.de

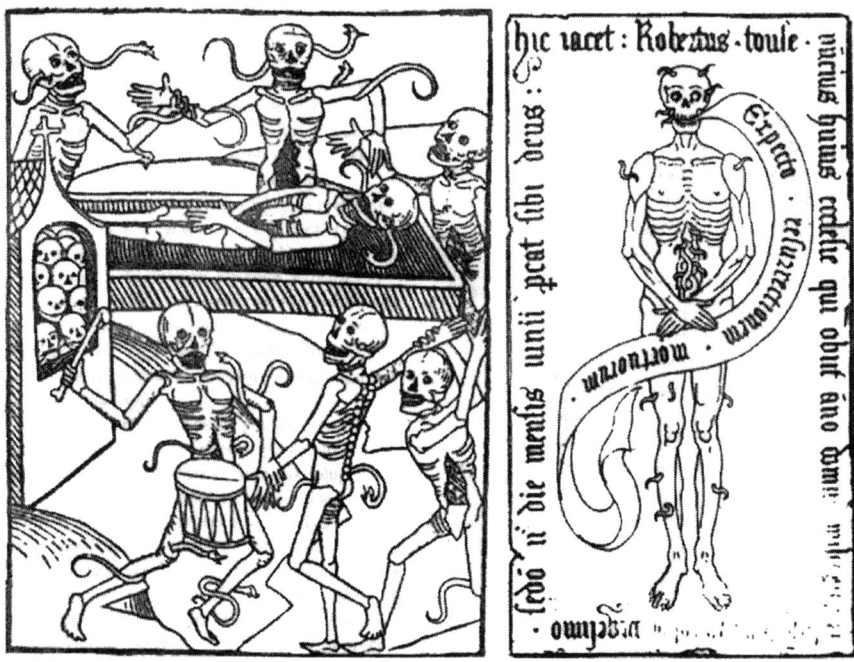

Fig. 10.1 Realistic descriptions of insect infestation from Medieval times (maggots with a snake-like shape) *Left*: "Dance of the dead" (ca. 1460, from: Stammler W.: Der Totentanz, München 1922); *Right*: Epitaph with banner, date unknown (after Grassberger and Amendt 2010)

attracted by corpses at a very early stage of decomposition. In 1829, Mende compiled a list of necrophagous insects, including flies and beetles as well as other taxa. In 1850, skeletonized remains of a child, infested by insects, were found behind a chimney by workmen during redecoration in France (Bergeret 1855). Even though the forensic expert at that time concluded that the development of the adult flies lasted about one year, and therefore his results would be questionable today, this case was a kind of breakthrough for forensic entomology in Europe.

In the following years, topics such as the grave fauna, the skeletonizing of corpses, or the modification of dead bodies caused by insects were explored (Greenberg and Kunich 2002), but data concerning the biology, ecology, and succession of necrophagous insects were not applied to criminal investigations since forensic examiners had no idea of insect biology. In a slow process, forensic entomology starts to move in Europe. Leclercq (Belgium), Nuorteva (Finland), and Schumann (East Germany) were the first to use forensic entomology in Europe in the 1950s and 1960s of the twentieth century for the determination of the time since death, the so-called postmortem interval (PMI).

The first comprehensive textbook of entomological methods for estimating time of death was provided by Smith (1986), and subsequently updated in books edited by Catts and Haskell (1990), Byrd and Castner (2010, 1st edition 2001), and

Amendt et al. (2010). The rate at which new methods for obtaining evidence are becoming available has escalated dramatically since 2000, and forensic entomology is, in a sense, coming of age.

10.2 Insects on Cadavers

Necrophagous insects are attracted to a body immediately after death, often within minutes, depending on the level of accessibility and environmental conditions (Smith 1986; Campobasso et al. 2001). Usually the first taxa to arrive on a body are flies (Diptera), mainly blowflies (Calliphoridae), which can locate an odor source with great spatial precision (Greenberg 1991). Larvae will hatch from the eggs and feed on the dead tissues. As they grow they shed their cuticle twice, and after each molt a new larval instar is formed. When the third and last instar finishes feeding it enters the postfeeding or wandering stage, when most species migrate away from the body to find shelter, either within soil or underneath objects, for example stones or leaf litter, or, if the scene of death is indoor, underneath furniture. Here, they form pupae within a protective outer case, the puparium (the hardened cuticle of the third instar larva), from which adult flies emerge at the completion of metamorphosis. Flies are the most important, but not the only group of insects found on cadavers. According to Smith (1986) four ecological categories can be recognized in a carrion species-community:

1. Necrophagous species, feeding on the carrion, like blowflies or silphid beetles.
2. Predators and parasites of necrophagous species, feeding on other insects or arthropods, like parasitic wasps or rope beetles. This group also contains schizophagous species, i.e., species that are feeding on the carrion at the beginning, but may become predaceous in late instar stages.
3. Omnivorous species such as wasps, ants, and some beetles feeding both on the corpse and its colonizers.
4. Other species, which use the corpse as an extension of their environment, e.g. springtails and spiders.

Many insects are attracted to certain stages of decomposition as the attractiveness of a decaying body differs to necrophilous insects (Fig. 10.2). They will not all occur simultaneously on a cadaver, but in a more or less predictable chronological sequence, known as insect succession (Anderson 2010). However, their occurrence on the body does not necessarily indicate that they have colonized the body. Some groups do not feed on dead tissue at all, but are attracted to a corpse to feed as predators on necrophagous insects.

For the purpose of forensic entomology the first two groups are the most important. They include mainly species from the orders Diptera (flies) and Coleoptera (beetles). Blowflies are attracted to the carrion by the odor, produced during decomposition, even over large distances. Besides those olfactory stimuli, vision, color, and the presence of other conspecific insects on the dead body play a role.

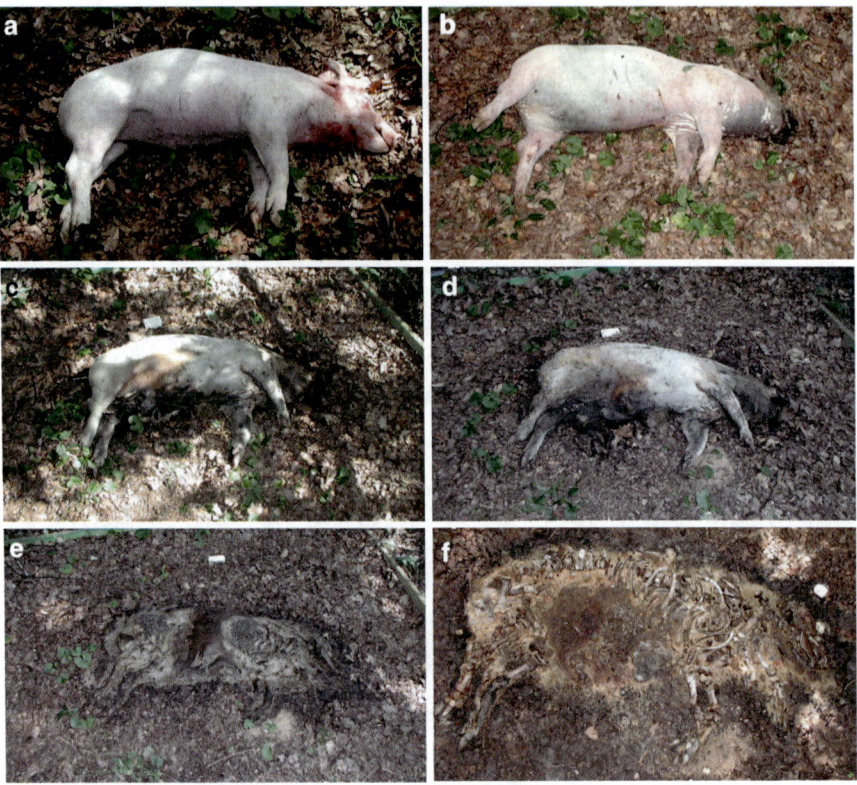

Fig. 10.2 Decomposition of a pig carcass in a forest in Germany during a period of 59 days (June and July); mean temperature 19.5°C (min 10.3°C and max 32.9°C); (**a**) postmortem interval of 1 day, (**b**) PMI of 4 days, (**c**) PMI of 9 days, (**d**) PMI of 21 days, (**e**) PMI of 34 days, (**f**) PMI of 59 days (after Amendt et al. 2005)

Last but not least ammonia-rich compounds and hydrogen sulfide are important stimulants for oviposition, as well as moisture and tactile stimulants. Female Diptera recognize dehydrated or mummified tissue as an inappropriate substrate, as eggs and larvae need moisture for successful development. Several Coleoptera, especially the larder beetles (Dermestidae), are adapted to utilize dry foods like skin, and even hairs and bones. They tend, therefore, to be attracted to the very last stages of decomposition.

10.3 Estimating the Time of Death

Supporting or disproving an alibi, linking a suspect with a scene of crime: The time of death is always crucial in crime investigations. However, the number of methods to achieve this goal is quite limited from a medical point of view.

10.3.1 Pathology

After death a human body undergoes many changes (Table 10.1) caused by autolysis of tissue, which is promoted by the internal chemical breakdown of cells, released enzymes as well as by the activity of bacteria and fungi from both the intestine and outer environment. The body temperature decreases (algor mortis) and the skin color turns to red (livor mortis or lividity) due to the gravitational pooling of blood in dependent body parts. The stiffening of the muscle fibers due to the breakdown of glycogen and the accumulation of lactic acid (rigor mortis) is another mark of death. Its duration depends on the metabolic state at death and on several factors such as body size, surrounding temperature, etc. Later, large quantities of putrefaction gasses cause the physical distortion of the body. Hydrogen sulfide (H_2S) reacts with hemoglobin and forms a green pigment which initially marks the superficial blood vessels, but later may also be seen as a green coloration in the gastrointestinal region and those portions of the body, where livor mortis was most prominent. All these parameters can be used to estimate the Postmortem interval (PMI), which refers to the time between the death and discovery of a corpse, but many of these are reciprocal functions and become inaccurate in

Table 10.1 Postmortal changes (21°C ambient temperature und 30% humidity); after Clark et al. (1997) and Amendt et al. (2004)

Time after death	Postmortem changes	Modifiers	Category
0 min	Circulation and breathing stop Pallor Early lividity Muscular relaxation Sphincters may relax	Temperature Humidity Outdoor location Indoor location Submerged in water	Early changes
2 h	Vascular changes in eye Rigor mortis begins Algor mortis begins Lividity easily seen		Late changes
4–5 h	Coagulation of blood Fixation of lividity		
24 h	Drying of cornea Reliquefication of blood		Putrid Tissue changes
48 h	Rigor disappears Intravascular hemolysis		
72 h	Loss of hair and nails		
96 h	Skin slippage and bulla formation Bacterial overgrowth	Insect activity Animal activity	Bloated
Days–Month	Green discoloration Bloating Release of gasses Release of liquefied internal organs Gradual loss of soft tissues Partial skeletonization Complete skeletonization	Mummification Adipocere formation	Destruction Skeleton

application very quickly. Moreover, they are limited to the first 72 h postmortem (Campobasso et al. 2001).

10.3.2 Entomology

During that 72 h and well beyond, insects can be a very powerful tool for estimating the time since death. However, it has to be stated clearly, that the forensic entomologist calculates not the time since death, but the "time of colonization" or "period of insect activity." This is simply due to the fact that there might be a lag between the onset of decomposition at death, the arrival of adult insects, and the deposition of the first immature insects. The clock for the entomologist is started only once eggs are laid, therefore the time given by the forensic entomologist is a so-called minimum postmortem interval (PMI_{min}) because it estimates the most recent date or time at which death may have occurred. The lag between death and the oviposition of insects may be of the magnitude of less than an hour up to several days (Amendt et al. 2007).

Contrary to this, there might be a prior-to-death colonization and the immature stages of flies infest humans or animals premortem (termed "Myiasis" according to Zumpt 1965: *The infestation of live human and vertebrate animals with dipterous larvae, which, at least for a certain period, feed on the host's dead or living tissue, liquid body-substances, or ingested food*). In veterinary cases, the period of infestation of livestock or pet animals in cases of cruelty or neglect must be clarified. Human infestations are most common in the young, the very elderly or debilitated, especially at sites of wounding such as ulcers (Hall and Wall 1995; Goff et al. 2010). It can occur in domestic situations as well as in a hospital. In a forensic context those cases of neglect of people who are in need of care needs to prove the period of infestation, using the same techniques as are used in estimating the PMI_{min} of dead bodies (Amendt et al. 2007).

10.3.2.1 Development

Insects are poikilotherms (their body temperatures fluctuate greatly), which means that ambient temperature has a great influence on their species-specific metabolic rates. Knowing those temperatures enables the forensic investigator to calculate the growth of the immature stages of necrophagous insects, and eventually establish a PMI_{min}.

Morphology and Metamorphosis

Within upper and lower lethal limits, warmer conditions accelerate insect metabolism, so that both developmental and ecological processes speed up. Finally, the age of specimens is positively correlated with the summed thermal input they accumulate during growth, which is calibrated usually in thermal accumulation units such as degree hours (hrs*°C) or degree days (d*°C), which are a summation of temperature (°C) above the lower threshold multiplied by time (hours or days) (Higley and Haskel 2010; Villet et al. 2010). By knowing the stage of development of the oldest immature insects on the body, and the environmental temperature of the crime scene while the body was in situ, it is possible to calculate an accurate PMI_{min} from this model. This method requires that one establishes the threshold temperature below which the physiological activity effectively ceases and therefore stops accumulating, which means that the physiological process is measured over a range of realistic temperatures for every species of interest (Donovan et al. 2006).

There are other less sophisticated methods for estimating the age of a fly specimen, like simply plotting the duration of developmental events (e.g., egg hatching or the onset of pupariation) against temperature. Each contour in this so-called isomorphen diagram (Fig. 10.3) represents one of these developmental events (Villet et al. 2010). This technique only models the timing of developmental events, and shows no gradation between events, and the estimate is only accurate if derived from live entomological evidence which is reared through to the next developmental event at a known, constant temperature. The isomegalen diagram (Fig. 10.4) can estimate PMI_{min} using dead larvae, and has the capability of accounting for and analyzing fluctuating environmental temperature data (Grassberger and Reiter 2001). It models larval size (length, weight, or width) as a measure of age, against temperature and time. Larval size is a measure of age with a higher resolution than developmental event and shows details of growth between two

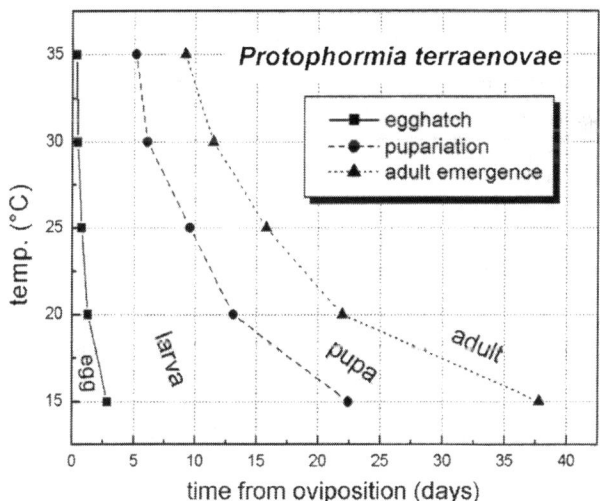

Fig. 10.3 Growth curves for the blowfly *Protophormia terraenovae* (so-called Isomorphen-Diagram), showing the time required to reach the larval, pupal, and adult stage at 15, 20, 25, 30, and 35°C; areas between lines represent identical morphological stages, e.g., pupa (Grassberger and Reiter 2002)

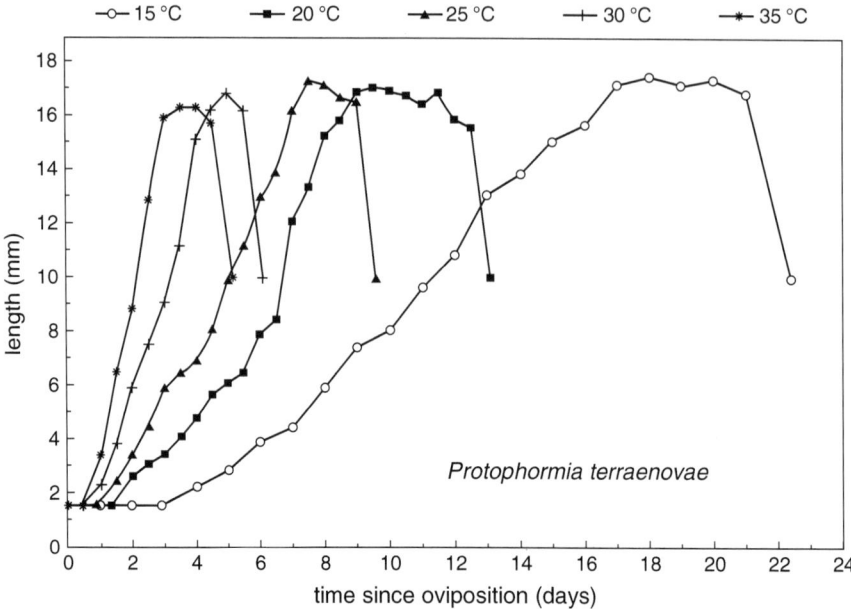

Fig. 10.4 Growth curves for the blowfly *Protophormia terraenovae* (so-called Isomegalen-Diagram), showing the time required for the maggots to reach the given larval length at 15, 20, 25, 30, and 35°C (Grassberger and Reiter 2002)

such developmental events. However, it is well documented that size is not always an accurate measure of age, particularly for the forensically important blowfly larvae (Villet et al. 2010). Larvae are known to shrink in size before entering the next developmental stage, particularly before pupariation. Moreover, stunted larval forms, smaller than normal, are quite common, for example as a result of competition for food. Therefore, it is possible for larvae of a small size to be older than larvae of a larger size. Last but not least this model is limited to larvae only, which accounts for only approximately half the total duration of development from egg to adult. The other half is egg and pupariation.

Blowflies and fleshflies have been used to provide estimates of time of death for decades, and support the estimation of a minimum PMI ranging from one day up to more than one month, depending on the insect species involved and the climatic conditions at the death scene (Amendt et al. 2007). Other taxa are less used due to a lack of knowledge regarding their development biology. Recently, Midgley et al. (2010) showed that the silphid beetle, *Thanatophilus micans*, arrives at a cadaver very soon after death occurred and has a longer development than flies, thus extending the period over which a PMI_{min} can be estimated. Similar advantages exist with hymenopteran parasitoids, which emerge from their hosts several days after unparasitized hosts have enclosed (Amendt et al. 2007). Development data are available for several cosmopolitan parasitoids and more species await detailed investigation.

Genes and Hormones

Age estimation of premature individuals is of special importance for pupae because here an age-dependent change in size cannot be used for. However, puparia can be opened to observe the physical changes associated with development of the pupae within, which correlate with age.

The growth of the maggot and much more the metamorphosis of the fly (lasting about 50% of the total time span of development), which takes place inside the puparia, is characterized by modifications such as cellular proliferation, tissue remodeling, cell migration, and apoptosis. Many genes are involved in these actions and display age-dependent expression and monitoring the increase or decrease of certain gene products, which are essential in specific development stages, can give a time scale for age estimation. This is performed by determining the amount of mRNA of specific genes through RNA extraction, reverse transcription into cDNA and quantification of this cDNA. Indeed, gene expression studies have demonstrated such changes in gene transcription profiles in *Drosophila* and mosquitoes of the genera *Aedes* and *Anopheles* (Tarone et al. 2007). Recently, this approach was applied to some blowfly species. Zehner et al. (2009) identified nine differentially expressed genes in the forensically important blowfly *Calliphora vicina* and were able to discriminate between pupae at the beginning and at the end of their metamorphosis and with an age up to at least 120 h (at 25°C) so far.

Gaudry et al. (2006) showed for the pupal stage the secretion of a hormone, 20-hydroxyecdysone, which fluctuates in a highly predictable way, reaching a peak 36–96 h after pupariation at about 20°C in the blowfly *Protophormia terraenovae*.

10.3.2.2 Succession

Several weeks after the arrival of the first wave of colonization, generally by flies, the first insects finish their development and hatch. Empty puparia may be found and identified, even on species level. However, despite several achievements there is no tool for estimating the age of those remnants, i.e., the time of hatching. This means that the first empty puparia shows the end of the estimations to the day by calculating the age of immature insects.

Studies on animal carcasses demonstrated that species composition and insect succession on a cadaver varies with respect to the geographical region and the season (Smith 1986; Anderson 2010). When knowing the chronology of insects colonizing a cadaver in a certain area, analysis of the fauna on a corpse can be used to estimate the time elapsed since death (Fig. 10.5). Succession data have been used to calculate a PMI_{min} up to 2 months and, if there exist adequate data, may be applied to a much longer time interval (Amendt et al. 2007). Several parameters like for example type of habitat, season, or the burial or wrapping of the corpse influence the composition of the species community. Even the degree of sun exposure can alter the pattern of insect colonization. However, the speed of decomposition is influenced by many internal and external parameters, and so the succession

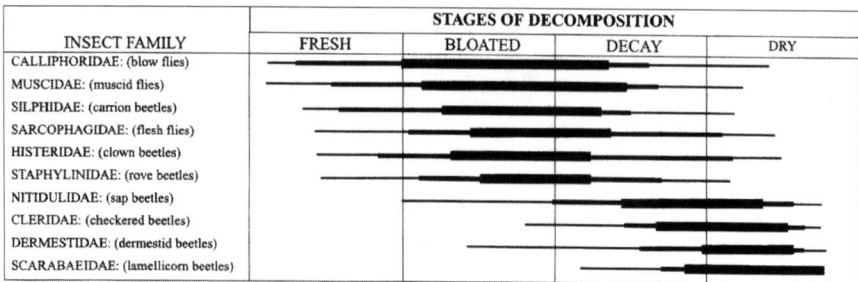

Fig. 10.5 Succession of adult arthropods on human cadavers in east Tennessee (during spring and summer); adapted from Rodriguez and Bass (1983), Hall (2010) and Amendt et al. (2004)

variations can be so large as to be of limited value. Nevertheless, in certain cases succession may be the only tool available to give an idea of the time since death, even estimates based on insect succession are given as windows of time around the event because of the uncertainties inherent in estimation (Matuszewski et al. 2010a, b).

10.4 Entomotoxicology

Insects which feed on corpses may sequester drugs and toxicants which had been ingested by the deceased person (Introna et al. 2001; Carvalho 2010). In badly decomposed bodies, their immature stages and remnants are therefore not only useful for estimating the PMI_{min}, but they can also be used as a reliable substrate for toxicological analysis. This analysis of carrion-feeding insects in order to detect toxic substances and to investigate the effects on the insect's development is named entomotoxicology.

10.4.1 Detection of Drugs

After a long period of time since death a decomposed body may be difficult to examine for toxic substances due to the lack of appropriate sources such as tissue, blood, or urine. Here, analysis of the necrophagous insects collected may enable toxicological assessment of the cause of death. Most of the substances involved in drug-related deaths are detectable through toxicological analyses of maggots and the potential use of insects as alternative samples for detecting drugs and toxins has been well documented in the literature. Sometimes those analyses can provide a more reliable biosample than those from cadaver tissues, as some studies demonstrated that drugs could be detected in larvae but not in associated soft tissues. Most of the substances involved in drug-related deaths (e.g., morphine, cocaine, amphetamines, benzodiazepines, etc.) are detectable through analyses of maggots (Kintz et al. 1990), but the method can also be applied to empty puparial cases and even beetle exuviae and fecal material (Gagliano-Candela and Aventaggiato 2001), which are often found at the death scene after years. This could help to resolve the history and identity of the dead body which was the source of food for the larvae that gave rise to the remnants.

However, despite the fact that some studies suggested a correlation between concentrations of drugs in for example larvae and the tissues on which the specimens had fed (Campobasso et al. 2004), it should be clearly stated that insects are certainly useful as qualitative toxicological specimens, but they are still of limited quantitative value. Last but not least, while a positive result indicates the presence of a drug or a toxin, the absence of a drug may not indicate that there are no drugs in the body. The accumulation of a drug by larvae is unpredictable, and the intake depends on the type of tissue upon which the maggots were feeding.

10.4.2 Influence of Drugs on Development

Insects metabolize and/or eliminate the substances ingested during their development. These processes may influence the development of the necrophagous insects, having a potential impact when estimating PMIs by calculating the rate of development (O'Brien and Turner 2004). In fact, drug-induced changes in the development rate of necrophagous insects can be large enough to significantly alter PMI estimates, leading to significant errors if overlooked and not taken into account during a death investigation. Goff et al. studied the effect of cocaine (Goff et al. 1989) and heroin (Goff et al. 1991) on the rate of development in Sarcophagidae and demonstrated that maggots develop more rapidly if reared on the liver or spleen of rabbits which had been killed by a lethal dose of cocaine or heroin and Bourel et al. (1999) showed that an underestimation of the postmortem interval up to 24 h is possible, if the presence of morphine in tissue is not considered when calculating the development of the blowfly *Lucilia sericata*. Hence, insects on corpses can be used in toxicological analyses, but it should be kept in mind that there is the risk of calculating an incorrect PMI_{min} due to the drug-modified rate of development of the immature stages.

10.5 Molecular Analysis of Insects

As already pointed out above, molecular tools could be very helpful in future in estimating the age of immature insects, provided that there will be substantial research. In contrast, DNA techniques for identifying forensically relevant species are firmly established in forensic entomology and routinely used (Wells and Stevens 2008).

10.5.1 Species Determination

Assuring the identity of a specimen is a crucial step in forensic entomology. However, species identification on a morphological basis may be hampered by limited expertise in taxonomy or the lack of morphological characters of the immature specimens, especially in the early larval stages. Within some genera (e.g., the flesh fly *Sarcophaga* spp.), discrimination of individual species is not possible from larval morphology alone (Zehner et al. 2004). Rearing up to the adult stage might allow determination, but this is not always possible due to rearing difficulties or because the immature stages were killed during or after collection for reasons of preservation. Here, molecular techniques as alternative tools for identification of forensically important species have been established within the last 15 years, based on species-specific nucleotide sequences of certain genes. Most important is the so-called subunit I of the mitochondrial cytochrome oxidase (CO I). Usually sequences of 350–650 base pairs (bp) are sufficient for analysis (Zehner et al. 2004; Wells and Stevens 2008). Reference sequences are available for a large variety of species, including blowflies and other groups of forensic interest. Even insect fragments or empty puparia may be assigned to certain species by using sensitive molecular techniques. As it cannot be ruled out that closely related species may exhibit a very similar nucleotide sequence (which may lead to incorrect species determination), extensive knowledge of all relevant species is mandatory, which includes the analysis of the nucleotide sequence of the used gene. Deviations between the sequence of a specimen in question and a reference sequence of up to 3% are possible. Higher values indicate different species.

Because of biological complexity surprising observations can be made. Blowflies originating from Hawaii which were morphologically identified as *Lucilia cuprina* showed a CO I sequence which assigned them to *L. sericata* (Stevens et al. 2002). Moreover, it has been demonstrated that cytochrome oxidase sequence-based classification of flies of the blowfly genus *Protocalliphora* cannot be performed reliably in specimens that are infested by the parasitic microbe *Wolbachia* (Whitworth et al. 2007). These examples highlight the fact that it has to be kept in mind that a CO I sequence has to be interpreted with caution and whenever it is possible should be supported by a morphological examination.

10.5.2 Detection and Typing of Human DNA Within Insect Larvae

Examination of human DNA from the digestive tract of insect larvae is recommended if for example the food source of the larvae sampled at the scene is in doubt because of the presence of alternative food sources and a possible contamination of the entomological samples (Campobasso et al. 2005). The detection of human DNA in larval guts may also indicate that a decomposing body was previously present at a scene of crime and has subsequently been relocated. If necessary, the identity of the person can be determined by forensic STR typing, by comparing it with existing genetic profiles. It seems that this kind of analysis can also be undertaken on nonfeeding stages, for example sheep-specific DNA were detected within two-day-old pupae of the Australian blowfly *Calliphora dubia* (Carvalho et al. 2005).

10.6 Forensic Entomology: A Silver Bullet?

It has been shown that entomology can act as a powerful tool in forensic science, being the most reliable method for establishing the minimum time since death in the first weeks postmortem. Especially precise estimation of the PMI as the most important value of forensic entomology has been achieved by refining the techniques used. However, there are several limitations which need further attention (Amendt et al. 2007; Villet et al. 2010).

Several parameters can lead to a delay in colonization for example covering corpses with branches, tight wrapping with blankets, carpets or plastic-bags etc., and indoor placement. Seasonal influences, such as cold and rainy weather, may inhibit or even prevent fly activity and delay oviposition as well. Last but not least blowflies are not active at night and generally do not lay eggs during nighttime. All these possibilities can lead to an underestimation of PMI_{min}.

In many forensically important blowflies diapause, the period during which growth and development of insects is suspended, is under maternal control – exposure of females to short day lengths induces diapause of the offspring. Beside day length, temperature may also influence its incidence. From a forensic point of view it remains still ambiguous which exact parameters induce diapause and which complex interactions terminate it; for the blowfly *Calliphora vicina* it was for example demonstrated that diapause ended earlier in larvae whose parents had been kept at 20°C than those kept at 15°C. Hence, calculating the time of development for diapausing specimens is still a challenge in forensic entomology.

When facing the huge masses of maggots on a decomposing corpse, it seems evident to expect competition in those situations. This may affect development and growth of immature Diptera, but is not really analyzed in forensic entomology. It is of special interest as all reference data dealing with the growth of forensically important blowflies are produced by examining monocultures of each species. Realizing those tens of thousands of larvae feeding gregariously on a dead body highlights another questionable parameter: Larval-generated heat must be considered as a potential factor influencing larval development. Values for larval masses of up to about 25°C above ambient temperature can be measured in extreme scenarios and it is still not clear whether and how to handle this fact when calculating the PMI_{min}. There are several ways in which larvae could regulate thermal stress, for example by evaporative cooling; and it seems impossible to account for all these factors.

A last pitfall when calculating the age of specimens collected on a corpse is the fact that laboratory-derived data which are used as a reference, may just reflect the developmental biology of the local population from where the culture of flies was initially established. It is debatable how well those developmental data relate to populations of the same species, but living in different geographic areas. Indeed variation in developmental time for geographically distinct populations is reported for areas of different scales (Tarone and Foran 2008; Hwang and Turner 2009; Gallagher et al. 2010). Reference data from different studies present different developmental rates for the same species.

All these examples highlight the need for a careful interpretation of the evidence, keeping in mind that every case has its own, very specific history due to the complexity of integrating all factors and parameters (diapause, competition, maggot-generated heat, etc.). Determining the PMI is extremely difficult, and accurate precision is quite impossible, even by using the entomological method, due to a wide biological variability.

References

Amendt J, Krettek R, Zehner R (2004) Forensic entomology. Naturwissenschaften 91:51–65

Amendt J, Krettek R, Zehner R (2005) Insekten auf Leichen. Biol unserer Zeit 35(4):232–240

Amendt J, Campobasso CP, Gaudry E, Reiter C, LeBlanc HN, Hall MJR (2007) Best practice in forensic entomology – standards and guidelines. Int J Legal Med 121:90–104

Amendt J, Goff ML, Campobasso CP, Grassberger M (2010) Current concepts in forensic entomology. Springer, Dordrecht

Anderson GS (2010) Factors that influence insect succession on carrion. In: Byrd JH, Castner JL (eds) Forensic entomology: the utility of arthropods in legal investigations. CRC, Boca Raton

Bergeret M (1855) Infanticide, momification naturelle du cadavre. Ann Hyg Publique Med Leg 4:442–452

Bourel B, Hédouin V, Martin-Bouyer L, Becart A, Tournel G, Deveaux M, Gosset D (1999) Effects of morphine in decomposing bodies on the development of *Lucilia sericata* (Diptera: Calliphoridae). J Forensic Sci 44:354–358

Byrd JH, Castner JL (2010) Forensic entomology: the utility of arthropods in legal investigations. CRC, Boca Raton

Campobasso CP, Di Vella G, Introna F (2001) Factors affecting decomposition and Diptera colonization. Forensic Sci Int 120:18–27

Campobasso CP, Gherardi M, Caligara M, Sironi L, Introna F (2004) Drug analysis in blowfly larvae and in human tissues: a comparative study. Int J Legal Med 118:210–214

Campobasso CP, Linville JG, Wells JD, Introna F (2005) Forensic genetic analysis of insect gut contents. Am J Forensic Med Pathol 26:161–165

Carvalho LML (2010) Toxicology and forensic entomology. In: Amendt J, Campobasso CP, Goff ML, Grassberger M (eds) Current concepts in forensic entomology. Springer, Dordrecht

Carvalho F, Dadour IR, Groth DM, Harvey ML (2005) Isolation and detection of ingested DNA from the immature stages of *Calliphora dubia* (Diptera: Calliphoridae): a forensically important blowfly. Forensic Sci Med Pathol 1:261–265

Catts EP, Haskell NH (1990) Entomology and death – a procedural guide. Joyce's Print shop, Clemson

Clark MA, Worrell MB, Pless JE (1997) Postmortem changes in soft tissues. In: Haglund WD, Sorg MH (eds) Forensic taphonomy: the postmortem fate of human remains. CRC, Boca Raton

Donovan SE, Hall MJR, Turner BD, Moncrieff CB (2006) Larval growth rates of the blowfly, *Calliphora vicina*, over a range of temperatures. Med Vet Entomol 20:106–114

Gagliano-Candela R, Aventaggiato L (2001) The detection of toxic substances in entomological specimens. Int J Legal Med 114:197–203

Gallagher MB, Sandhu S, Kimsey R (2010) Variation in developmental time for geographically distinct populations of the common green bottle fly, *Lucilia sericata* (Meigen). J Forensic Sci 55:438–442

Gaudry E, Blais C, Maria A, Dauphin-Villemant C (2006) Study of steroidogenesis in pupae of the forensically important blow fly *Protophormia terraenovae* (Robineau-Desvoidy) (Diptera: Calliphoridae). Forensic Sci Int 160:27–37

Goff ML, Omori AI, Goodbrod JR (1989) Effect of cocaine in tissues on the rate of development of *Boettcherisca peregrina* (Diptera: Sarcophagidae). J Med Entomol 26:91–93

Goff ML, Brown WA, Hewadikaram KA, Omori AI (1991) Effects of heroin in decomposing tissues on the development rate of *Boettcherisca peregrina* (Diptera: Sarcophagidae) and implications of this effect on estimation of postmortem intervals using arthropod development patterns. J Forensic Sci 36:537–542

Goff LM, Campobasso CP, Gherardi M (2010) Forensic implications of myiasis. In: Amendt J, Campobasso CP, Goff ML, Grassberger M (eds) Current concepts in forensic entomology. Springer, Dordrecht

Grassberger M, Amendt J (2010) Forensische Entomologie. In: Aspöck H (Hrsg) Krank durch Arthropoden. Denisia 30:843–861

Grassberger M, Reiter C (2001) Effect of temperature on *Lucilia sericata* (Diptera: Calliphoridae) development with special reference to the isomegalen- and isomorphen-diagram. Forensic Sci Int 120:32–36

Grassberger M, Reiter C (2002) Effect of temperature on development of the forensically important holarctic blowfly *Protophormia terraenovae* (Robineau-Desvoidy) (Diptera: Calliphoridae). Forensic Sci Int 128:177–182

Greenberg B (1991) Flies as forensic indicators. J Med Entomol 28:565–577

Greenberg B, Kunich J (2002) Entomology and the law. Cambridge University Press, Cambridge

Hall RD (2010) Perceptions and status of forensic entomology. In: Byrd JH, Castner JL (eds) Forensic Entomology – the utility of arthropods in legal investigations. CRC, Boca Raton

Hall MJR, Wall R (1995) Myiasis of humans and domestic animals. Adv Parasitol 35:257–334

Higley LG, Haskel NH (2010) Insect development and forensic entomology. In: Byrd JH, Castner JL (eds) Forensic entomology – the utility of arthropods in legal investigations. CRC, Boca Raton

Hwang CC, Turner BD (2009) Small-scaled geographical variation in life-history traits of the blowfly *Calliphora vicina* between rural and urban populations. Entomol Exp Appl 132:218–224

Introna F, Campobasso CP, Goff ML (2001) Entomotoxicology. Forensic Sci Int 120:42–47

Kintz P, Godelar A, Tracqui A, Mangin P, Lugnier AA, Chaumont AJ (1990) Fly larvae: a new toxicological method of investigation in forensic medicine. J Forensic Sci 35:204–207

Matuszewski S, Bajerlein D, Konwerski S, Szpila K (2010a) Insect succession and carrion decomposition in selected forests of Central Europe. Part 1: Pattern and rate of decomposition. Forensic Sci Int 194:85–93

Matuszewski S, Bajerlein D, Konwerski S, Szpila K (2010b) Insect succession and carrion decomposition in selected forests of Central Europe. Part 2: Composition and residency patterns of carrion fauna. Forensic Sci Int 195:42–51

Mende LJK (1829) Ausführliches Handbuch der gerichtlichen Medizin für Gesetzgeber, Rechtsgelehrte, Aerzte und Wundaerzte, Teil 5

Midgley JM, Richards CS, Villet MH (2010) The utility of Coleoptera in forensic investigations. In: Amendt J, Campobasso CP, Goff ML, Grassberger M (eds) Current concepts in forensic entomology. Springer, Dordrecht

O'Brien C, Turner B (2004) Impact of paracetamol on *Calliphora vicina* larval development. Int J Legal Med 118:188–189

Rodriguez WC, Bass WM (1983) Insect activity and its relationship to decay rates of human cadvers in East Tennessee. J Forensic Sci 28:423–432

Smith KGV (1986) A manual of forensic entomology. The Trustees, British Museum London

Stevens JR, Wall R, Wells JD (2002) Paraphyly in Hawaiian hybrid blowfly populations and the evolutionary history of anthropophilic species. Insect Mol Biol 11:141–148

Tarone AM, Foran DR (2008) Components of developmental plasticity in a Michigan population of Lucilia sericata (Diptera: Calliphoridae). J Forensic Sci 53:942–948

Tarone AM, Kimberley C, Jennings MS, Foran DR (2007) Aging blow fly eggs using gene expression: a feasibility study. J Forensic Sci 52:1350–1354

Villet MH, Richards CS, Midgley JM (2010) Contemporary precision, bias and accuracy of minimum post-mortem intervals estimated using development of carrion-feeding insects. In: Amendt J, Campobasso CP, Goff ML, Grassberger M (eds) Current concepts in forensic entomology. Springer, Dordrecht

Wells JD, Stevens JR (2008) Application of DNA-based methods in forensic entomology. Annu Rev Entomol 53:103–120

Whitworth TL, Dawson RD, Magalon H, Baudry E (2007) DNA barcoding cannot reliably identify species of the blowfly genus *Protocalliphora* (Diptera: Calliphoridae). Proc R Soc B 274:1731–1739

Zehner R, Amendt J, Schütt S, Sauer S, Krettek R, Povolný D (2004) Genetic identification of forensically important flesh flies (Diptera: Sarcophagidae). Int J Legal Med 118:245–247

Zehner R, Amendt J, Boehme P (2009) Gene expression analysis as a tool for age estimation of blowfly pupae. For Sci Int Genet Suppl 2:292–293

Zumpt F (1965) Myiasis in man and animals in the old world. Butterworths, London

Chapter 11
"Living Syringes": Use of Hematophagous Bugs as Blood Samplers from Small and Wild Animals

André Stadler, Christian Karl Meiser, and Günter A. Schaub

Abstract Sampling of blood from small and/or wild animals is very difficult since the animals are stressed strongly and can be severely affected by conventional sampling and/or anesthesia. Triatomines are the largest blood-sucking insects and feed on all warm-blooded animals including warm amphibia and reptiles. These insects develop through five larval instars which ingest increasing amounts of blood which is stored in the distensible stomach. There it is concentrated by the withdrawal of the fluid components and remains essentially undigested. Since the blood can be withdrawn easily with a syringe and used for determination of blood and physiological parameters and for the identification of pathogens, triatomines offer a noninvasive method to obtain blood samples. Especially larvae of *Rhodnius prolixus*, *Triatoma infestans*, and *Dipetalogaster maxima* are used as "living syringes." In this review we summarize the application of this methodology and its advantages and disadvantages.

11.1 Introduction

The collection of blood samples from small and/or wild animals is often stressful for the focus animal. Small animals have to be fixed to avoid injury and subsequent bleeding (Baer and McLean 1972). Large wild animals must be anesthetized, a procedure that is slightly risky for humans but more so for wild animals for which relevant parameters for the dose of anesthetics are not directly available but need to be estimated roughly. The use of bugs as samplers of blood is based on xenodiagnosis, i.e., the use of foreign animals to obtain blood samples for a proof of

---

A. Stadler
ZoologicalGarden, Wuppertal, Germany

C.K. Meiser and G.A. Schaub (✉)
Zoology/Parasitology Group, Department of Animal Ecology, Evolution and Biodiversity Group, Universitätsstr. 150, 44780 Bochum, Germany
e-mail: guenter.schaub@rub.de

parasites, and has a long tradition, especially in Chagas disease (summarized by Meiser and Schaub 2011). Such a diagnosis of infections with *Trypanosoma cruzi* is used if the concentration of the parasite in the human blood is too low to allow direct microscopic diagnosis, but it can be identified more easily after a period of multiplications of the parasite. This method was first used by Brumpt (1914). In xenodiagnosis usually groups of bugs originating from laboratory colonies are placed in a beaker and allowed to suck on the arm of the patient suspected of infection. To avoid allergic reactions, an artificial xenodiagnosis is recommended, offering the blood under a membrane (summarized by Meiser and Schaub 2011).

The first use of hematophagous arthropods to investigate components of the blood donor was suggested by Prof. Dr. W. Frank in Stuttgart–Hohenheim and performed by Will (1971, 1975, 1977), sampling blood of lizards after engorgement by ticks and then by triatomines. Independently, based on suggestions from Prof. Dr. J. Núñez, bugs were then used to obtain blood from bats (von Helversen and Reyer 1984; von Helversen et al. 1986). For about 7 years now, triatomines have been increasingly used whenever stress has to be avoided (Voigt et al. 2004; Stadler 2006). The blood-sucking bugs *Rhodnius prolixus, Triatoma infestans*, and *Dipetalogaster maxima* as "living syringes" offer a new and less invasive method for blood collection (Thomsen and Voigt 2006; Stadler 2007; Stadler et al. 2007, 2008a, b, 2009; Kruszewicz et al. 2009).

11.2 The "Living Syringes": Triatomines

The insect subfamily Triatominae of the family Reduviidae at present contains 140 species (Galvão et al. 2003; Schofield and Galvão 2009). The hemimetabolic insects develop through five larval instars, named nymphs by other scientists (Schaub 2008). The majority of species live between the Great Lakes in the USA and Southern Argentina and are vectors of the etiologic agent of Chagas disease, the protozoon *Trypanosoma cruzi* (Schaub 2009). Species of the monophyletic tribe Rhodniini prefer trees with nests of birds and glue the eggs to the feathers of birds or palm leaves. Using these leaves for the roofs of houses, one major vector of *T. cruzi*, *R. prolixus*, was transported into the houses and has adapted to living there. It occurs in the northern states of Latin America from Brazil to Mexico, often producing two generations per year (Zeledón and Rabinovich 1981). *R. prolixus* is one of the smaller species of Triatominae, adult males and females ranging up to 20 and 22 mm, respectively (Lent and Wygodzinsky 1979). The five larval instars increase in length from 3 to 4.5, 8.5, 10, and 15 mm (Schaub unpublished). The colonies are easy to breed in the laboratory, completing one generation within 3 months if the larvae are fed four days after the molt (Schaub 1988). Since they are also strict food specialists they became one of the most important laboratory insects for the study of insect physiology and were chosen as the first triatomine for a genome project (Wigglesworth 1972; Huebner 2007).

Even more important as a vector of *T. cruzi* is *T. infestans*. This species presumably originates from the Chaco, ranging from Bolivia to Argentina. There it sucked on guinea pigs and rodents. During domestication of the guinea pigs as meat animals it colonized houses. It is best adapted to life there and was dispersed by human trading and traveling to all southern states of Latin America from Ecuador to Argentina, usually developing one generation per year. Insecticide-spraying campaigns inside houses have reduced its distribution mainly to the sylvatic habitats (Dias and Schofield 1999; Schofield et al. 2006). This species is bigger than *R. prolixus*, males and females possessing lengths of up to 26 and 29 mm, respectively (Lent and Wygodzinsky 1979) and the five larval instars increase in length from 4 to 6, 9, 12, and 19 mm (Schaub unpublished). Under optimal conditions in the laboratory, the generation time lasts at least 4 months (Schaub 1988).

D. maxima only occurs in the fog desert of the peninsula Baja California Sur in Mexico and is the biggest species of triatomines (Ryckman and Ryckman 1963; Lent and Wygodzinsky 1979). Originally named *D. maximus* we now follow the current suggestions of triatomine taxonomists (Marsden 1986; Schofield and Galvão 2009). Adults of this univoltine species are usually present in June and July. Males develop to lengths of up to 35 mm, females of up to 42 mm (Lent and Wygodzinsky 1979). The five larval instars increase in length from 8 to 11.5, 15, 20, and 27 mm (Schaub unpublished). The generation time of laboratory colonies lasts at least about 5–6 months (Schaub and Breger 1988). Because of the hostile conditions this species needs to be very aggressive and begins blood sucking very fast in comparison to other species. *D. maxima* mainly feeds on small reptiles or mammals and on birds that live on the ground. In addition, they are diurnal and not strictly nocturnal like the other two species (Ryckman and Ryckman 1963, 1967; Lent and Wygodzinsky 1979). Because of the strict export regulations of Mexico, all non-Mexican laboratory colonies of *D. maxima* are based on the collections of Ryckman and Ryckman (1963) or Marsden et al. (1979).

11.3 The "Living Syringes": Adaptations of Triatomines to Hematophagy

Triatomines are classified as obligate blood-suckers, although some species like *T. klugi* ingest hemolymph of other insects (Carvalho-Pinto et al. 2000). In contrast to other blood-sucking insects like mosquitoes, all postembryonic developmental stages and not only females but also males suck blood (Schaub 2008). For host finding, triatomines use different stimuli. During warm nights, hungry adults fly to light sources, thereby covering great distances. Adults and larvae are attracted by the ammonia in the feces deposited by the engorged bugs. Also the increased proportion of CO_2 in expired air and components of sweat attract the bugs (Schofield 1979; Lorenzo Figueras et al. 1994; Guernstein and Guerin 2001;

Barrozo et al. 2004; Barrozo and Lazzari 2004, 2006; Ortiz and Molina 2010). At short distances, they react mainly to thermal differences, respectively infrared stimuli (Schmitz et al. 2000; Ferreira et al. 2007). The receptors for these stimuli are mainly located on the antennae but also on the mouthparts (Guernstein and Guerin 2001).

On the host, triatomines swing the mouthparts forward from a resting position below the thorax. The mouthparts of triatomines consist of the labrum, labium, mandibles, and maxillae. The labrum and labium form the proboscis and protect the internal mandibles and these in turn the maxillae. The left and right maxilla are moveable against each other in the longitudinal axis. Between left and right maxilla the ventral salivary tube and the dorsal food tube are formed. At the proximal end of the maxillary bundle the two maxillae are separated (Geigy and Kraus 1952; Wirtz 1987; Wenk et al. 2010). These fine mouthparts indicate that triatomines are not pool-feeders like horse-, stable- and tsetse flies, which ingest the blood directly from a wound, but vessel-feeders, like fleas, bedbugs and mosquitoes, all ingesting blood directly from a blood vessel (Lavoipierre 1965). Upon contact with the skin, this is punctured by a jerky movement of the head. The serrated proximal ends of the mandibles open the upper layers of skin and anchor there. Then the maxillae delve deeper into the skin and whip until a blood vessel is found. During this phase minor hematomas are caused. Several times the bugs probe to suck blood on a trial basis, but in this probing phase no blood is ingested from the hematomas. When a suitable blood vessel is found, the left maxilla attaches on the vessel wall while the right maxilla penetrates into the vessel. This may occur by chance due to waving movements of the maxillae or by perception of the minimally higher temperature of the capillaries (Lavoipierre et al. 1959; Wirtz 1987; Ferreira et al. 2007). The duration of the probing phase can last only a few seconds up to several minutes. If the bug is disturbed or no suitable vessel is found, the bug starts new attempts elsewhere (Lavoipierre et al. 1959; Sant'Anna et al. 2001). The duration of feeding depends on host and triatomine species as well as the developmental stage and lasts about 5–50 min (Soares et al. 2000). During feeding, the diameter of the blood vessel is affected by the rhythm of the aspiration of the food pump and the injection of saliva (Lavoipierre et al. 1959; Wirtz 1987; Soares et al. 2006).

Compared to other hematophagous insects, triatomines are the biggest and ingest very large amounts of blood. The volumes vary – depending on species and developmental stage – for larvae between 3 and 23 times and for adults 1–3 times their own body weight (summarized by Meiser 2009). In *D. maxima*, the females ingest up to 4.3 g blood (Barreto et al. 1981). Each larval instar requires in the respective stage several smaller or at least one full engorgement. The period of time from the full engorgement to ecdysis depends on the climatic conditions, the larval stage, and the species. At $26 \pm 1°C$ and 50–60% relative humidity, the five larval stages of *R. prolixus* need 9, 9, 10, 13, and 14 days before the first larvae of the respective stage molt, those of *T. infestans* 10, 10, 12, 17, and 28 days, respectively, and those of *D. maxima* 14, 14, 16, 21, and 51 days, respectively (Schaub 1988; Schaub and Breger 1988).

11.4 The "Living Syringes": Factors of Triatomines Interacting with the Host Blood

Blood is an energy-rich food source, but focusing on this food causes many problems for ectoparasites. The blood vessel must be found and the blood made available without attracting the attention of the host and thus mechanical defense reactions. After the blood is made available, the hematophagous ectoparasite is confronted with the different factors of the defense of the host, especially the complement system and blood clotting (Andrade et al. 2005). In the initial phase, salivary compounds are relevant, then factors from the stomach.

The saliva of triatomines contains hundreds of compounds enabling the hematophagous lifestyle (Ribeiro et al. 2004; Santos et al. 2007; Assumpção et al. 2008; Meiser et al. 2010b). Some compounds cause allergic reaction in humans (see above). In animals, less abundant compounds in the saliva induce the strongest immune reactions (Schwarz et al. 2009). In the majority of species the bite is painless; one exception is *Triatoma rubrofasciata* (Hase 1932; Ryckman and Bentley 1979). However, so far only in the saliva of *T. infestans* has a protein been identified that inhibits Na^+ channels and acts similarly to local anesthetics (Dan et al. 1999). Thereby, the blood ingestion, which usually lasts 5–20 min, is not recognized by the host. In the saliva of *R. prolixus*, nitrophorins carry NO radicals that induce a vasodilatation of capillaries, enabling a more rapid blood engorgement (Kaneko et al. 1999; Weichsel et al. 2000; Martí et al. 2008). During the probing phase salivary apyrases seem to be relevant (Ribeiro and Garcia 1981a; Sarkis et al. 1986). Furthermore, these apyrases inhibit the primary hemostasis (Ribeiro et al. 1998). The collagen-induced adhesion of platelets is inhibited by at least two proteins in the saliva of *R. prolixus* and thromboxane A_2-mediated aggregation by another compound (Ribeiro and Garcia 1981b). In *R. prolixus,* aggregation inhibitor-1 (RPAI-1) binds ADP, thus preventing platelet aggregation (Francischetti et al. 2000, 2002). The saliva of *Triatoma pallidipennis* contains another inhibitor of collagen-induced activation of platelets, pallidipin, and that of *T. infestans* has triplatin-1 and triplatin-2, both possessing high sequence similarities to pallidipin (Noeske-Jungblut et al. 1994; Morita et al. 2006). Other compounds interact by a direct interference with components of the coagulation cascade (Noeske-Jungblut et al. 1995; Ribeiro et al. 1998). Additionally, the saliva of different triatomines inhibits the classical and/or alternative pathway of the complement system (Cavalcante et al. 2003; Barros et al. 2009). An antibacterial protein in the saliva of triatomines has been described so far only from *T. infestans*, namely the 30-kDa trialysin (Amino et al. 2002) and in addition a sialidase with suggested anti-inflammatory function (Amino et al. 1998). A 39-kDa protein in the saliva of *R. prolixus* induces a similar effect but also inhibits platelet aggregation (Ribeiro and Garcia 1981b; Ribeiro 1982; Ribeiro and Walker 1994).

Although the compounds of the saliva are ingested together with the blood, for their use as "living syringes" especially the modifications of blood in the intestine seem to be relevant. The digestive tract of triatomines is a simple tube without any lateral diverticula. Considering the three major regions, foregut, midgut, and hindgut, especially the midgut is relevant. It can be divided into an anterior part, consisting of the short cardia and a strongly distensible stomach, and the posterior midgut, named small intestine, into which small portions of stomach content are passed for digestion (Kollien and Schaub 2000). After blood ingestion, the abdomen of fully engorged bugs is nearly completely round, and the larvae can hardly move. Therefore, triatomines possess the most effective excretion system of the animal kingdom (Maddrell 1969; Maddrell et al. 1991). Already during blood engorgement, the excretion of blood compounds without nutritional value starts, in *R. prolixus* 3 min after the beginning of ingestion (Maddrell 1963). Within this short period of time, ions are transported via the wall of the stomach into the hemolymph and the water follows also into the Malpighian tubules, which end at the midgut/hindgut transition. Secretory activity of the isolated Malpighian tubules increases to about 3 μl min^{-1} mm^{-1} in *R. prolixus* and *T. infestans* (Maddrell 1969; Schnitker et al. 1988). Thereby *R. prolixus* excretes about 45% of the weight of the ingested blood within the first 4 h and *T. infestans* about 60% of it within the first 24 h (Wigglesworth 1931; Maddrell 1969; Eichler and Schaub 1998). In the first few days after ingestion, the concentrated blood is stored essentially undigested except for a lysis of the erythrocytes and some modifications induced by the increasing activity of some glycosidases and lipases (Bauer 1981; Azambuja et al. 1983; Ribeiro and Pereira 1984; Canavoso et al. 2004; Schaub 2009; Garcia et al. 2010; Schaub et al. 2011). According to the host species, the hemoglobin may crystallize (Bauer 1981; Oliveira et al. 2007), but usually the blood takes on a jelly-like consistency (Lehane 2005).

Whereas the excretion affects the hematocrit levels only after a delayed withdrawal of the blood, concentrations of the factors of the complement system and of the coagulation cascade are immediately changed, not only by the compounds in the saliva of triatomines, but also by those present in the stomach. The classical and the alternative activation pathway of the complement system are inhibited by the stomach contents of *R. prolixus*, *Triatoma brasiliensis*, and *T. infestans* (Barros et al. 2009). The most potent anticoagulatory proteins are secreted into the stomach of the triatomine. These are mainly Kazal-type inhibitors, directly interacting with thrombin and other serine proteases of the coagulation cascade, and have been characterized from the stomach of *R. prolixus*, *D. maxima*, *T. infestans*, *T. brasiliensis*, and *Panstrongylus megistus* (summarized by Meiser et al. 2010a).

11.5 The "Living Syringes": Rearing of Triatomines

The rearing of triatomines is not very difficult, but there are some pitfalls. In endemic regions bugs can be maintained under normal laboratory conditions without any regulation of temperature and humidity (Dias 1955; Wood 1964).

However, regulated conditions enable a production of triatomines without fluctuations of the development due to environmental changes. Specific descriptions for the rearing of *R. prolixus*, *T. infestans*, and *D. maxima* are available (e.g., Gomez-Núnez 1964; Neghme et al. 1967; Gardiner and Maddrell 1972; Barreto et al. 1981). The majority of triatomines can be well reared in a temperature range of 26–28°C (summarized by Ryckman and Ryckman 1966; Garcia et al. 1984; Núñez and Segura 1987). The relative humidity is critical! Below 50% rH the bugs have to spend energy to avoid the loss of body water and die during molting, while above 90% relative humidity fungi develop on the feces (Ryckman and Ryckman 1966). A light/dark rhythm is better than complete darkness, and long-day conditions are recommended (Gomez-Núnez 1964; Ryckman and Ryckman 1966).

Special containers have been developed for maintenance of triatomines (e.g., Dias 1938; Deane 1948; Corrêa 1954; Ryckman and Ryckman 1966). According to our experience it is less money- and labor-consuming if normal 2-l plastic beakers of 15×18 cm are used, covered by a nylon cloth which is fixed by commercial rubber rings. For the collection of eggs, the rubber rings can be detached in a short region and the eggs poured out. Another possibility is the use of screens at the bottom (Ryckman and Ryckman 1966). Both methods cannot be used for *R. prolixus*, since the females glue the eggs to all surfaces, for example cardboards or other adults. Crossed cardboards are necessary to increase the area (Schaub 1988); long folded filter papers are not stiff enough. Filter paper at the bottom of the beakers soaks up the feces and urine. If the enormous amount of urine excreted after feeding does not evaporate from the bottom of the beakers, especially young larvae will be drowned (Wood 1964). In our colonies, 100–200 first instar larvae are placed in the 2-l beakers which will not be opened until the bugs have developed to the adult stage. The exuviae remain in the beakers and are usually compressed at the bottom by feces and urine. Inspections for counting should be avoided since triatomines react very sensitively to handling stress (summarized by Schaub 1988; Schaub and Breger 1988). The group size is important to avoid overcrowding and isolation effects (Schaub 1990b). Using *T. infestans*, up to 40 larvae can be maintained in a 1-l beaker and up to 200 in a 2-l beaker without difficulties if crossed cardboards are included (Schaub 1988, 1989). If the feeding area and the period of time offered to suck blood are limited, crowding effects may affect feeding and thereby the development (Gomez-Núnez 1964). The hosts are fixed on a wooden device possessing a central opening under which the beaker with the bugs is placed (Ryckman and Ryckman 1966). In our laboratory, the bugs get the opportunity to suck blood of cocks or hens for 1 h or, if a uniform development of the group is required, for 2 h.

The feeding of bugs varies strongly. Since the bugs are usually searching for hosts during night, with the highest activities in the evening (Núñez 1987), complete darkness is often recommended (e.g., Neghme et al. 1967). However, in the field *D. maxima* is also active during daylight (Marsden et al. 1979), and bugs can be fed during the day without difficulties. Critical is the source of blood (summarized by Martínez-Ibarra et al. 2003). The feeding on living mice is optimal; rabbits seem to cause difficulties in the rearing of *R. neivai* (Cabello et al. 1987). According to our experience, also cocks and hens are optimal, especially if groups of 200 old larvae or *D. maxima* have to be fed. In vitro feeding via artificial membranes is possible (Garcia et al. 1984) but causes many difficulties in a long-term rearing (Gomez-Núñez 1964). A contamination of the blood with air-borne bacteria is deleterious; the bugs die within several days. Even with the use of very thin silicone membranes fewer bugs engorge than on a living mouse (Schaub 1990a). In addition, the clotting of the offered blood has to be avoided. Feeding of *R. prolixus* with defibrinated sheep and cow blood increases the mortality rate or reduces the reproduction rate but is successful with defibrinated pig blood (Gardiner and Maddrell 1972; Langley and Pimley 1978). However, usually after about 2 years the vitality of the colonies declines (Schaub unpublished).

Critical is the supplementation with symbionts, so far all belonging to the actinomycetes (summarized by Vallejo et al. 2009). Aposymbiotic bugs rarely develop to adults; the majority dies in the fourth and fifth instar (summarized by Vallejo et al. 2009). The symbionts are obtained by coprophagy, usually after feeding (Schaub et al. 1989). Therefore, a clean maintenance of bugs acts against this supply. Starting colonies with the first instar, old larvae or adult males should be added as symbiont donors (Schaub 1989; Jensen and Schaub 1991). Using the males is advantageous in order to have a standardized stage of the cohorts. Another possibility is the in vitro culture of the actinomycete symbionts and the distribution of a suspension in the beaker after feeding (Kollien and Schaub 1998). In *T. infestans* and *R. prolixus* the symbionts have been identified and are not known from other sources. In *P. megistus*, a *Rhodococcus equi*-like isolate has full symbiotic action. Since some strains of this actinomycete cause nocardiosis (Schaal 1992), this species should not be used as a "living syringe" or only after rearing with other actinomycetes as symbionts. The symbiont of *D. maxima* is unknown, but a supplementation with symbionts of different triatomines is sufficient for a good development (Schaub unpublished).

Are other bacteria relevant for the rearing of triatomines and can pathogens be transmitted by the use of triatomines as "living syringes"? Bugs possess an innate immunity and antimicrobial homeostasis mechanisms in the gut. However, the contact of the colonies with other insects should be excluded to avoid an infection with insect pathogenic bacteria. Such infections are usually indicated by the appearance of blood in the feces (Schaub unpublished). Small numbers of different bacteria have been isolated from bugs from the field and from laboratory colonies (summarized by Vallejo et al. 2009). All these bacteria did not affect the bugs and were not human pathogenic. In addition, within the last nearly 100 years of xenodiagnoses no human infections were evident. Since the surface of eggs can be disinfected, aposymbiotic bugs can be produced. A supplementation of the blood with vitamins of the vitamin B complex in the in vitro feeding enables a normal development of the bugs (Eichler and Schaub 1998, 2002). However, the axenic maintenance is very difficult, and easily triatomines become contaminated (Schaub unpublished).

11.6 The "Living Syringes": Comparison of Species of Triatomines for the Suitability as "Living Syringes"

Generally all species are suitable as "living syringes" but so far three species have been used. In comparing the three species for their suitability for use as "living syringes," one important aspect is the ease of rearing (Table 11.1). A disadvantage of *D. maxima* is the long generation time and the time required for each larval stage (see above). A disadvantage of *R. prolixus*, which has the shortest developmental times, is the behavior of females in gluing all the eggs to surfaces, whereas the other two species deposit eggs on the ground. In addition, adults of *R. prolixus* possess adhesive modifications on the legs and are easily able to climb up even glass walls (Lent and Wygodzinsky 1979).

Correlated to the sizes of the larvae is the volume of the ingested blood. In *D. maxima*, the first, second, third, fourth, and fifth instar larvae ingest in a single meal about 0.1, 0.2, 0.6, 1.2, and 2.5 g blood, respectively (Schaub unpublished). Therefore, the disadvantageous long-lasting development of *D. maxima* can be classified as advantageous, since the young instars of *D. maxima* are usually large enough to acquire the required volumes of blood. If less than 0.1 g blood is sufficient, young instars of the other species can be used or first instars of *D. maxima* before being fully engorged. The optimal instar can be chosen according to the volume of blood required for the ensuing investigations.

The induction of allergic reactions by the saliva of the triatomines may be relevant if bugs suck repeatedly on the same host. Data are only available for humans. *R. prolixus* induces such marked skin reactions that artificial xenodiagnosis is recommended (summarized by Meiser and Schaub 2011). The allergic reactions in humans are less intensive for *T. infestans* than *R. prolixus* (Lapierre and Lariviere 1954; Hoffman 1987), and *D. maxima* produces only minor skin reactions (Marsden 1986). In the use of second and third instar larvae of *D. maxima* as "living syringes" no response of the bird *Sterna hirundo* to the bite of *D. maxima* could be detected (Becker et al. 2005). Also in intensive use in the Zoological Garden Wuppertal no allergic reactions were evident (Stadler unpublished).

To be used as living syringes, not only the size is relevant. Even more important is the aggressiveness, since this reduces the period of time required for the

Table 11.1 Comparison of the advantages and disadvantages of *R. prolixus*, *T. infestans*, and *D. maxima* in their use as "living syringes"

	Species of triatomines		
	R. prolixus	*T. infestans*	*D. maxima*
Generation time	Shortest	Medium	Longest
Collection of eggs	Difficult[a]	Easy	Easy
Allergenicity of saliva	Highest	Medium	Lowest
Aggressiveness	Highest	Lowest	Medium[b]

[a]Eggs are glued to the cardboards or the surface of other bugs, and the adults can climb up glass walls since they possess attachment modifications on the legs
[b]Only slightly less than *R. prolixus*

animal–triatomine contact. In a ranking, *R. prolixus* seems to be most aggressive, followed by the slightly less aggressive *D. maxima* and the much less aggressive *T. infestans* (Marsden 1986; Schaub unpublished). However, the aggressiveness depends on the state of starvation. Shortly after ecdysis, the stomachs of the larvae still contain remnants of blood and some bugs refuse to suck. Usually, the larvae can be fed/used 1 week after the molt of the respective stage, although a 2-week period is recommended to increase the aggressiveness. With increasing periods of starvation, aggressiveness increases, but after the optimal phase, increasing numbers of bugs refuse to suck. Since the starvation capacities of the different instars differ (summarized by Schaub and Lösch 1989) the number of totally engorged bugs should be determined after different periods of starvation for each instar and each species in order to use the larvae during the optimal period of time.

Summarizing the advantages and disadvantages (Table 11.1) we recommend the aggressive *D. maxima*, especially since the second and third instar larvae, which will deliver sufficient blood for the majority of investigations, can already be used 3 and 6 weeks after hatching of first instar larvae (if the second instar is fed 1 week after molting). In addition, sterile first and second instar larvae can also be obtained more easily from this species.

11.7 The Use of "Living Syringes": Modes of Applications, Advantages and Disadvantages

Using triatomines as "living syringes," the application has to be modified according to the biology of the host species and whether the host animal has to be captured in the field or is maintained in human care. Using small mammals in the field or in the zoo, the animals can be fixed and the free active bugs placed near them. Investigating bats, the blood-sucking bugs were placed on the inner side of the wings of the bats, and the bugs recognized the higher temperature of the wing compared to the surroundings (von Helversen and Reyer 1984; von Helversen et al. 1986). Blood samples up to 70 µl were obtained by this method from fifth and fourth instar larvae of *R. prolixus* and *T. infestans*, respectively, and up to 200 µl from third instars of *D. maxima*. Using fourth instars of *T. infestans* and third instars of *D. maxima* the bugs ingested within 2–12 min about 100 µl blood, of which 70–80 µl could be withdrawn from the stomach (Voigt et al. 2005). Free active triatomines are also advantageous if in the zoo large aggressive animals have to be anesthetized, but are sometimes in cages free of litter, for example during feeding of the big carnivores *Panthera leo persica* (Asiatic lion), *Neofelis nebulosa* (clouded leopard), and *Canis lupus hudsonicus* (Canadian wolf) (Stadler 2007; Stadler et al. 2007, 2008b, 2009; Kruszewicz et al. 2009) (Figs. 11.1–11.4). The triatomines have to be watched to avoid an escape of the bugs which freely approach the host and leave it after engorgement. Then the zoo animals are removed from the cage and the bug is captured to collect the blood. To enable easier location of the bugs, a white thread of

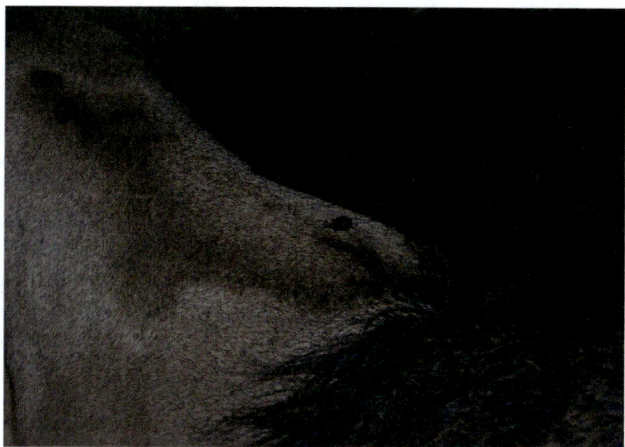

Fig. 11.1 Engorged *Dipetalogaster maxima* on the back of a male Asiatic lion (*Panthera leo persica*) (photo: André Stadler)

Fig. 11.2 *Dipetalogaster maxima* before engorgement on a Zambian common molerat (*Fukomys anselli*) (photo: Stephan Gatzen)

cotton can be placed around the thorax of the bug. The free active triatomine has the big advantage that the hosts do not need to be touched and the bugs can even be thrown into the cage. The risk to lose even a single bug is minimal. If this occurs, an insecticide can be applied in the region of disappearance.

In shy animals, for example okapis (*Okapia johnstoni*), which can be attracted for a short time but not for the period necessary for a full engorgement, the bugs are placed on the skin in a beaker. Before applying the beaker to the host, it is advantageous to exhale a few times into the beaker to stimulate the bugs and thereby to reduce the application time (Stadler unpublished). After the first bugs

Fig. 11.3 *Dipetalogaster maxima* before engorgement on the back of a hippopotamus (*Hippopotamus amphibius*) (photo: André Stadler)

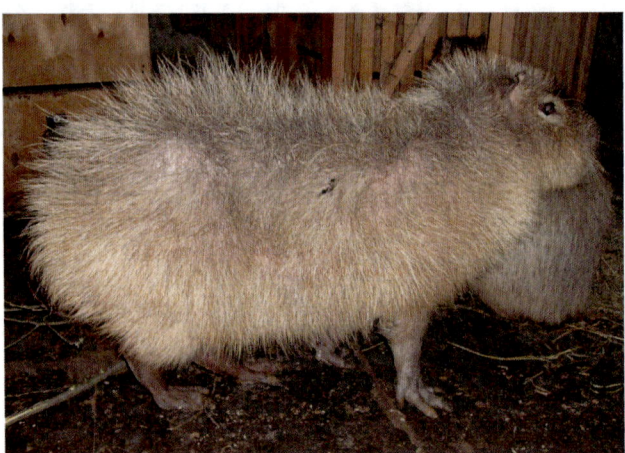

Fig. 11.4 *Dipetalogaster maxima* before engorgement on a capivara (*Hydrochoerus hydrochaeris*) (photo: André Stadler)

have started blood ingestion, the beaker is removed. The position of the bug can easily be observed. This minimizes the stress for the zoo animal since the period of time during which a person has direct contact with the animal is reduced (Stadler et al. 2007, 2008b, 2009).

A beaker containing the triatomines and closed by a metal or cotton gauze – similar to xenodiagnosis – should be used if keepers are familiar with the zoo animals and can handle them. Feeding or ruffling the animal the keepers place the beaker with bugs on the skin for about 15–20 min (Figs. 11.5–11.7). These beakers can either be small, containing a single bug, or bigger, containing more bugs which

Fig. 11.5 Use of a container with *Dipetalogaster maxima* for a juvenile Baird's tapir (*Tapirus bairdii*) (photo: Arne Lawrenz)

Fig. 11.6 Use of a container with *Dipetalogaster maxima* for an African elephant (*Loxodonta africana*) (photo: Arne Lawrenz)

are displaced if the first one is engorged totally. In the containers, a small piece of paper enables a better attack of the bugs (Hoffmann et al. 2005; Thomsen and Voigt 2006, 2008; Stadler 2007; Stadler et al. 2007, 2008b, 2009; Kruszewicz et al. 2009). In a specific, patented container, which resembles a syringe, the engorged bug can be pushed to and fixed on the gauze by a moveable part (Hoffmann et al. 2005; Thomsen and Voigt 2008). This allows an easier handling to obtain the blood from the bug, but a small risk that the fragile bug bursts. A further variation is a collar which is fixed around the neck of medium-sized mammals, suricates

Fig. 11.7 Use of a container with *Dipetalogaster maxima* for a reticulated giraffe (*Giraffa camelopardis reticulata*) (photo: André Stadler)

(*Suricata suricatta*) (Habicher 2009). This method is more complicated and restricted to specific animals, since initially the suricates were guided in a positive enforcement training, i.e., by offering small rewards, to tolerate the collar, initially without the insect and then with an insect. A second reward had to be offered to take the collar after the engorgement of the bugs, *R. prolixus* and *D. maxima*. This investigation indicated some of the limitations in the use of triatomines as "living syringes" in the field. At ambient temperatures below 20°C the bugs refuse to suck blood. High temperatures, especially direct sun, kill the bugs rapidly. In addition, in the tropics and subtropics, during applications in the field, triatomines may escape and establish. The sterilization of bugs seems to be a solution to this problem, but this needs to be standardized. In addition, outside of Latin America, a specific permission is required to introduce a nonindigenous species. In such countries the use of indigenous blood-sucking insects, for example tsetse flies, may be an easier solution (Habicher 2009).

Another variation is preferably used for animals which sleep or rest in caves or boxes. Under these circumstances, a second floor can be included as a resting place for the zoo animal and the triatomines are placed in beakers below the second floor under areas covered with metal gauze. This method was used for suricates and afterwards for small cat species and lemurs (Stadler 2006; Thomsen and Voigt 2006; Stadler et al. 2007, 2008b, 2009).

In ornithological studies the "living syringes" were used also without trapping the animals. Two different methods have been described to obtain blood from breeding animals, either the placement of artificial eggs or of small bags into the nests (Becker et al. 2005; Arnold et al. 2008; Bähnisch 2011). The latter was unsuccessful, since within less than 10 min the great tits, *Parus major,* removed the foreign material containing third instars of *D. maxima* from their nests, even if the bags were lightly attached (Bähnisch 2011). Only small amounts of blood were obtained, but not from sufficient numbers of birds to enable an analysis of paternity and parasitemia. Artificial eggs of such a tiny size were not available. In the common terns (*S. hirundo*), artificial eggs were produced and a hole in them closed with gauze; later the eggs were perforated. The artificial egg containing the bug was placed into the nest, and the triatomine sucked the blood at the brood patch of breeding adults (Becker et al. 2005; Arnold et al. 2008; Bauch et al. 2010). This mode of application strongly reduced the risk of an escape of bugs. Using third instar larvae of *D. maxima*, about 190 μl were obtained (Becker et al. 2005). Initially, the method was successful in only 34% of the trials, but after modification of the artificial eggs this rate was increased to 86 and 82% (Becker et al. 2005; Bauch et al. 2010). Most recently, these artificial eggs were also successfully used for Montagus' harriers (*Circus pygargus*) (Janowski 2010). The majority of harriers ignored the artificial egg, but four out of 48 birds threw the egg out of the nest. Seven females changed the position of the eggs in the nest unfavorably, excluding a contact possibility with the bug. Therefore, more specific adaptations of the eggs and an anchorage of eggs in the nest are necessary. Small artificial eggs were produced for common swifts (*A. apus*) (Will unpublished).

If the bugs are engorged, the blood has to be taken from the stomach. Initially bugs were decapitated and the abdomen perforated with a microcapillary tube (von Helversen et al. 1986), but it is much easier to hold the bugs between the fingers or with forceps and to withdraw the blood from the stomach directly using normal syringes and a thick needle, for example 16G (Voigt et al. 2004) (Fig. 11.8). The majority of bugs will survive this procedure but should be killed to avoid a mixture of samples. If the patented container is used, the needle is pushed through the gauze into the bug (Hoffmann et al. 2005). If the blood cannot be withdrawn directly after engorgement, the larvae should be placed at low temperatures to slow down the temperature-dependent excretion and digestion processes.

Fig. 11.8 Insertion of a drain tube into the abdomen of *Dipetalogaster maxima* to withdraw the blood (photo: Andreas Fischer)

11.8 The Use of "Living Syringes": Fields of Application

In the first use of "living syringes" the protein profiles of the blood of geckoes collected by ticks and fifth instar larvae of *R. prolixus* or by a decapitation of tail tips were similar (Will 1971, 1975, 1977). Then energy and water budgets of nectar-feeding bats were investigated using doubly labeled water and fourth instars of *T. infestans* (von Helversen and Reyer 1984; von Helversen et al. 1986; Voigt et al. 2003). In a verification of the methodology, data of samples obtained by the bugs were compared with those taken conventionally from veins, and the isotope enrichment in the latter was significantly higher (Voigt et al. 2003). This was suggested to be caused by the dilution with intestinal liquids or hemolymph of the bug. Since the doubly labeled water methodology depends on the rate of isotope decrease over time in a comparison of several samples, an identical dilution does not affect the conclusions. In a follow-up investigation with the bats and third instars of *D. maxima* and fourth instars of *T. infestans*, the doubly labeled water method and the energy balance method yielded similar results. Since conventional blood collection is stressful and may cause body mass losses in experimental animals, the bugs are a gentle method to minimize the stress in investigations of the energy and water budget (Voigt et al. 2005). On the basis of these studies, bugs were also used for doubly labeled water investigations of suricates (Habicher 2009).

In endocrinological studies, concentrations of the sex hormones testosterone and progesterone in the blood of domestic rabbits, *Oryctolagus cuniculus*, were determined after collection of blood with syringes or fifth instar larvae of *D. maxima* (Voigt et al. 2004). Hormone concentrations did not statistically significantly differ between the two methods. Drawing the blood from the stomach of the bug immediately after blood ingestion and 1, 2, 4, 8, and 24 h later, levels of progesterone significantly increased between 8 and 24 h after blood ingestion. Presumably, this increase was caused by the transport of water out of the stomach and the concentrating of the blood. Samples collected via blood-sucking bugs, *R. prolixus* and *D. maxima*, were also successfully used for gestation hormone analyses of Iberian lynxes (*Lynx pardinus*) (Vargas 2008) and in Dublin Zoo for elephants (Stadler unpublished). However, if the concentrations of adrenocorticotropic hormone metabolites in blood samples of elands (*Taurotragus oryx*) collected by third and fourth instars of *D. maxima* were compared to those obtained by the conventional method via needle and syringe and feces analyses, the blood samples collected with bugs significantly possessed sixfold higher concentrations than the samples collected conventionally (Hubmer et al. 2010, in press). The same happened with Indian rhinos, *Rhinoceros unicornis* (Stadler unpublished). Since the methodology of hormone analysis of rabbits and elands differs, it must be investigated whether or not components of the triatomines react with the compounds used in the analysis or if elands and rhinos are species in which triatomines cannot be used for gestation hormone analyses.

The investigation of rabbits also focused on the concentrations of the stress hormone hydrocortisone, which remained unchanged within 24 h after feeding

(Voigt et al. 2004). Hydrocortisone levels demonstrated the advantage of the "living syringes" since the standard deviations of the concentrations were lower than in samples obtained by the conventional method (Voigt et al. 2004). Baseline stress hormone levels of common terns *S. hirundo* collected by the use of artificial eggs did not statistically significantly differ from levels obtained by the conventional methods if in the latter, the samples were drawn within 3 min after trapping the bird (Arnold et al. 2008). In a sample drawn between 4 and 5 min after trapping the concentration was higher than in the sample obtained from the bug.

Analyses of blood metabolites offer information about the condition of an animal and illnesses. Using three fourth instars of *D. maxima* and a clouded leopard to determine the concentrations of 22 clinically relevant blood parameters (e.g., sodium, calcium, potassium, chloride, total carbon dioxide, bicarbonate, phosphate, creatinine, urea, glucose, hematocrit, hemoglobin, albumin, alkaline phosphatase, alanine aminotransferase, amylase, total bilirubin, globulin, total protein), the data of the three bugs differed by 1–13.6% (Stadler et al. 2007). Comparing the number of leucocytes in samples obtained by the "living syringes" and the conventional method from five zoo animals, in three animals the numbers differed by about 7%, but in two animals by 17 and 28% (Stadler et al. 2007). In common terns, cholesterol values of both methods were strongly correlated. Also in herring gulls, these concentrations and also those of triglyceride and uric acid did not differ statistically significantly (Bauch et al. 2010).

A fourth field of investigations, in which blood samples are required, is the determination of antibody titers. In a study of rabbit hemorrhagic disease virus, blood samples of 20 rabbits were collected by the conventional method (needle and syringe from the vena auricularis) or via fifth instars of *D. maxima* (Voigt et al. 2006). Virus antibody titers were determined in three different dilutions using standard test kits and were identical in 56 paired samples and differed in only four cases. Using NMRI-mice (*Mus musculus*), vaccinated against rabies, the virus-neutralizing antibody titers determined in the stomach contents of the fourth and fifth instars of *R. prolixus* and *D. maxima* were equivalent to those collected by retro-orbital bleeding from the same mice (Vos et al. 2010). Antibody titers in samples taken from different bugs after feeding on the same mouse did not differ, and were also not affected by storage in the stomach for 4 h.

Another field of application is the collection of blood for the isolation of lymphocytes, which were used in karyological analyses to identify species of bats (von Helversen et al. 1986). Genetics of populations of Montagus' harriers, *Circus pygargus*, were investigated using the blood for the isolation of DNA and the PCR amplification of microsatellite DNA (Janowski 2010). Although the "living syringe" technique is based on xenodiagnosis, so far it has not been used to obtain blood parasites.

Since the data of samples collected conventionally or by bugs are identical, the advantages of the "living syringe" technique support a more widespread use. The puncture of the needles is much bigger than the puncture caused by mouthparts of final instars of *R. prolixus*, *T. infestans*, or of *D. maxima* (Voigt et al. 2004, 2006). Whereas needles sometimes cause bleeding after blood collection (Bähnisch 2011), the capillaries close directly after retraction of the mouthparts of the triatomines, sometimes leaving a little hemorrhage in the tissue (Lavoipierre et al. 1959). The major advantages of the "living syringe" technique are the possibility to obtain blood without stress and from animals for which veins are inaccessible and the fact that anesthesia of focus animals is not necessary (Voigt et al. 2006).

11.9 The Use of "Living Syringes": Field Studies

"Living syringes" have been successfully used in several studies in the field (Table 11.2). In such projects, often host mammals are captured initially. Why then use bugs instead of syringes? It is advantageous if little animals are probed. In bats, a sampling of blood by cardiac puncture or from the orbital sinus requires good training (Baer 1966; Baer and McLean 1972) and is not possible for the investigation of tiny species (von Helversen and Reyer 1984). In these first field studies in which bugs were used as "living syringes" for the doubly labeled water method, the measurements of nectar intake and energy expenditure indicate that the energy turnover of a flower-visiting bat is high compared with that of other bat species, small mammals, and birds (von Helversen and Reyer 1984; von Helversen et al. 1986). The doubly labeled water method requires at least two blood samplings, a difficulty of investigations in the field. However, after training suricates to wear collars, it was also possible with these mammals (Habicher 2009).

The noninvasive technique to obtain blood samples was especially successful in investigating colonies of the common tern *S. hirundo* (Becker et al. 2005; Arnold et al. 2008; Bauch et al. 2010). Conventional blood sampling disturbs the birds greatly and can induce a leaving of the nest. In addition, baseline stress hormone

Table 11.2 Use of "living syringes" in the field

Host	Vector	Instar	Aim of research	Reference
Tailed tailless bat (*Anoura caudifer*)	*R. prolixus*	L5	Energy metabolism, Methodology	von Helversen and Reyer (1984), von Helversen et al. (1986)
Pallas's long-tongued bat (*Glossophaga soricina*)	*T. infestans*/*D. maxima*	L4, L5	Energy metabolism	Voigt et al. (2005)
Common tern (*Sterna hirundo*)	*D. maxima*	L2, L3	Methodology	Becker et al. (2005)
Common tern (*Sterna hirundo*)	*D. maxima*	L3	Corticosterone level	Arnold et al. (2008)
Suricate (*Suricata suricatta*)	*R. prolixus*	L2, L3	Energy metabolism	Habicher (2009)
Montagus' harrier (*Circus pygargus*)	*D. maxima*	L2	Genetic variability	Janowski (2010)
Common tern (*Sterna hirundo*)	*D. maxima*	L3	Blood metabolite level	Bauch et al. (2010)

titers in blood can only be determined if the capture of a bird and withdrawal of blood is finished within 3 min (Arnold et al. 2008). This difficulty was excluded using bugs. The "living syringes" are optimal for determining the baseline stress hormone titers that are indicators of the fitness, health, rank, and disturbance levels of wild organisms (Arnold et al 2008; Bähnisch 2011). Since the collection of blood induced no stress in the common terns, blood metabolite levels over the course of incubation could be determined, indicating that breeding experience affects the condition of these seabirds (Bauch et al. 2010).

Blood of 13 species of bats was also collected by triatomines to establish lymphocyte cultures for taxonomic purposes (von Helversen et al. 1986). Most recently, "living syringes" were used for an investigation of the genetic composition of populations of Montagus' harriers in Germany (Janowski 2010). In about 50 trials about 25 blood samples were obtained.

11.10 The Use of "Living Syringes": Zoo and Laboratory Studies

"Living syringes" have been used more often in zoos and laboratories than in the field (Tables 11.3 and 11.4). The main goal in zoos is not to disturb the animals since they are very valuable and can easily be damaged or even killed in the course of anesthesia. In addition, zoological staff are also endangered if they try to catch zoo animals for anesthesis. In laboratory animals, "living syringes" are advantageous if the animals are small, rendering the collection of blood more difficult, or if they are more strongly affected. Since blood of geckoes was usually obtained by decapitation of the tip of the tail, triatomines were used for the first time as "living syringes," investigating the protein profile (Will 1971, 1975, 1977)). Seven years later, the method was reinvented for use in the field (see above) and used also for captive colonies of bats (Voigt et al. 2003, 2005). Investigating also the energy metabolism with doubly labeled water, *R. prolixus* and tsetse flies were used for suricats, and six samples were obtained out of 47 trials (Habicher 2009). The application of "living syringes" to collect blood for endocrinological investigations has been successful in rabbits, lynxes, and elephants, but not in captive elands and rhinos (see above).

Beside these energy budget and endocrinological analyses, the "living syringe" method was intensively tested to determine whether or not *D. maxima* offers an application in regular zoo veterinary practice for the determination of blood parameters (Stadler 2007; Stadler et al. 2007, 2008b, 2009; Kruszewicz et al. 2009) (Table 11.4). This work was carried out inter alia with black panthers (*Panthera pardus*), Asiatic lions, Grant's zebras (*Equus quagga boehmi*), three species of tapirs, seals (*Phoca vitulina*), okapis, arctic wolves, and both species of elephants (Genus *Elephas* and *Loxodonta*), in total in over 120 samples from 44 species (Table 11.3). Blood values up to 1,100 µl were obtained from zoo animals. Already one bug was sufficient to obtain all relevant clinical parameters via photo-, potentiometric or via

Table 11.3 Use of "living syringes" in zoo and laboratory animals for evaluations of methodology, energy metabolism, endocrinological studies, and diagnoses of diseases

Host	Instar	Aim of research	Reference
17 species of scaled reptiles (Squamata)	L5	Methodology	Will (1971, 1975, 1977)
Sumatran Orang utan (*Pongo abelii*)	L5	Methodology	Thomsen and Voigt (2006)
Bonobo (*Pan paniscus*)	L5	Methodology	Thomsen and Voigt (2006)
Gray mouse lemur (*Microcebus murinus*)	L5	Methodology	Thomsen and Voigt (2008)
Suricate (*Suricata suricatta*)	?	Energy metabolism	Habicher (2009)
Rabbit (*Oryctolagus cuniculus*, dom)	?	Stress hormones	Voigt et al. (2004)
Eland antilope (*Taurotragus oryx*)	L3	Reproductive endocrinology	Hubmer et al. (2010, in press)
Iberian Lynx (*Lynx pardinus*)	?	Reproductive endocrinology	Vargas (2008)
Malayan tapir (*Tapirus indicus*)	L3, L4	Tuberculosis diagnosis	Stadler et al. (2008a)
South american sea lion (*Otaria flavescens*)	L3, L4	Tuberculosis diagnosis	Stadler et al. (2008a)
Lowland tapir (*Tapirus terrestris*)	L3, L4	Tuberculosis diagnosis	Stadler et al. (unpublished)
Indian rhinoceros (*Rhinoceros unicornis*)	L3, L4	Tuberculosis diagnosis	Stadler et al. (unpublished)
Giraffe (3 subspecies and hybrids) (*Giraffa camelopardalis*)	L3, L4	Tuberculosis diagnosis	Stadler et al. (unpublished)
Harbor seal (*Phoca vitulina*)	L3, L4	Tuberculosis diagnosis	Stadler et al. (unpublished)
Baird's tapir (*Tapirus bairdii*)	L3, L4	Tuberculosis diagnosis	Stadler et al. (unpublished)
Black faced sheep (*Ovis ammon f. aries*)	L3, L4	Blue tongue virus diagnosis	Stadler et al. (unpublished)
Yellow-backed duiker (*Cephalophus silvicultor*)	L3, L4	Blue tongue virus diagnosis	Stadler et al. (unpublished)
Dromedary (*Camelus dromedarius*, dom.)	L3, L4	Blue tongue virus diagnosis	Stadler et al. (unpublished)
Siberian ibex (*Capra ibex sibirica*)	L3, L4	Blue tongue virus diagnosis	Stadler et al. (unpublished)
White-lipped deer (*Cervus albirostris*)	L3, L4	Blue tongue virus diagnosis	Stadler et al. (unpublished)
Giraffe (2 subspecies) (*Giraffa camelopardalis*)	L3, L4	Blue tongue virus diagnosis	Stadler et al. (unpublished)
Lowland tapir (*Tapirus terrestris*)	L3, L4	Blue tongue virus diagnosis	Stadler et al. (unpublished)
Red river hog (*Potamochoerus porcus*)	L3, L4	Brucellosis diagnosis	Stadler et al. (unpublished)
Giraffe (2 subspecies and hybrids) (*Giraffa camelopardalis*)	L3, L4	Brucellosis diagnosis	Stadler et al. (unpublished)
Mice (*Mus musculus*)	L4, L5	Rabies diagnosis	Vos et al. (2010)

Table 11.4 Use of third and fourth instar larvae of *D. maxima* as "living syringes" in the zoo for blood chemistry analyses[a]

Host	Scientific name
Red-necked wallaby	*Macropus rufogriseus*
Dusky langur	*Presbytis obscura*
Columbian black spider monkey	*Ateles fusciceps robustus*
Golden headed tamarin	*Leontopithecus chrysomelas*
Red handed tamarin	*Saguinus midas*
Ansell's Mole-rat	*Cryptomys ansellii*
Capybara	*Hydrochoerus hydrochaeris*
Giant anteater	*Myrmecophaga tridactyla*
Canadian wolf	*Canis lupus ssp.*
Suricate	*Suricata suricatta*
Brown hyaena	*Hyaena brunnea*
Black-footed cat	*Felis nigripes*
Sandcat	*Felis margarita*
Geoffroy's cat	*Oncifelis geoffroyi*
Clouded leopard	*Neofelis nebulosa*
Cheetah	*Acinonyx jubatus*
Black panther	*Panthera pardus*
Asiatic lion	*Panthera leo persicus*
Harbor seal	*Phoca vitulina*
South american sea lion	*Otaria flavescens*
Californian sea lion	*Zalophus californianus*
African elephant	*Loxodonta africana*
Asiatic elephant	*Elephas maximus*
Baird's tapir	*Tapirus bairdii*
Malayan tapir	*Tapirus indicus*
Lowland tapir	*Tapirus terrestris*
White rhinoceros	*Ceratotherium simum*
Indian rhinoceros	*Rhinoceros unicornis*
African donkey	*Equus asinus*, dom.
Horse	*Equus ferus caballus*
Grant's Zebra	*Equus quagga boehmi*
Collared peccary	*Tayassu tajacu*
Babirusa	*Babyrousa babyrussa*
Red river hog	*Potamochoerus porcus*
White-lipped deer	*Cervus albirostris*
Siberian ibex	*Capra ibex sibirica*
Takin	*Budorcas taxicolor*
Okapi	*Okapia johnstoni*
Giraffe	*Giraffa camelopardalis* (3 subspecies)
Bongo	*Taurotragus euryceros*
Yellow-backed duiker	*Cephalophus silvicultor*
Zebu	*Bos taurus*, dom.
Banteng	*Bos javanicus*
Bactrian camel	*Camelus bactrianus*, dom.
Hippopotamus	*Hippopotamus amphibius*
African goat	*Capra hircus*, dom.
Capivara	*Hydrochoerus hydrochaeris*.

[a]Stadler et al. (2007) or Stadler (unpublished)

counting methods. Samples obtained from blood-sucking bugs were qualified for the analyses of 22 clinically relevant parameters, for example blood sugar, leukocyte counts, or kidney values. Sometimes concentrations of potassium, glucose, or iron differed from samples obtained conventionally, but this mainly occurred in hemolytic samples. Hemolysis could be caused in the stomach of the bug or by a too rapid withdrawal of the blood from the stomach. In addition, bugs suck blood from capillaries, and some larvae require >30 min. Then initial absorption processes might have changed the composition. In bugs finishing engorgement in <30 min, the blood parameter did not change significantly in the first 6 h after collection sampling. The concentrations of potassium are a good indicator to judge the hemolytic process in the sample collected via blood-sucking bugs and to conclude whether or not the sample can be used for the other relevant clinical parameters obtained via photometrics or potentiometrics or via counting methods (Stadler 2007; Stadler et al. 2007, 2008b, 2009; Kruszewicz et al. 2009). Especially primates are easily stressed by the conventional veterinary blood sampling routine, and blood parameters cannot be regularly monitored in big apes without difficulties. The fifth instar larvae of *D. maxima* were successfully used on gray mouse lemurs (*Microcebus murinus*) and two species of great apes (*Pongo abelii, Pan paniscus*) as an alternative, noninvasive technique for bleeding primates. Within 6–62 min, 0.01–2.4-ml blood was obtained in 11 out of 12 trials from all three species (Thomsen and Voigt 2006).

Another currently significant field of blood analyses is the identification of diseases, especially tuberculosis, blue-tongue disease, and brucellosis (Table 11.4). At the Zoological Garden Wuppertal the application of blood-sucking bugs as a method for tuberculosis screening by the rapid test was investigated. To run this test, only a small amount of blood is required. In other zoos, blood samples from South American sea lions (*Otaria flavescens*) and Malayan tapirs (*Tapirus indicus*) were taken by common venous puncture as well as by third and fourth instar larvae of *D. maxima*. In total, 15 paired samples were compared, and all positive and negative results were identical in both collection methods (Stadler et al. 2008a, unpublished). In a screening of 27 and eight samples for blue tongue disease and brucellosis, respectively, two samples contained antibodies against the blue tongue virus, none were positive for brucellosis. False positive or negative samples did not occur. Also in one giraffe at the Cologne Zoo an infection with the Blue tongue virus was verified using *D. maxima* (Habicher 2009).

11.11 Conclusion

Triatomines offer a good alternative to the conventional veterinary blood sampling routine with great advantages, for example that they are noninvasive and so do not disturb the vector. Normally the animals need to be immobilized for blood sampling. This is unnecessary using "living syringes." Adapting the collection methodology to the respective species of mammals and birds, so far a handful of species have been successfully considered in the wild and nearly 40 species in zoos.

Altogether three species of Triatominae were used, but *D. maxima* seems to be most advantageous. The samples obtained by "living syringes" could be used for example for hormone analysis, testing for diseases, and blood chemistry values. There are still some problems which occur, for example in hormone analysis or with a few blood chemistry values, but especially the values of potassium are a good indicator to judge the hemolytic process in the sample collected via blood-sucking bugs and therefore to conclude whether this gentle method can be used for all relevant clinical parameters via photometric, potentiometric or counting methods or not.

Acknowledgements We thank Dr. R. Cassada for correcting the English style. We are deeply indebted to Dr. Ulrich Schürer and Dr. Arne Lawrenz from the Zoological Garden Wuppertal for permission and support of the studies. We also thank them, Andreas Fischer and Stephan Gatzen for the permission to use their photos.

References

Amino R, Porto RM, Chammas R, Egami MI, Schenkman S (1998) Identification and characterization of a sialidase released by the salivary gland of the hematophagous insect *Triatoma infestans*. J Biol Chem 273:24575–24582

Amino R, Martins RM, Procopio J, Hirata IY, Juliano MA, Schenkman S (2002) Trialysin, a novel pore-forming protein from saliva of hematophagous insects activated by limited proteolysis. J Biol Chem 277:6207–6213

Andrade BB, Teixeira CR, Barral A, Barral-Netto M (2005) Haematophagous arthropod saliva and host defense system: a tale of tear and blood. Ann Acad Bras Cienc 77:665–693

Arnold JM, Oswald SA, Voigt CC, Palme R, Braasch A, Bauch C, Becker PH (2008) Taking the stress out of the blood collection; comparison of field blood-sampling techniques for analysis of baseline corticosterone. J Avian Biol 39:588–592

Assumpção TCF, Francischetti IMB, Andersen JF, Schwarz A, Santana JM, Ribeiro JMC (2008) An insight into the sialome of the bood-sucking bug *Triatoma infestans*, a vector of Chagas' disease. Insect Biochem Mol Biol 38:213–232

Azambuja P, Guimarães JA, Garcia ES (1983) Haemolytic factor from the crop of *Rhodnius prolixus*: evidence and partial characterization. J Insect Physiol 29:833–837

Baer GM (1966) A method for bleeding small mammals. J Mammal 47:340–341

Baer GM, McLean RG (1972) A new method of bleeding small and infants bats. J Mammal 53:231–232

Bähnisch E (2011) Parasitämie, Stress-Parameter und Fremdgehrate von Kohlmeisen (*Parus major*) in unterschiedlichen Habitaten. Dissertation, Universität Duisburg-Essen

Barreto AC, Prata AR, Marsden PD, Cuba CC, Trigueira CP (1981) Aspectos biológicos e criação em massa de *Dipetalogaster maximus* (Uhler, 1894) (Triatominae). Rev Inst Med Trop São Paulo 23:18–27

Barros VC, Assumpção JG, Cadete AM, Santos VC, Cavalcante RR, Araújo RN, Pereira MH, Gontijo NF (2009) The role of salivary and intestinal complement system inhibitors in the midgut protection of triatomines and mosquitoes. PLoS One 4(6):e6047

Barrozo RB, Lazzari CR (2004) The response of the blood-sucking bug *Triatoma infestans* to carbon dioxide and other host odours. Chem Senses 29:319–329

Barrozo RB, Lazzari CR (2006) Orientation response of haematophagous bugs to CO_2: the effect of the temporal structure of the stimulus. J Comp Physiol A Neuroethol Sens Neural Behav Physiol 192:827–831

Barrozo RB, Minoli SA, Lazzari CR (2004) Circadian rhythm of behavioural responsiveness to carbon dioxide in the blood-sucking bug *Triatoma infestans* (Heteroptera: Reduviidae). J Insect Physiol 50:249–254

Bauch C, Kreutzer S, Becker PH (2010) Breeding experience affects condition: blood metabolite levels over the course of incubation in a seabird. J Comp Phys B 180:835–845

Bauer PG (1981) Ultrastrukturelle und physiologische Aspekte des Mitteldarms von *Rhodnius prolixus* Stal (Insecta, Heteroptera). Dissertation, Universität Basel

Becker PH, Voigt CC, Arnold JM, Nagel R (2005) A non-invasive technique to bleed incubating birds without trapping: a blood-sucking bug in a hollow egg. J Ornithol 147:115–118

Brumpt E (1914) Le xénodiagnostic. Application au diagnostic de quelques infection parasitaires et in particulier à la trypanosome de Chagas. Bull Soc Path Exot 77:706–710

Cabello DR, Lizano E, Valderrama A (1987) Estadísticas vitales de *Rhodnius neivai* Lent, 1953 (Hemiptera: Reduviidae) en condiciones experimentales. Mem Inst Oswaldo Cruz 82:511–524

Canavoso LE, Frede S, Rubiolo ER (2004) Metabolic pathways for dietary lipids in the midgut of hematophagous *Panstrongylus megistus* (Hemiptera: Reduviidae). Insect Biochem Mol Biol 34:845–854

Carvalho-Pinto CJ, Grisard EC, Loroza ES, Steindel M (2000) Ecological and behavioral aspects of *Triatoma klugi*, a new triatomine species recently described from Rio Grande do Sul State, Brazil. Mem Inst Oswaldo Cruz 95(Suppl 2):336–337

Cavalcante RR, Pereira MH, Gontijo NF (2003) Anti-complement activity in the saliva of phlebotomine sand flies and other haematophagous insects. Parasitology 127:87–93

Corrêa RR (1954) Alguns dados sôbre a criação de Triatomíneos em laboratório. (Hemiptera Reduviidae). Folia Clin Biol 22:51–56

Dan A, Pereira MH, Pesquero JL, Diotaiuti L, Beirão PS (1999) Action of the saliva of *Triatoma infestans* (Heteroptera: Reduviidae) on sodium channels. J Med Entomol 36:875–879

Deane MP (1948) Um método para manter colônias de triatomídeos em laboratórios. Rev Serv Esp Saúde Públ 2:493–500

Dias E (1938) Criação de triatomideos no laboratorio. Mem Inst Oswaldo Cruz 33:407–412

Dias E (1955) Notas sôbre o tempo de evolução de algumas espécies de triatomíneos em laboratório. Rev Bras Biol 15:157–158

Dias JCP, Schofield CJ (1999) The evolution of Chagas disease (American trypanosomiasis) control after 90 years since Carlos Chagas discovery. Mem Inst Oswaldo Cruz 94(Suppl 1):103–121

Eichler S, Schaub GA (1998) The effects of aposymbiosis and of an infection with *Blastocrithidia triatomae* (Trypanosomatidae) on the tracheal system of the reduviid bugs *Rhodnius prolixus* and *Triatoma infestans*. J Insect Physiol 44:131–140

Eichler S, Schaub GA (2002) Development of symbionts in triatomine bugs and the effects of infections with trypanosomatids. Exp Parasitol 100:17–27

Ferreira RA, Lazzari CR, Lorenzo MG, Pereira MH (2007) Do haematophagous bugs assess skin surface temperature to detect blood vessels? PLoS ONE 2(9):e932

Francischetti IMB, Ribeiro JMC, Champagne D, Andersen J (2000) Purification, cloning, expression, and mechanism of action of a novel platelet aggregation inhibitor from the salivary gland of the blood-sucking bug, *Rhodnius prolixus*. J Biol Chem 275:12639–12650

Francischetti IMB, Andersen JF, Ribeiro JMC (2002) Biochemical and functional characterization of recombinant *Rhodnius prolixus* platelet aggregation inhibitor 1 as a novel lipocalin with high affinity for adenosine diphosphate and other adenine nucleotides. Biochemistry 41:3810–3818

Galvão C, Carcavallo R, da Silva RD, Jurberg J (2003) A checklist of the current valid species of the subfamily Triatominae Jeannel, 1919 (Hemiptera, Reduviidae) and their geographical distribution, with nomenclatural and taxonomic notes. Zootaxa 202:1–36

Garcia ES, de Azambuja P, Contreras VT (1984) Large-scale rearing of *Rhodnius prolixus* and preparation of metacyclic trypomastigotes of *Trypanosoma cruzi*. In: Morel CM (ed) Genes and antigens of parasites, 2nd edn. Fundação Oswaldo Cruz, Rio de Janeiro, pp 43–46

Garcia ES, Genta FA, de Azambuja P, Schaub GA (2010) Interactions of intestinal compounds of triatomines and *Trypanosoma cruzi*. Trends Parasitol 26:499–505

Gardiner BOC, Maddrell SHP (1972) Techniques for routine and large-scale rearing of *Rhodnius prolixus* Stål (Hem., Reduviidae). Bull Entomol Res 61:505–515

Geigy R, Kraus C (1952) Rüssel und Stechakt von *Rhodnius prolixus*. Acta Trop 9:272–276

Gómez-Núñez JC (1964) Mass rearing of *Rhodnius prolixus*. Bull World Health Org 31:565–567

Guernstein PG, Guerin PM (2001) Olfactory and behavioural responses of the blood-sucking bug *Triatoma infestans* to odours of vertebrate hosts. J Exp Biol 204:585–597

Habicher A (2009) Behavioural cost minimisation and minimal invasive blood-sampling in meerkats (*S. suricatta*, Herpestidae). Dissertation, Universität Köln

Hase A (1932) Beobachtungen an venezolanischen *Triatoma*-Arten, sowie zur allgemeinen Kenntnis der Familie der Triatomidae (Hemipt.-Heteropt.). Z Parasitenkd 4:585–652

Hoffman DR (1987) Allergy to biting insects. Clin Rev Allergy 5:177–190

Hoffmann H, Voigt CC, Thomsen R (2005) Patentschrift: DE102004004066B32005.06.09 Vorrichtung zur minimal-invasiven Blutentnahme bei Tieren mittels blutsaugender Raubwanzen. Deutsches Patent- und Markenamt

Hubmer I, Kotrba R, Stadler A, Schwarzenberger F (2010) Minimally invasive pregnancy monitoring in captive elands (*Taurotragus oryx*) – faecal steroid hormone metabolites and blood sucking bugs (*Dipetalogaster maxima*). In: Wibbelt G (ed) Proceedings of the International Conference on Diseases of Zoo and Wild Animals 2010, Madrid, pp 200–203

Hubmer I, Kotrba R, Stadler A, Schwarzenberger F (in press) Hormonal study of captive female elands *(Taurotragus oryx)*: analysis of the steroid hormones in plasma and steroid hormone metabolites in feces. Theriogenology

Huebner E (2007) The *Rhodnius* Genome Project: the promises and challenges it affords in our understanding of reduviid biology and their role in Chagas' transmission. Comp Biochem Physiol 148(Suppl 1):S130

Janowski S (2010) Erste Ansätze zur populationsgenetischen Untersuchung von mainfränkischen Wiesenweihen (*Circus pygargus*) mit genetischen Markern unter Einsatz von Raubwanzen zur Blutgewinnung. Diploma thesis, Universität Heidelberg

Jensen C, Schaub GA (1991) Development of *Blastocrithidia triatomae* (Trypanosomatidae) in *Triatoma infestans* after vitamin B-supplementation of the blood-diet of the bug. Eur J Protistol 27:17–20

Kaneko Y, Shojo H, Yuda M, Chinzei Y (1999) Effects of recombinant nitrophorin-2 nitric oxide complex on vascular smooth muscle. Biosci Biotechnol Biochem 63:1488–1490

Kollien AH, Schaub GA (1998) *Trypanosoma cruzi* in the rectum of the bug *Triatoma infestans*: effects of blood ingestion by the starved vector. Am J Trop Med Hyg 59:166–170

Kollien AH, Schaub GA (2000) The development of *Trypanosoma cruzi* in Triatominae. Parasitol Today 16:381–387

Kruszewicz AG, Grothmann P, Czujkowska A, Stadler A, Lawrenz A, Schaub GA (2009) Use of kissing bugs for blood sampling of exotic animals. Zycie Weteryn 84:405–407

Langley PA, Pimley RW (1978) Rearing triatomines in the absence of a live host and some effects of diet on reproduction in *Rhodnius prolixus* Stål (Hemiptera: Reduviidae). Bull Entomol Res 68:243–251

Lapierre J, Lariviere M (1954) Réaction allergique aux piqûres de réduvidés (*Rhodnius prolixus*). Bull Soc Pathol Exot Filiales 47:563–566

Lavoipierre MMJ (1965) Feeding mechanism of blood-sucking arthropods. Nature 208:302–303

Lavoipierre MMJ, Dickerson G, Gordon RM (1959) Studies on the methods of feeding of blood-sucking arthropods. Ann Trop Med Parasitol 53:235–250

Lehane MJ (2005) Managing the blood meal. In: Lehane MJ (ed) The biology of blood-sucking in insects, 2nd edn. Cambridge University Press, Cambridge, pp 84–115

Lent H, Wygodzinsky P (1979) Revision of the Triatominae (Hemiptera, Reduviidae), and their significance as vectors of Chagas disease. Bull Am Museum Nat Hist 163:123–520

Lorenzo Figueras AN, Kenigsten A, Lazzari CR (1994) Aggregation in the haematophagous bug *Triatoma infestans*: chemical signals and temporal pattern. J Insect Physiol 40:311–316

Maddrell SHP (1963) Excretion in the bloodsucking bug, Rhodnius prolixus Stål. I. The control of diuresis. J Exp Biol 40:247–256

Maddrell SHP (1969) Secretion by the Malpighian tubules of *Rhodnius*. The movements of ions and water. J Exp Biol 51:71–97

Maddrell SHP, Herman WS, Mooney RL, Overton JA (1991) 5-Hydroxytryptamine: a second diuretic hormone in *Rhodnius prolixus*. J Exp Biol 156:557–566

Marsden PD (1986) *Dipetalogaster maxima* or *D. maximus* as a xenodiagnostic agent. Rev Soc Bras Med Trop 19:205–207

Marsden PD, Cuba CC, Alvarenga NJ, Barreto AC (1979) Report on a field collection of *Dipetalogaster maximus* (Hemiptera, Triatominae) (Uhler, 1894). Rev Inst Med Trop São Paulo 21:202–206

Martí MA, González Lebrero MC, Roitberg AE, Estrin DA (2008) Bond or cage effect: how nitrophorins transport and release nitric oxide. J Am Chem Soc 130:1611–1618

Martínez-Ibarra JA, Novelo López M, Hernández Robles MR, Guillén YG (2003) Influence of the blood meal source on the biology of *Meccus picturatus* Usinger 1939 (Hemiptera: Reduviidae: Triatominae) under laboratory conditions. Mem Inst Oswaldo Cruz 98:227–232

Meiser CK (2009) Bacteriolytic and anticoagulant proteins in the saliva and intestine of blood sucking bugs (Triatominae, Insecta). Dissertation, Ruhr-Universität Bochum

Meiser CK, Schaub GA (2011) Xenodiagnosis. In: Mehlhorn H (ed) Nature helps – How plants and other organisms contribute to solve health problems. Springer, Heidelberg

Meiser CK, Piechura H, Werner T, Dittmeyer-Schäfer S, Meyer HE, Warscheid B, Schaub GA, Balczun C (2010a) Kazal-type inhibitors in the stomach of *Panstrongylus megistus* (Triatominae, Reduviidae). Insect Biochem Mol Biol 40:345–353

Meiser CK, Piechura H, Meyer HE, Warscheid B, Schaub GA, Balczun C (2010b) A salivary serine protease of the haematophagous reduviid *Panstrongylus megistus*: sequence characterization, expression pattern and characterization of proteolytic activity. Insect Mol Biol 19:409–421

Morita A, Isawa H, Orito Y, Iwanaga S, Chinzei Y, Yuda M (2006) Identification and characterization of a collagen-induced platelet aggregation inhibitor, triplatin, from salivary glands of the assassin bug, *Triatoma infestans*. FEBS J 273:2955–2962

Neghme A, Alfaro E, Beyes H, Schenone H (1967) Método para la crianza de laboratorio de *Triatoma infestans* (Klug, 1934) (Hemiptera, Reduviidae). Bol Chil Parasitol 22:107–112

Noeske-Jungblut C, Krätzschmar J, Haendler B, Alagon A, Possani L, Verhallen P, Donner P, Schleuning W-D (1994) An inhibitor of collagen-induced platelet aggregation from the saliva of *Triatoma pallidipennis*. J Biol Chem 269:5050–5053

Noeske-Jungblut C, Haendler B, Donner P, Alagon A, Possani L, Schleuning WD (1995) Triabin, a highly potent exosite inhibitor of thrombin. J Biol Chem 270:28629–28634

Núñez JA (1987) Behavior of triatominae bugs. In: Brenner RR, Stoka A de la M (eds) Chagas' disease vectors, vol. II Anatomy and physiological aspects. CRC Press, Boca Raton, pp 1–29

Núñez JA, Segura EL (1987) Rearing of Triatominae. In: Brenner RR, Stoka A de la M (eds) Chagas' disease vectors, vol. II Anatomy and physiological aspects. CRC, Boca Raton, pp 31–40

Oliveira MF, Gandara AC, Braga CMS, Silva JR, Mury FB, Dansa-Petretski M, Menezes D, Vannier-Santos MA, Oliveira PL (2007) Heme crystallization in the midgut of triatomine insects. Comp Biochem Physiol C Toxicol Pharmacol 146:168–174

Ortiz MI, Molina J (2010) Preliminary evidence of *Rhodnius prolixus* (Hemiptera: Triatominae) attraction to human skin odour extracts. Acta Trop 113:174–179. Erratum in Acta Trop (2010) 115:165

Ribeiro JMC (1982) The antiserotonin and antihistamin activities of salivary secretion of *Rhodnius prolixus*. J Insect Physiol 28:69–75

Ribeiro JMC, Garcia ES (1981a) The role of the salivary glands in feeding in *Rhodnius prolixus*. J Exp Biol 94:219–230

Ribeiro JM, Garcia ES (1981b) Platelet antiaggregating activity in the salivary secretion of the blood sucking bug *Rhodnius prolixus*. Experientia 37:384–386

Ribeiro JMC, Pereira MEA (1984) Midgut glycosidases of *Rhodnius prolixus*. Insect Biochem 14:103–108

Ribeiro JMC, Walker FA (1994) High affinity histamine-binding and antihistaminic activity of the salivary nitric oxide-carrying heme protein (nitrophorin) of *Rhodnius prolixus*. J Exp Med 180:2251–2257

Ribeiro JMC, Schneider M, Isaias T, Jurberg J, Galvão C, Guimarães JA (1998) Role of salivary antihemostatic components in blood feeding by triatomine bugs (Heteroptera). J Med Entomol 35:599–610

Ribeiro JMC, Andersen J, Silva-Neto MAC, Pham VM, Garfield MK, Valenzuela JG (2004) Exploring the sialome of the blood-sucking bug *Rhodnius prolixus*. Insect Biochem Mol Biol 34:61–79

Ryckman RE, Bentley DG (1979) Host reactions to bug bites (Hemiptera, Homoptera): a literature review and annotated bibliography (parts I and II). CA Vector Views 26:1–49

Ryckman RE, Ryckman AE (1963) Loma Linda University's 1962 expedition to Baja California. Med Arts Sci 17:65–76

Ryckman RE, Ryckman AE (1966) Reduviid bugs. In: Smith CN (ed) Insect colonization and mass production. Academic, New York, pp 183–200

Ryckman RE, Ryckman AE (1967) Epizootiology of *Trypanosoma cruzi* in Southwestern North America. Part X: The biosystematics of *Dipetalogaster maximus* in Mexico (Hemiptera: Reduvidae) (Kinetoplastida: Trypanosomidae). J Med Entomol 4:180–188

Sant'Anna MRV, Diotaiuti L, de Figueiredo GA, de Figueiredo GN, Pereira MH (2001) Feeding behaviour of morphologically similar *Rhodnius* species: influence of mechanical characteristics and salivary function. J Insect Physiol 47:1459–1465

Santos A, Ribeiro JMC, Lehane MJ, Gontijo NF, Veloso AB, Sant'Anna MRV, Araujo RN, Grisard EC, Pereira MH (2007) The sialotranscriptome of the bood-sucking bug *Triatoma brasiliensis* (Hemiptera, Triatominae). Insect Biochem Mol Biol 37:702–712

Sarkis JJF, Guimarães JA, Ribeiro JMC (1986) Salivary apyrase of *Rhodnius prolixus*. Kinetics and purification. Biochem J 233:885–891

Schaal KP (1992) Pathogene aerobe Aktinomyceten. In: Burkhardt F (ed) Mikrobiologische Diagnostik. Thieme, Stuttgart, pp 258–268

Schaub GA (1988) Developmental time and mortality in larvae of the reduviid bugs *Triatoma infestans* and *Rhodnius prolixus* after coprophagic infection with *Blastocrithidia triatomae* (Trypanosomatidae). J Invertebr Pathol 51:23–31

Schaub GA (1989) *Trypanosoma cruzi*: quantitative studies of development of two strains in small intestine and rectum of the vector *Triatoma infestans*. Exp Parasitol 68:260–273

Schaub GA (1990a) Membrane feeding for infection of the reduviid bug *Triatoma infestans* with *Blastocrithidia triatomae* (Trypanosomatidae) and pathological effects of the flagellate. Parasitol Res 76:306–310

Schaub GA (1990b) The effect of *Blastocrithidia triatomae* (Trypanosomatidae) on the reduviid bug *Triatoma infestans*: influence of group size. J Invertebr Pathol 56:249–257

Schaub GA (2008) Kissing bugs. In: Mehlhorn H (ed) Encyclopedia of parasitology, vol 1, 3rd edn. Springer, Heidelberg, pp 684–686

Schaub GA (2009) Interactions of trypanosomatids and triatomines. Adv Insect Physiol 37:177–242

Schaub GA, Breger B (1988) Pathological effects of *Blastocrithidia triatomae* (Trypanosomatidae) on the reduviid bugs *Triatoma sordida*, T. pallidipennis and Dipetalogaster maxima after coprophagic infection. Med Vet Entomol 2:309–318

Schaub GA, Lösch P (1989) Parasite/host-interrelationships of the trypanosomatids *Trypanosoma cruzi* and *Blastocrithidia triatomae* and the reduviid bug *Triatoma infestans*: influence of starvation of the bug. Ann Trop Med Parasitol 83:215–223

Schaub GA, Böker CA, Jensen C, Reduth D (1989) Cannibalism and coprophagy are modes of transmission of *Blastocrithidia triatomae* (Trypanosomatidae) between triatomines. J Protozool 36:171–175

Schaub GA, Meiser CK, Balczun C (2011) Interactions of *Trypanosoma cruzi* and triatomines. In: Mehlhorn H (ed) Progress in the fight against parasitic diseases. Springer, Heidelberg

Schmitz H, Trenner S, Hofmann MH, Bleckmann H (2000) The ability of *Rhodnius prolixus* (Hemiptera; Reduviidae) to approach a thermal source solely by its infrared radiation. J Insect Physiol 46:745–751

Schnitker A, Schaub GA, Maddrell SHP (1988) The influence of *Blastocrithidia triatomae* (Trypanosomatidae) on the reduviid bug *Triatoma infestans*: in vivo and in vitro diuresis and production of diuretic hormone. Parasitology 96:9–17

Schofield CJ (1979) The behaviour of Triatominae (Hemiptera: Reduviidae): a review. Bull Entomol Res 69:363–379

Schofield CJ, Galvão C (2009) Classification, evolution and species groups within the Triatominae. Acta Trop 110:88–100

Schofield CJ, Jannin J, Salvatella R (2006) The future of Chagas disease control. Trends Parasitol 22:583–588

Schwarz A, Sternberg JM, Johnston V, Medrano-Mercado N, Anderson JM, Hume JCC, Valenzuela JG, Schaub GA, Billingsley PF (2009) Antibody responses of domestic animals to salivary antigens of *Triatoma infestans* as biomarkers for low-level infestation of triatomines. Int J Parasitol 39:1021–1029

Soares RPP, das Graças Evangelista L, Laranja LS, Diotaiuti L (2000) Population dynamics and feeding behavior of *Triatoma brasiliensis* and *Triatoma pseudomaculata*, main vectors of Chagas disease in Northeastern Brazil. Mem Inst Oswaldo Cruz 95:151–155

Soares AC, Carvalho-Tavares J, Gontijo NF, dos Santos VC, Teixeira MM, Pereira MH (2006) Salivation pattern of *Rhodnius prolixus* (Reduviidae; Triatominae) in mouse skin. J Insect Physiol 52:468–472

Stadler A (2006) Einfluss des Geschlechtes und psychoneuroimmunologischer Faktoren auf die Parasitierung von Zootieren. Diploma thesis, Ruhr-Universität Bochum

Stadler A (2007) Non invasive use of *Dipetalogaster maxima* for obtaining a blood sample from zoo animals. Proc Br Vet Zool Soc, Abstracts:96–97

Stadler A, Lawrenz A, Schaub GA (2007) Der Einsatz von Raubwanzen zur Gewinnung von Blutproben bei Zootieren. Zeitschr Kölner Zoo 50:163–173

Stadler A, Lawrenz A, Schaub GA (2008a) A minimaly-invasive technique for blood sampling suitable for TB screening – preliminary results. Proc Eur Zoo Wildl Vet Abst:143

Stadler A, Lawrenz A, Schaub GA (2008b) Nicht-invasiver Einsatz von Raubwanzen zur Gewinnung von Blutproben bei Zootieren. 26. Arbeitstagung der Zootierärzte, Frankfurt Tagungsbericht Abst:110–112

Stadler A, Lawrenz A, Schaub GA (2009) Der Einsatz der südamerikanischen Raubwanze *Dipetalogaster maxima* in Zoologischen Gärten zur Gewinnung von Blutproben. Tierärztliche Umschau 64:147–153

Thomsen R, Voigt CC (2006) Non invasive blood sampling from primates using laboratory-bred blood-sucking bugs (*Dipetalogaster maximus*; Reduviidae, Heteroptera). Primates 47:397–400

Thomsen R, Voigt CC (2008) Patent Application Publication: US 2008/0221536 A1 Device for carrying out the minimally invasive withdrawal of blood from animals by using blood-sucking assassin bugs. United States of America Patent Office

Vallejo GA, Guhl F, Schaub GA (2009) Triatominae – *Trypanosoma cruzi/T. rangeli*: vector-parasite interactions. Acta Trop 110:137–147

Vargas A (2008) Diagnósticos de gestacion. Bol Programa de conservacion ex-situ del lince ibérico 48:1–2

Voigt CC, von Helversen O, Michener RH, Kunz TH (2003) Validation of a non invasive blood sampling technique for doubly-labelled water experiments. J Exp Zool A Comp Exp Biol 296:87–97

Voigt CC, Fassbender M, Denhard M, Wibbelt G, Jewegenow K, Hofer H, Schaub GA (2004) Validation on a minimally invasive blood-sampling technique for the analysis of hormones in domestic rabbits, *Oryctolagus cuniculus* (Lagomorpha). Gen Comp Endocrinol 135:100–107

Voigt CC, Michener R, Wibbelt G, Kunz TH, von Helversen O (2005) Blood-sucking bugs as a gentle method for blood collection in water budget studies using doubly labelled water. Comp Biochem Phys 142:318–324

Voigt CC, Peschel U, Wibbelt G, Frölich K (2006) An alternative, less invasive blood sample collection technique for serologic studies utilizing triatomine bugs (Heteroptera; Insecta). J Wildl Dis 42:466–469

von Helversen O, Reyer HU (1984) Nectar intake and energy expenditure in a flower visiting bat. Oecologia 63:178–184

von Helversen O, Volleth M, Núñez J (1986) A new method for obtaining blood from a small mammal without injuring the animal: use of triatomid bugs. Experientia 42:809–810

Vos AC, Müller T, Neubert L, Voigt CC (2010) Validation of less invasive blood technique in rabies serology using reduviid bugs (*Triatominae*, Heminoptera). J Zoo Wildl Med 41:63–68

Wenk P, Lucic S, Betz O (2010) Functional anatomy of the hypopharynx and the salivary pump in the feeding apparatus of the assassin bug *Rhodnius prolixus* (Reduviidae, Heteroptera). Zoomorphology 129:225–234

Weichsel A, Andersen JF, Roberts SA, Montford WR (2000) Nitric oxid binding to nitrophorin 4 induces complete distal pocket burial. Nat Struct Biol 7:551–554

Wigglesworth VB (1931) The physiology of excretion in a blood-sucking insect, *Rhodnius prolixus* (Hemiptera, Reduviidae) I. Composition of the urine. J Exp Biol 8:411–427

Wigglesworth VB (1972) The principles of insect physiology, 7th edn. Chapman and Hall, London

Will R (1971) Serologische Normalwerte und deren krankhafte Veränderungen bei Reptilien (Squamata). Diploma thesis, Universität Hohenheim

Will R (1975) Die Verschiebungen des Bluteiweißbildes (Dysproteinämien) bei Lebererkrankungen von Reptilien (Boidae, Pythonidae, Varanidae). Zbl Vet Med B 22:635–655

Will R (1977) Hämatologische und serologische Untersuchungen bei Lacertiden (Reptila, Squamata). Dissertation, Universität Hohenheim

Wirtz HP (1987) Eindringen der Mundwerkzeuge von Raubwanzen durch eine Membran (Hemiptera: Reduviidae). Entomol Gen 12:147–153

Wood SF (1964) The laboratory culture of *Triatoma* (Hem., Reduviidae). Bull World Health Org 31:579–581

Zeledón R, Rabinovich JE (1981) Chagas' disease: an ecological appraisal with special emphasis on its insect vectors. Annu Rev Entomol 26:101–133

Chapter 12
Xenodiagnosis

Christian Karl Meiser and Günter A. Schaub

Abstract In xenodiagnosis, the vector of a pathogenic agent is infected to make identification of the parasite easier after a period of multiplications. This enables a diagnosis if the strain of the pathogen possesses a low virulence or if very low numbers of parasites occur in the blood, for example *Trypanosoma cruzi* in the chronic phase of Chagas disease. In 1914, 5 years after the discovery of *T. cruzi* in the triatomines, its blood-sucking vectors, these were used for the first time as a diagnostic tool of this disease. For many decades thereafter xenodiagnosis represented the golden standard. Meanwhile serological and molecular biological methods are preferably used for diagnosis today, but xenodiagnosis and in vitro cultivation are still the best methods for a proof of the living parasites or for the isolation of the flagellate. In this context, the different methodological variations and the disadvantages and advantages of xenodiagnosis are discussed in our review.

12.1 Introduction

Xenodiagnosis originates from the Greek words xenos (= foreign) and diagnosi (= diagnosis). This classical indirect parasitological method is a proof of parasites using another (foreign) host species, in which the pathogenic agent can be identified more easily after incubating for a long enough time for several multiplications. This is necessary if the concentration of the parasite in the original host is too low to allow a direct microscopic diagnosis. Usually the vectors are used as the "foreign" host. This method was first used by Brumpt (1914) to identify infections of a Brazilian watersnake (*Helicops modestus*) with *Trypanosoma brazili* after the development of the trypanosomes in leeches. It was also used for the detection of *Bartonella quintana*, the etiological agent of trench fever: body lice were fed on suspected infected people and about 1 week after infection, after multiplication in

---

C.K. Meiser and G.A. Schaub (✉)
Zoology/Parasitology Group, Department of Animal Ecology, Evolution and Biodiversity Group, Universitätsstr. 150, 44780 Bochum, Germany
e-mail: guenter.schaub@rub.de

the digestive tract of this vector, the bacteria could be detected (Töpfer 1916). Today, xenodiagnosis is a scientific tool to isolate and/or identify *Borrelia* sp. with ticks or arboviruses with mosquitoes, for example dengue virus (Donahue et al. 1987; summarized by Mourya et al. 2007). However, Brumpt (1914) already emphasized the importance of using the vector, triatomines, in the diagnosis of an infection with *Trypanosoma cruzi*, the etiologic agent of Chagas disease. Following its introduction, xenodiagnosis was the golden diagnostic standard for decades, but then it was displaced by serological and molecular biological tools. Serology gives rapid results but also has some difficulties, for example false positive results by cross reaction of antibodies produced against other pathogens and false negative results with antigen preparations which do not react with antibodies against all strains of *T. cruzi* (Malchiodi et al. 1994; Caballero et al. 2007; Verani et al. 2009). Meanwhile, xenodiagnosis is mainly used for the isolation and/or identification of the parasite and also as a low-cost method for the diagnosis of an infection with *T. cruzi*. In any case, like all diagnostic tools in Chagas disease, it has advantages and disadvantages that will be summarized in this review.

12.2 The Diagnostic Tool: Triatomines

The current taxonomy of Triatominae recognizes 140 species in five or six tribes – Alberproseniini, Bolboderini, Cavernicolini, Rhodniini, and Triatomini, with or without Linshcosteini – and 18 genera (Galvão et al. 2003; Schofield and Galvão 2009), but this number will increase with the use of molecular biological methods (Mas-Coma and Bargues 2009). The majority of species occur on the American continent between the Great Lakes of North America and Argentina (Lent and Wygodzinsky 1979; Schofield 1994; Gorla et al. 1997). Species of the genus *Linshcosteus* are confined to the Indian subcontinent. The eight species forming the Rubrofasciata group originate from the New World and have been distributed by sailing ships to many ports in tropical and subtropical regions (Haridass and Ananthakrishnan 1980; Gorla et al. 1997).

Triatomines prefer to stay near the host and therefore colonize burrows of rodents, nests of birds, and caves of bats. These habitats are often occupied by specific groups of triatomines, for example species of the genus *Triatoma* are often found in rocky habitats and rodent burrows, and species of the genus *Panstrongylus* in tree cavities and burrows, while species of the tribe Rhodniini – the only monophyletic tribe of Triatominae – often prefer palm trees with nests of birds and burrows of rodents (Gaunt and Miles 2000; Schofield and Galvão 2009). The transition from animals to humans seems to have taken place when the dwellings of the indigenous humans of Latin America were caves or were made of material from the forest. Until now, the construction of houses as a wooden frame covered with adobe or mud still offers a good habitat for triatomines since the cracks in the adobe or mud offer optimal hiding places during the day for the night-active bugs. In addition, since species of the genus *Rhodnius* glue their eggs to the leaves of palm trees, the use of this natural material for the roof of houses provides a direct access to the house. One species of Rhodniini, *Rhodnius prolixus,* is rarely found on palm trees, but has strongly adapted to houses and therefore is classified as being domestic. It may have evolved from the almost morphologically indistinguishable but nearly strictly sylvatic *Rhodnius robustus* (Feliciangeli et al. 2007). An even more important domestic species is *Triatoma infestans,* which presumably originates from sylvatic populations in the Andean valleys of Bolivia, colonizing burrows of guinea pigs and being transferred to the houses during domestication of this small mammal (Schofield 1994; Noireau et al. 2005). This species was distributed over wide parts of the South American continent and has successfully displaced other species of Triatominae, for example *Panstrongylus megistus* (Pereira et al. 2006). Thanks to control campaigns using insecticides, *T. infestans* has been eliminated from many countries (Schofield and Dias 1999), but other species are now invading the houses, for example *P. megistus* in the eastern states of Brazil (Villela et al. 2005). These bugs originate from peridomestic habitats where they suck blood of domestic animals or from sylvatic habitats where the majority of triatomine species live. One of these mainly sylvatic triatomines is the biggest species of triatomines, *Dipetalogaster maxima,* occurring only in the fog desert of the peninsular Baja California Sur in Mexico (Lent and Wygodzinsky 1979) (Fig. 12.1).

All postembryonic stages of triatomines are obligatorily hematophagous, developing from eggs through five larval instars to the adults (Schaub 2008). (Some scientists call all five instars or the first four instars of Hemiptera "larvae" and the fifth a nymph, while others use the term "nymph" for all five preadult instars.) In each instar, the larvae require blood meals, several small ones or a minimum of one full engorgement of about 6–12 times their own body weight. The blood is stored essentially undigested in the strongly distensible stomach in which it is concentrated. After lysis of erythrocytes it is passed into the small intestine for digestion and absorption. Remains of blood digestion are stored in the rectum. The full engorgement activates distension receptors and thereby the secretion of hormones and the development of the new cuticle and the molt to the next larval instar or to the adult (Wigglesworth 1940; Anwyl 1972; Chiang and Davey 1988). The duration of the different developmental stages is a characteristic of the respective species and is affected by the relative humidity, the temperature and the presence of symbiotic bacteria, which colonize the anterior regions of the midgut and deliver essential compounds (Eichler and Schaub 2002). About 1 week after the molt, the larvae can ingest the next blood meal, but usually hosts are not continuously present, and the availability of blood determines the generation time. If no host is available, bugs can survive starvation for long periods of time. The starvation capacity is lower at higher temperatures and lower relative humidities and increases in older larvae. In *T. infestans* the fourth instars can survive 8–13 months after feeding in the third instar and with maintenance under optimal climatic conditions (Schaub and Lösch 1989). Availability of blood, abiotic factors and species characteristics determine not only the duration of larval development but also the longevity of adults, which live for about 3 months to 1 year (summarized by Schaub 2009).

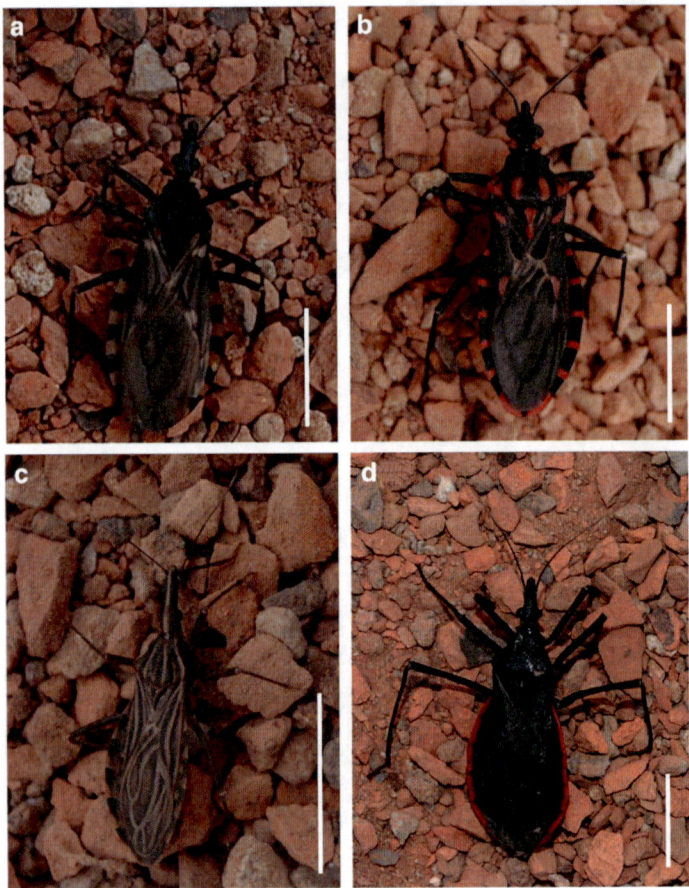

Fig. 12.1 Adults of four species of triatomines: (**a**) *Triatoma infestans*, (**b**) *Panstrongylus megistus*, (**c**) *Rhodnius prolixus*, (**d**) *Dipetalogaster maxima*. (**a**–**c**) from Schaub et al. (2011); permission from Springer is gratefully acknowledged

12.3 The Diagnosed Parasite: *Trypanosoma cruzi*

12.3.1 Taxonomy

All species of triatomines are assumed to be capable of transmitting *T. cruzi*, which belongs to the family Trypanosomatidae and the order Kinetoplastida. The different strains of *T. cruzi* show very diverse characteristics in their multiplication time, development in vector and mammalian hosts and pathogenicity for humans. In an attempt to correlate some of these peculiarities to biochemical characteristics, the strains were classified according to their isoenzyme patterns into two major groups and one less frequently represented, named zymodemes 1, 2, and 3 (Z1, Z2, Z3)

(Miles et al. 1978; Ebert and Schaub 1983). Similar groupings were obtained using restriction enzymes for kinetoplast DNA and molecular biological tools, denominated *T. cruzi* I and *T. cruzi* II (Anonymous 1999). Then subpopulations of *T. cruzi* I and *T. cruzi* II were classified (summarized by Vallejo et al. 2009). Most recently, the nomenclature of *T. cruzi* strains was changed to six discrete typing units based on their genetic similarity, abbreviated TcI–TcVI (Zingales et al. 2009). However, so far the new classification does not allow a correlation of these groups to specific regions, pathology or other biological parameters.

Evolutionarily, *T. cruzi* I seems to originate from an association with opossums and predominates north of Amazonia also in humans, while *T. cruzi* II arose from an association with armadillos and predominates in Southern Cone countries of South America also in humans (summarized by Schaub 2009). In some regions, these subpopulations are associated with infections in wild mammals and sylvatic vectors, humans and sylvatic vectors, and humans and domestic vectors (summarized by Vallejo et al. 2009), but under natural conditions often double infections with parasites belonging to both groups are found (summarized by Carneiro et al. 1990; Solari et al. 1998). In Brazil, *T. cruzi* I and *T. cruzi* II are often involved in the sylvatic and domestic transmission cycles, respectively. However, in Chile both groups of *T. cruzi* occur in sylvatic areas but circulate in different species of rodents (Galuppo et al. 2009). The increasing contact of humans to sylvatic cycles, especially in the Amazon region will introduce originally sylvatic *T. cruzi* strains into the domestic cycle and vice versa. Since the mammalian host and also the triatomine act as "filters," enabling or retarding the development of specific and sometimes quite different strains (Araújo et al. 2009), diagnosis and also xenodiagnosis will have to identify infections of a wide and very diverse number of strains of *T. cruzi*.

12.3.2 Developmental Cycle

During the developmental cycle of this protozoon, there is a host change between mammals and insects, and a change of forms between a-, trypo-, epi- and spheromastigotes (summarized by Schaub 2009; Garcia et al. 2010) (Fig. 12.2). In the stomach, different vector-derived factors interact with the flagellates, regulating the development of *T. cruzi* (summarized by Garcia et al. 2010). If *T. cruzi* is ingested together with the blood, the blood-trypomastigotes are immediately aggregated in the stomach and shorten and develop to a- and spheromastigotes (Brack 1968; Brener 1973; Alvarenga 1974). The flagellates are passed together with the concentrated blood into the small intestine, where epimastigotes multiply and attach to the perimicrovillar membranes of the midgut cells (Gonzalez et al. 1999). According to the flow of the intestinal contents, epi- and spheromastigotes arrive within 2–3 days in the rectum, where they attach to the rectal cuticle and also multiply intensively. In regularly fed fifth instars of *T. infestans* 1.5 million flagellates/bug develop in the rectum, about three to four times more parasites than in the small intestine and in *T. brasiliensis* up to 10 times more (Schaub and Lösch 1988; Schaub 1989; Kollien

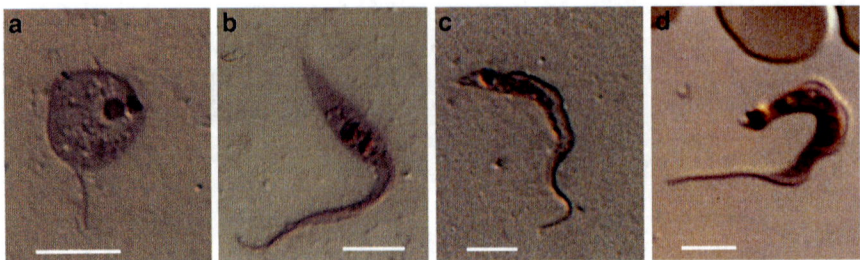

Fig. 12.2 Different developmental stages of *Trypanosoma cruzi* from the vector (**a–c**) or the mammalian host (**d**): (**a**) spheromastigote, (**b**) epimastigote, (**c**) metacyclic trypomastigote, (**d**) blood trypomastigote. (Scale bar: 5 µm); from Schaub et al. (2011); permission from Springer is gratefully acknowledged

and Schaub 1998a, b; Araújo et al. 2008). In only one parasite/vector system, the isolate mainly colonized the small intestine (Araújo et al. 2007). In many isolates, the infection extends to the ampullae (the final enlargements of the Malpighian tubules) and then to the tubules (Schaub and Lösch 1988; Schaub et al. 1989b). The non-replicative metacyclic trypomastigotes develop almost exclusively in the rectum. This infectious form is excreted in the feces of the bug, initiating the development in the mammal (Kollien and Schaub 1998b, 2000).

The metacyclic trypomastigote invades the mammalian host via mucous membranes or small lesions of the skin, often caused by the mouthparts of the vector for blood ingestion or by scratching as a reaction to the saliva of the vector (Schuster and Schaub 2000). First it infects cells in the invasion area, especially macrophages. In the phagosome of a macrophage, the parasite transforms to the amastigote form and secretes pore-forming proteins to evade into the cytosol of the immune cell, before phagosome and lysosomes fuse (Burleigh and Andrews 1995; Contreras et al. 2002). In the cytosol of the host cell, the amastigotes divide repeatedly, and cystic nests, so-called pseudocysts, are formed in the tissue (Sacks and Sher 2002). When the resources of the host cell are depleted, the amastigotes develop via pro- and epimastigote stages to trypomastigotes and are released upon bursting of the host cell. The blood-trypomastigote can be detected for only a short time in the blood, where it is protected against the immune system of the host by a surface coat of glycoproteins (Hall and Joiner 1991). Via the blood these trypomastigotes gain access to cells of other organs, in which the flagellate changes again to the amastigote form and multiplies.

12.4 The Diagnosed Disease: Chagas Disease

Chagas disease is one of the "Big Six," that is one of the six most important tropical parasitic diseases selected by the World Health Organization (WHO) as major topics in the campaign against tropical diseases (Schaub and Wülker 1984). The etiological agent of the disease, *T. cruzi*, was described by Chagas for the first time

Fig. 12.3 Old banknote of Brazil showing Dr. Carlos Chagas and the developmental cycle of *Trypanosoma cruzi* (currency until 1989)

in 1909 (Chagas 1909) (Fig. 12.3). During the investigation of a malaria epidemic in Minas Gerais, a Federal State in southeast Brazil, he first found the flagellate in the intestine of a blood-sucking assassin bug, *P. megistus* (Chagas 1909, 1922). Chagas disease occurs mainly in Latin America, but also in the south of North America. While Colombia and Bolivia, with about 30 and 24%, had the highest infection rates and still remain strongly affected, in the southern United States only sporadic infections occurred (WHO 2002).

The course of the disease in humans is divided into two or three barely distinguishable phases (WHO 2008). After the infection, in the acute phase, nonspecific symptoms of infections develop, such as fever and the swelling of local lymph nodes. These symptoms remain for 1–2 weeks after the infection and are caused by the increased number of blood stages of the pathogen. In this acute phase, less than 5% of infected people die, especially children (Guimarães et al. 1968).

In the following latent and chronic phases of the disease, which are grouped together as the chronic phase, hardly any blood stages can be detected and the pathogen develops almost exclusively intracellularly. In the initial latent phase, which can last from several years to several decades, hardly any symptoms are noticeable. The late chronic phase is characterized by the damage caused by the intracellular development of the pathogen. Frequently, megaorgan syndromes occur in large hollow muscular organs, such as the colon or the heart. Damage of the myocardium often results, due to the deteriorated electrical conduction, in pathological cardiac arrhythmia (Coura and de Castro 2002; Coura 2007).

12.5 Diagnosis of Chagas Disease

In the acute phase of infection the number of parasites is high enough to observe them microscopically in a blood smear or in a thick drop blood test directly or after concentration in microhematocrit tubes (La Fuente et al. 1984). However, some low

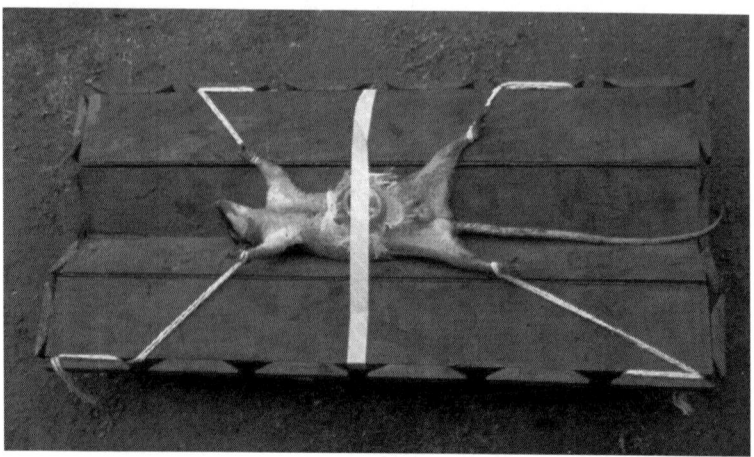

Fig. 12.4 Xenodiagnosis of an anaesthetized opossum in the field

virulent strains develop no visible parasitemia. In the chronic phase of the disease, diagnosis uses indirect proof via xenodiagnosis or hemoculture and serological methods, the latter are based on the appearance of IgG antibodies. Today molecular techniques are on the verge of becoming standardized (Gomes et al. 2009).

In xenodiagnosis of *T. cruzi*, laboratory-reared uninfected instars of triatomines are used. Groups of 5–40 larvae are fed on the patient – often on the skin of the arm or upper limb – in a cylindric pot or wooden box closed with gauze (Vega Chirinos and Náquira Velarde 2006). The triatomines are also used in xenodiagnoses in the field on anesthetized animals (Fig. 12.4). Within about 30 min the bugs can engorge. Afterwards, the insects are maintained under standardized conditions for an incubation period, often about 30 days. Then some laboratories homogenize the bugs or the intestinal tract, others obtain rectal content by applying pressure to the abdomen of the bug. Finally, the samples are microscopically examined for the presence of *T. cruzi*.

12.6 Disadvantages of Xenodiagnosis of Chagas Disease

12.6.1 Disadvantages of Xenodiagnosis of Chagas Disease: Time Involvement

The first major disadvantage is the long period of time until the final result. The period of incubation between infection and examination is necessary, because *T. cruzi* must have enough time to multiply in the rectum. Three to four days after experimental infection of *R. prolixus*, first flagellates were passed to the rectum, but in *T. infestans* an attached population was not present at 7 days after infection (Brack 1968; Böker and Schaub 1984). If the rectum was well colonized about 3 months after infection and monthly feeding, starvation reduced the population density, but even after starvation periods of 4 or 11 months some parasites remained attached to the rectal cuticle (Schaub and Lösch 1989; Kollien and Schaub 1998b). One week and 17 days after xenodiagnosis, the rectal content of five and six groups out of seven groups of *Rhodnius neglectus*, respectively, contained trypanosomes, but the highest infection rates were detectable 3 weeks after xenodiagnosis (Forattini et al. 1976). Also in seven out of 16 groups of *P. megistus*, *T. infestans*, *R. neglectus*, and *R. prolixus* fed on guinea pigs in the chronic phase, the percentages of positive bugs increased between 15 and 30 days after feeding (Perlowagora-Szumlewicz and Muller 1987). However, the infection rates in groups of first instars of *P. megistus*, *T. infestans*, and *R. prolixus* fed on naturally infected opossums and examined 7–10 and 25–28 days after feeding did not increase with the longer incubation period (Minter et al. 1977).

Also the microscopic examinations are very time- and work-consuming – the second disadvantage. Especially slow-growing strains of *T. cruzi* require a careful examination. In diagnosis centers at least 30 min of examinations without seeing any trypanosomes is the limit before classifying a sample as *T. cruzi*-negative. Having to view several samples without parasites is tiring for the eyes, increasing the risk that an infection is overlooked. Since triatomines may also possess other trypanosomatids (see below), a positive result requires a verification in stained smears.

12.6.2 Disadvantages of Xenodiagnosis of Chagas Disease: Allergic Reactions

A third disadvantage is the direct exposure of the patient to the bite of a triatomine. In addition to the psychological effect, which can be reduced by using pots in which the bugs cannot be seen, the triatomine continuously injects saliva during blood ingestion, and salivary proteins act as agents of allergic reactions. Anaphylactic reactions caused by the bite of a triatomine are often reported (Hoffman 1987). After xenodiagnosis, cutaneous allergic reactions are common, and sometimes more than 80% of patients are affected (Mott et al. 1980). These reactions often include localized urticaria or more severe cutaneous reactions, for example the so-called Romaña sign, a local edema on the eye lid, often assumed to be a sure symptom of a fresh *T. cruzi* infection (Dias 1968; Mott et al. 1980). Cutaneous reactions to triatomine bites can be divided into immediate and delayed reactions. While immediate reactions appear within minutes after the bite and in xenodiagnosis are often established at the end of blood ingestion of the bugs, delayed reactions develop afterwards and can last for several hours up to some days, in single cases up to weeks (Marsden et al. 1969). Only a minority of probands (about 20%) showed immediate reactions while up to 90% had a delayed reaction (Mott et al. 1980). Comparing the species *T. infestans*, *P. megistus*, and *R. prolixus*, which are common local vectors and are frequently used in xenodiagnosis, the saliva of *R. prolixus* seems to induce reactions more often and more strongly (Marsden et al. 1969; Dias 1968). Applications of corticosteroids and antihistamines do not always exclude such reactions (Dias 1968; Mott et al. 1980). In addition to cutaneous allergic reactions and itching, in single cases generalized systematic reactions such as anaphylaxis are reported after bites of *R. prolixus* and *T. infestans*, but also from triatomine species in the south of North America, where anaphylactic reactions to triatome bites exceed those from bee stings and other Hymenoptera (Lapierre and Lariviere 1954; Klotz et al. 2010). A repeated exposure to triatomine bites seems to sensitize the patient (summarized by Mott et al. 1980).

The problems of allergic reactions induced the search for a less allergenic triatomine species. Since the allergic reaction could be caused by previous exposures to the respective species of triatomine at home, Mott et al. (1980) introduced *T. infestans* for xenodiagnosis in an area where *P. megistus* is the only domestic vector, and Costa et al. (1981) used the Mexican *D. maxima* in an area in Brazil. Both studies showed allergic reactions after the bite of the foreign species, emphasizing a cross-reactivity. This result is also indicated by a previous investigation in which a low-concentration salivary protein reacted with sera of chicken immunized against different species of triatomines (Schwarz et al. 2009).

In another approach, an artificial xenodiagnosis, originally reported by Romaña and Gil (1947), blood of the patient is obtained by venous puncture and offered under an artificial membrane. In this "in vitro" technique, contacts with the saliva of the bug and thereby allergic reactions are avoided. In comparisons of "in vivo" and "in vitro" xenodiagnosis, diagnostic sensitivity was similar or sporadically improved in the latter (Nussenzveig and Sonntag 1952; dos Santos et al. 1995; Freitas et al. 1955; Pineda et al. 1998). Success of artificial xenodiagnosis is affected by several factors, for example anticoagulants and membranes. Blood clotting has to be avoided, but EDTA strongly reduces the survival rate of the triatomines (Campos et al. 1988). Another anticoagulant, sodium citrate solution, reduces the amount of ingested blood but not the detection efficiency (Cedillos et al. 1982b; Merks and Werner 1983; Christophel et al. 1988). Membranes commonly used are household plastic wrap or uncoated condoms and in earlier publications preparations of the gut wall of animals (Cedillos et al. 1982b; Campos et al. 1988). The type of membrane strongly affects the volume of ingested blood (Cedillos et al. 1982b). Thin silicone membranes are very superior to polyvinylchloride household wrap, but some bugs even refuse to suck through this membrane (Schaub 1990). This difficulty can be overcome by using more larvae. A factor that has been proven to directly affect the sensitivity of artificial xenodiagnosis is the amount of offered blood and the number of triatomines: using a four times greater blood volume and a higher number of bugs increases the percentage of positive patients from 19% to 44% (Franco et al. 2002). In addition to the avoidance of a direct contact with the triatomine, this technique opens a time window of up to 4 h without impairing diagnostic results (Castro et al. 2004). Even storage for 12 h at room temperature followed by 24 h at 4 or 27°C and a subsequent storage for another day at room temperature did not kill all *T. cruzi* and allowed a development in 50% of *D. maxima* (Werner and Merks 1984). This offers the chance of a postmortem xenodiagnosis some hours after death (Lopes et al. 1986).

12.6.3 Disadvantages of Xenodiagnosis of Chagas Disease: False Positive Results

Beside *T. cruzi* other trypanosomes are common in the endemic areas of Chagas disease and develop in triatomines. They can only be distinguished by well-trained staff and sometimes require further molecular or biochemical characterizations (Schaub 2009). The homoxenous flagellate *Blastocrithidia triatomae* owed its discovery to extensive surveys of xenodiagnosis (Cerisola et al. 1971; Schaub 1988b). When the BAYER company proved the efficiency of the first chemotherapy compound for Chagas disease, the nitrofuran Lampit®, the treatment seemed to be ineffective in some cases. However, this occurred only using bugs from an Argentine laboratory colony. Detailed inspection of these bugs proved that they had been infected with *B. triatomae* before being used for xenodiagnosis. Although only very low numbers of triatomines were found to be infected in natural populations – in Argentina 0.42% of *T. infestans*, 2% of *Triatoma guasayana*, and 1.5% of *Triatoma garciabesi*, in Brazil 0.03% and 0.08% of *P. megistus*, 0.36% of *Triatoma sordida*, and 0.02% of different triatomines (mainly *T. sordida*), and in Venezuela 5–10% of *Triatoma maculata* (Corrêa et al. 1977; da Rocha e Silva et al. 1977; de Hubsch et al. 1977; Luz and Silveira 1984; Schijman et al. 2006; Marti et al. 2009) – in laboratory populations high infection rates can occur (Schaub 1988a). In such populations the transmission via coprophagy – a behavior that occurs regularly in triatomines to obtain essential symbionts – seems to occur more often or is more efficient and also results in higher infection rates with the *Triatoma* virus (Schaub et al. 1989a; Muscio et al. 1997). Using higher magnifications in the microscopic examinations, *B. triatomae* can be distinguished from *T. cruzi* because during unequal divisions a cyst stage develops, which often remains as a "straphanger" at the flagellum of the mother flagellate and is still infective for the bug in dried feces. Using low magnification microscopy, the identification is very difficult, and this insect flagellate was erroneously classified as *T. cruzi* before the first description of *B. triatomae* (Silva 1958). Therefore, laboratory colonies used for xenodiagnosis should be constantly monitored for the presence of *B. triatomae*.

In addition to this insect trypanosomatid, beside *T. cruzi* two heteroxenous trypanosomatids of the genus *Trypanosoma* develop in triatomines. Although *Trypanosoma conorhini* has been reported from primates in Asia, it is mainly restricted to rats and is found naturally only in the triatomine *Triatoma rubrofasciata* (Hoare 1972; Dennig and Karcher 1986), which is closely associated to rats in the harbors of Africa,

Asia, and South America (Schaub 2009). Also in the endemic areas of Chagas disease it seems to be present only in rats and cannot interfere with xenodiagnosis of humans. More widely distributed and always in regions endemic for Chagas disease is *Trypanosoma rangeli*, which is subpathogenic for humans (D'Alessandro 1976). It develops nearly exclusively in species of the genus *Rhodnius*, only rarely colonizes the rectum of the bug and is transmitted by the saliva in contrast to the two other *Trypanosoma* sp. (Schaub 2009). In single infections of humans, usually parasitemia is low and persists only for 3–18 months (Hoare 1972). The blood trypomastigotes possess a less prominent kinetoplast than *T. cruzi*, but this can be recognized only in stained smears (Hoare 1972). In the metacyclic trypomastigotes in the triatomine, this difference is less pronounced, but the kinetoplast of *T. rangeli* is more terminal (Hoare 1972). Also in the triatomine epimastigotes of both species can hardly be distinguished, but only *T. rangeli* develops long epimastigotes of up to 80 μm (Hoare 1972). The major differences between *T. cruzi* and *T. rangeli* are the triatomine hosts since the latter mainly develops in species of the genus *Rhodnius*, often but not always invading the hemocoel (Vallejo et al. 2009). Therefore, flagellates in the hemolymph can solely be attributed to *T. rangeli*, but a missing infection there does not exclude this species.

12.6.4 Disadvantages of Xenodiagnosis of Chagas Disease: Sensitivity

The major disadvantage of xenodiagnosis is its low sensitivity causing false negative results. In most studies, in <50% of seropositive patients *T. cruzi* is present in the blood and develops in the vector (Fistein and Chowdhury 1980; Schenone 1999). Out of 10,297 patients with a positive Machado-Guerreiro test, a complement fixation test for *T. cruzi* infection, only 29.3% were positive in xenodiagnosis (summarized by Urribarri 1970). Using three groups of 40 third and fourth instars of *T. infestans* per patient in a xenodiagnosis of 206 chronically infected patients, *T. cruzi* was detected in 49% of patients. However, after feeding on 26.7% of patients only less than 2% of bugs became infected, on 13.1% up to 7% and only xenodiagnosis of 9.2% of patients resulted in 7–20% infected *T. infestans* (Pereira et al. 1989).

Comparing the two parasitological techniques, Minter-Goedbloed (1976) found no qualitative differences, but hemoculture is emphasized to take much longer. However, results of comparative studies are contradictory (Table 12.1). In vitro cultivation is of similar sensitivity, that is in both proofs of living parasites in the blood, the number of parasites in the blood ingested by the vector or in the sample used for the inoculation of the in vitro culture medium is too low or in both methods blood trypomastigotes do not develop (Table 12.1). The often low sensitivity of xenodiagnosis and in vitro cultivation affect the therapy. While a positive xenodiagnosis of a suspected individual is an indisputable proof for a *T. cruzi* infection,

Table 12.1 Sensitivity of different diagnosis methods for Chagas disease in detecting *T. cruzi* in the blood of serologically positive chronically infected humans

Sensitivity of diagnostic method[a]			Reference
Xenodiagnosis	In vitro culture	PCR[b]	
36–54	2–7	n.d.[c]	Cerisola et al. (1974)
2–32	17	n.d.	Minter-Goedbloed et al. (1978)
42	24	n.d.	Bronfen et al. (1989)
54	n.d.	100[d]	Avila et al. (1993)
36	26	59	Junqueira et al. (1996)
n.d.	37	84	Gomes et al. (1999)
n.d.	70	87	Castro et al. (2002)

[a]Proportion of serologically positive patients
[b]Proof for DNA, not living parasites
[c]Not determined
[d]Slightly higher sensitivity than serology

a negative result does not necessarily indicate the absence of the protozoon and repeated examinations are required. In a negative xenodiagnosis, but with positive serology and pathology a therapy should be started, especially in infected children.

As xenodiagnosis includes the labor-intensive microscopical examination of the bugs' feces, any factor increasing the number of trypanosomes in the rectum will improve its efficiency. Several attempts focused on improvements of xenodiagnosis. Since triatomines are mainly night-active, a coevolution of parasite and vector might have resulted in a day–night rhythm of parasitemia. Xenodiagnosis at the optimal hours would then increase the sensitivity. Initial studies performing xenodiagnosis in a 2 h rhythm seemed to indicate a daily rhythm in the concentration of *T. cruzi* in the peripheral blood (Almeida et al. 1976). However, after feeding third instar larvae of *T. infestans* on the same chronic chagasic patient during the day or the night an examination of the bugs in monthly intervals up to 120 days after xenodiagnosis showed positivity of 24–38% for the insects fed during the day and 20–36% for those fed during the night (Schenone et al. 1977). Using the same developmental stage of *T. infestans* for xenodiagnosis every 3 h also similar percentages of infected bugs were evident. Therefore, there seems to be no circadian rhythm in the presence of *T. cruzi* in the peripheral bloodstream (Castro and Prata 2000).

However, the technique used to obtain the examined material influences the sensitivity of xenodiagnosis. In microscopy of the rectal contents obtained by abdominal compression of 2,016 larvae and adults of *T. infestans* from the field an infection rate of 46.7% was found; further investigation of 625 previously negative individuals showed an infection in an additional 100 bugs (Guedes 1952). In another analysis of the rectal contents obtained by abdominal compression 3.5 days after infection, one out of seven groups of *R. neglectus* was classified as positive, but after dissection of the posterior intestine six out of seven groups were positive (Forattini et al. 1976). Examination of pooled rectal contents of third instars of *T. infestans* instead of an individual examination of rectal contents required two or three times more bugs to obtain similar or superior results

(Cuba et al. 1979). The microscopy of pooled homogenate of whole insects followed by cotton filtration and centrifugation steps was twofold more sensitive than that of individual intestinal contents of *R. prolixus* (Maekelt 1964). In a later investigation of the same species and using a similar procedure no difference in sensitivity was evident (Cedillos et al. 1982a). The chance to initiate an infection even at low numbers of parasites in the blood is increased by the use of more triatomines. Using one, two, four and six units – each consisting of seven to ten third instars of *T. infestans* – the percentage of identified positive cases increased from 46.1% to 54.7%, 61.8%, and 69.1%, respectively (Schenone et al. 1974, 1991). Similarly, sensitivity was increased from 59.6% to 75.2% by an additional xenodiagnosis in cases in which an infection had not been recognized in the first xenodiagnosis (Schenone 1998). The use of more bugs is advisable either in the individual examination or in the use of pooled material that reduces the amount of time used in the final step of xenodiagnosis.

The amount of blood ingested by the triatomines can also be increased by the use of older larvae. Comparing first and third instars of *D. maxima*, percentages of diagnosed cases increased from 31.5 to 65.7% and the number of infected bugs from 14.6 to 36% in the older larvae (Cuba et al. 1978). In a follow-up investigation of first and third instars of *D. maxima*, also in four times more older larvae an infection developed (Cuba et al. 1979). Comparing first and fifth instars of *R. prolixus*, the percentages increased from 21% to 59% and between first and fourth instars of *P. megistus* from 7% to 49% (Minter et al. 1978). However, only 29% of the fifth instar larvae were infected. Therefore, the percentages of infection are not simply correlated to the amount of ingested blood (Minter et al. 1978; Perlowagora-Szumlewicz and Müller 1982). The disadvantage of young instars can be overcome by the use of more bugs. The use of first instars seems to be advantageous, as it is cost reducing, for example Marsden (1986) calculated the costs of a single xenodiagnosis kit containing 40 first instars of *D. maxima* to be as low as 4 US$. Another advantage especially in field studies is the easier transport of eggs (Marsden et al. 1979; Minter-Goedbloed and Minter 1987).

In the different approaches performed in the last nine decades to increase the sensitivity of xenodiagnosis, most studies focused on the vector, searching for the most effective species of triatomines. Differences in the efficiency of detecting *T. cruzi* in chronically infected humans using different vector species were already described by Dias (1936). Comparing only published data does not give conclusive information since the infection rates of one species varied strongly, in *T. infestans* from 2 to 95% (Table 12.2). A comparison of the efficiency of species is only possible after xenodiagnosis on the same host (Tables 12.2 and 12.3). A comparison of three of the major domestic vectors of Chagas disease – *R. prolixus*, *T. infestans*, and *P. megistus* – using a monkey in a stage of infection comparable to the human chronic stage and a Peruvian strain of *T. cruzi* resulted in infection rates of 56.8% for *R. prolixus* and about 75% for the two other species (Miles et al. 1975). However, another *Rhodnius* species, the sylvatic *R. neglectus*, was more susceptible than *T. infestans* after feeding on a naturally infected opossum (Forattini et al. 1976). Using a fresh *T. cruzi* isolate originating from *Triatoma vitticeps* and

Table 12.2 Percentages of infected triatomines using different instars of *Triatoma infestans*, *Rhodnius prolixus*, and *Panstrongylus megistus* and different hosts in the chronic phase of infection

Triatomine	Instar	Host[a]	% Positive[b]	Reference	ID[c]
T. infestans	I	Opossum	34	Minter et al. (1977)	A
	I	Opossum	37	Minter et al. (1978)	B
	I	Monkey[d]	95	Minter et al. (1977)	A
	III	Human	7	Marsden et al. (1979)	C
	III	Human	11	Cuba et al. (1978)	D
	III	Human	3	Cuba et al. (1979)	E
	III	Human	16	Bronfen et al. (1989)	F
	IV	Guinea-pig	53	Perlowagora-Szumlewicz and Muller (1987)	G
	IV	Guinea-pig	6–94	Perlowagora-Szumlewicz et al. (1990)	H
	IV	Human	2	Moreira and Perlowagora-Szumlewicz (1997)	I
	IV/V	Human	20	Barretto et al. (1978)	J
	V	Monkey[d]	77	Miles et al. (1975)	K
	V	Human	26	Minter et al. (1978)	B
R. prolixus	I	Opossum	25	Minter et al. (1977)	A
	I	Opossum	21	Minter et al. (1978)	B
	I	Monkey[d]	65	Minter et al. (1977)	A
	IV	Guinea-pig	60	Perlowagora-Szumlewicz and Muller (1987)	G
	IV	Guinea-pig	6–81	Perlowagora-Szumlewicz et al. (1990)	H
	V	Opossum	59	Minter et al. (1978)	B
	V	Monkey[d]	57	Miles et al. (1975)	K
	V	Human	11	Minter et al. (1978)	B
P. megistus	I	Opossum	43	Minter et al. (1977)	A
	I	Opossum	42	Minter et al. (1978)	B
	I	Monkey[d]	69	Minter et al. (1977)	A
	I	Human	2	Minter et al. (1977)	A
	I	Human	7	Minter et al. (1978)	B
	III	Human	17	Bronfen et al. (1989)	F
	IV	Guinea-pig	89	Perlowagora-Szumlewicz and Muller (1987)	G
	IV	Guinea-pig	19–100	Perlowagora-Szumlewicz et al. (1990)	H
	IV	Human	49	Minter et al. (1978)	B
	IV	Human	5	Moreira and Perlowagora-Szumlewicz (1997)	I
	V	Monkey[d]	76	Miles et al. (1975)	K
	V	Human	29	Minter et al. (1978)	B

[a] Hosts were serologically positive or experimentally infected
[b] Examination 20–45 days after feeding of triatomines, average according to the respective publication or minimum and maximum using seven *T. cruzi* strains from Brazil
[c] Identical investigations are indicated by the same letter
[d] Rhesus monkey (*Macaca mulatta*)

acute phase mice, all *R. neglectus* but only 39% of *T. infestans* and 83% of *P. megistus* became infected (Kollien et al. 1998). However, *T. infestans* verified an infection in all three patients and 83% of the 12 groups of 10 third or fourth

Table 12.3 Percentages of infected triatomines using different instars of different triatomines and different hosts in the chronic phase of infection

Triatomine	Instar	Host[a]	% Positive[b]	Reference	ID[c]
R. neglectus	IV	Guinea-pig	83	Perlowagora-Szumlewicz and Muller (1987)	G
	IV	Guinea-pig	6–94	Perlowagora-Szumlewicz et al. (1990)	H
	IV	Human	2	Moreira and Perlowagora-Szumlewicz (1997)	I
D. maxima	I	Human	15	Cuba et al. (1978)	D
	I	Human	5	Cuba et al. (1979)	E
	I	Human	10	Marsden et al. (1979)	C
	I	Human	12	Bronfen et al. (1989)	F
	III	Human	36	Cuba et al. (1978)	D
	III	Human	19	Cuba et al. (1979)	E
	IV/V	Human	49	Baretto et al. (1978)	J
T. brasiliensis	IV	Guinea-pig	38	Perlowagora-Szumlewicz and Muller (1987)	G
	IV	Guinea-pig	19–100	Perlowagora-Szumlewicz et al. (1990)	H
T. sordida	IV	Guinea-pig	72	Perlowagora-Szumlewicz and Muller (1987)	G
	IV	Guinea-pig	19–100	Perlowagora-Szumlewicz et al. (1990)	H
T. pseudo-maculata	IV	Guinea-pig	73	Perlowagora-Szumlewicz and Muller (1987)	G
	IV	Guinea-pig	13–100	Perlowagora-Szumlewicz et al. (1990)	H
T. rubrovaria	IV	Guinea-pig	76	Perlowagora-Szumlewicz and Muller (1987)	G
	IV	Guinea-pig	32–100	Perlowagora-Szumlewicz et al. (1990)	H
T. dimidiata	IV	Guinea-pig	0	Perlowagora-Szumlewicz and Muller (1987)	G
	IV	Guinea-pig	0–69	Perlowagora-Szumlewicz et al. (1990)	H
T. vitticeps	IV	Guinea-pig	38–94	Perlowagora-Szumlewicz et al. (1990)	H
	IV	Human	6	Moreira and Perlowagora-Szumlewicz (1997)	I
T. pessoia	IV	Guinea-pig	25–94	Perlowagora-Szumlewicz et al. (1990)	H

[a]Hosts were serologically positive or experimentally infected
[b]Examination 20–45 days after feeding of triatomines, average according to the respective publication or minimum and maximum using seven *T. cruzi* strains from Brazil
[c]Identical investigations are indicated by the same letter

instars were infected, whereas *T. pallidipennis* and *R. prolixus* identified the infection only in two and one of these patients, respectively, and only 50 and 18% of the groups, respectively, contained *T. cruzi*. *T. dimidiata*, *Panstrongylus herreri*, and *Rhodnius pallescence* as well as another strain of *R. prolixus* were refractory to the *T. cruzi* strain of these patients (Cerisola et al. 1974). After feeding third instars of the domestic *T. infestans* and the sylvatic Mexican *D. maxima* on chronically infected patients from Brazil sensitivity of both species were similar, but third instar larvae of *D. maxima* had a threefold higher infection rate than those of *T. infestans* (Cuba et al. 1978). Comparing the efficiency of fourth and fifth instar larvae of these two species in an examination at 30–45 days after feeding, both species identified similar numbers of chronic cases, 84.4% of 32 cases for *D. maxima* and 75% for *T. infestans*. However, after 55–65 days 52.3% of *D. maxima* were infected and only 31% of *T. infestans* (Barretto et al. 1978). Further studies comparing these species in xenodiagnosis of patients from the same area indicate a lower but still significant efficiency of *D. maxima* (Cuba et al. 1979). However, the use of this day-active and aggressive triatomine species is restricted because it cannot be reared so easily in the laboratory (Marsden 1986; Perlowagora-Szumlewicz et al. 1988).

Whereas the majority of investigations used only up to three species, the group of Perlowagora-Szumlewicz compared nine species of triatomines using guinea pigs infected with the Y strain of *T. cruzi* in the acute stage of infection. The sylvatic or peridomestic species *P. megistus*, *T. rubrovaria*, *T. pseudomaculata*, and *R. neglectus* possessed infection rates of >90%, the peridomestic species *T. sordida* and *T. brasiliensis* of 76.9% and 80%, respectively, and the domestic *T. infestans*, *T. dimidiata*, and *R. prolixus* <55% (Perlowagora-Szumlewicz and Müller 1982). Not only the prevalence but also the population density indicated a higher efficiency of sylvatic species for xenodiagnosis (Perlowagora-Szumlewicz and Müller 1982). A follow-up investigation of the same system, but for the chronic phase, emphasized the value of *P. megistus* in xenodiagnoses since <50% of larvae of the other domestic species were infected (Perlowagora-Szumlewicz and Muller 1987) (Tables 12.2 and 12.3). In an upscaling of the experiments the group included seven Brazilian strains of *T. cruzi* (Perlowagora-Szumlewicz et al. 1988). In the acute phase, all strains of *T. cruzi* infected all nine species of triatomine with a slight tendency to higher efficiencies in sylvatic and peridomestic species. Interestingly the Gávea strain of *T. cruzi* originally isolated from *P. megistus* established less successfully in this species. One *T. cruzi* strain did not establish in *T. dimidiata*, the others up to 87% (Perlowagora-Szumlewicz et al. 1988). Only in this species a loss of natural infections occurred in Costa Rica (Vargas and Zeledón 1985). Also from the chronic phase of infection, prevalence and density of *T. cruzi* supported the use of sylvatic species as xenodiagnosis agents. Therefore, the authors adopted *P. megistus* as the insect model of choice in xenodiagnosis of Chagas disease, based on its availability, simplicity in breeding under laboratory conditions, short developmental time, and slow locomotion (Perlowagora-Szumlewicz et al. 1990). The superiority of *P. megistus* over the generally used *T. infestans* had also been shown under diagnostic conditions using 264 Brazilian patients (Moreira and Perlowagora-Szumlewicz 1997).

Especially the detailed investigations of the group of Perlowagora-Szumlewicz also emphasize the importance of the strain of *T. cruzi* (Table 12.4). Although she focused on material originating from Brazil, an interspecific difference in the susceptibility of triatomines to different *T. cruzi* strains can be generalized. The different infection rates of triatomines after infection with different strains indicate that intrinsic factors of the insect as well as the strain of *T. cruzi* determine whether or not the ingested parasites develop and establish (Garcia et al. 2010). Although single investigations report an initial concentration of *T. cruzi* in the

Table 12.4 Prevalence of *T. cruzi* in groups of different triatomines after xenodiagnosis using different strains or isolates of the flagellate

Triatomine	% Infected by the *T. cruzi* strain/isolate[a, b]											ID[c]
	Ber	CL	Col	CVP	FL	Gáv	Per	SF	SM6	SM7	Y	
T. infestans	35	19	5		29	21		56			19	H
	69	50	84		72	72		47			56	M
											40	G
											52	L
							95					A
							77					K
				100					39	61		N
		100									67	O
P. megistus	84	56	25		69	88		96			71	H
	100	100	97		100	66		97			100	M
											90	G
											98	L
							69					A
							76					K
				100					83			N
R. prolixus	38	31	23		50	74					29	H
	91	97	97		81	100		72			88	M
											47	G
											53	L
							65					A
							57					K
R. neglectus	75	44	13		53	73		95			26	H
	100	100	100		100	100		100			100	M
											74	G
											94	L
				100					100			N

[a] Average according to the respective publication independent of time after infection; if different times after feeding were investigated maximal values are given

[b] Abbreviations of the strains/isolates of *T. cruzi*: *Ber* Berenice; *Col* Columbiana; *CVP* CV. Pl.1106.94; *Gáv* Gávea; *Per* Peru; *SF* São Felipe; *SM6* SMM106; *SM7* SMM107

[c] Identical investigations are indicated by the same capital letter (for details see Table 12.2), the follow-up investigations are indicated by additional letters: *L* Perlowagora-Szumlewicz and Müller (1982); acute infection; *M* Perlowagora-Szumlewicz et al. (1988); acute infection; *N* Kollien et al. (1998), acute infection; *O* Alvarenga and Bronfen (1984)

stomach of *R. prolixus* directly after feeding on experimentally infected mice compared to blood obtained from the tail tip (Fistein and Chowdhury 1980), the major development of the protozoon occurs in the following weeks. Between 30 and 90 days after infection, the numbers of metacyclic trypomastigotes of the Y strain of *T. cruzi* increased 18- and 33-fold in *T. infestans* and *D. maxima*, respectively. However, about 33% of *T. infestans* were refractory to an infection with this *T. cruzi* strain, but only 2% and 4% of *P. megistus* and *D. maxima*, respectively (Alvarenga and Bronfen 1984). The comparison of the development of two different strains of *T. cruzi* in *R. neglectus*, *R. robustus*, and *T. infestans* showed a similar development of the Y strain in all triatomines, reaching at 10–15 days after infection the highest and later on a constant prevalence in all triatomine species. However, the strain AMJM reached the highest infection rates about 20 days after infection only in *R. neglectus*, while in the two other species infection rates peaked between 10 and 20 days after infection and decreased afterwards, indicating a loss of infection (Martins et al. 2000). Comparing two fresh isolates, both originating from *T. vitticeps*, one isolate infected 61% and the other one 39% of *T. infestans* (Kollien et al. 1998). Clone or strain specificities are indicated not only by different numbers of parasites in the gut but also by different percentages of metacyclic trypomastigotes (e.g., Schaub 1989; Perlowagora-Szumlewicz et al. 1990; Perlowagora-Szumlewicz and Moreira 1994; Alvarenga and Bronfen 1997; Pinto et al. 1998; Lana et al. 1998; Lima et al. 1999). Using two *T. cruzi* strains from the same locality and *T. infestans* also originating from there, the number of metacyclics differed strongly (Schaub 1989). Summarizing these strain differences of *T. cruzi* and the distribution of different strains in the endemic areas of South America, there seems to be no optimal triatomine species for xenodiagnosis in all countries.

12.7 Future Impact of Xenodiagnosis in Diagnosis of Chagas Disease

Today xenodiagnosis is practically restricted to the isolation of parasites for research in the more highly developed areas of Latin America (Sosa-Estani et al. 2009). In xenodiagnosis of animals in the field, the difficulties are the transport and maintenance conditions of the insects outside the insectarium, the long anesthesia (especially disadvantageous if the animal should be released afterwards), and the unknown vector in the transmission cycle in the field, causing a "filter" effect of the species chosen for xenodiagnosis. However, in combination with PCR (polymerase chain reaction), higher infection rates of a small rodent were found (Campos et al. 2007). Xenodiagnosis in humans is restricted to wide parts of endemic areas, where the infrastructure is poor and where sophisticated laboratory facilities often are absent. The current state of the art of Chagas disease diagnosis involves serology, but this method also shows disadvantages. First, some patients fail to produce sufficient antibodies and give negative serological results, although a presence of the parasite could be proven by xenodiagnosis (Luquetti 1987). Second, after chemotherapeutical treatment conventional serological tests remain positive up to years after successful therapy so that a possible parasite persistence or new infection in a number of patients is not detectable. In these cases PCR has to be used, but this requires well-equipped laboratories: 20 years after treatment in the acute phase 27% and 14% of 37 patients were positive in PCR or xenodiagnosis, respectively, and of 48 patients treated in the chronic phase, 35% and 17%, respectively, were proven to be still infected (Britto et al. 2001). Since PCR needs to be standardized, even nearly 100 years after the first use of xenodiagnosis, this and the in vitro cultivation still remain the only proofs for the presence or absence of living *T. cruzi* in the blood, for the transmissibility of the parasite, and for the infectivity of the vector.

Acknowledgments We thank Dr. R. Cassada for correcting the English style. We very much appreciate helpful comments from Dr. Ana Jansen, Instituto Oswaldo Cruz, Rio de Janeiro, Brazil; and Dr. Nora Medrano Mercado, Univ. San Simon, Cochabamba, Bolivia.

References

Almeida SP, Sherlock IA, Fahel E (1976) Novo procedimento de xenodiagnóstico na forma crônica da doença de Chagas. Mem Inst Oswaldo Cruz 74:285–288

Alvarenga NJ de (1974) Evolução do *Trypanosoma cruzi* no trato digestivo de *Triatoma infestans*. Dissertation, Universidade Federal de Minas Gerais, Brazil

Alvarenga NJ, Bronfen E (1984) Interação do *Trypanosoma cruzi* com diferentes vetores: uso para o xenodiagnóstico. Rev Soc Bras Med Trop 17:145–149

Alvarenga NJ, Bronfen E (1997) Metaciclogênese do Trypanosoma cruzi como parâmetro de interação do parasita com o triatomíneo vetor. Rev Soc Bras Med Trop 30:247–250

Anonymous (1999) Recommendations from a satellite meeting. Mem Inst Oswaldo Cruz 94:429–432

Anwyl R (1972) The structure and properties of an abdominal stretch receptor in *Rhodnius prolixus*. J Insect Physiol 18:2143–2154

Araújo CAC, Cabello PH, Jansen AM (2007) Growth behaviour of two *Trypanosoma cruzi* strains in single and mixed infections: in vitro and in the intestinal tract of the blood-sucking bug, *Triatoma brasiliensis*. Acta Trop 101:225–231

Araújo CAC, Waniek PJ, Jansen AM (2008) Development of a *Trypanosoma cruzi* (TcI) isolate in the digestive tract of an unfamiliar vector, *Triatoma brasiliensis* (Hemiptera, Reduviidae). Acta Trop 107:195–199

Araújo CAC, Waniek PJ, Jansen AM (2009) An overview of Chagas disease and the role of triatomines on its distribution in Brazil. Vector-borne Zoonotic Dis 9:227–234

Avila HA, Pereira JB, Thiemann O, De Paiva E, DeGrave W, Morel CM, Simpson L (1993) Detection of *Trypanosoma cruzi* in blood specimens of chronic chagasic patients by polymerase chain reaction amplification of kinetoplast minicircle DNA: comparison with serology and xenodiagnosis. J Clin Microbiol 31:2421–2426

Barretto AC, Marsden PD, Cuba CC, Alvarenga NJ (1978) Estudo preliminar sobre o emprego de *Dipetalogaster maximus* (Uhler, 1894) (Triatominae) na técnica do xenodiagnóstico em forma crônica de Doença de Chagas. Rev Inst Med Trop São Paulo 20:183–189

Böker CA, Schaub GA (1984) Scanning electron microscopic studies of *Trypanosoma cruzi* in the rectum of its vector *Triatoma infestans*. Z Parasitenkd 70:459–469

Brack C (1968) Elektronenmikroskopische Untersuchungen zum Lebenszyklus von *Trypanosoma cruzi*. Acta Trop 25:289–356

Brener Z (1973) Biology of *Trypanosoma cruzi*. Annu Rev Microbiol 27:347–382

Britto C, Silveira C, Cardoso MA, Marques P, Luquetti A, Macêdo V, Fernandes O (2001) Parasite persistence in treated chagasic patients revealed by xenodiagnosis and polymerase chain reaction. Mem Inst Oswaldo Cruz 96:823–826

Bronfen E, de Assis Rocha FS, Machado GBN, Perillo MM, Romanha AJ, Chiari E (1989) Isolamento de amostras do *Trypanosoma cruzi* por xenodiagnóstico e hemocultura de pacientes na fase chrônica do doença de Chagas. Mem Inst Oswaldo Cruz 84:237–240

Brumpt E (1914) Le xénodiagnostic. Application au diagnostic de quelques infection parasitaires et in particulier à la trypanosome de Chagas. Bull Soc Path Exot 77:706–710

Burleigh BA, Andrews NW (1995) The mechanisms of *Trypanosoma cruzi* invasion of mammalian cells. Annu Rev Microbiol 49:175–200

Caballero ZC, Sousa OE, Marques WP, Saez-Alquezar A, Umezawa ES (2007) Evaluation of serological tests to identify *Trypanosoma cruzi* infection in humans and determine cross-reactivity with *Trypanosoma rangeli* and *Leishmania* spp. Clin Vaccine Immunol 14:1045–1059

Campos R, Amato Neto V, Matsubara L, Moreira AAB, Pinto PLS (1988) Estudos sobre o xenodiagnóstico "in vitro". 1. Escolha de anticoagulante e de membrana. Rev Hosp Clin Fac Med São Paulo 43:101–103

Campos R, Botto-Mahan C, Ortiz S, Acuña M, Cattan PE, Solari A (2007) *Trypanosoma cruzi* detection in blood by xenodiagnosis and polymerase chain reaction in the wild rodent *Octodon degus*. Am J Trop Med Hyg 76:324–326

Carneiro M, Chiari E, Gonçalves AM, da Silva Pereira AA, Morel CM, Romanha AJ (1990) Changes in the isoenzyme and kinetoplast DNA patterns of *Trypanosoma cruzi* strains induced by maintenance in mice. Acta Trop 47:35–45

Castro C, Prata A (2000) Absence of both circadian rhythm and *Trypanosoma cruzi* periodicity with xenodiagnosis in chronic chagasic individuals. Rev Soc Bras Med Trop 33:427–430

Castro AM, Luquetti AO, Rassi A, Rassi GG, Chiari E, Galvão LM (2002) Blood culture and polymerase chain reaction for the diagnosis of the chronic phase of human infection with *Trypanosoma cruzi*. Parasitol Res 88:894–900

Castro C, Santos MCA, Silveira CA (2004) Estudo comparativo entre o xenodiagnóstico artificial realizado imediatamente e quatro horas após a coleta de sangue. Rev Soc Bras Med Trop 37:128–130

Cedillos RA, Hubsch R, Tonn RJ, Escalante P, Carrasquero B, Liendo H (1982a) Comparison of two laboratory procedures for xenodiagnostic examination. Bull Pan Am Health Organ 16:255–260

Cedillos RA, Torrealba JW, Tonn RJ, Mosca W, Ortegón A (1982b) El xenodiagnóstico artificial en la Enfermedad de Chagas. Bol Oficina Sanit Panam 93:240–249

Cerisola JA, Del Prado CE, Rohwedder R, Bozzini JP (1971) *Blastocrithidia triatomae* n. sp. found in *Triatoma infestans* from Argentina. J Protozool 18:503–506

Cerisola JA, Rohwedder R, Segura EL, Del Prado CE, Alvarez M, De Martini GJW (1974) El xenodiagnostico. Ministerio de Bienstar Social, Buenos Aires, Argentina

Chagas C (1909) Novo tripanozomiaze humana. Über eine neue Trypanosomiasis des Menschen. Mem Inst Oswaldo Cruz 1:159–281

Chagas C (1922) Descoberta do *Tripanozoma cruzi* e verificação da Tripanozomiase Americana. Retrospecto histórico. Mem Inst Oswaldo Cruz 15:67–76

Chiang R, Davey KG (1988) A novel receptor capable of monitoring applied pressure in the abdomen of an insect. Science 241:1665–1667

Christophel E-M, Scheede S, Bommer W (1988) Massenfütterung von Reduviiden mit aussortierten menschlichen Blutkonserven in Beuteln aus Haushaltsfolie: zugleich eine Möglichkeit der In-vitro-Xenodiagnose der Chagas-Krankheit. Mitt Österr Ges Tropenmed Parasitol 10:23–32

Contreras VT, Navarro MC, De Lima AR, Duran F, Arteaga R, Franco Y (2002) Early and late molecular and morphologic changes that occur during the in vitro transformation of *Trypanosoma cruzi* metacyclic trypomastigotes to amastigotes. Biol Res 35:47–58

Corrêa R de, Alves UP, da Cunha JT (1977) Observações sobre insetos e protozoários que podem danificar a criação de triatomíneos e a diagnose do *Trypanosoma cruzi* nas fezes desses hemípteros. Rev Bras Malariol Doencas Trop 29:23–31

Costa CHN, Costa MT, Weber JN, Gilks GF, Castro C, Marsden PD (1981) Skin reactions to bug bites as a result of xenodiagnosis. Trans R Soc Trop Med Hyg 75:405–408

Coura JR (2007) Chagas disease: what is known and what is needed – a background article. Mem Inst Oswaldo Cruz 102(Suppl 1):113–122

Coura JR, de Castro SL (2002) A critical review on Chagas disease chemotherapy. Mem Inst Oswaldo Cruz 97:3–24

Cuba CC, Alvarenga NJ, Barreto AC, Marsden PD, Chiarini C (1978) Nuevos estudios comparativos entre *Dipetalogaster maximus* y *Triatoma infestans* en el xenodiagnóstico de la infección chagásica crónica humana. Rev Inst Med Trop São Paulo 20:145–151

Cuba CC, Alvarenga NJ, Barreto AC, Marsden PD, Macedo V, Gama MP (1979) *Dipetalogaster maximus* (Hemiptera, Triatominae) for xenodiagnosis of patients with serologically detectable *Trypanosoma cruzi* infection. Trans R Soc Trop Med Hyg 73:524–527

da Rocha e Silva EO, Pattoli DBG, Corrêa R de, de Andrade JCR (1977) Observações sobre o encontro de tripanossomatídeos do gênero *Blastocrithidia*, infetando naturalmente triatomíneos em insetário e no campo. Rev Saúde Pub 11:87–96

D'Alessandro A (1976) Biology of *Trypanosomas* (*Herpetosoma*) *rangeli* Tejera, 1920. In: Lumsden WHR, Evans DA (eds) Biology of the Kinetoplastida, vol 1. Academic, London, pp 327–403

de Hubsch R, Nuñez V, Mora E, Carrasquero B (1977) *Blastocrithidia triatomae torrealbai* n. subsp. encontrado en *Triatoma maculata* (Hemiptera, Reduviidae) de Venezuela. Bol Dir Malariol Saneam Ambient 17:14–19

Dennig HK, Karcher F (1986) *Trypanosoma* (*Megatrypanum*) *conorhini* im Javaneraffen (*Macaca fascicularis*) auf den Philippinen. Mitt Österr Ges Tropenmed Parasitol 8:153–162

Dias E (1936) Xenodiagnóstico e algumas verificações epidemiológicas na doença de Chagas. IX Reuniao de la Soc Argentina Pat Regional 1:89–119

Dias JCP (1968) Manifestações cutâneas na prática do xenodiagnóstico. Rev Bras Malariol Doenças Trop 20:247–258

Donahue JG, Piesman J, Spielman A (1987) Reservoir competence of white-footed mice for Lyme disease spirochetes. Am J Trop Med Hyg 36:92–96

dos Santos AH, da Silva IG, Rassi A (1995) Estudo comparativo entre o xenodiagnóstico natural e o artificial, em chagásicos crônicos. Rev Soc Bras Med Trop 28:367–373

Ebert F, Schaub GA (1983) The characterization of Chilean and Bolivian *Trypanosoma cruzi* stocks from *Triatoma infestans* by isoelectrofocusing. Z Parasitenkd 69:283–290

Eichler S, Schaub GA (2002) Development of symbionts in triatomine bugs and the effects of infections with trypanosomatids. Exp Parasitol 100:17–27

Feliciangeli MD, Sanchez-Martin M, Marrero R, Davies C, Dujardin J-P (2007) Morphometric evidence for a possible role of *Rhodnius prolixus* from palm trees in house re-infestation in the State of Barinas (Venezuela). Acta Trop 101:169–177

Fistein B, Chowdhury MNH (1980) *Trypanosoma cruzi*. A suggested adjunct to xenodiagnosis. Ann Trop Med Parasitol 74:251–253

Forattini OP, Cotrim M das D, Galati EAB, Sarzana SB, Cruz CF, Van Dinteren NHS, Gotlieb SLD (1976) Estudo sobre a utilização de *Rhodnius neglectus* para xenodiagnóstico realizados em marsupiais (*Didelphis*). Rev Saude Publica 10:335–343

Franco YBA, da Silva IG, Rassi A, Rocha ACRG, da Silva HHG, Rassi GG (2002) Correlação entre a positividade do xenodiagnóstico artificial e a quantidade de sangue e triatomíneos utilizados no exame, em pacientes chagásicos crônicos. Rev Soc Bras Med Trop 35:29–33

Freitas JLP, Nussenzweig V, Amato Neto V, Sontag R (1955) Estudo comparativo entre o xenodiagnóstico praticado "in vivo" e "in vitro" em formas crônicas da moléstia de Chagas. Hospital 47:181–188

Galuppo S, Bacigalupo A, García A, Ortiz S, Coronado X, Cattan PE, Solari A (2009) Predominance of *Trypanosoma cruzi* genotypes in two reservoirs infected by sylvatic *Triatoma infestans* of an endemic area of Chile. Acta Trop 111:90–93

Galvão C, Carcavallo R, da Silva RD, Jurberg J (2003) A checklist of the current valid species of the subfamily Triatominae Jeannel, 1919 (Hemiptera, Reduviidae) and their geographical distribution, with nomenclatural and taxonomic notes. Zootaxa 202:1–36

Garcia ES, Genta FA, de Azambuja P, Schaub GA (2010) Interactions of intestinal compounds of triatomines and *Trypanosoma cruzi*. Trends Parasitol 26:499–505

Gaunt M, Miles M (2000) The ecotopes and evolution of triatomine bugs (Triatominae) and their associated trypanosomes. Mem Inst Oswaldo Cruz 95:557–565

Gomes ML, Galvao LM, Macedo AM, Pena SDJ, Chiari E (1999) Chagas' disease diagnosis: comparative analysis of parasitologic, molecular, and serologic methods. Am J Trop Med Hyg 60:205–210

Gomes YM, Lorena VMB, Luquetti AO (2009) Diagnosis of Chagas disease: what has been achieved? What remains to be done with regard to diagnosis and follow up studies? Mem Inst Oswaldo Cruz 104(Suppl 1):115–121

Gonzalez MS, Nogueira NFS, Mello CB, De Souza W, Schaub GA, Azambuja P, Garcia ES (1999) Influence of brain and azadirachtin on *Trypanosoma cruzi* development in the vector, *Rhodnius prolixus*. Exp Parasitol 92:100–108

Gorla DE, Dujardin JP, Schofield CJ (1997) Biosystematics of Old World Triatominae. Acta Trop 63:127–140

Guedes A da S (1952) Determinação do índice de infecção de triatomíneos por *Schizotrypanum cruzi* pelo exame simples de fezes obtidas por expressão e por dissecção do intestine posterior do inseto. Dados comparativos preliminares. Rev Bras Malariol Doencas Trop 4:433–436

Guimarães FN, da Silva NN, Clausell DT, de Mello AL, Rapone T, Snell T, Rodrigues N (1968) Um surto epidêmico de doença de Chagas de provável transmissão digestiva, ocorrido em Teutonia (Estrêla – Rio Grande Do Sul). Hospital (Rio J) 73:1767–1804

Hall BF, Joiner KA (1991) Strategies of obligate intracellular parasites for evading host defences. Immunol Today 12:A22–A27

Haridass ET, Ananthakrishnan TN (1980) Bionomics and behaviour of the reduviid bug, *Triatoma rubrofasciata* (de Geer), the vector of *Trypanosoma* (*Megatrypanum*) *conorhini* (Donovan), in India (Insecta: Heteroptera). Proc Indian Natl Sci Acad B Biol Sci 46:884–891

Hoare CA (1972) The trypanosomes of mammals. Blackwell Scientific, Oxford

Hoffman DR (1987) Allergy to biting insects. Clin Rev Allergy 5:177–190

Junqueira ACV, Chiari E, Wincker P (1996) Comparison of the polymerase chain reaction with two classical parasitological methods for the diagnosis of Chagas disease in an endemic region of north-eastern Brazil. Trans R Soc Trop Med Hyg 90:129–132

Klotz JH, Dorn PL, Logan JL, Stevens L, Pinnas JL, Schmidt JO, Klotz SA (2010) "Kissing bugs": potential disease vectors and cause of anaphylaxis. Clin Infect Dis 50:1629–1634

Kollien AH, Schaub GA (1998a) The development of *Trypanosoma cruzi* (Trypanosomatidae) in the reduviid bug *Triatoma infestans* (Insecta): influence of starvation. J Eukaryot Microbiol 45:59–63

Kollien AH, Schaub GA (1998b) *Trypanosoma cruzi* in the rectum of the bug *Triatoma infestans*: effects of blood ingestion by the starved vector. Am J Trop Med Hyg 59:166–170

Kollien AH, Schaub GA (2000) The development of *Trypanosoma cruzi* in Triatominae. Parasitol Today 16:381–387

Kollien AH, Goncalves TCM, de Azambuja P, Garcia ES, Schaub GA (1998) The effect of azadirachtin on fresh isolates of *Trypanosoma cruzi* in different species of triatomines. Parasitol Res 84:286–290

La Fuente C, Saucedo E, Urjel R (1984) The use of microhaematocrit tubes for the rapid diagnosis of Chagas disease and malaria. Trans R Soc Trop Med Hyg 78:278–279

Lana M, Pinto A da S, Barnabé C, Quesney V, Noël S, Tibayrenc M (1998) *Trypanosoma cruzi*: compared vectorial transmissibility of three major clonal genotypes by *Triatoma infestans*. Exp Parasitol 90:20–25

Lapierre J, Lariviere M (1954) Réaction allergique aux piqûres de réduvidés (*Rhodnius prolixus*). Bull Soc Pathol Exot Filiales 47:563–566

Lent H, Wygodzinsky P (1979) Revision of the Triatominae (Hemiptera, Reduviidae) and their significance as vectors of Chagas' disease. Bull Am Mus Nat Hist 163:123–520

Lima VS, Mangia RHR, Carreira JC, Marchewski RS, Jansen AM (1999) *Trypanosoma cruzi*: correlations of biological aspects of the life cycle in mice and triatomines. Mem Inst Oswaldo Cruz 94:397–402

Lopes ER, Chapadeiro E, Brener Z, Franciscon JU, Cardoso JE, Adad SJ, Tostes Júnior S (1986) Xenodiagnóstico artificial "post-mortem" em chagásicos crónicos. Rev Soc Bras Med Trop 19:259–262

Luquetti AO (1987) Megaesôfago a anticorpos anti-*Trypanosoma cruzi*. Rev Goiana Med 33:1–16

Luz FCO, Silveira AC (1984) Prevalência da infecção em triatomíneos por tripanosomatídeos monogenéticos (*Blastocrithidia* sp), parasitos encontrados no conteúdo intestinal de hemípteros do gênero *Zelus* e outros. Rev Soc Bras Med Trop 17(Suppl):70

Maekelt GA (1964) A modified procedure of xenodiagnosis for Chagas disease. Am J Trop Med Hyg 13:11–15

Malchiodi EL, Chiaramonte MG, Taranto NJ, Zwirner NW, Margni RA (1994) Cross-reactivity studies and differential serodiagnosis of human infections caused by *Trypanosoma cruzi* and *Leishmania* spp; use of immunoblotting and ELISA with a purified antigen (Ag163B6). Clin Exp Immunol 97:417–423

Marsden PD (1986) *Dipetalogaster maxima* or *D. maximus* as a xenodiagnostic agent. Rev Soc Bras Med Trop 19:205–207

Marsden PD, Prata A, Sarno P, Sherlock IA, Mott K (1969) Some observations on xenodiagnosis with *Rhodnius prolixus* and *Triatoma infestans* in human infections with Bahian strains of *Trypanosoma cruzi*. Trans R Soc Trop Med Hyg 63:425–426

Marsden PD, Barreto AC, Cuba CC, Gama MB, Ackers J (1979) Improvements in routine xenodiagnosis with first instar *Dipetalogaster maximus* (Uhler 1894) (Triatominae). Am J Trop Med Hyg 28:649–652

Marti GA, Echeverria MG, Susevich ML, Becnel JJ, Pelizza SA, García JJ (2009) Prevalence and distribution of parasites and pathogens of triatominae from Argentina, with emphasis on *Triatoma infestans* and *Triatoma virus* TrV. J Invertebr Pathol 102:233–237

Martins LPA, da Rosa JA, Castanho REP, Sauniti GL, Medeiros H (2000) Susceptibilidade de *Rhodnius neglectus*, *Rhodnius robustus* e *Triatoma infestans* (Hemiptera, Reduviidae, Triatominae) à infecção por duas cepas de *Trypanosoma cruzi* (Kinetoplastida, Trypanosomatidae) utilizando xenodiagnóstico artificial. Rev Soc Bras Med Trop 33:559–563

Mas-Coma S, Bargues MD (2009) Populations, hybrids and the systematic concepts of species and subspecies in Chagas disease triatomine vectors inferred from nuclear ribosomal and mitochondrial DNA. Acta Trop 110:112–136

Merks C, Werner H (1983) Xenodiagnose "in vitro" – Möglichkeit der Chagas-Diagnose. Mitt Österr Ges Tropenmed Parasitol 5:13–16

Miles MA, Patterson JW, Marsden PD, Minter DM (1975) A comparison of *Rhodnius prolixus*, *Triatoma infestans* and *Panstrongylus megistus* in the xenodiagnosis of a chronic *Trypanosoma* (*Schizotrypanum*) *cruzi* infection in a rhesus monkey (*Macaca mullatta*). Trans R Soc Trop Med Hyg 69:377–382

Miles MA, Souza A, Povoa M, Shaw JJ, Lainson R, Toyé PJ (1978) Isoenzymic heterogeneity of *Trypanosoma cruzi* in the first autochthonous patients with Chagas' disease in Amazonian Brazil. Nature 272:819–821

Minter DM, Minter-Goedbloed E, Vela CF (1977) Quantitative studies with first-instar triatomines in the xenodiagnosis of *Trypanosoma* (*Schizotrypanum*) *cruzi* in experimentally and naturally infected hosts. Trans R Soc Trop Med Hyg 71:530–541

Minter DM, Minter-Goedbloed E, de C Marshall TF de C (1978) Comparative xenodiagnosis with three triatomine species of different hosts with natural and experimental chronic infections with *Trypanosoma* (*Schizotrypanum*) *cruzi*. Trans R Soc Trop Med Hyg 72:84–91

Minter-Goedbloed E (1976) Hemoculture compared with xenodiagnosis for the detection of *T. cruzi* infection in man and in animals. In: New Approaches in American Trypanosomiasis Research. Proceedings of an International Symposium Belo Horizonte, Minas Gerais, Brazil, 18th–21st March, 1975, Scientific Publication No. 318, Pan American Health Organization, Washington, pp 245–252

Minter-Goedbloed E, Minter DM (1987) Value of first-instar triatomines (Hemiptera; Reduviidae) in comparative xenodiagnosis of *Trypanosoma* (*Schizotrypanum*) *cruzi*. Parasitol Res 73:565–567

Minter-Goedbloed E, Minter DM, Marshall TF de C (1978) Quantitative comparison between xenodiagnosis and haemoculture in the detection of *Trypanosoma* (*Schizotrypanum*) *cruzi* in experimental and natural chronic infections. Trans R Soc Trop Med Hyg 72:217–225

Moreira CJ de C, Perlowagora-Szumlewicz A (1997) Attempts to improve xenodiagnosis: comparative test of sensibility using *Rhodnius neglectus*, *Panstrongylus megistus*, *Triatoma vitticeps* and *Triatoma infestans* in endemic areas of Brazil. Mem Inst Oswaldo Cruz 92:91–96

Mott KE, Franca JT, Barrett TV, Hoff R, de Oliveira TS, Sherlock TA (1980) Cutaneous allergic reactions to *Triatoma infestans* after xenodiagnosis. Mem Inst Oswaldo Cruz 75:3–10

Mourya DT, Gokhale MD, Kumar R (2007) Xenodiagnosis: use of mosquitoes for the diagnosis of arboviral infections. J Vector Borne Dis 44:233–240

Muscio OA, La Torre JL, Bonder MA, Scodeller EA (1997) *Triatoma* virus pathogenicity in laboratory colonies of *Triatoma infestans* (Hemiptera:Reduviidae). J Med Entomol 34:253–256

Noireau F, Cortez MGR, Monteiro FA, Jansen AM, Torrico F (2005) Can wild *Triatoma infestans* foci in Bolivia jeopardize Chagas disease control efforts? Trends Parasitol 21:7–10

Nussenzveig V, Sonntag R (1952) Xenodiagnóstico artificial. Novo processo. Primeiros resultados positivos. Rev Paul Med 40:41–43

Pereira JB, Willcox HPF, Marcondes CB, Coura JR (1989) Parasitemia em pacientes chagásicos crônicos avaliada pelo índice de triatomíneos infectados no xenodiagnóstico. Rev Soc Bras Med Trop 22:39–44

Pereira MH, Gontijo NF, Guarneri AA, Sant'Anna MRV, Diotaiuti L (2006) Competitive displacement in Triatominae: the *Triatoma infestans* success. Trends Parasitol 22:516–520

Perlowagora-Szumlewicz A, Moreira CJC (1994) In vivo differentiation of *Trypanosoma cruzi*. – 1. Experimental evidence of the influence of vector species on metacyclogenesis. Mem Inst Oswaldo Cruz 89:603–618

Perlowagora-Szumlewicz A, Müller CA (1982) Studies in search of a suitable experimental insect model for xenodiagnosis of hosts with Chagas' disease. 1 – Comparative xenodiagnosis with nine triatomine species of animals with acute infections by *Trypanosoma cruzi*. Mem Inst Oswaldo Cruz 77:37–53

Perlowagora-Szumlewicz A, Muller CA (1987) Studies in search of a suitable experimental insect model for xenodiagnosis of hosts with Chagas' disease. 2 – Attempts to upgrade the reliability and the efficacy of xenodiagnosis in chronic Chagas' disease. Mem Inst Oswaldo Cruz 82:259–272

Perlowagora-Szumlewicz A, Muller CA, Moreira CJC (1988) Studies in search of a suitable experimental insect model for xenodiagnosis of hosts with Chagas' disease. 3 – On the interaction of vector species and parasite strain in the reaction of bugs to infection by *Trypanosoma cruzi*. Rev Saude Publica 22:390–400

Perlowagora-Szumlewicz A, Muller CA, Moreira CJC (1990) Studies in search of a suitable experimental insect model for xenodiagnosis of hosts with Chagas' disease. 4 – The reflection of parasite stock in the responsiveness of different vector species to chronic infection with different *Trypanosoma cruzi* stocks. Rev Saude Publica 24:165–177

Pineda JP, Luquetti A, Castro C (1998) Comparação entre o xenodiagnóstico clássico e artificial na fase crônica da doença de Chagas. Rev Soc Bras Med Trop 31:473–480

Pinto AS, de Lana M, Bastrenta B, Barnabé C, Quesney V, Noël S, Tibayrenc M (1998) Compared vectorial transmissibility of pure and mixed clonal genotypes of *Trypanosoma cruzi* in *Triatoma infestans*. Parasitol Res 84:348–353

Romaña C, Gil L (1947) Xenodiagnóstico artificial. An Inst Med Regional (Tucumán) 2:57–60

Sacks D, Sher A (2002) Evasion of innate immunity by parasitic protozoa. Nat Immunol 3:1041–1047

Schaub GA (1988a) Developmental time and mortality in larvae of the reduviid bugs *Triatoma infestans* and *Rhodnius prolixus* after coprophagic infection with *Blastocrithidia triatomae* (Trypanosomatidae). J Invertebr Pathol 51:23–31

Schaub GA (1988b) Parasite-host interrelationships of *Blastocrithidia triatomae* and triatomines. Mem Inst Oswaldo Cruz 83(Suppl 1):622–632

Schaub GA (1989) *Trypanosoma cruzi*: quantitative studies of development of two strains in small intestine and rectum of the vector *Triatoma infestans*. Exp Parasitol 68:260–273

Schaub GA (1990) Membrane feeding for infection of the reduviid bug *Triatoma infestans* with *Blastocrithidia triatomae* (Trypanosomatidae) and pathogenic effects of the flagellate. Parasitol Res 76:306–310

Schaub GA (2008) Kissing bugs. In: Mehlhorn H (ed) Encyclopedia of parasitology, vol 1, 3rd edn. Springer, Heidelberg, pp 684–686

Schaub GA (2009) Interactions of trypanosomatids and triatomines. Adv Insect Physiol 37:177–242

Schaub GA, Lösch P (1988) *Trypanosoma cruzi*: origin of metacyclic trypomastigotes in the urine of the vector *Triatoma infestans*. Exp Parasitol 65:174–186

Schaub GA, Lösch P (1989) Parasite/host-interrelationships of the trypanosomatids *Trypanosoma cruzi* and *Blastocrithidia triatomae* and the reduviid bug *Triatoma infestans*: influence of starvation of the bug. Ann Trop Med Parasitol 83:215–223

Schaub GA, Wülker W (1984) Tropische Parasitosen im Programm der Weltgesundheitsorganisation. Universitas 39:71–80

Schaub GA, Böker CA, Jensen C, Reduth D (1989a) Cannibalism and coprophagy are modes of transmission of *Blastocrithidia triatomae* (Trypanosomatidae) between triatomines. J Protozool 36:171–175

Schaub GA, Grünfelder CG, Zimmermann D, Peters W (1989b) Binding of lectin-gold conjugates by two *Trypanosoma cruzi* strains in ampullae and rectum of *Triatoma infestans*. Acta Trop 46:291–301

Schaub GA, Meiser CK, Balczun C (2011) Interactions of *Trypanosoma cruzi* and triatomines. In: Mehlhorn H (ed) Progress in the fight against parasitic diseases. Springer, Heidelberg

Schenone H (1998) Tratamiento etiológico en la fase crónica de la enfermedad de Chagas en niños de Chile. Rev Pat Trop 27(Suppl):33–34

Schenone H (1999) Xenodiagnosis. Mem Inst Oswaldo Cruz 94(Suppl 1):289–294

Schenone H, Alfaro E, Rojas A (1974) Bases y rendimiento del xenodiagnóstico en la infección chagásica humana. Bol Chil Parasitol 29:24–26

Schenone H, Rojo M, Rojas A, Concha L (1977) Positividad diurna y nocturna del xenodiagnóstico en un paciente con infección chagásica crónica de parasitemia permanente. Bol Chil Parasitol 32:63–66

Schenone H, Contreras MC, Rojas A (1991) Rendimiento del xenodiagnóstico, según el número de cajas utilizadas en 1.181 personas con infección chagásica crónica diagnosticada mediante la reacción de hemaglutinación indirecta. Bol Chil Parasitol 46:58–61

Schijman AG, Lauricella MA, Marcet PL, Duffy T, Cardinal MV, Bisio M, Levin MJ, Kitron U, Gürtler RE (2006) Differential detection of *Blastocrithidia triatomae* and *Trypanosoma cruzi* by amplification of 24S- ribosomal RNA genes in faeces of sylvatic triatomine species from rural northwestern Argentina. Acta Trop 99:50–54

Schofield CJ (1994) Triatominae. Biology and control. Eurocommunica, West Sussex

Schofield CJ, Dias JCP (1999) The Southern Cone Initiative against Chagas disease. Adv Parasitol 42:1–27

Schofield CJ, Galvão C (2009) Classification, evolution and species groups within the Triatominae. Acta Trop 110:88–100

Schuster JP, Schaub GA (2000) *Trypanosoma cruzi*: skin-penetration kinetics of vector-derived metacyclic trypomastigotes. Int J Parasitol 30:1475–1479

Schwarz A, Sternberg JM, Johnston V, Medrano-Mercado N, Anderson JM, Hume JCC, Valenzuela JG, Schaub GA, Billingsley PF (2009) Antibody responses of domestic animals to salivary antigens of *Triatoma infestans* as biomarkers for low-level infestation of triatomines. Int J Parasitol 39:1021–1029

Silva II (1958) Forma quistica del *Trypanosoma (Schizotrypanum) cruzi*. Rev Fac Med Tucuman 1:39–66

Solari A, Wallace A, Ortiz S, Venegas J, Sanchez G (1998) Biological characterization of *Trypanosoma cruzi* stocks from Chilean insect vectors. Exp Parasitol 89:312–322

Sosa-Estani S, Viotti R, Segura EL (2009) Therapy, diagnosis and prognosis of chronic Chagas disease: insight gained in Argentina. Mem Inst Oswaldo Cruz 104(Suppl 1):167–180

Töpfer H (1916) Zur Ursache und Übertragung des Wolhynischen Fiebers. Münch Med Wochenschr 63:1495–1496

Urribarri RS (1970) El xenodiagnóstico. Experiencia personal en 100 casos de enfermedad de Chagas crónica. Kasmera 3:167–225

Vallejo GA, Guhl F, Schaub GA (2009) Triatominae-*Trypanosoma cruzi/T. rangeli*:vector-parasite interactions. Acta Trop 110:137–147

Vargas LG, Zeledón R (1985) Effect of fasting on *Trypanosoma cruzi* infection in *Triatoma dimidiata* (Hemiptera: Reduviidae). J med Entomol 22:683

Vega Chirinos S, Náquira Velarde C (2006) Capitulo 7 Xenodiagnóstico. In: Ministerio de Salud, Instituto Nacional de Salud (eds) Manual de procedimientos de laboratorio para el diagnóstico de la trypanosomiosis americana (enfermedad de Chagas), 2nd edn, Serie de Normas Técnicas; 26, Ministerio de Salud, Instituto Nacional de Salud, Lima, pp 47–50

Verani JR, Seitz A, Gilman RH, LaFuente C, Galdos-Cardenas G, Kawai V, de LaFuente E, Ferrufino L, Bowman NM, Pinedo-Cancino V, Levy MZ, Steurer F, Todd CW, Kirchhoff LV, Cabrera L, Verastegui M, Bern C (2009) Geographic variation in the sensitivity of recombinant antigen-based rapid tests for chronic *Trypanosoma cruzi* infection. Am J Trop Med Hyg 80:410–415

Villela MM, Souza JB, Mello VP, Azeredo BVM, Dias JCP (2005) Vigilância entomológica da doença de Chagas na região centro-oeste de Minas Gerais, Brasil, entre os anos de 2000 e 2003. Cad Saude Publica 21:878–886

Werner H, Merks C (1984) Neue Möglichkeiten zum Nachweis einer Chagas-Krankheit mittels künstlicher Xenodiagnose. In: Diesfeld HJ (ed) Medizin in Entwicklungsländern, vol 16. Boch J (ed) Tropenmedizin, Parasitologie. Verlag Peter Lang, Frankfurt, pp 101–102

WHO (2002) Control of Chagas disease. Second Report from the Committee of Experts. WHO, Series of Technical Reports 905. WHO, Geneva, Switzerland

WHO (2008) Chagas disease: control and elimination, Report of the Secretariat EB124/17:1–4

Wigglesworth VB (1940) The determination of characters at metamorphosis in *Rhodnius prolixus* (Hemiptera). J Exp Biol 17:201–222

Zingales B, Andrade SG, Briones MR, Campbell DA, Chiari E, Fernandes O, Guhl F, Lages-Silva E, Macedo AM, Machado CR, Miles MA, Romanha AJ, Sturm NR, Tibayrenc M, Schijman AG (2009) Second Satellite Meeting. A new consensus for *Trypanosoma cruzi* intraspecific nomenclature: second revision meeting recommends TcI to TcVI. Mem Inst Oswaldo Cruz 104:1051–1054

Chapter 13
Blowfly Strike and Maggot Therapy: From Parasitology to Medical Treatment

Heike Heuer and Lutz Heuer

Abstract Patients, especially elderly and diabetic ones, may develop chronic wounds on the leg and foot, so called ulcers, which are open sores that go through the skin. These often tend not to heal due to insufficient circulation, will eventually get infected, and might result in serious consequences such as amputation. Physicians all over the world are involved in the daily conflict as to how to treat such wounds, which are even when not life threatening very unpleasant for these patients as they usually have a strong smell and produce continuous pain. Within the last 20 years the treatment of wounds has not been based on a dry dressing but on a wet dressing (products such as hydrogel) and healing as the primary goal is obtained in some cases. Chronic wounds will heal only when the insufficient circulation is stopped. Most often it is required that all the dead material is removed from a wound, therefore a debridement of the wound is undertaken. Although there is no clear evidence that debridement is useful for wound healing at all, several techniques are in use. One of these is forced myiasis, MDT. Maggot debridement therapy (MDT) is a very efficient debridement technique and otherwise necessary amputations are avoided in some cases. For this therapy *Lucilia* spec. are the least invasive fly larvae tested for MDT and *Lucilia sericata* is in use in most places of the world. Other, more frequently found species of *Lucilia* are in use in some countries as well.

As maggots are fly larvae and they live on the patients wound they usually do not distinguish between dead and vital tissue and must be used in correct doses. Therapy should be stopped when pain occurs or bleeding is observed. The fly larvae combat some multiresistant bacteria, for example MRSA, but many gram-negative bacteria are contra-indicated as they might kill the fly larvae. The spit of the fly larvae contains a powerful mixture of enzymes and protein-based antibiotics, both of which are under evaluation for their use as a pharmaceutical drug. Either these compounds isolated from maggots or well-dosed fly larvae can be used for debridement of chronic wounds. Acute dehiscent wounds can be debrided

H. Heuer (✉) and L. Heuer
Fa. Agiltera GmbH, Am Krausberg 31, 41542 Dormagen, Germany
e-mail: agiltera@t-online.de

efficiently by maggots and closed afterwards by other standard techniques for fast healing.

Larval treatment costs are in the same range or slightly higher compared to standard treatment when used for leg ulcers. Compared to products such as hydrogel no additional benefit was found in all available randomized clinical studies on fly larvae treatment on chronic wounds. Therefore, MDT should be used for selected patients only, especially as there is no full market authorization in Europe or anywhere else in the world apart from USA.

13.1 Introduction

Medical interventions including medication often are not man-invented but man-made only. In case of technologies "invented" by man but created by nature we call these *bionics*, a synthetic word formed by the terms *bio*logy and tech*nic*. Bio-inspiration is a very important tool for humans as nature has optimized life for billions of years and man for only some thousands of years. So whenever man observes something consciously, he decides if he can use that technology for his own purposes and if so for what benefit. So it happened that ancient societies observed myiasis, the infection of living animals by fly larvae and evidence exists that larvae have been used for the last thousand years by some cultures (Zumpt 1965; Gupta 2008), such as by the *Iranian* scientist Avicenna (Mirabzadeh et al. 2010), the aboriginal *Ngemba tribe* of New South Wales, the *Hill people* of Northern Myanmar (Burma), and the *Mayan healers* of Central America (Whitaker et al. 2007). Later the French surgeons, Ambroise Pare (1510–1590) and Baron Dominique-Jean Larrey (1766–1842), or the American physicians John Forney Zacharias (1837–1901) and William Williams Keen (1837–1932) reported independently from each other on myiasis on man and on their potential use to clean wounds (Fleischmann et al. 2004). If a wound was infected by fly larvae they did not remove them. Pare, Larrey, Zacharias, and Keen were war surgeons, treating wounds inflicted by military action. Also William S Baer, a military surgeon in France, observed in 1917 the effect of fly larvae cleaning ragged wounds, gashes and acute wounds of dead tissue, something which is used even today and described in the US Army Special Forces Medical Handbook (US Army 1982) for the use of fly larvae in military life-threatening situations.

Baer was the first who decided to use fly larvae deliberately to treat wounds. In 1929, during his appointment as Professor of Orthopedic Surgery at the Johns Hopkins University, he chose 21 patients with failed primary treatment for *osteomyelitis*. He exposed the wounds to maggots and found that, 2 months after the initiation of treatment, all of the patients' wounds had healed (Whitaker et al. 2007). As Baer could carefully supervise his patients under clinical conditions, he also was the first who observed that maggots, collected from the wild, could transfer *Clostridium perfringens* or *Clostridium tetani* to the patient and that this is fatal in most cases.

Therefore, he developed a method of sterilizing eggs of flies and breeding disinfected fly larvae. More than 300 US American hospitals introduced maggots into their program of wound healing between 1930 and 1940, and in this period more than 100 publications appeared (Whitaker et al. 2007). Larval therapy was subsequently the fastest and most successful mode of treatment for chronic *osteomyelitis*. As disinfected fly larvae do not have a shelf life longer then some hours, this *Maggot Therapy* disappeared when Penicillin was invented by Fleming some years later.

Nowadays *Maggot Therapy*, *Larval Therapy*, or *Maggot Debridement Therapy* has been reintroduced into hospital routine by Sherman (USA), Thomas (Wales), Fleischman (Germany), and Mitsui (Japan; Mitsui et al. 2010). They all are inspired by John Church, a retired orthopedic surgeon, who reported in lectures in Britain on the use of maggots after his observations in Africa and by an enthusiastic report by Bunkis (Bunkis et al. 1985). With Stephen Thomas, he set up the Biosurgical Research Unit in Bridgend, South Wales in the early 1990s (Church 2010).

Why has such an old and most complicated therapy been reinvented? MDT has been shown to be successful in diabetic ulcers (Sherman 2003; Tantawi et al. 2007), osteomyelitis (Sherman et al. 1996; Wayman et al. 2000), pressure ulcers (Sherman 2002), necrotizing fasciitis (Dunn et al. 2002; Steenvoorde et al. 2007), postsurgical wound infections and burns (Namias et al. 2000), perineal gangrene and Fournier's gangrene (Dossey 2002), and other wounds and conditions (Chan et al. 2007). And maggots do not care if the bacteria they swallow during feeding on human flesh are multiresistant to modern antibiotics. They liquefy all organic matter by their juices and grow on this powerful diet rapidly. And since MRSA (*m*ultiresistant *Staphylococcus aureus*) is a global hazard in hospitals against which physicians can only compete using the most inventive antibiotics, maggots are the last resort for many patients, which otherwise would lose their foot, leg, or arm to the saw of a surgeon.

13.2 Using Maggots for Wound Healing

Physicians and surgeons both handle patients with wounds every day. Besides open and closed wounds, chronic wounds are relevant in clinical practice. Closed wounds are contusions, bruises, hematomas, and crush injuries, all representing damage to tissue and/or blood vessels under the skin without breaking the skin. These wounds are usually not treated by debridement as they are not infected in most cases and the body will remove the dead tissue on its own if the total damage is not fatal. If on the other hand the tissue has been disrupted the open wounds provide a surface for the attack of bacteria and acute and chronic wounds are usually infected by some germs that have specialized in living on dead or dying tissue. Often surgical debridement of the dead tissue is needed. Early hopes that using MDT instead would improve the healing of the wounds (Wollina et al. 2002; Courtenay et al. 2000) have been disappointed by a recent multicenter randomized clinical study in

Great Britain (Dumville et al. 2009a, b) and now leading experts are calling for a new study to prove healing (Sherman 2010b).

13.3 Acute Wound: Definition and Background

Acute or traumatic wounds are the result of injuries that disrupt the tissue. If the patient is not malnourished, very old, weakened by other maladies or when not too much blood or tissue is lost, the human body itself is capable of closing the open wound within some days or weeks depending on the severity of the wound. As dehiscent wounds formed during accidents or in times of war usually have a large surface and foreign matter such as wood, metal etc. is enclosed by the remaining tissue, these wounds might easily be attacked by bacteria. When a larger amount of tissue is infected wound healing might be too slow and the infection might spread to yield sepsis. Sepsis is always life threatening, even in modern times with antibiotics available. Surgical cleaning of the wound is urgently needed. This is called debridement. In addition a systemic antibiotic is used to kill the bacteria that have already spread. In these cases MDT has been shown to be very useful (Fleischmann et al. 2004).

13.4 Chronic Wound: Definition and Background

Chronic wounds are caused by a relatively slow process that leads to tissue damage, for example pressure ulcers, venous ulcers, and diabetic ulcers. Insufficiency in the circulation or other systemic support of the tissue causes failure and disintegration of tissue and infections take hold of the site. As with traumatic wounds the infection might spread locally or become systemic (sepsis). We must consider that until now any debridement of a chronic wound has not shown a significant benefit over standard treatment without debridement (Steenvoorde 2008; Paul et al. 2009; Dumville et al. 2009a, b; Edwards and Stapley 2010; Bell-Syer 2010).

13.5 Debridement: Is It a Useful Intervention?

In case sepsis is assumed or has already been observed the physician must act. If antibiotics are available systemic sepsis can be avoided for some time and he can decide to remove some tissue (debridement) or to perform minor (toe, finger) or major amputations (food, hand, leg, or arm). As amputations cripple the patient, debridement of patient's dead, damaged, or infected tissue is always the first choice. Removal may be surgical, mechanical, chemical, autolytic (self-digestion), and by MDT. We will concentrate on the last, the fly larvae.

13.6 Selection of Species for Larval Therapy

All blowflies are feared in their habitat by the local shepherds. *Lucilia sericata* and *Lucilia cuprina* destroy an unnumbered amount of sheep every year in Australia (Sneddon et al. 2010), England, New Zealand, and other countries depending on sheep. *L. cuprina* is the predominant pathogenic species in subtropical and warm temperate habitats (e.g., Australia and South Africa), *L. sericata* in cool temperate habitats (e.g., Europe and New Zealand) and *Lucilia caesar* and *Lucilia illustris* become more common in sheep myiasis in more northerly Palearctic regions (Stevens and Wall 1997; Amendt et al. 2004). *Protophormia terraenovae* causes myiasis of cattle, sheep, and reindeer in the northern holarctic region. Fly larvae of *L. sericata* are usually found in humans in nasal myiasis (Poetker et al. 2006) or in wounds.

Although the fly larvae are small, usually white and live on dead matter, they are not equipped with hooks or other mechanisms to cling on to their pray. That's why the female flies deposit their eggs in natural orifices [nose holes, ear holes (Çetinkaya et al. 2008), and so on (Hsiao et al. 2008)], in wounds or on skin when the fleece is soaked by fluids and causes myiasis. As one female fly lays eggs in masses of up to 200 the hatched fly larvae are usually not alone. The maggots immediately start to feed on whatever they can find by spitting their juices onto the organic matter the female fly has selected. As the little maggots have only a very little juice when freshly hatched, they try to cling to each other and add their enzymatic fluids to boost their impact. On limited food source grown fly larvae might attack each other and cannibalize themselves so that some of the fly larvae grow to full size to pupate and to emerge as adults. The life cycle: adult fly → egg → fly larvae first instar → second instar → third instar → pupa and again adult fly differs in the different fly species and strongly depends on temperatures during each stage.

In case of myiasis prompt treatment is indicated. As in some myiasis causing fly larvae the maggots have spines located on their body, the fly larvae should not be forcibly removed from the wound. Surgical debridement with wide local excision of the fly larvae is then the best course. A foreign body reaction may occur if parts of the fly larvae remain in the wound (Poetker et al. 2006). Antiseptic dressings and oral antibiotics are recommended. Vaccination for *C. tetani* should be considered (Poetker et al. 2006) and infection by *C. perfringens* should be tested. Even scrapie can be transmitted during myiasis (Post et al. 1999).

In addition it is of extraordinary importance for fly larvae used in larval therapy that the selected fly species should lay eggs which eventually hatch after some time. Eggs are the only way to disinfect these organisms as hatched fly larvae cannot be disinfected (Sherman et al. 2000; Sherman 2009). This procedure is described in detail for eggs of *L. sericata* (Wolff and Hansson 2005) but is most properly in use for other *Lucilia* spec. also.

After a period of experimentation in which the black blow fly *Phormia regina* (Lederle 1932), the blue-bottle fly *Protophormia terraenovae* (Nuesch et al. 2002) and other flies were tested on a flesh diet, nearly all over the world green-bottle flies *Lucilia (Phaenicia)* spec. are the favorite fly family for larval therapy. All of these are in the kingdom of *Animalia*, phylum of *Arthropoda*, class of *Insecta*, order of *Diptera*, and family of *Calliphoridae* and are known to cause myiasis. *Myiasis* is defined as the infestation of live human and vertebrate animals with dipterous larvae, which, at least for a certain period, feed on the host's dead or living tissue, liquid body-substances, or ingested food by fly larvae hatched from fly eggs laid by a female fly. Therefore, for larval therapy use the fly species should be selected carefully. A review of myiasis in man is given by Derraik et al. (2010) and by the National History Museum (2010). Over time several species of flies have been tested and by a careful screening of the available literature *L. sericata* was selected as it was the least invasive species (Sherman et al. 2000; Grassberger 2002). Meanwhile only *L. sericata*, *L. cuprina* (Paul et al. 2009; Tantawi et al. 2010; Gohar et al. 2010), *Lucilia eximia* (Wolff Echeverri et al. 2010), and *L. caesar* (Susan 2008) are still in use for larval therapy, whereas *Calliphora vincina*, *Chrysoyma rufifacies*, *Phormia regina*, *Protophormia terraenovae*, *Wohlfahrtia nuba*, and *Musca domestica* (Park et al. 2010) have been used in the past (Sherman et al. 2000). Which *Lucilia* species is used for MDT most often depends on the fly species available in the local habitat and not on a specific action some *Lucilia* fly larvae might have. In Europe, North America, Japan, and New Zealand *L. sericata* is the most used species for MDT, *L. cuprina* in Egypt (Tantawi et al. 2010) and Indonesia, and *L. eximia* in South America. In total numbers *L. sericata* is the most used species, as this species has been used and therefore recommended by all pioneers in MDT.

L. sericata is now one of the best-studied flies besides *Drosophila melanogaster* and more and more producers concentrate on this species as they need a full market authorization for the production of this pharmaceutical drug (Heuer et al. Chap. 15).

Although most cases of myiasis in humans are self-limiting, some fly larvae are able to invade paranasal sinuses, orbital structures, and the cranial vault and brain tissue (Poetker et al. 2006). Therefore, when the choice for debridement is maggot therapy the used fly larvae must be controlled by the physician. The routine controls must not be time consuming and easily done, the fly larvae should not bore into the tissue, should not destroy any viable tissue or separate products from their metabolism which can lead to further maladies in the patient.

It must taken into account that the early observations on MDT using *L. sericata* for only removing dead tissue (Vistnes et al. 1981) were doubted by other authors (Sherman et al. 2000; Vilcinskas 2011) and recently disproved (Heuer and Heuer 2010b). Physicians need to keep an eye on their patients and need to set the right dose of maggots onto the patients wound (Welt-online 2010).

13.7 Action of Fly Larvae

The primary action of fly larvae is almost completely local. By their crawling, using their tiny mouth hooks or their spicules (fine, dorsally projecting, hook-like appendages (spicules) on each body segment) they are not able to cause anything but a tickling sensation on the patients wound, which indeed is felt by some patients (Nigam et al. 2010). The debridement is done chemically by the juices of the fly larvae, which they spit into the close environment and a short time later they will be sucked in again. Over the time of about 3 days the larvae will remove all dead tissue and yield a clean wound in many cases. When only vital tissues are left these juices will continuously itch and will cause pain. This type of external digestion is typical for these animals and is mostly done by enzymes included in the larvae's spit. The composition of the spit is discussed later. On the next pages we concentrate on the question, how the wanted action of the fly larvae is enforced and any unwanted action is suppressed, for example when the fly larvae leave the wound and crawl onto the patient, into his bed, or into his natural orifices.

13.8 Fly Larvae Cages: A Market Overview

For medicinal purposes the fly larvae of *Lucilia* spec. must be forced by some kind of cage to crawl only on parts of the body needing debridement. During war time or when myiasis infected a patient, the fly larvae were bound to the wound by bandages or in the natural orifices of a human body by a dirty cloth. Fly larvae from *Lucilia* spec. cannot necessarily hold themselves on a chronic wound as the larvae are leg- and spineless, and so the physician must provide help so that the larvae stay where they are needed. For this several constructions have been invented, and the most commonly used ones are discussed and depicted herein (Figs. 13.1–13.5).

To minimize the disgust of patients, as well as staff, and to avoid the escape of fly larvae, Fine and Alexander published posthumously the creation of Baer's net-cage bandages to cover and hide the fly larvae. To reduce the strong itching effect that fly larvae caused on healthy skin, he covered the border of wounds with a special bandage (Fine and Alexander 1934). These cages have been optimized over the years.

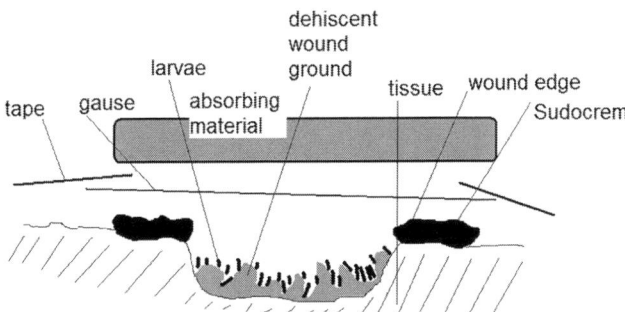

Fig. 13.1 Picture of a standard cage

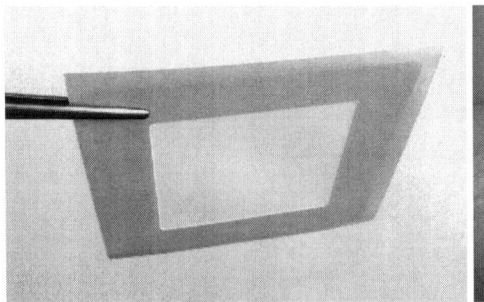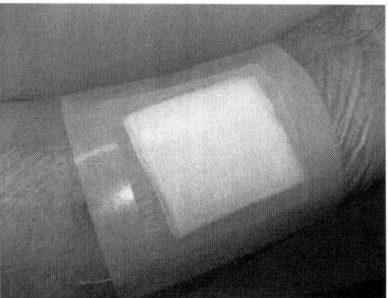

Fig. 13.2 Pictures of AgilPad, sterile and in use

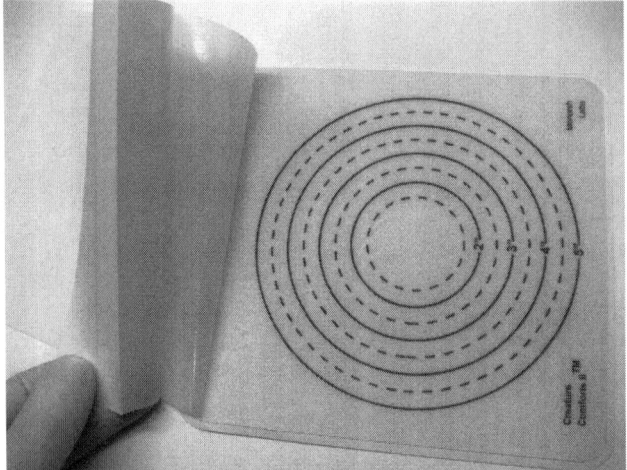

Fig. 13.3 Picture of LeFlap, new

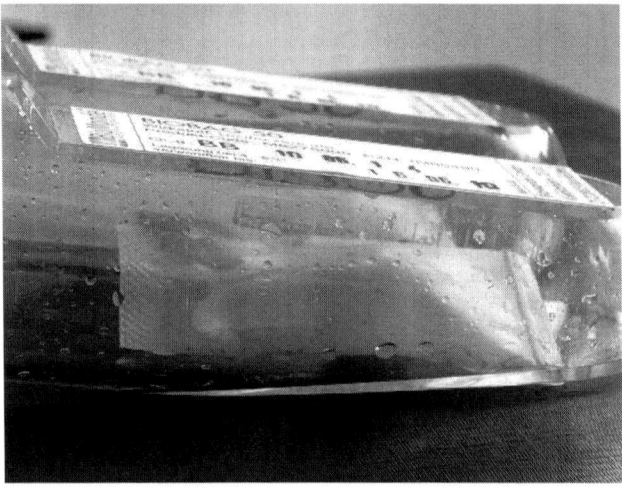

Fig 13.4 Picture of BioBag in its primary packaging

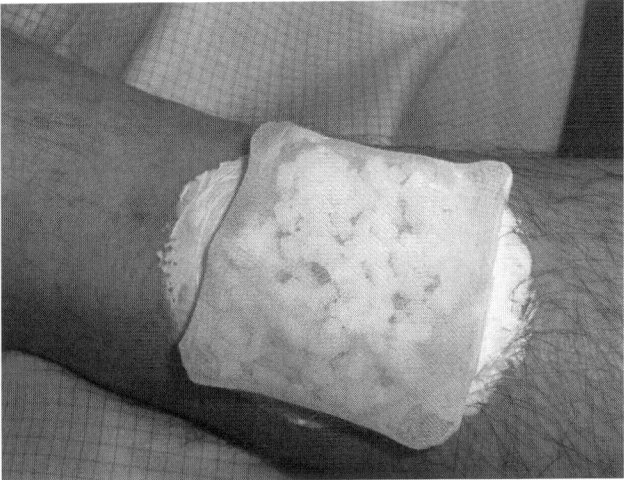

Fig. 13.5 Picture of BioFOAM on a wound protected by Sudocrem

A *standard cage* for the so-called "free-range" fly larvae is a fine net such as from curtain or gauze. The wound is surrounded by some kind of ointment mostly based on Zink oxide, for example *Sudocrem*, and the gauze is fixed by some adhesive tape (Fig. 13.1). The maggots are placed within the wound. After wetting the whole construction with a little physiological sodium chloride solution and bandaging by a standard dressing the maggots will do their work and the wound will eventually be debrided. This construction is very cheap, but quite complicated to apply and usually not tight for the crawling maggots as the ointment loosens the adhesive tape. Eventually escaping maggots will frighten the patient, the physician, the nurse, and everybody else.

These standard cages are still recommended by AGILTERA, BioMonde, MonarchLabs, and ZooBiotic, and they all have medicinal gauzes for this purpose available (the gauzes are medical devices).

An even better cage for free-range fly larvae is also a medical device. AGILTERA recommends a cage called *AgilPad*™ (Fig. 13.2), which is formed from medicinal gauze which has a framed sticky edge. To use this pad, the wound is first surrounded by a commercial tape made of polyurethane, for example *Fixomull*™ *Transparent*. As this material can easily be cut into pieces even a sinuate wound can rapidly be covered by a nontrained nurse. AgilPad can then be fixed by one nurse without a helping hand. The standard dressing and wetting is done within seconds and the maggots cannot escape. In case the size of AgilPad is smaller than the wound surface, two or more AgilPads can easily be glued to each other as AgilPad is equipped with a framed sticky edge on both sides of the gauze.

Finally, another piece of *Fixomull Transparent* is put onto the second sticky surface of AgilPad, leaving a gauze surface in the middle for passing exudates from the wound and oxygen needed for the action of the maggots. Sterile absorbing material is placed onto this and *Fixomull stretch* might be additionally used to fix the whole construction (Fig. 13.2). The procedure takes less than 2 min and gives an extraordinarily strong cage, so that no maggots can escape. As the polyurethane film and the gauze both prevent the wound from occlusion, this type of cage building is the most developed at this time.

Something very similar was invented by MonarchLabs recently. *LeFlap*™ (Fig. 13.3) or *Creatures Comforts*™ is also built from medicinal gauze but combined with a hydrocolloid dressing as the base for the cage (medical device to FDA (US Food and Drug Administration) rules). The physician or nurse must cut a hole into the hydrocolloid plate more or less the shape and the size of the wound. This can be done by scissors or a scalpel. After this step the hydrocolloid base is glued around the wound edges, the free-range maggots are placed into the wound and the sticky gauze is glued to the hydrocolloid. Finally, the dressing is finished with standard material and wetted as before. To cover different sizes of wound LeFlap is available in two sizes. Cutting the hole without a blueprint shape of the wound is challenging for this construction. In addition hydrocolloids are an occlusive material and there are some complaints on this filed with the FDA (Maude reports 2009).

All dressings discussed so far require the building of a cage onto the patient's wound. That always takes some time for the nurse and additionally gives reason for eventually escaping fly larvae. To overcome this problem, Dr. Fleischmann (1999), a German surgeon, invented a "teabag"-like cage into which the maggots are filled and totally enclosed. As the action of the fly larvae is mostly chemical, they spit through the gauze holes their juices and the debridement is performed "indirectly". These fly larvae are "contained" or "bagged" fly larvae and two products are available on the market. Both suffer from the fact, that the action of the maggots is limited to the distance of the "inner" gauze of the bag and the wound surface and that restricts the action of the fly larvae. In addition these bags are no longer medical devices but pharmaceutical drugs and need a market license which they lack until now (they are on the market anyway, based on temporary permission from the British respectively German authorities).

BioMonde is on the market with the *BioBag*™ available in four sizes and ZooBiotic (owner of BioMonde and recently renamed also to BioMonde) in roughly a dozen sizes, named *BioFOAM*™. Besides the names, both "teabags" differ in quite a lot of characteristics.

The BioBag is a small, quite rigid tube-like gauze including a piece of sponge to maintain distance between the two sides (Fig. 13.4). The amount of fly larvae (min. 50, min. 100, min. 200, or min. 300) is placed into this pouch by BioMonde and it is heat sealed afterwards. The physician can use this bag after picking up the pouch with sterile tweezers and placing it on the wound. Standard wetted dressings are applied finally. The sharp edges from the heat-sealing process of the pouch only rarely give reason for extra pain.

ZooBiotic is on the market with their BioFOAM (Fig. 13.5), which is not as rigid as the BioBag. BioFOAM uses gauze with much larger perforations and much larger volume. In addition the inner volume of the BioFOAM is almost filled with little pieces of a sponge material which itself has a foraminous structure. As the heat-sealing of the maggot-filled bag is done only at one little edge, the whole pouch feels very soft like a pillow. As a consequence of the larger perforation the fly larvae must develop for longer periods so that they become bigger and cannot escape. Eventually smaller, less-developed fly larvae will use that possibility and escape from the pouch. In addition the BioFOAM has a large inner surface which ensures that the pouch is wet over the whole time of usage. On the other hand many more fly larvae must be used (up to 900) to compensate for the large volume. Recently, ZooBiotic developed BioFOAM Maintenance, a BioFOAM of the same type and structure but using less fly larvae to maintain the debridement. This addresses the fact that some wounds are covered by a bacterial film forming a slough very soon after removing the maggots. ZooBiotic did not give numbers of fly larvae used for any product.

Contained maggots are produced, bagged, delivered, and eventually stored for additional days. The effective number of maggots cannot be seen without destroying the device. It is known, that maggots cannot be stored without loss of activity. Numbers of live maggots decreases during cold periods (Andersen et al. 2010).

"Free-range" fly larvae and contained fly larvae showed quite a different picture upon usage. When a wound is dehiscent (Golinko et al. 2008), if it is a burning or other kind of acute or traumatic wound, free-range fly larvae can be better dosed, more flexibly applied, and are the cheaper solution. Contained fly larvae on the other hand can be used when cages cannot be built up easily, for example on a toe, heel or finger, or when natural orifices are close and escaping fly larvae might be dangerous, for example close to the ear or nose. Also when the physician needs to debride only little wounds as in angiology from time to time, bagged maggots might be better. Contained fly larvae are much more expensive, and about half as effective as free-range maggots (Steenvoorde 2008). As the total number of maggots in a BioFOAM is not indicated by the producer ("optimized number") and the amount of maggots in a BioBag usually differs to higher numbers (Andersen 2010) effect and side effect cannot easily be estimated using contained fly larvae.

13.9 Biosurgical Debridement

When taking a special therapy like using fly larvae into consideration, especially when the pharmaceutical drug substance has not been investigated in full, it must have some major advantages over the alternatives, for example surgical, mechanical, chemical, or autolytic (self-digestion) debridement.

In the following table the pharmacological effects of fly larvae are reviewed and stated.

Pharmacological effect	Maggots influence on it	Statement	Literature
Primary effects	Debridement	Debridement is done in an efficient way and useful for selected patients. Debridement is not specific to dead tissue only and depends on doses = on the number of fly larvae used.	Sherman et al. (2000) Grassberger (2002a, b) Gray (2008) Steenvoorde (2008) Gupta (2008) Dumville et al. (2009a, b) Vilcinskas (2011)
	Promotion of wound healing	Wound healing is shown in cell culture only and depends on doses. Overdoses show negative effects. There is no indication, that healing can be observed under clinical conditions. Even in artificial in vitro testing the effect of larval secretions depends on doses.	Prete (1997) Petherick et al. (2006) Horobin et al. (2006) Gray (2008) Dumville et al. (2009a, b) van der Plas et al. (2009a, b) Edwards and Stapley (2010)
	Disinfection of wound	Disinfection from bacteria is shown in laboratory testing only. Reduction of bacteria load is seen on wounds in some cases, but there is no clear proof that disinfection is of clinical evidence.	Thomas et al. (1999) Steenvoorde (2008) Dumville et al. (2009a, b) Cazander et al. (2009) Margolin and Gialanella (2010) Nenoff et al. (2010)
Side effects	Debridement	One side effect is removal of viable tissue and bleeding, another pain. Strong pain is observed in 30–40% of all patients.	Sherman et al. (2000) Dumville et al. (2009a, b) Mumcuoglu et al. (2010a, b) Sherman (2010b, c)
	Promotion of wound healing	As there is no promotion of wound healing, no side effects are known.	Petherick et al. (2006) Dumville et al. (2009a, b)
	Disinfection of wound	Like most antibiotics *Lucilia sericata*'s action on bacteria is specific to some bacteria. When a specific antibiotic is used, unaffected bacteria or even fungi will spread eventually. This is seen for MDT as well.	Thomas et al. (1999) van der Plas et al. (2008) MAUDE Adverse Event Report (2009) Andersen et al. (2010) Sherman (2010a)
	Pain	Pain is observed in nearly all wounds, which are not diabetic. For diabetic patients is it known that they suffer from polyneuropathy which depresses pain through destroyed nerve cells. Pain can only sometimes be controlled by paracetamol, other patients	Mumcuoglu et al. (1999) Steenvoorde (2008) Dumville et al. (2009a, b) Rufli and Rudin (2002) Sherman et al. (2000) Mumcuoglu et al. (2010a, b) Sherman (2010b, c) Bekins (2010) Nenoff et al. (2010)

(*continued*)

Pharmacological effect	Maggots influence on it	Statement	Literature
		receive morphine and still felt pain. In these cases high numbers of fly larvae = high doses are used.	
	Bleeding	Bleeding happens often and is observed in both, free-range and bagged maggots. It occurs by corroding blood vessels by proteases and other enzyme action.	Dumville et al. (2009a, b) Steenvoorde (2008)
	Allergic reaction	Fly larvae include p-hydroxybenzoic acid, a known allergic reagent. Allergic reactions might also occur from rearing media on which the larvae have been fed before their action on the wound.	Hubermann et al. (2007a, b)
Contraindications	Debridement	Debridement by fly larvae should not be undertaken, when tissue is not well supplied with blood as these tissues, although still viable are removed and destroyed. See also ischemia.	BioMonde, Germany, information and directions for use.
	Disinfection of wound	Like most other antibiotics *Lucilia sericata* action on bacteria is specific to some bacteria only. When a specific antibiotic is used, unaffected bacteria or even fungi will spread eventually, e.g., *Pseudomonas aeruginosa* infection will spread, as maggots are themselves killed by these bacteria. *P. aeruginosa* is a well-known wound infection bacteria, which can be found in nearly all wounds. *Proteus* infection will spread, as these bacteria live in symbiosis with *L. sericata*. *E. coli* as other Gram-negative bacteria infections cannot be stopped by *L. sericata* and might spread. *Candida albicans* as	Pavillard and Wright (1957) Thomas et al. (1999) van der Plas et al. (2008) Steenvoorde (2008) Jaklič et al. (2008) MAUDE Adverse Event Report (2009) Andersen et al. (2010) Andersen et al. (2010) BioMonde, Germany, information and directions for use

(*continued*)

Pharmacological effect	Maggots influence on it	Statement	Literature
		a fungus known for wound infections cannot be eradicated by *L. sericata*.	
	Pain	Pain can only be controlled by analgesics, most preferred is morphine.	Steenvoorde (2008) Sherman (2010b, c) Heuer and Heuer (2010b)
		Local anesthetics cannot be used because they have a strong influence on maggots.	Sherman (2010b)
	Bleeding	Bleeding occurs most often. Patients using anticoagulants must not be treated. Contained fly larvae might show less bleeding.	Steenvoorde (2008)
	Ischemia	Patients suffering from ischemia did not benefit from larval therapy.	Steenvoorde (2008)
	Escaping fly larvae	Escaped fly larvae might penetrate into the visceral cavity. This will give further complications and must not happen. This is so for fly larvae treatment independently of whether they are contained or free-range.	Probst and Vasel-Biergans (2010)
	Death	Although no direct influence from larval therapy and death of the patient has been reported, an unusual high number of patients died during or soon after treatment (23.8%).	Steenvoorde (2008)

13.10 Active Ingredients of *Lucilia sericata*

It has been claimed that fly larvae are effective for wound healing, antimicrobial action, and debridement. Only the latter could be proven in randomized clinical studies (VenUS II) (Petherick et al. 2006; Dumville et al. 2009a, b). In any case, fly larvae's spit did show quite a lot of different actions in the laboratory and these have been studied in the last years. Although some literature claims that maggots always secrete the same compounds in their spit (Barnes and Gennard 2010), other authors found a strong dependency of kind and amount of compounds up to a factor of 10x if bacteria are present (Hubermann et al. 2007a, b; BingHong et al. 2007;

Altincicek and Vilcinskas 2009; Sherman 2010a, b, c, Kawabata et al. 2010) and that maggot's secretion might adapt to the patients microbiology. In addition the maggots secrete a huge amount of different compounds (Simmons 1935; Heuer and Heuer 2010a; Gohar et al. 2010), most of which are unknown (Bexfield et al. 2004, 2008; Kerridge et al. 2005; van der Plas et al. 2007, 2008, 2009a, b, 2010; Jaklič et al. 2008a, b; Barnes et al. 2010; Arora et al. 2010). That might explain the differences in outcomes seen by the MDT.

Known active ingredients	Pharmaceutical effect	Comment	Literature
10-Oleic acid	Angiogenic activity	Fatty acid	Zhen et al. (2010)
3-Guanidinopropionic acid	Proliferative effectors on growth of human endothelial cells	Amino acid derivative	Bexfield et al. (2010)
5,8,11,14-Arachidonic acid	Angiogenic activity	Fatty acid	Zhen et al. (2010)
7,10,13-Eicosatrienoic acid	Angiogenic activity	Fatty acid	Zhen et al. (2010)
7,10-Oleic acid	Angiogenic activity	Fatty acid	Zhen et al. (2010)
9,12-Linoleic acid	Angiogenic activity	Fatty acid	Zhen et al. (2010)
9-Hexadecenoic acid	Angiogenic activity	Fatty acid	Zhen et al. (2010)
9-Oleic acid	Angiogenic activity	Fatty acid	Zhen et al. (2010)
Allantoin	Skin softener	Intermediate	Fleischmann et al. (2004)
Ammonia		Toxic intermediate	Fleischmann et al. (2004)
Carboxypeptidase	Enzyme	Enzyme needed for the debridement.	Probst and Vasel-Biergans (2010)
Cecropoine	Bactericides	Insect defensive Two Lucifensins (4,113.89 and 4,063.118 Da) and other proteins	Bexfield et al. (2008) Čeřovský et al. (2010) Heuer and Heuer (2010a) Takac et al. (2010)
Chymotrypsin	Enzyme	Enzyme needed for the debridement.	Pritchard et al. (2010) Telford et al. (2010)
Collogenase	Enzyme	Enzyme needed for the debridement.	Ziffren et al. (1952)
Dermaseptine	Bactericides	Dermaseptin-similar propeptides	Berger (2006)
Histidine	Proliferative effectors on growth of human endothelial cells	Amino acid derivative	Bexfield et al. (2010)

(*continued*)

Known active ingredients	Pharmaceutical effect	Comment	Literature
Hypothiocyanate	Bactericides	In situ production by the enzymatic catalyzed reaction on thiocyanates and hydrogen peroxide	Daeschlein et al. (2006)
INF-γ	Cell growth factor, cytokine	Highly active cytokine	Fleischmann et al. (2004)
Interleukin-10	Cell growth factor, cytokine	Highly active cytokine	Probst and Vasel-Biergans (2010)
Leucinamino peptidase	Enzyme	Enzyme needed for the debridement	Probst and Vasel-Biergans (2010)
Lucifensin; Seraticin	Bactericide	Protein of the defensin-class	Čeřovský et al. (2010)
Nucleases	Enzyme	Enzyme needed for the debridement	Pritchard et al. (2009)
Octadecoic acid	Angiogenic activity	Fatty acid	Zhen et al. (2010)
Octahydro-dipyrrolo [1,2-a;1′,2′-d] pyrazine-5,10-dione	Bactericides	cyclo(Pro, Pro)	Hubermann et al. (2007a, b)
p-Hydroxy phenylacetic acid	Bactericides	Phenol derivative	Hubermann et al. (2007a, b)
p-Hydroxybenzoic acid	Bactericides	Phenol derivative, Parabene, potent allergic agent	Hubermann et al. (2007a, b)
Palmic acid	Angiogenic activity	Fatty acid	Zhen et al. (2010)
Phenyl acetaldehyde	Bactericides	Intermediate	Fleischmann et al. (2004)
Phenylacetic acid		Intermediate	Fleischmann et al. (2004)
Protease	Enzyme	Enzyme needed for the debridement	Pritchard (2001) Chambers et al. (2003)
Proteinase	Enzyme	Enzyme needed for the debridement	Pritchard (2006)
Serinprotease	Enzyme	Enzyme needed for the debridement	Probst and Vasel-Biergans (2010)
Tetradecanoic acid	Angiogenic activity	Fatty acid	Zhen et al. (2010)
Trypsin	Enzyme	Enzyme needed for the debridement	Probst and Vasel-Biergans (2010)
Urea	Wetting agent	Intermediate	Fleischmann et al. (2004)
Valinol	Proliferative effectors on growth of human endothelial cells	Amino acid derivative	Bexfield et al. (2010)

(*continued*)

Known active ingredients	Pharmaceutical effect	Comment	Literature
β-Interferon	Cell growth factor, cytokine	Highly active cytokine	Probst and Vasel-Biergans (2010)

13.11 Larval Therapy and Pharmaceutical Law

Fly larvae for medicinal uses are classified as a medical device in the USA, as a pharmaceutical drug in Europe and Canada and have not been finally classified in most other countries of the world. As fly larvae clearly showed a pharmaceutical interaction with the human body which is not only based on physical but on chemical action, they most properly will be classified as pharmaceutical drugs worldwide sooner or later. At the moment there is a full market authorization only for USA (medical device) and some temporary permissions for Europe (Heuer et al. Chap. 15, Schmidt 2009). In Europe prescriptions for maggots for medicinal use are based on the last resort right of physicians (Heuer et al. Chap. 15), whereas in the US even cancer patients are treated with maggots (Blum et al. 2010).

Fly larvae are grown on synthetic media. These include soy, brewer's yeast, chicken, and other organic food ingredients. From all of these components allergic reactions in the patient might occur (Mehlhorn et al. 2005; Monarch Labs product information) or when meat is used as food for the fly larvae, scrapie might be transferred (Post et al. 1999). As the digestion of fly larvae does not fully destroy the biological material of their food they might transfer DNA from the last food source (Kondakcia et al. 2009) to the wound. With respect to the rules of pharmaceutical drugs, all of these potential risks as well as the transfer of any virus from the food must be ruled out. To overcome all these problems in Nottingham, UK, the enzyme believed to be responsible for the debridement action has been isolated and produced without maggots (Pritchard 2010). This single active ingredient might much more easily fulfill all the aspects of a pharmaceutical drug. The same is ongoing for the bactericide Lucifensin (Takac et al. 2010) in Slovakia.

13.12 Market Size and Growth

The treatment by medicinal maggots, MTD, is as cost effective as the standard therapy (Soares et al. 2009). The market of fly larvae for MDT is estimated to be 9 Mio€ worldwide (2010), mostly in Europe, some in USA, a little in Japan, Australia (Geary and Russell 2010), Malaysia, Hong Kong (Chan et al. 2007), Singapore (Arora et al. 2010), New Zealand, Israel, Egypt, Near and Far East, and South America. 70 to 90% of treated wounds are chronic wounds from insufficient circulation in one way or another, the remainder consists of treatment of burns, acute wounds, or in angiology. Contained maggots have a market share of 70 to 80%. The market grows by 15% per year.

13.13 Addresses

AGILTERA GmbH & Co. KG, Am Krausberg 31, 41542 Dormagen, Germany.
BioMonde GmbH, Kiebitzhörn 33–35, 22885 Barsbüttel, Germany.
Department of Medical Entomology, ICPMR, Westmead Hospital, Westmead NSW 2145, Australia.
Japan Maggot Company, ジャパンマゴットカンパニー、マゴットセラピー事業部, ビー・　フライ事業部, 岡山市北区野田三丁目16-24.
Monarch Labs, 17875 Sky Park Circle, Suite K, Irvine, CA 92614, USA.
ZooBiotic Ltd., Units 2–4 Dunraven Business Park, Coychurch Road, Bridgend CF31 3BG, United Kingdom.

Products mentioned in the text are given and named for comparison reasons. They are marked as trademarks most carefully. In case a product is not marked by a trademark sign, this does not implicate that this product is not a registered trademark. All trademarks are owned by companies and not by the authors of this article.

References

Altincicek B, Vilcinskas A (2009) Septic injury-inducible genes in medicinal maggots of the green blow fly Lucilia sericata. Insect Mol Biol 18(1):119–125. doi:10.1111/j.1365-2583.2008.00856.x

Amendt J, Krettek R, Zehner R (2004) Forensic entomology. Naturwissenschaften 91:51–65. doi:10.1007/s00114-003-0493-5

Andersen AS (2010) personnel communication to the authors, Copenhagen Wound Healing Center, Bispebjerg Hospital & ABMP, Statens Serum Institut, Denmark

Andersen AS, Joergensen B, Bjarnsholt T, Johansen H, Karlsmark T, Givskov M, Krogfelt KA (2010) Quorum-sensing-regulated virulence factors in Pseudomonas aeruginosa are toxic to Lucilia sericata maggots. Microbiology 156:400–407

Arora S, Sing LC, Babtista C (2010) Antibacterial activity of Lucilia cuprina maggot extracts and its extraction techniques. Int J Integr Biol 9:43–48

Barnes KM, Gennard DE (2010) The effect of bacterially-dense environments on the development and immune defences of the blowfly Lucilia sericata. Physiol Entomol. doi:10.1111/j.1365-3032.2010.00759.x. Accessed 2010-11-30

Barnes KM, Gennard DE, Dixon RA (2010) An assessment of the antibacterial activity in larval excretion/secretion of four species of insects recorded in association with corpses, using Lucilia sericata Meigen as the marker species. Bull Entomol Res. doi:10.1017/S000748530999071X

Bekins L (2010) Thesis, Maggot Therapy for Removal of Non-healing Wounds, School of Physician Assistant Studies, Pacic University

Bell-Syer S (2010) Managing Editor for the Cochrane Wounds Group, personnel communication to the authors.

Berger M (2006) Thesis, Identifizierung biologisch aktiver Peptide und Proteine in den Sekreten von Lucilia sericata im Wundheilungsgeschehen, University of Cologne

Bexfield A, Nigam Y, Thomas S, Ratcliffe NA (2004) Detection and partial characterisation of two antibacterial factors from the excretions/secretions of the medicinal maggot Lucilia sericata and their activity against methicillinresistant Staphylococcus aureus (MRSA). Microbes Infect 6:1297–1304

Bexfield A, Bond AE, Roberts EC, Dudley E, Nigam Y, Thomas S, Newton RP, Ratcliffe NA (2008) The antibacterial activity against MRSA strains and other bacteria of a <500 Da fraction from maggot excretions/secretions of Lucilia sericata (Diptera: Calliphoridae). Microbes Infect 10:325–333

Bexfield A, Bond AE, Morgan C, Wagstaff J, Newton RP, Ratcliffe NA, Dudley E, Nigam Y (2010) Amino acid derivatives from Lucilia sericata excretions/secretions may contribute to the beneficial effects of maggot therapy via increased angiogenesis. Br J Dermatol 162 (3):554–562

BingHong X, LiPing Z, YuChuan A, MingZhu S, XianJun Y, JinFen L (2004) Induction and characterization of antibacterial substances in the blowfly Lucilia sericata. Chin J Zoonoses, 2004, 06-017

Blum K, Mendez S, Miller-Cox D (2010), Uncommon applications of maggot therapy for common an problematic wounds. In: 8th International Conference on Biotherapy, Los Angeles, CA, USA, 12 Nov 2010

Bunkis J, Gherini S, Walton RL (1985) Maggot therapy revisited. West J Med 142:554–556

Cazander G, van Veen KEB, Bernards AT, Jukema GN (2009) Do maggots have an influence on bacterial growth? A study on the susceptibility of strains of six different bacterial species to maggots of Lucilia sericata and their excretions/secretions. J Tissue Viab 18:80–87. doi:10.1016/j.jtv.2009.02.005

Čeřovský V, Zdarek J, Fučík V, Monincová L, Voburka Z, Bém R (2010) Lucifensin, the long-sought antimicrobial factor of medicinal maggots of the blowfly Lucilia sericata. Cell Mol Life Sci 67:455–466. doi: 10.1007/s00018-009-0194-0

Çetinkaya M, Özkan H, Köksal N, Coşkun SZ, Hacımustafaoğlu M (2008) Neonatal myiasis: a case report. Turk J Pediatr 50:581–584

Chambers L, Woodrow S, Brown AP, Harris PD, Phillips D, Hall M, Church JCT, Pritchard DI (2003) Degradation of extracellular matrix components by defined proteases from the green-bottle larva Lucilia sericata used for the clinical debridement of non-healing wounds. Br J Dermatol 148:14–23

Chan DCW, Fong DHF, Leung JYY, Patil NG, Leung GKK (2007) Maggot debridement therapy in chronic wound care. Hong Kong Med J 13:382–386

Church JC (2010) personnel communication to the authors

Courtenay M, Church JC, Ryan TJ (2000) Larva therapy in wound management. J R Soc Med 93:72–74

Daeschlein G, Hoffmeister B, Below H, Kramer A (2006) GMS Krankenhaushygiene Interdisziplinär, 1(1), ISSN 1863-5245

Derraik JGB, Heath ACG, Rademaker M (2010) Human myiasis in New Zealand: imported and indigenously-acquired cases; the species of concern and clinical aspects. NZMJ 123(1322). doi: http://www.nzma.org.nz/journal/123-1322/4333/

Dossey L (2002) Maggots and leeches: when science and aesthetics collide. Altern Ther Health Med 8:12–17

Dumville JC, Worthy G, Bland JM, Cullum N, Dowson C, Iglesias CP, Mitchell JL, Nelson EA, Soares MO, Torgerson DJ (2009a) Fly larvae therapy for leg ulcers (VenUS II) randomized controlled trial. BMJ 338:b773. doi:10.1136/bmj.b773

Dumville JC, Worthy G, Soares MO, Bland JM, Cullum N, Dowson C, Iglesias C, McCaughan D, Mitchell JL, Nelson EA, Torgerson DJ on behalf of the VenUS II team (2009b) VenUS II: a randomized controlled trial of larval therapy in the management of leg ulcers Health Technology Assessment 2009; Vol. 13: No. 55 DOI: 10.3310/hta13550.

Dunn C, Raghavan U, Pfleiderer AG (2002) The use of maggots in head and neck necrotizing fasciitis. J Laryngol Otol 116:70–72

Edwards J, Stapley S (2010) Debridement of diabetic foot ulcers. Cochrane Database of Systematic Reviews, Issue 1. Art. No.: CD003556. doi: 10.1002/14651858.CD003556.pub2

Fine A, Alexander H (1934) Maggot therapy: technique and clinical application. J Bone Joint Surg 16:572–582

Fleischmann W (1999) Verbandmaterial mit dem Sekret von Fliegenlarven, EP 1020197B1 (1999-12-31)

Fleischmann W, Grassberger M, Sherman R (2004) Maggot therapy. Thieme, Stuttgart, New York. ISBN 3-13-136811-x

Geary MJ, Russell RC (2010) Maggot debridement therapy (MDT)- not a 'fly by night' therapy, vol 17. Centre of Infectious Diseases and Microbiology, pp 1–2

Gohar YM, Tantawi TI, El-Ghaffar HA, El-Shazly BMA (2010) The antibacterial acidity of medicinal maggots of the blow fly Lucilia cuprina against multidrug-resistance bacteria frequently infected diabetic foot ulcers in Alexandria, Egypt: a preliminary in vitro study. In: 8th International Conference on Biotherapy, Los Angeles, CA, USA, 2010-11-12

Golinko MS, Joffe R, Maggi J, Cox D, Chandrasekaran EB, Tomic-Canic RM, Brem H (2008) Operative debridement of diabetic foot ulcers. J Am Coll Surg 207:E1–E6

Grassberger M (2002) Entomologie und Parasitologie, Fliegenmaden: Parasiten und Wundheiler, Denisia 6. Neue Folge 184:507–534

Gray P (2008) Is larval (Maggot) debridement effective for removal of necrotic tissue from chronic wounds? J Wound Ostomy Continence Nurs 35:378–384. doi:10.1097/01.WON.0000326655.50316.0e

Gupta A (2008) A review of the use of maggots in wound therapy. Ann Plastic Surg 60:224–227. doi:10.1097/SAP.0b013e318053eb5e

Heuer H, Heuer L (2010a) Deutsche Apotheker Zeitung 150:394–395

Heuer H, Heuer L (2010b) Pain release drugs in MTD – Uncover the physiological interaction. In: 8th International Conference on Biotherapy, Los Angeles, CA, USA, 2010-11-12

Horobin AJ, Shakesheff KM, Pritchard DI (2006) Promotion of human dermal fibroblast migration, matrix remodelling and modification of fibroblast morphology within a novel 3D model by Lucilia sericata larval secretions. J Invest Dermatol 126:1410–1418

Hsiao FC, Chen Y, Chang LW (2008) Umbilical myiasis in a healthy adult. South Med Assoc 101 (10):1054–1055

Hubermann L, Gollop N, Mumcuoglu KY, Block C, Galun R (2007a) Antibacterial properties of whole body extracts and haemolymph of Lucilia sericata maggots. J Wound Care 16:123–127

Hubermann L, Gollop N, Mumcuoglu YK, Breuer E, Bhusare SR, Sha Y, Galun R (2007b) Antibacterial substances of low molecular weight isolated from the blowfly, Lucilia sericata. Med Vet Entomol 21:127–131. doi:10.1111/j.1365-2915.2007.00668.x

Jakličč D, Lapanje A, Zupančič K, Smrke D, Gunde-Cimerman N (2008) Selective antimicrobial activity of maggots against pathogenic bacteria. J Med Microbiol 57:617–625. doi:10.1099/jmm.0.47515-0

Kawabata T, Mitsui H, Yokota K, Ishino K, Oguma K, Sano S (2010) Induction of antibacterial activity in larvae of the blowfly Lucilia sericata by an infected environment. Med Vet Entomol 24:375. doi:10.1111/j.1365-2915.2010.00902.x

Kerridge A, Lappin-Scott H, Stevens JR (2005) Antibacterial properties of larval secretions of the blowfly, Lucilia sericata. Med Vet Entomol 19:333–337

Kondakcia GO, Bulbula O, Shahzada MS, Polatb E, Cakana H, Altuncula H, Filoglu G (2009) STR and SNP analysis of human DNA from Lucilia sericata fly larvae's gut contents. Forensic Sci Int Genet Suppl 2(1):178–179

Lederle Laboratories (1932) Council on Pharmacy and Chemistry, Surgical Maggots-Lederle, Journal of the American Association (JAMA), advert 1932

Margolin L, Gialanella P (2010) Assessment of the antimicrobial properties of maggots. Int Wound J 7(3):202–204

Maude Adverse Event Report (2009) Monarch labs, LLC. Medical maggots with Leflap dressing. Lot Number MM-090406/CCII-090 Event Date 04/13/2009 Event Type Injury Patient Outcome Hospitalization

Maude Reports (2009) http://www.accessdata.fda.gov/scripts/cdrh/cfdocs/cfmaude/detail.cfm?mdrfoi__id=1394596, http://www.accessdata.fda.gov/scripts/cdrh/cfdocs/cfmaude/detail.cfm?mdrfoi__id=1394598

Mehlhorn H, Schmidt J, Walldorf V (2005) Zucht von Lucilia-, Phormia-, Sarcophaga- und Calliphora-Larven und Puppen auf Pflanzenextrakten, DE10328102A1 20.01.2005

Mirabzadeh A, Ladani MJN, Brojerdi SS, Imani B (2010) Maggot therapy in Iran. In: 8th International conference on biotherapy, Los Angeles, CA, USA, 12 Nov 2010

Mitsui H, Kawabata T, Ugaki S, Fujii Y, Sakrai S, Sano S (2010) Maggot debridement therapy for treating diabetic foot ulcers in Japan. In: 8th international conference on biotherapy, Los Angeles, CA, USA, 12 Nov 2010

Mumcuoglu KY, Ingber A, Gilead L (1999) Maggot therapy for the treatment of intractable wounds. Int J Dermatol 38(8):623–627

Mumcuoglu K, Davidson E, Gilead L (2010a) Pain related to maggot debridement therapy. In: 8th international conference on biotherapy, Los Angeles, CA, USA, 12 Nov 2010

Mumcuoglu K, Gilead L, Ingber A (2010b) The use of maggot debridement therapy in the treatment of chronic and acute wounds in hospitalized and ambulatory patients of the Hadassah University Hospital Jerusalem. In: 8th international conference on biotherapy, Los Angeles, CA, USA, 12 Nov

Namias N, Varela JE, Varas RP, Quintana O, Ward CG (2000) Biodebridement: a case report of maggot therapy for limb salvage after fourthdegree burns. J Burn Care Rehab 21:254–257

National History Museum (2010) http://www.nhm.ac.uk/print-version/?p=/research-curation/research/projects/myiasis-larvae/intro-myiasis/index.html. Accessed 05 Dec 2010

Nenoff P, Herrmann A, Gerlach C, Herrmann J, Simon JC (2010) Biochirurgisches Débridement mittels Lucilia sericata-Maden – ein Update, Wien Med Wochenschr 1–8. doi: 10.1007/s10354-010-0806-1

Nigam Y, Dudley E, Bexfield A, Bond AE, Evans J, James J (2010) The physiology of wound healing by the medicinal maggot, Lucilia sericata. Adv Insect Physiol 39:39–81

Nuesch R, Rahm G, Rudin W, Steffen I, Frei R, Rufli T, Zimmerli W (2002) Clustering of bloodstream infections during maggot debridement therapy using contaminated fly larvae of Protophormia terraenovae. Infection 30:306–309

Park SO, Shin JH, Choi WK, Park BS, Seok Oh J, Jang A (2010) Antibacterial activity of house fly-maggot extracts against MRSA (Methicillin-resistant Staphylococcus aureus) and VRE (Vancomycin-resistant enterococci). J Environ Biol 31(5):865–871

Paul AG, Ahmad NW, Lee HL, Ariff AM, Saranum M, Naicker AS, Osman Z (2009) Maggot debridement therapy with Lucilia cuprina: a comparison with conventional debridement in diabetic foot ulcers. Int Wound J 6(1):39–46. http://www.onlinelibrary.wiley.com/doi/10.1111/j.1742-481X.2008.00564.x/pdf. Accessed 19 Sept 2010

Pavillard ER, Wright EA (1957) An antibiotic from maggots. Nature 180:916–917

Petherick ES, O'Meara S, Spilsbury K, Iglesias CP, Nelson EA, Torgerson DJ (2006) Patient acceptability of larval therapy for leg ulcer treatment: a randomised survey to inform the sample size calculation of a randomised trial. BMC Med Res Methodol 6:43. doi:10.1186/1471-2288-6-43

Poetker DM, Cristobal R, Smith TL (2006) Head & Neck Surgery – Otolaryngology, 4th edn. Lippincott Williams & Wilkins, Baltimore

Post K, Riesner D, Walldorf V, Mehlhorn H (1999) Fly fly larvae and pupae as vectors for scrapie. Lancet 354(9194):1969–1970

Prete PE (1997) Growth effects of Phaenicia sericata larval extracts on fibroblasts: mechanism for wound healing by maggot therapy. Life Sci 60(8):505–510

Pritchard DI (2001) Protease from *Lucila sericata* and its use in treatment of wounds, WO/2001/031033

Pritchard DI (2006) Treatment of wounds, US 7144721 B1
Pritchard D (2010) The Greenbottle Pharmacy Project: Next generation wound debridement products. In: 8th international conference on biotherapy, Los Angeles, CA, USA, 12 Nov 2010
Pritchard DI, Horobin AJ, Brown A (2009) Larval polypeptides having a nuclease activity, US020090304668A1
Pritchard DI, Horobin AJ, Brown A (2010) Chymotrypsin from Lucilia sericata fly larvae and its use for the treatment of wounds, US20100008898
Probst W, Vasel-Biergans A (2010) Wundmanagement. WVG, Stuttgart, p 337 ff.
Rufli T, Rudin W (2002) Biochirurgie: Bewährtes Verfahren in der Wundbehandlung. Dtsch Arztebl 99:A2038–A2039
Schmidt M (2009) Madentherapie statt Amputation, ph Nr. 23 Dezember 2009, pp 30–32
Sherman RA (2002) Maggot versus conservative debridement therapy for the treatment of pressure ulcers. Wound Rep Reg 10:208–214
Sherman RA (2003) Maggot therapy for treating diabetic foot ulcers unresponsive to conventional therapy. Diabetes Care 26:446–451
Sherman RA (2009) Maggot therapy takes us back to the future of wound care: new and improved maggot therapy for the 21st century. J Diabetes Sci Technol 3:336–344
Sherman RA (2010a) Antimicrobially-primed medicinal maggot therapy, WO 2010/011611 A2
Sherman RA (2010b) In search of pain-free MDT: Effects of lidocaine on the debridement capacity of medicinal maggots. In: 8th international conference on biotherapy, Los Angeles, CA, USA, 12 Nov 2010
Sherman RA (2010c) In search of pain-free MDT: Healing properties of maggot therapy – What is the evidence? In: 8th international conference on biotherapy, Los Angeles, CA, USA, 12 Nov 2010
Sherman RA, Tran JMT, Sullivan R (1996) Maggot therapy for venous stasis ulcers. Arch Dermatol 132:254–256
Sherman RA, Hall MJR, Thomas S, Maggots M (2000) An ancient remedy for some contemporary afflictions. Annu Rev Entomol 45:55–81
Simmons S (1935) A bacteriocidal principle in excretions surgical maggots which destroys important etiological agents of pyrogenic infections. J Bacteriol 30:253–267
Sneddon J, Lee JA, Soutar GN (2010) An exploration of ethical consumers' response to 'animal friendly' apparel labelling. J Res Consumers 18:1–10
Soares MO, Iglesias PC, Bland JM, Cullum N, Dumville JC, Nelson EA, Torgerson DJ, Worthy G on behalf of the VenUS II team (2009) Cost effectiveness analysis of larval therapy for leg ulcers. BMJ 338:b825. doi:10.1136/bmj.b825
Steenvoorde P (2008) Maggot debridement therapy Surgery, Thesis, 9 Jan 2008
Steenvoorde P, Jacobi C, Wong C, Jukema G (2007) Maggot debridement therapy in necrotizing fasciitis. Methods Wounds 19:73–78
Stevens J, Wall R (1997) The evolution of ectoparasitism in the genus Lucilia (Diptera: Calliphoridae). Int J Parasitol 27:51–59
Susan SV (2008) In: Capinera JL (ed) Encyclopedia of entomology, Band 4, Maggot Therapy, page 2257
Takac P, Majtan J, Novak P, Bohova J, Cambal M, Kozanek M (2010) Antimicrobial Factors – Lucilia sericata. In: 8th international conference on biotherapy, Los Angeles, CA, USA, 12 Nov 2010
Tantawi TI, Gohar YM, Kotb MM, Beshara FM, El-Naggar MM (2007) Clinical and microbiological efficacy of MDT in the treatment of diabetic foot ulcers. J Wound Care 16:379–383
Tantawi TI, Williams KA, Villet MH (2010) An accidental but safe and effective use of Lucilia cuprina (Wiedemann) (Diptera: Calliphoridae) in maggot debridement therapy in Alexandria, Egypt. In: 8th international conference on biotherapy, Los Angeles, CA, USA, 12 Nov 2010
Telford G, Brown AP, Seabra RAM, Horobin AJ, Rich A, English JSC, Pritchard DI (2010) Degradation of eschar from venous leg ulcers using a recombinant chymotrypsin from *Lucilia sericata*. Br J Dermatol. doi:10.1111/j.1365-2133.2010.09854.x

Thomas S, Andrews AM, Hay NP, Bourgoise S (1999) The anti-microbial activity of maggot secretions: results of a preliminary study. J Tissue Viab 9(4):127–131

US Army (1982) Special Forces Medical Handbook, ST 31-91B, United States Army Institute for Military Assistence, 1982-03-01. http://stealthsurvival.blogspot.com/2009/11/free-downloads-us-army-special-forces.html. Accessed 10 Oct 2010

van der Plas MJA, van der Does AM, Baldry M, Dogterom-Ballering HCM, van Gulpen C, van Dissel JT, Nibbering PH, Jukema GN (2007) Maggot excretions/secretions inhibit multiple neutrophil pro-inflammatory responses. Microb Infect 9:507–514

van der Plas MJ, Jukema GN, Wai SW, Dogterom-Ballering HC, Lagendijk EL, van Gulpen C, van Dissel JT, Bloemberg GV, Nibbering PH (2008) Maggot excretions/secretions are differentially effective against biofilms of Staphylococcus aureus and Pseudomonas aeruginosa. J Antimicrob Chemother 61:117–122

van der Plas MJA, Baldry M, van Dissel JT, Jukema GN, Nibbering PH (2009a) Maggot secretions suppress pro-inflammatory responses of human monocytes through elevation of cyclic AMP. Diabetologia 52:1962–1970

van der Plas MJA, van Dissel JT, Peter H, Nibbering PH (2009b) Maggot secretions skew monocyte-macrophage differentiation away from a pro-inflammatory to a pro-angiogenic type. PLoS One 4(11):e8071

van der Plas MJA, Dambrot C, Dogterom-Ballering HC, Kruithof S, van Dissel JT, Nibbering PH (2010) Combinations of maggot excretions/secretions and antibiotics are effective against Staphylococcus aureus biofilms and the bacteria derived therefrom. J Antimicrob Chemother 65:917–923

Vilcinskas A (2011) Insect biotechnology, biologically-inspired systems, vol 2, Part 1. From traditional maggot therapy to modern biosurgery. Springer, Heidelberg, pp 67–75. doi: 10.1007/978-90-481-9641-8_4; http://www.springerlink.com/content/978-90-481-9640-1/#section=801283&page=1&locus=4. Accessed 01 Dec 2010

Vistnes L, Lee R, Ksander A (1981) Proteolytic activity of blowfly fly larvae secretions in experimental burns. Surgery 90:835–841

Wayman J, Nirojogi V, Walker A, Sowinski A, Walker MA (2000) The cost effectiveness of larval therapy in venous ulcers. J Tissue Viab 10(3):91–94

Welt-online (2010) Maden zerfressen Bettlägerigen bei lebendigem Leib. http://www.welt.de/vermischtes/weltgeschehen/article8703194/Maden-zerfressen-Bettlaegerigen-bei-lebendigem-Leib.html

Whitaker IS, Twine C, Whitaker MJ, Welck M, Brown CS, Shandall A (2007) Larval therapy from antiquity to the present day: mechanisms of action, clinical applications and future potential. Postgrad Med J 83(980):409–413. doi:10.1136/pgmj.2006.055905

Wolff H, Hansson C (2005) Rearing fly larvae of Lucilia sericata for chronic ulcer treatment–an improved method. Acta Derm Venereol 85(2):126–131

Wolff Echeverri MI, Rivera Álvarez C, Herrera Higuita SE, Wolff Idárraga JC, Escobar Franco MM, Lucilia eximia (2010) (Diptera: Calliphoridae), una nueva alternativa para la terapia larvaly reporte de casos en Colombia. IATREIA 23(2)

Wollina U, Liebold K, Schmidt WD, Hartmann M, Fassler D (2002) Biosurgery supports granulation and debridement in chronic wounds – clinical data and remittance spectroscopy measurement. Int J Dermatol 41:635–639

Zhen Z, Shouyu W, Yunpeng D, Jianing Z, Decheng L (2010) Fatty acid extracts from Lucilia sericata fly larvae promote murine cutaneous wound healing by angiogenic activity. Lipids Health Dis 9:24. doi:10.1186/1476-511X-9-24

Ziffren SE, Heist HE, May SC, Womack NA (1952) The secretion of collagenase by maggots and its implication. Ann Surg 1953:932–934

Zumpt F (1965) Myiasis in man and animals in the old world. Butterworths, London, UK

Chapter 14
Extracts from Fly Maggots and Fly Pupae as a "Wound Healer"

Heinz Mehlhorn and Falk Gestmann

Abstract On one hand the population in good old Europe and North America is decreasing in number, yet on the other there is an enormous increase in age. This leads to an increase in typical diseases of the elderly such as diabetes, decubitus and/or "nonhealing wounds" of other origin as consequences of incorrect diet and/or being constantly bedridden. There are many approaches to clear the situation of those people with "nonhealing wounds". However, although huge amounts of money are spent on different therapies thousands of amputations have to be carried out per year in most of the so-called industrialized countries. For example, in 2004 about 42,000 amputations were carried out in Germany as consequences of the typical diabetic-foot-syndrome that occurred among the 6-8 million people suffering in Germany from diabetes – worldwide there are more than 300 million humans involved in diabetic diseases. The chapter describes a new approach to heal "non-healing" wounds by use of a patented extract from larvae of the fly *Lucilia sericata*, which is finally lyophilyzed and thus can be stored for long before use as "wound cover".

14.1 Introduction

On one hand the population in good old Europe and North America is decreasing in number, yet on the other there is an enormous increase in age. This leads to an increase in typical diseases of the elderly such as diabetes, decubitus and/or "non-healing wounds" of other origin as consequences of incorrect diet and/or being constantly bedridden. There are many approaches to clear the situation of those people with "nonhealing wounds." However, although huge amounts of money are spent on different therapies thousands of amputations have to be carried out per year in most of the so-called industrialized countries. For example, in 2004 about 42,000 amputations were carried out in Germany as consequences of the typical diabetic-foot-syndrome that occurred among the 6–8 million people suffering in

---

H. Mehlhorn (✉) and F. Gestmann
Department of Zoology and Parasitology, Heinrich Heine University, 40225 Düsseldorf, Germany
e-mail: mehlhorn@uni-duesseldorf.de

Germany from diabetes – worldwide there are more than 300 million humans involved in diabetic diseases.

Therefore, there is an increasing urgency to alleviate the suffering of the mostly very old patients living with such very painful and perspectiveless wounds.

One of the recent approaches besides different applications of mechanical and chemotherapeutic dressing of the wounds is the use of fly maggots as "biosurgeons" to clean wounds and thus to stimulate them to heal again (Baer 1931; Bonn 2000; Beasley and Hirst 2004; Bexfield et al. 2004, 2008, 2010; Evans 2002; Fleischmann and Grassberger 2002; Grassberger and Frank 2003; Hobson 1931; Horobin et al. 2003, 2005; Mumcuoglu 2001; Robinson and Norwood 1933; Sherman et al. 2000; Sherman 2003). Another approach entails the evaluation of different extracts of fly larvae for their ability to do the same as living fly maggots do, but with less risk of the transmission of bacterial superinfections and/or with less psychological stress for the mostly very old patients.

These approaches, although not new, but already known in various forms for centuries, are in a way astonishing as flies have been known for a very long time as vectors of agents of disease (Tables 14.1–14.3; Mehlhorn et al. 2010). Thus, contact of humans and their animals with flies has been intensively avoided by the development of skilful hygiene methods and a broad spectrum of insecticides that should reduce the number of flies in the surroundings of humans and their animals (Mehlhorn 2008).

Considering the fact, that flies which feed on living or dead bodies and/or live in or on decomposing material, had already solved problems with wound healing and defense against attacking microbes millions of years ago, it becomes more likely, that their "abilities and capacities" might be helpful to humans, too. Thus, it is therefore not surprising that the activities of flies had excited the interest of scientists as was seen 80 years ago by the discovery of fungi that produce antibiotics.

14.2 The Process of Wound Healing

The wound-healing process is based on many components that have to be active at particular phases during normal wound healing. Therefore, it is not surprising that – if one or several of these factors are disturbed or even completely blocked – wound healing might become severely affected. Principally two basic types of wound healing are distinguished (Knapp and Hansis 1999; Protz 2009).

14.2.1 The Primary Healing Wound

This type is seen in aseptic surgical wounds that form fresh (4–6 h old), noninfected wounds with rather close borders (e.g., knife cuts). These wounds are mostly closed at present within 8–10 days and only small scars remain visible.

14 Extracts from Fly Maggots and Fly Pupae as a "Wound Healer"   327

Table 14.1 Developmental data of flies attacking man and animals

Species	Size/adult (mm)	Eggs/place of deposition	Hatch of larvae within	Larval development	Pupal rest	Life span of adults
Musca domestica, house typhoid fly	6–7	600–1,000 of 0.25 mm on feces	15°C: 50 h 20°C: 23 h 30°C: 10 h	15°C: 10 Days 20°C: 8 Days 30°C: 4 Days	15°C: 18 Days 20°C: 10 Days 30°C: 4 Days	60–70 days, 6–9 generations/year pupal hibernation
Musca autumnalis, face fly	5–7	600–900 on feces	Temperature dependent	4–7 Generations per year	4–7 Generations per year	♀ of last generations hibernates
Fannia canicularis, small sink fly	4–6	Feces and putrescent material	25°C: 20–48 h	6 Days	7–10 Days	6–7 Generations per year
Muscina stabulans, false stable fly	6–8	Eggs, larvae on chicken feces	Life cycle in summer about 2–3 weeks	Life cycle in summer about 2–3 weeks	Life cycle in summer about 2–3 weeks	4–5 Generations per year
Stomoxys calcitrans, biting stable fly	6–7	800 in groups of 25–50 in silage, in stables with urine and feces	1–2 Days, temperature dependent: 14 days up to months	6–8 Days, temperature dependent: 14 days up to months	6–8 Days, temperature dependent: 14 days up to months	♀ Live about 70–90 days
Haematobia irritans, small stable fly	4.5–4.5	In fresh cattle dung	Temperature dependent: 24 days up to months, optimum 27–30°C	Temperature dependent: 24 days up to months, optimum 27–30°C	Temperature dependent: 24 days up to months, optimum 27–30°C	3–4 Generations per year
Calliphora spp., blue blowfly	9–14	Eggs on feces with cadavers	Temperature dependent	Temperature dependent	10–40 Days	1–2 Months
Sarcophaga carnaria, gray flesh fly	10–19	Eggs on cadavers, agent of myiasis	After deposition	Temperature dependent	10–40 Days	1–2 Months
Lucilia sericata, green blowfly	5–11	1,000–2,000 eggs in batches of 250 on feces, wounds, meat	6–24 h	4–7 Days	1–3 Weeks on the ground	1–2 Months
Oestrus ovis, sheep nose bot fly	8–15	8–10 larvae are deposited at the nose or eyes	Immediately after laying	Larvae hibernate in the nose	2–4 Weeks on the ground	4 Weeks
Hypoderma bovis, warble fly	13–15	600–800 eggs on hair	4–7 days, then invading the skin	Inside the body until March	15–65 Days on the ground	3–5 Days
Hypoderma lineatum, cattle grub fly	11–13	5–20 eggs per hair	3–6 days, then invading the skin	Inside the body until March	23–28 Days on the ground	3–5 Days

Table 14.2 Transmission of agents of diseases and/or introduction of diseases by flies and other ectoparasitic insects (examples)

Ectoparasites	Symptoms of disease
House flies: *Musca domestica*, etc.	Mechanical vector for more than 100 animal and human pathogenic viruses, bacteria, and parasites. In case of ruminants, the following symptoms may occur besides restlessness, reduced food uptake, and even loss of weight: mastitis, diarrhea, and eye diseases due to bacteria and *Thelazia* worms
Cadaver flies: *Calliphora*, *Lucilia* species, *Sarcophaga*	Myiasis in wounds or in the hair and nostrils; transportation of viruses, bacteria, or parasitic eggs/larvae to mouth, eye, udder, or wounds
Stable flies: *Stomoxys calcitrans*	Inflammation of biting sites, restlessness, hypersensitivity, anemia, loss of weight, potential mechanical transmission of viruses or bacteria (e.g., paratyphus)
Bot flies: *Hypoderma* species, *Oestrus ovis*	Restlessness, hypodermosis, inner edema, paralysis during larval wandering, possible death, loss of the leather due to skin bots and nose bots, nose and eye problems, wrong turning syndrome in sheep, general loss of weight
Simuliids: *Simulium*, *Odagmia*, *Boophthora* species	Painful, burning bite sites with considerable subcutaneous hemorrhages, shock reaction in case of numerous bites, heart and blood circulation problems, paralysis of breathing activities, eventual death in case of mass infestation, transmission of filarial worms
Tabanids: *Tabanus*, *Haematopota* species	Painful bites introduce restlessness and severe itching and therefore loss of weight follows. Mechanical transmission of *Anaplasma* stages, bacteria, filarial worms, and probably also viruses
Blood-sucking lice: *Haematopinus*, *Linognathus* species	Restlessness, itching, loss of hair and weight, anemia, reduced activity
Mallophaga: *Bovicola*, *Lepikentron* species	Restlessness, itching, loss of hair, skin infections, loss of weight
Louse flies: *Lipoptena*, *Melophagus* species	Restlessness, itching, loss of hair and weight, dermal myiasis
Mosquitoes: *Aedes*, *Cules*, *Anopheles* species etc.	Itching, potential transmission of viruses, bacteria, protozoans and worms, restlessness and skin edema
Midges: *Culicoides* species	Vectors of Bluetongue and other viruses, possible death, painful bites, skin edema, especially at the udder, loss of weight, restlessness

14.2.2 The Secondary Healing Wound

This process occurs in superinfected, larger wounds and includes also all "nonhealing wounds" such as decubitus, diabetic foot, ulcus cruris etc.

In many cases the *primary* and *secondary wound-healing* process covers three, sometimes overlapping phases.

Table 14.3 List of bacteria and fungi found on the mouthparts and other regions of the body of different flies caught close to farms (according to Förster et al. 2007)

Microorganisms isolated	Test flies (No)	Percent ($n = 56$)
Bacteria		
Acinetobacter wolfii	1	1.8
Aerobic spore formers	49	87.5
Corynebacteria	7	12.5
E. coli	15	26.8
EAEC	1	1.8
EPEC	1	1.8
ETEC	2	3.6
Enterobacter aerogenes	3	5.4
Enterobacter group	9	16.1
Enterococci	3	5.4
Enterococcus faecium	1	1.8
Klebsiella spp.	3	5.4
Morganella morganii	7	12.5
Meisseria sp.	4	7.1
Nonfermenter group	1	1.8
Pantoea agglomerans	1	1.8
Proteus sp.	20	35.7
Providencia rettgeri	4	7.1
Pseudomonas sp.	1	1.8
Sphingomonas paucimobilis	1	1.8
Staphylococci, coagulase-negative	20	35.7
Staphylococcus aureus	5	8.9
Streptococcus sp.	5	8.9
Streptococcus viridans group	10	17.9
Fungi		
Aspergillus fumigatus	1	1.8
Mucor sp.	1	1.8

No. Number of test flies carrying microorganism; *Percent* percentages of total test flies transferring microorganisms ($N = 56$)

14.2.2.1 Phase of Cleaning, Exudation, and Inflammatory Reactions

This phase starts seconds after the injury by bleeding which removes remnants of cells and/or bacteria etc. The body contracts blood vessels in order to decrease the loss of blood. Contact with oxygen and release of tissue factors activating factor VII introduces the chain of the blood coagulation cascade, which is accompanied by the production of growth factors, cytokines, and installation of the arachidonic acid pathway. The visible symptoms of these processes are occurrence of heat, redness, and swelling along the wound region and feeling of pain (classically described as *rubor*, *calor*, *tumor*, and *dolor*). These phenomena and the final formation of a fibrin clot closing the wound support the *hemostasis*. Besides their main function during hemostasis the thrombocytes (= platelets) release many proinflammatory peptides (e.g., TGF-β, PDGF), which act as chemoattractants for monocytes, neutrophilic cells, and fibroblasts. At the same time the damaged tissue releases arachidonic acids (= fatty acids from membranes) and eicosanoids. Both stimulate the immune reaction. The release of Interleukin-2-cytokine by T-cells thus promotes the mitosis and formation of monocytes, which have their maximum between 48 and 60 h after the wound appeared and which are finally transformed into macrophages. During the phagocytosis of bacteria and cell remnants the macrophages release a factor (bFGF), which acts as a chemoattractant for fibroblasts and endothelial cells, in addition to IL-1, which stimulates the division of many cells that are needed during angiogenesis. Finally, the macrophages excrete MDGF (Macrophage-Derived Growth Factor), which increases the migration of keratinocytes, fibroblasts, and endothelial cells into the wound. This phase is mostly finished within 3 days after the injury.

14.2.2.2 Phase of Granulation and Proliferation

This phase starts with the migration of the fibroblasts from the wound's border in the direction of the center of the wound, being attracted by PDGF, TGF-β, and bFGF factors in the wound's fluid. These fibroblasts release proteoglycans and glucosaminoglycans as a basis for a new extracellular matrix formed by collagen and granulation tissue. The number of fibroblasts reaches its maximum between the 7th and 14th day having started as early as on the 2nd day after injury. The newly biosynthesized collagen is then gathered together extracellularly by fibroblasts forming cross-linked fibers. This formation of a new connective tissue goes on continuously for 6 weeks, while endothelial cells and keratinocytes are also growing continuously. In parallel to these processes angiogenesis occurs developing new blood vessels from the edges of the wound into its center. The new vessels produce collagenase and plasminogen activators, which support the slow digestion of fibrin clots and of the temporary matrix, while further new granulation tissue (collagen, capillaries, and extracellular matrix) is formed to cover the whole wounded area. The degrading provisional matrix sets free hyaluronic acid, while the chondroin sulfate level decreases, which slows down fibroblast migration and proliferation. Since the tissues are well supplied with blood, the wound appears reddish and granulated. This phase therefore is also called the "granulation phase." The surface of this wound appears smooth, but it must be protected from mechanical injury and kept humid at any time.

14.2.2.3 Phase of Maturation, Regeneration, and Epithelization

This phase may start in wounds as early as on the 4th day and may need 21–25 days. In chronic wounds, however, it may take weeks, months or even years. The main process is the maturation of collagen. The primarily formed collagen type I is gradually replaced by collagen type III, which forms the scar. The granulation tissue loses water and epithelial cells protrude into the center of the wound and this layer becomes thickened. Then fibroblasts differentiate into myofibroblasts including α-smooth muscle actin fibrils, which bind cells together causing a buckling of the wound's edges which move closer together. These myofibrils are responsible for wound contraction and apparently play a role as precursors of apoptosis, since finally the wound tissue is replaced by healthy tissue.

14.3 The "Nonhealing Wounds": The Problem

Skin is the most important barrier in protecting humans and animals from external environmental influences, for example it offers shelter against loss of temperature and toxic solutions as well as hindering or even blocking penetration of agents of diseases including ectoparasites. Openings in the body surface are entrance doors called wounds, which may lead to outflow of blood and lymph fluid. Thus, all living beings have developed sophisticated methods to close such wounds as soon as possible. In general – under normal conditions even without medical support – a wound heals at the latest within 13–30 days during which a cascade of events occurs including the healing phases of blood clotting, inflammation, tissue formation (= cell proliferation), and final remodeling. These different stages involve processes such as coagulation, prevention of infection, exudation, angiogenesis, collagen biosynthesis, and finally epithelization. This leads at the beginning to the formation of rather weak scar tissue without hair, follicles and glands, which later, however, will be re-formed.

According to the standards of the so-called "school medicine" there are several types or origins of wounds (Protz 2009; Knapp and Hansis 1999):

1. Mechanical wounds
2. Thermic wounds
3. Chemical wounds
4. Ulcus wounds

If these wounds do not heal within 4–12 weeks – even under strict clinical treatment – they are considered *chronic wounds*. Especially in group 4 of the wounds cited above three basic types can be diagnosed.

14.3.1 Decubitus Wounds

These wounds have their origin in pressure afflicting certain regions of the body due to immobility and due to several diseases such as diabetes, disturbances in metabolism, damage to blood vessels, drug abuse, dementia etc. According to the European Pressure Ulcer Advisory Panel (EPUAP) there are four grades with the following main characteristics:

1. Phase one: Skin reddening
2. Phase two: Partial loss of the skin

3. Phase three: Deep hollows in tissues
4. Phase four: Involvement of destruction of bones and muscles

14.3.2 Ulcus Cruris (venosum) Wounds

This disease is due to a chronic deficiency of the walls of the veins (very often the veins along the legs are afflicted). Such an ulcus is described as therapy resistant, if there is no healing within 1 year under use of approved clinical methods. Dermatologists differentiate three different grades:

1. Occurrence of edema on the legs (in the region of the knuckles)
2. Edema along the tibia, pigmentation of the skin, atrophies along the skin of the legs
3. Ulcus formation

14.3.3 Diabetic Foot Syndrome

About 20% of all patients suffering from the disease *diabetes mellitus* are finally hit by *diabetic foot syndrome*. This syndrome has as its origin in more than 50% of cases a so-called polyneuropathy, which is described as a deficiency of the sensoric, motoric, and autonomous nerves leading to the following symptoms:

1. *Sensoric neuropathy*:
   Appears as loss of sensitivity and/or occurrence of sudden itching of the sole of the foot.
2. *Motoric neuropathy*:
   The deformation of the muscles of the feet leads to a changing of the movement of the whole body thus introducing the increase of the stratum corneum along the sole of the feet (= appearance of a thick horny skin).
3. *Autonomous neuropathy*:
   This stage is introduced by a dilation of peripheral blood vessels. This leads to reddening and heating up of the skin. About 15% of cases of diabetic foot syndrome are due to an occlusion of the peripheral arterial blood vessels (the risk for diabetes patients is about six times higher than in normal persons). In general the diabetic foot syndrome occurs in more than 50% of cases due to the coexistence of all three forms of neuropathy.

The classical differentiations of the diabetic foot syndrome of *Armstrong* respectively *Wagner* (see Protz 2009) consider groups A–D (Armstrong) respectively 0–5 (Wagner). The most severe stage in both grading lists are characterized by a complete necrosis of the foot combined with local loss of blood supply and additional bacterial superinfection (Table 14.4).

Table 14.4 Factors that may make wounds "nonhealing wounds"

Type of factor	Effects
Necrotic tissue	Formation of a growth medium for bacteria increasing their number to more than 10^5/g tissue
Bacterial infection	Slow down of restoration of the wound
Degradation of cells	Slow down of cell migration
Increase of proteases	Toxic effects on healthy cells, which may die or reduce migration into wounds, granulation is retarded
Destruction of blood vessels	Macrophages and own proteolytic enzymes do not arrive in wounds, oxygen cannot arrive and decreases in wounds thus slowing down formation of collagen, lack of oxygen (less than 40 mmHg), blocks linking of fibrils due to blocking of hydralization of lysine and proline, which are the main amino acids of collagen
Lack of nutrients due to insufficient transport	Lack of fatty acids leads to reduced membrane formations, lack of amino acids reduce enzyme and protein formation, lack of vitamin K causes coagulopathy lack of metals (calcium, copper, iron, zinc etc.), reduce protein synthesis

14.4 Available Approaches to Heal "Nonhealing Wounds"

14.4.1 Mechanical Remedies

In traditional wound care the following remedies are used (Table 14.5):

(a) Compressions
   1. Mull covers
   2. Fleece covers
   3. Plaster, pavements
   4. Covers containing elements of polyurethane, silver, alginates, hydrocolloids, active charcoal, collagen, betaisadonna etc. (Hoffman 2002)
(b) Medicinal honey covering the wound

14.4.2 Chemotherapeutics

The use of wound-rinsing baths (hydrotherapy) is common, but depends on the stage of the wound. Sterile fluids containing chemotherapeutically active substances are used (e.g. polyhexanid, antibiotics, jodids, or even H_2O_2). Furthermore, enzymes are used to obtain a debridement of the borders of the wound – often in combination with systemically acting antibiotics. Recently also wound dressing containing growth factors have been in use.

The *enzyme therapy* is rather old, since in former times enzymes used were often obtained from fruit juices (e.g. lemon). However, although the efficacy of such juices had been approved in tests since 1960, for the past 10 years proteolytic enzymes have been preferred and were adapted as standard methods for necrotic

Table 14.5 Examples of wound dressings that claim to influence wound healing

Type of dressing	Mechanism of action	Authors
Addition of honey (Manuka honey)	Changing of pH on the wound acts via lowering activity of wound proteases	Gethin et al. (2008)
Alginate containing dressings with uron acids and polysaccharides	Stimulated granulation processes of the wound	Lee et al. (2009), Murakami et al. (2010)
Alginate containing wound dressing	Binding of proinflammatory factors	Wiegand et al. (2009a, b)
Natural extracts and Medihoney	Claim influences of antibacterial compounds	Jull et al. (2008), Robson et al. (2009), Blair et al. (2009), George and Cutting (2007)
Polymer and gel dressings	Managing of wound exudate levels by protecting against wound dehydration and external bacterial contamination, producing cushioning and absorption	Many products; e.g. Principelle Matrix® Fa. Principelle BV, The Netherlands

tissue removal. Furthermore, streptokinase (originating from nonpathogenic *Streptococcus* group C) is used to cut proteins and peptides. This enzyme also catalyzes the production of plasmine from plasminogen. The plasmine degrades blood clots by fibrin digestion and prevents the new formation of fibrin by decomposing fibrinogen, factor V and VII into peptides and amino acids. Another enzyme used is streptodornase which is produced by fermentation of *Streptococcus haemolyticus* products. Other enzymes in use are desoxyribonuclease, fibronilysine, krillase, collagenases of human type (elastase, hydroxylase, myeloperoxidase), and bacterial collagenases [e.g., from *Clostridium histolyticum*] belong also to the spectrum of remedies (Huberman et al. 2007; König et al. 2005; Lappin-Scott 1998).

14.4.3 Debridement by Surgery

Surgical debridement is carried out with the help of the good old scalpel and leads to the removal of necrotic layers. This result can also be obtained by an ultrasound therapy in a waterbath or via a hydrogel dressing. Furthermore, a laser and beam may be used, too.

14.4.4 Maggot Biotherapy

Maggot therapy (see Chap. 13 of this book) is very old and was already used by the early high cultures of the Maya up to the Australian aborigines of today. In the

Medieval wars of the knights and in the European wars (around 1800) with and against the French usurper Napoleon Bonaparte, medical doctors observed that the wounds of soldiers had different consequences depending on the situation, whether they were infested with fly larvae or not. Unexpectedly those soldiers survived whose wounds contained fly larvae, while those without larvae often died. This knowledge was rediscovered in the First World War (1914–1918) and used until the penicillin antibiotics were discovered (Baer 1931).

In the early 1980s, this form of biosurgery was again rediscovered and is used today in Germany as a so-called *"individually prescribed medicament."* About 5–8 larvae per centimeter are placed within a gauze-bag onto the wound. After 2–4 days the bag is taken away and the wound is rinsed to remove potential bacterial superinfections. It is claimed that the excreted proteinases of the salivary glands lead to disinfection of the wound and to its debridement by dissolving dead cells and thus allowing healthy intact tissues to grow again and to cover the wound surface (e.g., comments of the producers – Fa. Biomonde and Fa. Agiltera, Germany). However, many of the different efficacies of the maggots are not very clearly documented, but in any case wound healing works in many cases of so-called "nonhealing wounds." If this interpretation is true, the effects are due to pharmacological reactions and thus the flies are *"living medicaments"* [Figs. 14.1–14.9 (for figures 14.2, 14.3, 14.7 and 14.8 see end of this chapter)].

14.4.5 Osmotic Debridement

This method is one of the oldest ones, which applies hyperosmolar sugar derivates such as diextranomer (e.g., as honey) twice a day. This procedure binds fluids and acts antiseptically. However, there may be allergic reactions and gluing of the wound surface (Table 14.5, Manuka- and Medo-honey).

14.5 The "Other" Approaches: Extracts to the Fore

14.5.1 The Idea – Why Use Extracts of Fly Larvae

When looking at the present situation of increasing numbers of diseased elderly people suffering from "nonhealing wounds" due to different supporting factors and the fact that severe resistances against antibiotics have been developed by many bacteria, the method involving the use of fly maggots for biosurgery represents important progress. However, this method has also some weak points:

1. Many people are disgusted with the idea that living fly larvae are feeding on their body.
2. The larvae move constantly on the wounds. Thus, there is a permanent irritation due to their movements and attachment of their mouth hooks [Figs. 14.3–14.5

(for figure 14.3 see end of this chapter)], although they are embedded in bags of gauze. These movements remind patients all day and night of the presence of the fly larvae.
3. Living fly larvae digest and thus excrete feces, which in combination with the fluid on the surface of the wounds produces a "somewhat muddy smell." Of course these feces lead also to a certain nonsterility.
4. The pads containing the fly larvae cannot be stored but must be constantly reproduced, since larvae two of *Lucilia sericata* are used in the bags, which will pupate after another molt at the latest within 3–5 days. This inevitable process may lead to logistic problems, since the pads have to arrive always just in time in order to be put to use for a period of about 3–5 days at the maximum.

Therefore, the idea to examine extracts of fly larvae was born in several groups of researchers (Bowles et al. 1988, 1992, 1996; Casu et al. 2000; Cerovský et al. 2010; Chambers et al. 2003; Chapman 1997; Elkington et al. 2009; Foti et al. 2007; Greener et al. 2005; Harris et al. 2009; Kerlin and East 1991; Light et al. 2000; Nemoto et al. 2003; Prete 1997; Tellam et al. 2001; Wicke et al. 2000; Wollina et al. 2002; Ziffren et al. 1953). This idea was primarily based on the morphology of these fly larvae. Fly larvae have neither chewing mouthparts, that would allow them to bite off pieces of dead tissues (as was formerly erroneously believed) nor do they possess any other true mouthparts. The dense structure seen at the anterior end of the larvae (Fig. 14.4) represent two hooks (Fig. 14.5), which are used as holdfast systems while the larva migrates forward by body stretching movements on the surface of wounds or meat (in cultures). The feeding process of fly larvae occurs by repeated sucking of portions of the fluid from the surface of wounds after they had secreted large amounts of saliva and intestinal fluid onto the wound. The larval excretions become mixed with fluid from the surface containing apparently lysed tissue materials.

This behavior and the effects seen along the wounds nourished the suspicion that the salivary excretions contain defined active compounds that might be extractable by various standard laboratory methods in order to become stored, for example after lyophilization. Therefore, it was not surprising that several groups started research in this direction. However, this turned out to become a promising path, since there are no animal models for testing the wound-healing process that should go on in the "nonhealing wounds" of humans.

14.5.2 The Life Cycle of the Fly Lucilia sericata

The so-called gold-fly or green-bottle *L.* (syn. *Phaenicia*) *sericata* [Fig. 14.8 (for figure see end of this chapter)] belongs to the family of Calliphoridae (blow-flies). *L. sericata* occurs in many temperate climates including Australia, New Zealand, Northern Europe, and North America. The body of the adults appears robust and has a length of about 6–9 mm. Thorax and abdomen appear bright metallic green.

The palps are yellowish. The dorsal thorax has no stripes and the bristles are stout. The dorsal side of the abdomen is characterized by a faint longitudinal line along the middle [Fig. 14.8 (for figure see end of this chapter)]. After copulation the females deposit about 10–15 times batches of 100–200 whitish eggs on substrates [Fig. 14.2 (for figure see end of this chapter)]. Apparently this species prefers sticky wounds for its egg deposition, since Martini (1946) reported that in 43 cases of *Lucilia* myiasis larvae were found on the following wounds:

17 cases in tuberculosis open bones
5 cases of infected ulcera
6 cases of syphilitic wounds
5 cases of lupus
5 cases of infected skin lesions
5 cases of chancroids
2 cases of oriental sore (due to *Leishmania* spp.)
1 case of skin cancer
1 case of lichen infection
1 case of mastitis
1 case of hemorrhoidal sore
1 case of blastomycosis
2 cases of framboesian ulcer
But only 1 case of a normal wound due to a knife-cut

In the laboratory (we use horse-meat in our laboratory) the development from egg laying until pupation takes about 6–7 days under standardized temperatures of 27°C. The larva one (maggot) has a length of about 1–2 mm when hatching within 24 h from the egg. After feeding sufficiently larva three gains 100 times more weight and reaches a length of 8–12 mm [Fig. 14.3 (for figure see end of this chapter)]. Then it leaves the feeding substrate and creeps for pupation to dry and hidden places, where the formation of the adult female or male takes place inside the pupal cocoon [Fig. 14.7 (for figure see end of this chapter)]. Depending on the temperature this process needs in general about 2 weeks. The adults have a life span of about 30–45 days.

14.5.3 Types of Extracts of Fly Maggots

14.5.3.1 Extracts of Salivary and Intestinal Excretions of *L. sericata* Maggots

As listed in Table 14.6 there have been many attempts to use different extracts from fly maggots for products to heal "nonhealing-wounds." Since wound healing is a very complex process of different steps in a broad cascade of events, the idea is that the addition of some blocked factors or the "deblocking" of blocked factors as well as the stimulation of particular activities may be successful with respect to causing a return to the events and steps of normal wound healing. Because of the complexity of the wound-healing system, of course several approaches will lead to more or less significant healing effects. The effects described when using the different extracts listed in Table 14.6 vary often considerably. This fact is very probably based on the great variation of "nonhealing-wounds": not one is identical to the other (Table 14.4).

Table 14.6 Compounds that are or might be used in would healing

Type of extract	Mechanism of activity	Authors
Excretory/secretory extracts in PBS	Motogenic activity on fibroblasts, determined in cell culture	Smith et al. (2006)
Extract contains essential amino acids	Change of microenvironment	Cassino and Ricci (2010)
Extract contains collagen, elastin, glycosaminglycans, glycoproteins, proteoglycans (OASIS product)	Formation of a cover, change of microenvironment	Romanelli et al. (2010)
Extract contains collagen type 1	Binds and inactivates proinflammatory cytokines and proteases	Wiegand et al. (2009a, b)
Extract contains collagen type 1	Deactivation of proteases in the fluid of the wound	Smeets et al. (2008), Lobmann et al. (2006)
Extract contains collagen	Change of microenvironment, better microcirculation due to covering the wound	Andriessen et al. (2009)
Extract interacts with angiostatin and other proteases	Binding and inactivation of angiostatin (= Kringle 1–3 from plasminogen), that acts negatively on formation of blood vessels and growth factors	Takahashi et al. (2010), Patthy et al. (1984), Aisina et al. (2009)
Extract of salivary glands	Prohibition of the formation of proinflammatory factors by neutrophilic granulocytes	Pecivova et al. (2008)
Secretions of fly maggots	Support of the development of monocytes into macrophages, change of microenvironment from proinflammatory to proangionenic	van der Plas et al. (2008, 2009, 2010)
Full extract of squeezed pupae (heated on 65°C)	Forms together with the fluid on the wounds a fine film which becomes attached to the wound dressing and thus removes inflammatory factors from the wound	Patents of the researcher group of Fa. Alpha-Biocare, Germany
Full extracts of squeezed larvae (heated on 65°C)	Forms together with the fluid on the wounds a fine film which becomes attached to the wound dressing and thus removes inflammatory factors and potential bacteria from the wound's surface	Patents of the researcher group of Fa. Alpha-Biocare, Germany

14.5.3.2 Extracts of Whole Pupae and Larvae of *L. sericata*

In our laboratory aqueous extracts were made from larvae respectively pupae after intense cleaning of the surfaces and after homogenization of the whole individuals.

After filtration this homogenate was heated for x min at 65°C leading to a fall-out of rough material. The remnant fine homogenate was stepwise ultrafiltrated, sterile filtered, and subsequently transferred into vials that were subjected to lyophilization and firmly closed. Prior to use this ultrafiltrate was diluted in a physiological fluid and then used for different trials. For exclusion of any secondary infection the vials had been radiated previously with γ-rays. The whole system was submitted to the process of obtaining patents. Some of which had been granted in the mean time. While testing the extract – besides many other trials and tests – it was seen that the extract upon contact with the fluids of the surface of "nonhealing-wounds" formed a fine film [Figs. 14.10–14.15 (for figures see end of this chapter)], which will be surely attached to the wound dressing and thus will be removed, as the normal wound dressing is regularly changed. This effect was apparently the basis of the debridement seen in some cases of so-called "healing trials" in persons, which had not reacted to any other medication. After repeated use of this procedure the debridement was successful in that the wounds were nearly closed 6–8 weeks after the first application of the extract [Figs. 14.16 and 14.17 (for figures see end of this chapter)]. This lyophilized extract is now registered under the name "Larveel®." The product removes after application apparently mechanically the proinflammatory substances from the wound's surface thus changing the microenvironment of the wound, so that normal autolysis of necrotic tissues by the wound itself may start again. Of course also the number of bacteria, which may have developed on the wound, is constantly reduced during each changing of the wound dressing, since these bacteria are apparently included in the fine films formed on the wound.

14.5.4 Characteristics of Larveel®

The extract has the following properties:

1. It is an aqueous, ultrafiltrated extract of whole maggots of larvac and/or fresh pupae of *L. sericata*.
2. It is heated at 65°C leading to a fall out of dense material.
3. It is absolute sterile.
4. It does not contain compounds for conservation.
5. It does not contain chemical ingredients other than those coming from the fly stage.
6. Thus it is a completely natural product.
7. In skin tests with daily doubling of doses no irritation was noted on the skin of test persons.

8. It produces a fine film on the wound exudates, which becomes attached to the normal wound dressing and thus apparently removes substances that block the normal debridement of the wound.
9. In healing trials very promising results were seen in finally healing "nonhealing wounds."

14.5.5 Advantages of Larveel® Versus Living Maggot Therapy

1. Patients treated with living maggots feel psychological stress when thinking that "insects are feeding on their wounds."
   Larveel® is used like a normal wound dressing.
2. Living larvae move and this is noted by the patient leading to restlessness.
   In Larveel® treatment no movements occur on the wound.
3. Living larvae have intestinal excretions (since they increase their body weight up to 100 times). These excretions introduce a somewhat "muddy" smell to the wound.
   Larveel® has no smell.
4. Living larvae grow every day. Therefore, they can only be used at a certain stage and thus they cannot be stored in pharmacies.
   Larveel® can easily be stored.
5. Because of the need that living larvae must be used at a certain stage, logistic problems during transportation may occur.
   Larveel® can be stored for long periods at adequate temperatures, it is available on request.
6. Living larvae – although widely produced sterile – will become infected with bacteria/viruses from the wound and perhaps support an increase in their numbers.
   Larveel® diminishes the amount of bacteria, which are normally on a wound, by including them into the fine film formed on the wound, which is removed during the changing of the wound dressing.
7. The costs of the use of living larvae must be rather high due to the sterile production and fresh transportation.
   Larveel® is more economical to its rather easy production in larger lots.

14.6 Conclusions

The present and the preceding chapters show that nature has developed many methods to protect individuals from death. All these methods were developed for survival in the struggle for life during evolution. Today we only see those organisms that were successful in this fight for survival. Therefore, it is the task of scientists to determine what can be used from these successful "experiments of

14 Extracts from Fly Maggots and Fly Pupae as a "Wound Healer" 341

nature." Some of these "natural inventions" had been noted by mankind even a thousand years ago by empirical observation. Such experiments as the obvious benefits of otherwise mostly nasty flies with respect to wound healing led to their use already in the early history of mankind. However, this knowledge was often forgotten, but also often rediscovered, so that today the use of larvae and/or the use of extracts of larvae and pupae really offer progress in the treating of "nonhealing wounds" in times of society-based diseases.

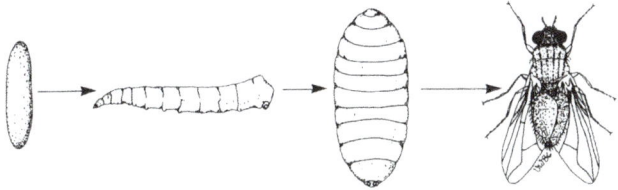

Fig. 14.1 Diagrammatic representation of the life cycle stages of flies (there are three larval stages, which grow via two molts)

Fig. 14.2 Light micrograph of a batch of *Lucilia sericata* eggs

Fig. 14.3 Light micrograph of a third stage larva of *L. sericata*. Note the dense appearing mouth hooks

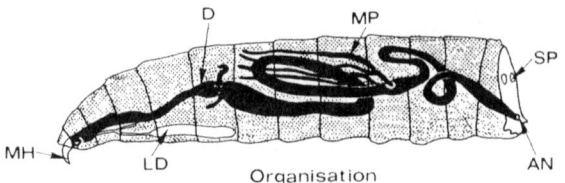

Fig. 14.4 Diagrammatic representation of a fly larva showing intestine and labial = salivary glands. AN = anus, D = gut, LD = labial glands, MH = mouth hooks, MP = malpighi ducts, SP = spiracles

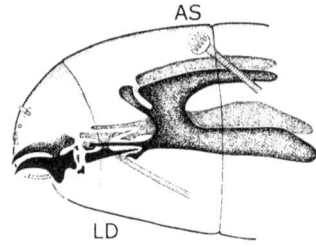

Fig. 14.5 Diagrammatic representations of the mouth hooks. AS = anterior spiracle, LD = labial duct starting from mouth

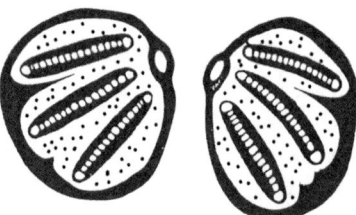

Fig. 14.6 Diagrammatic representation of the appearance of the two abdominal spiracles of larva three

Fig. 14.7 Light micrograph of a pupae of *L. sericata*

Fig. 14.8 Adult *Lucilia sericata*

Fig. 14.9 Diagrammatic representation of a wing of an adult *Lucilia sericata* (letters M, R, Sc mark important "veins" of the wing)

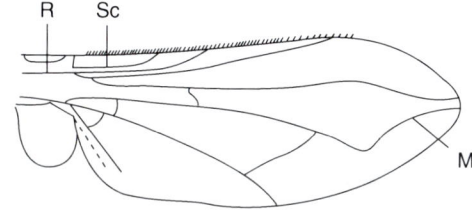

Fig. 14.10 Formation of a fine film on a wound exudate after application of the larval extract in six steps (step 1)

Fig. 14.11 Formation of a fine film on a wound exudate after application of the larval extract in six steps (step 2)

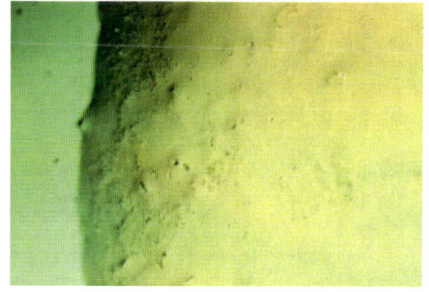

Fig. 14.12 Formation of a fine film on a wound exudate after application of the larval extract in six steps (step 3)

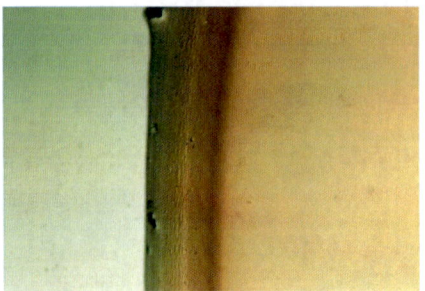

Fig. 14.13 Formation of a fine film on a wound exudate after application of the larval extract in six steps (step 4)

Fig. 14.14 Formation of a fine film on a wound exudate after application of the larval extract in six steps (step 5)

Fig. 14.15 Formation of a fine film on a wound exudate after application of the larval extract in six steps (step 6)

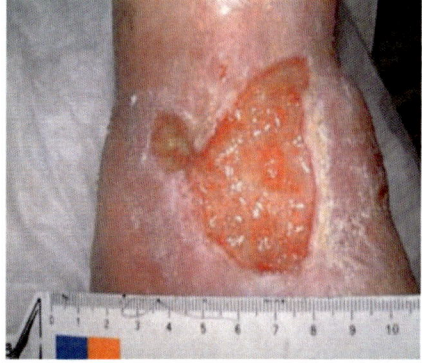

Fig. 14.16 Nonhealing wound of a patient before the application of the extract (photo by Prof. Dr. Stege from a patient application, given to Fa. Alpha-Biocare)

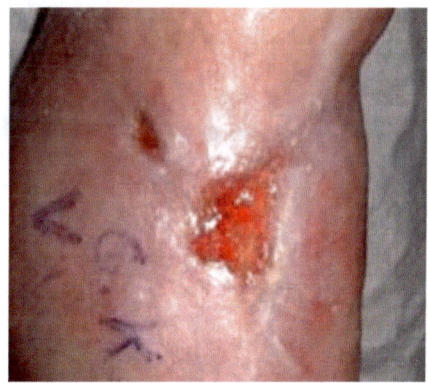

Fig. 14.17 Wound 6 weeks after application of the extract. Note the tendency to close the surface (photo by Prof. Dr. Stege given to Fa. Alpha-Biocare for a common patent)

References

Aisina RB, Mukhametova LI, Gulin DA, Levashov MY, Prisyazhnaya NV, Gershkovich KB, Varfolomeyev SD (2009) Inhibitory effect of angiostatins on activity of the plasminogen/plasminogen activator system. Biochemistry 74:1104–1113

Andriessen A, Polignano R, Abel M (2009) Monitoring the microcirculation to evaluate dressing performance in patients with venous leg ulcers. J Wound Care 18:145–150

Baer WS (1931) The treatment of chronic osteomyelitis with the maggot (larva of the blowfly). J Bone Jt Surg 13:438–475

Beasley WD, Hirst G (2004) Making a meal of MRSA – the role of biosurgery in hospital-acquired infection. J Hosp Infect 56:6–9

Bexfield A, Nigam Y, Thomas S, Ratcliffe NA (2004) Detection and partial characterisation of two antibacterial factors from the excretion/secretion of the medicinal maggot *Lucilia sericata* and their activity against methicillin-resistant *Staphylococcus aureus* (MRSA). Microbes Infect 6:1297–304

Bexfield A, Bond AE, Roberts EC, Dudley E, Nigam Y, Thomas S, Newton S, Newton RP, Ratcliffe NA (2008) The antibacterial activity against MRSA strains and other bacteria of a <500 Da fraction from maggot excretions/secretions of *Lucilia sericata* (Diptera: Calliphoridae). Microbes Infect 10:325–333

Bexfield A, Bond AE, Morgan C, Wagstaff J, Newton RP, Ratcliffe NA, Dudley E, Nigam Y (2010) Amino acid derivatives from *Lucilia sericata* excretions/secretions may contribute to the beneficial effects of maggot therapy via increased angiogenesis. Br J Dermatol 162:554–562

Blair SE, Cokcetin NN, Harry EJ, Carter DA (2009) The unusual antibacterial activity of medical-grade *Leptospermum* honey: antibacterial spectrum, resistance and transcription analysis. Eur J Clin Microbiol Infect Dis 28:1199–1208

Bonn D (2000) Maggot therapy: an alternative for wound infection. Lancet 356:1174

Bowles VM, Carnegie PR, Sandeman RM (1988) Characterization of proteolytic and collagenolytic enzymes from the larvae of *Lucilia cuprina*, the sheep blowfly. Aust J Biol Res 41:269–278

Bowles VM, Grey ST, Brandon MR (1992) Cellular immune response in the skin of sheep infected with larvae of *Lucilia cuprina*, the sheep blowfly. Vet Parasitol 44:151–162

Bowles VM, Meeusen EN, Young AR, Andrews AE, Nash AD, Brandon MR (1996) Vaccination of sheep against larvae of the sheep blowfly (*Lucilia cuprina*). Vaccine 14:1347–1352

Cassino R, Ricci E (2010) Effectiveness of tropical applications of amino acids to chronic wounds: a prospective observational study. J Wound Care 19:29–34

Casu RE, Jarmey JM, Elvin CM, Eisemann CH (2000) Isolation of a trypsin-like serine protease gene family from the sheep blowfly *Lucilia cuprina*. Insect Mol Biol 3:159–170

Cerovský V, Zdárek J, Fucík V, Monincová L, Voburka Z, Bém R (2010) Lucifensin, the long-sought antimicrobial factor of medicinal maggots of the blowfly *Lucilia sericata*. Cell Mol Life Sci 67:455–466

Chambers L, Woodrow S, Brown AP, Harris PD, Phillips D, Hall M, Church JC, Pritchard DI (2003) Degradation of extracellular matrix components by defined proteinases from the greenbottle larva *Lucilia sericata* used for the clinical debridement of non-healing wounds. Br J Dermatol 148:14–23

Chapman HA (1997) Plasminogen activators, integrins, and the coordinated regulation of cell adhesion and migration. Curr Opin Cell Biol 9:714–724

Elkington RA, Humphries M, Commins M, Maugeri N, Tierney T, Mahony TJ (2009) *Lucilia cuprina* excretory-secretory protein inhibits the early phase of lymphocyte activation and subsequent proliferation. Parasite Immunol 31:750–765

Evans P (2002) Larvae therapy and venous leg ulcers: reducing the "yuk factor". J Wound Care 11:407–408

Fleischmann W, Grassberger M (2002) Maden-Therapie. Trias Verlag, Germany

Förster M, Klimpel S, Mehlhorn H, Sievert K, Messler S, Pfeffer K (2007) Pilot study on synanthropic flies (e. g. *Musca, Sarcophaga, Calliphora, Tannia. Lucilia, Stomoxys*) as vectors of pathogenic microorganisms. Parasitol Res 101:243–246

Foti C, Conserva A, Casullo C, Scrimieri V, Pepe ML, Quaranta D (2007) Contact dermatitis with clostridiopeptidase A contained in Noruxol ointment. Contact Dermatitia 56:361–362

George NM, Cutting KF (2007) Antibacterial Honey (Medihoney^{TM}): in-vitro activity against clinical isolates of MRSA, VRE, and other multiresistant gram negative organisms including *Pseudomonas aeruginosa*. Wound 19:231–236

Gethin GT, Cowman S, Conroy RM (2008) The impact of Manuka honey dressings on the surface pH of chronic wounds. Int Wound J 5:185–194

Grassberger M, Frank C (2003) Wundheilung durch sterile Fliegenlarven: mechanische, biochemische und mikrobiologische Grundlagen. Schwarz, Urban, München

Greener B, Hughes A, Banister NP, Douglass J (2005) Proteases and pH in chronic wounds. J Wound Care 14:59–61

Harris LG, Bexfield A, Nigam Y, Rohde H, Ratcliffe NA, Mack D (2009) Disruption of *Staphylococcus epidermidis* biofilms by medicinal maggot *Lucilia sericata* excretions/secretions. Int J Artif Organs 32(9):555–564

Hobson RP (1931) On an enzyme from blowfly larvae (*Lucilia sericata*) which digests collagen in alkaline solution. Biochem J 25:1458

Hoffman AS (2002) Hydrogels for biomedical applications. Adv Drug Deliv Rev 43:3–12

Horobin AJ, Shakesheff KM, Woodrow S, Robinson C, Pritchard DI (2003) Maggots and wound healing: the effect of *Lucilia sericata* larval secretions upon human dermal fibroblasts. Br J Dermatol 148:923–933

Horobin AJ, Shakesheff KM, Pritchard DI (2005) Maggots and wound healing: an investigation of the effects of secretions from *Lucilia sericata* Larvae upon the migration of human dermal fibroblasts over a fibronectin-coated surface. Wound Rep Reg 13:422–433

Huberman L, Gollop N, Mumcuoglu KY, Breuer E, Bhusare SR, Shai Y, Galun R (2007) Antibacterial substances of low molecular weight isolated from the blowfly, *Lucilia sericata*. Med Vet Entomol 21(2):127–131

Jull A, Walker N, Parag V, Molan P, Rodgers A (2008) Randomized clinical trial of honey-impregnated dressings for venous leg ulcers. Br J Surg 95:175–182

Kerlin RL, East IJ (1991) Potent immunosuppression by secretory/excretory products of larvae from the sheep blowfly *Lucilia cuprina*. Parasite Immunol 14:595–604

Knapp U, Hansis M (1999) Die Wunde. Thieme, Stuttgart, New York

König M, Vanscheidt W, Augustin M, Kapp H (2005) Enzymatic versus autolytic debridement of chronic leg ulcers: a prospective randomised trial. J Wound Care 14:320–323

Lappin-Scott HM (1998) Bacterial–maggot interactions in wound therapy. In: Third international Conference on Biotherapy, Jerusalem, Israel

Lee WR, Park JH, Kim KH, Kim SJ, Park DH, Chae MH, Suh SH, Jeong SW, Park KK (2009) The biological effects of tropical alginate treatment in an animal model of skin wound healing. Wound Repair Regen 17(4):505–510

Light RW, Nguyen T, Mulligan ME, Sasse SA (2000) The in vitro efficacy of varidase versus streptokinase or urokinase for liquefying thick purulent exudative material from loculated empyema. Lung 178(1):13–18

Lobmann R, Zemlin C, Motzkau M, Reschke K, Lehnert H (2006) Expression of matrix metalloproteinases and growth factors in diabetic foot wounds treated with a protease absorbent dressing. J Diabetes Complications 20(5):329–335

Martini E (1946) Medizinische Entomologie. G. Fischer, Jena

Mehlhorn H (2008) Encyclopedia of parasitology, 3rd edn. Springer, New York

Mehlhorn H, Al-Rasheid KAS, Abdel-Ghaffar F, Klimpel S, Pohle H (2010) Life cycle and attacks of ectoparasites on ruminants during the gear in Central Europe. Parasitol Res 107:425–431

Mumcuoglu KY (2001) Clinical applications for maggots in wound care. Am J Clin Dermatol 2001(2):219–227

Murakami K, Aoki H, Nakamura S, Nakamura S, Takikawa M, Hanzawa M, Kishimoto S, Hattori H, Tanaka Y, Kiyosawa T, Sato Y, Ishihara M (2010) Hydrogel blends of chitin/chitinosan, fucoidan and alginate as healing-impaired wound dressings. Biomaterials 31(1):83–90

Nemoto K, Hirota K, Murakami K, Taniguti K, Murata H, Viducic D, Miyake Y (2003) Effect of Varidase (streptodornase) on biofilm formed by *Pseudomonas aeruginosa*. Chemotherapy 49(3):121–125

Patthy L, Trexler M, Váli Z, Baányai L, Váradi A (1984) Kringles: modules specialized for protein binding. Homology of the gelatine-binding region of fibronectin with the kringle structures of proteases. FEBS Lett 171(1):131–136

Pecivova J, Macickova T, Takac P, Kovacsova M, Cupanikova D, Kozanek M (2008) Effect of the extract from salivary glands of *Lucilia sericata* on human neurophils. Neuroendocrinol Lett 29(5):794–797

Prete PE (1997) Growth effects of *Phaenicia sericata* larval extracts on fibroblasts: mechanism for wound healing by maggot therapy. Life Sci 60(8):505–510

Protz K (2009) Moderne Wundversorgung, 5th edn. Urban & Fischer, München

Robinson W, Norwood VH (1933) The role of surgical maggots in the disinfection of osteomyelitis and other infected wounds. J Bone Joint Surg 15:409–412

Robson V, Dodd S, Thomas S (2009) Standardized antibacterial honey (Medihoney) with standard therapy in wound care: randomized clinical trial. J Adv Nurs 65(3):565–575

Romanelli V, Dini V, Bertone MS (2010) Randomized comparison of OASIS wound matrix versus moist wound dressing in the treatment of difficult-to-heal wounds of mixed arterial/venous etiology. Adv Skin Wound Care 23(1):34–38

Sherman RA (2003) Maggot therapy for treating diabetic foot ulcers unresponsive to conventional therapy. Diabetes Care 2003(26):446–451

Sherman RA, Hall MJ, Thomas S (2000) Medicinal maggots: ancient remedy for some contemporary applications. Annu Rev Entomol 45:55–81

Smeets R, Ulrich D, Unglaub F, Wöltje M, Pallua N (2008) Effect of oxidised regenerated cellulose/collagen matrix on proteases in wound exudates of patients with chronic venous ulceration. Int Wound J 5:195–203

Smith AG, Powis RA, Pritchard DI, Britland ST (2006) Greenbottle (*Lucilia sericata*) larval secretions delivered from a prototype hydrogel wound dressing accelerate the closure of model wounds. Biotechnol Prog 22:1690–1696

Takahashi S, Shinya T, Sugiyama A (2010) Angiostatin inhibition of vascular endothelial growth factor-stimulated nitric oxide production in endothelial cells. J Pharmacol Sci 112:432–437

Tellam RL, Eisemnn CH, Vuocolo T, Casu R, Jarmey J, Bowles V, Pearson R (2001) Role of oligosaccharides in the immune response of sheep vaccinated with *Lucilia cuprina* larval glycoprotein, peritrophin-95. Int J Parasitol 8:798–809

van der Plas MJ, Jukema GN, Wai SW, Dogterom-Ballering HC, Lagendijk EL, van Gulpen C, van Dissel JT, Bloemberg GV, Nibbering PH (2008) Maggot excretions/secretions are differentially effective against biofilms of *Staphylococcus aureus* and *Pseudomonas aeruginosa*. J Antimicrob Chemother 61(1):117–122

van der Plas MJ, van Dissel JT, Nibbering PH (2009) Maggot secretions skew monocyte-macrophage differentiation away from a pro-inflammatory to a proangiogenic type. PLoS One 30;4(11):e8071

van der Plas MJ, Dambrot C, Dogterom-Ballering HC, Kruithof S, van Dissel JT, Nibbering PH (2010) Combinations of maggot excretions/secretions and antibiotics are effective against *Staphylococcus aureus* biofilms and the bacteria derived therefrom. J Antimicrob Chemother 65:917–923

Wicke C, Halliday B, Allen D, Roche NS, Scheuenstuhl H, Spencer MM, Roberts AB, Hunt TK (2000) Effects of steroids and retinoids on wound healing. Arch Surg 135:1265–1270

Wiegand C, Heinze T, Hipler UC (2009a) Comparative in vitro study on cytotoxicity, antimicrobial activity, and binding capacity for pathophysiological factors in chronic wounds of alginate and silver-containing alginate. Wound Repair Regen 17(4):511–521

Wiegand C, Schönfelder U, Abel M, Ruth O, Kraatz M, Hipler UC (2009b) Protease and proinflammatory cytokine concentrations are elevated in chronic compared to acute wounds and can be modulated by collagen type I in vitro. Arch Dermatol Res 302(6):419–428

Wollina U, Liebold K, Schmidt WD, Hartmann M, Fassler D (2002) Biosurgery supports granulation and debridement in chronic wounds-clinical data and remittance spectroscopy measurement. Int J Dermatol 42:635–639

Ziffren SE, Heist HE, May SC, Womack NA (1953) The secretion of collagenase by maggots and its implication. Ann Surg 138:932–934

Chapter 15
Living Medication: Overview and Classification into Pharmaceutical Law

Heike Heuer, Lutz Heuer, and Valentin Saalfrank

Abstract This contribution summarises the knowledge of pharmaceutical drugs based on living organisms, living drugs such as leaches, the worm *Trichuris suis*, fly larvae form *Lucilia sericata* or bacteria such as *E. coli*, their application in medicine and their legal status in many countries. As more and more patients in the world suffer from diseases to which modern science does not have appropriate interventions, alternative therapies are welcome as a cure. Nature has many secrets and some are revealed and discussed herein. Man, *Homo sapiens*, is not a single species but an ecological system in itself. The complex interaction of two or more organisms is described by different paradigms such as symbiosis or parasitism and in this way are claimed to be helpful or dangerous. Looking more closely these paradigms provide two extreme positions of the same idea: nature is not an elaborated world but a wonderful creation within which man may find another organism to help him stay healthy. When other living creatures are used for man's health these organisms acquire the status of a medicinal drug in most cases and must be treated by pharmaceutical law.

15.1 Introduction

The world made by God has been partitioned [subdivided into parts] by man. Inorganic matter and life belong to these manmade partitions, plants and animals to another. The truth is that all that we can see, taste or feel and even that which we cannot see, taste or feel is part of the same world and everything interacts with everything else. And as man's intelligence and comprehension is not able and never will be to understand all these interactions, we are proud of the tiny portions and pieces of

---

This work is dedicated to Professor The Lord Lewis of Newnham on the occasion of his 83rd birthday.

H. Heuer (✉) and L. Heuer
Fa. Agiltera GmbH, Am Krausberg 31, 41542 Dormagen, Germany
e-mail: agiltera@t-online.de

V. Saalfrank
Berrenrather Str. 393, 50937 Köln, Germany

knowledge and understanding that we have collected within the roughly 4,500 years of reasonably well-documented scientific history within the 4.6 billion years of Earth.

Nature is the word we use for that complex interaction and nature gives man so many questions so that we, man, will be busy in science forever. Some of what has been discovered on useful interaction between the organism *Homo sapiens* and other living organisms will be part of this contribution.

Therefore, it is interesting to look into the manmade partitions on life. We will not discuss plants but animals only and we have selected animals that are or have been used for man's health. Further we will give an explanation as to why some of the known pharmaceutically useful interactions are still in use whereas others have been long forgotten or cannot be used within our modern world. This introduction covers more than 300 years of documents and the knowledge of about 4,500 years of human history.

15.2 Quality of a Pharmaceutical Drug

When it comes to man's health in all highly civilised countries strict and rigorous rules have been given to those who want to produce, import, sell, or use compounds or intervention of any kind to help patients. This is needed because patients are ill, sometimes doomed to death and in this situation patients would do anything to recover their health. This has been misused since man began helping man and civilised countries must provide help especially to those that are ill and in need of help in that most vulnerable group of people: the patients. Nowadays physicians trust interventions and pharmaceutical drugs that have been shown as reliable and reproducibly helpful within clinical trials. These trials should be randomised by a third party (patients and intervention) and wherever possible blinding is needed for both, physicians and patients. Both physicians and patients should not be aware of which persons receive the real thing and which are treated just for comparison in order to avoid what man will always do and science must avoid: *believe*. To all that lots of statistics must be added, pure naked statistics which will not worry about the anguished individual but only about the overall healthy effect. When a new (or old) intervention or compound has been shown in this type of clinical trial to overcome the reason for the illness (root cause treatment) or only the symptoms of it, it is worth stepping further along the decision tree towards it becoming a useful remedy for man. We call this *evidence-based medicine*.

Pharmaceutical drugs must not only address the pain, they should also not introduce as side effects other diseases. And finally it must be possible to produce these compounds to a high and long-lasting quality. These are the rules: a pharmaceutical drug must show a conclusive pharmaceutical effect, must not have unacceptable side effects and must be produced using Good Manufacturing Practise (GMP). In Europe, the United States, Canada, Japan and in many other countries these rules must all be addressed and after a complicated and most often failing procedure the new pharmaceutical drug or intervention brings a new cure for so far untreatable or less-favourably cured patients.

On the other hand in many countries physicians do have the right to test new interventions or to use old obsolete ones, if the physician had done everything for his patient that modern science and evidence-based medicine indicates him to do, but his patient remained ill. This is the compromise civilised countries still allow – if everything known fails, an experimental trial on a human patient is allowed. In all other cases experimental trials on humans are forbidden or need the special permission of an ethics committee (World Medical Association 2008). There is a reason for these strict and rigorous rules since any physician swears:

> § 4: "The health of my patient will be my first consideration," and the International Code of Medical Ethics declares that, "A physician shall act in the patient's best interest when providing medical care."
> § 7: Even the best current interventions must be evaluated continually through research for their safety, effectiveness, efficiency, accessibility and quality.
> § 12: Medical research involving human subjects must conform to generally accepted scientific principles, be based on a thorough knowledge of the scientific literature, other relevant sources of information, and adequate laboratory and, as appropriate, animal experimentation. The welfare of animals used for research must be respected.
> § 32: The benefits, risks, burdens and effectiveness of a new intervention must be tested against those of the best current proven intervention, except in the following circumstances: The use of placebo, or no treatment, is acceptable in studies where no current proven intervention exists.
> § 35: In the treatment of a patient, where proven interventions do not exist or have been ineffective, the physician, after seeking expert advice, with informed consent from the patient or a legally authorized representative, may use an unproven intervention if in the physician's judgement it offers hope of saving life, re-establishing health or alleviating suffering. Where possible, this intervention should be made the object of research, designed to evaluate its safety and efficacy. In all cases, new information should be recorded and, where appropriate, made publicly available.

Medical research involving human subjects is based on trust in the physician. He should consider this special treatment only when all other proved interventions have been ineffective. If the patient suffers from an experimental approach, the physician is responsible for his action. In Europe the law says that if a drug could possibly introduce a disease, it is supposed that this disease was the consequence of the drug. Whoever is responsible for that drug is also responsible for the side effects of this drug.

By these rules research using pharmaceutical drugs based on living organisms is of extraordinary complexity. Even if both, physician and patient agree to conduct this trial in the hope of recovering the patient's health, a reliable source of the pharmaceutical-grade organism must be found. Living fly larvae, leaches, *Plasmodium*-bearing mosquitoes, phage cultures, eggs of worms or living *Nematoda* will not be available even in a Chinese pharmacy, and in addition they are much less common in European or North American drug stores. And there is no way to produce these living drugs in a pharmacy, if there is no help from specialised companies producing these pharmaceuticals. In most countries of the western world such specialists in living drugs need a licence to produce these "compounds".

Apart from some microorganisms none of the organisms cited on the next pages are allowed to be produced, sold or used for medical interventions before they get a full market authorisation. Thus, only those indicated in the following text are allowed. Here it should be pointed out that the market is changing and the following might have changed as time goes by. It also should be noted that these rules cover living drugs using whole organisms and do not cover drugs based on separated cells, organs or fluids etc. These living parts are treated differently by law (Boergen et al. 2008; Schmidt 2010) and are not in the focus of this contribution.

Finally, living drugs cannot be sterile by definition. If they would be sterile, they would not be alive. And organisms are always unpredictable as life is capricious. Leaches for example need special bacteria in their gut (*Aeromonas hydrophila*), which might be transferred during biting of the patient. Furthermore, leaches big enough for treatment have sucked blood from sources before, which might be contaminated with agents of disease. That is why the patient-named prescription of every leach sold in a German pharmacy must be documented for 30 years. It is difficult and nowadays no longer allowed to import leaches from for example Turkey and to use them as a drug (there are temporary exemptions in Europe). It is better to breed the leaches, which proves to be a very complicated business.

It is also difficult, but not impossible to get high-quality eggs from the nematode *Trichuris suis*, with the worms mostly not isolated from the excrement of pigs but separated from the swine's intestine. And it remains tricky to determine the viability of these eggs within the patient's gut afterwards [Figs. 15.1–15.6 (for figures see end of the chapter)].

On one hand the physicians will need help from specialised companies and on the other hand these companies are usually not allowed to produce or sell these organisms for this purpose. Although many reports have been published, clinical trials are rare and because of this dilemma it is complicated to gain a market authorisation (Heuer et al. 2010). The North-America FDA (U.S. Food and Drug Administration) has decided to allow the production and use of leaches and fly larvae as a medical device. The European EMA (European Medicine Agency) has decided to classify larvae as a pharmaceutical drug but has not decided yet on a market authorisation. Japan and Australia are still waiting for decisions. Canada has decided, but since this decision has been taken, leaches and fly larvae are no longer on the market (only with special permission for experimentation on *an unproven intervention*).

15.3 Living Pharmaceutical Drugs

Scientists know two domains and six kingdoms of life on Earth (Cavalier-Smith 2004), which are used here as a guide to living pharmaceutical drugs.

The first domain includes only the kingdom bacteria. The other domain covers the kingdoms of animalia, fungi, plantae, chromista and protozoa. The group of viruses is placed between inorganic matter and living stages, as they are neither fully alive nor consist of only abiotic material and thus not able to reproduce themselves.

15.3.1 Kingdom Bacteria

Although bacteria usually are treated as agents of diseases they are very useful as pharmaceutical drugs. Many of the antibacterial drugs that we use are produced by bacteria, for example streptomycin by *Streptomyces griseus* to heal infections with *Mycobacterium tuberculosis* (tuberculosis). Furthermore, man has been using antibiotics from bacteria for centuries, for example the ancient Nubian people used *Streptomyces* in millet-seed brewed beer (contains tetracycline) (Nelson et al. 2010) or mouldy bread (contains penicillin) for healing was in use in Germany. Also bacteria live in and on man and help us to overcome potential infections from other bacteria, help us digest or produce vitamins. Our body is not our body alone. By a factor of 10 bacteria outnumber our body cells (Sagan and Margulis 1993) and live in symbiosis inside our gut system, on our skin, under our tongue (Kroes et al. 1999), in the corners of our eye, in our nose, in ear holes or within the vagina. Even under the extremely hard conditions of our stomach with a high concentration of hydrochloric acid more than 100 different phylotypes of bacteria have been identified (Bik et al. 2006), one of which is the probably cancer-inducing *Helicobacter pylori*. Bacteria are fighting in and on ourselves for their lives and are defending their ecological niche: our body. Man is an ecological system in itself and each individual of *Homo sapiens* is most properly a composition of many hundred different species and many thousands of subspecies, most of which are in our gut (Ley et al. 2006a, b). This is called a *super organism*. In a very recent study of the European population 1,000–1,150 different species of bacteria have been found and up to 160 in each individual (Qin et al. 2010). While in oceans live about 10^{29} bacteria cells the guts of all humans together contain 10^{23}–10^{24} bacteria. Therefore, together with other mammals, the human gut constitutes a substantial microbial habitat in our biosphere (Ley et al. 2006a, b).

When we are born we get infected by the first bacteria during the passage through the vagina (Ehrenberg 2010; Goho 2007) and we collect more and more useful species during our childhood and in the process of our life. When we are treated with antibiotics, we not only get rid of the unwanted species, but also of many others and this loss might lead to other illnesses (Czichos 2010) or obesity (Ley et al. 2006a, b; Turnbaugh et al. 2006; DiBaise et al. 2008). Looking back this is nothing of a surprise and has been used since more than half a century: antibiotics in the diet of pigs, broilers and other animal crops are used to increase their life-weight – obesity is the goal [Lehrer et al. 1953; Robinson et al. 1954]. In addition, Infectious bacteria might be spread such as *Klebsiella pneumoniae*, *Clostridium difficile* and *Proteus mirabilis*, leading for example to inflammatory bowel disease as *Colitis ulcerosa* (Garrett et al. 2010). In any case the gut microflora becomes changed as a reaction to repeated antibiotic perturbation and this change is irreversible (Dethlefsena and Relman 2010). Some species might be lost forever and consequences might follow years after the antibiotic therapy. The complexity of this ecological system is just about to be discovered by scientists, because only for a couple of years have technologies been available to obtain information on bacteria, which will grow easily in or on man but not in a Petri dish used in classical microbiology. Very recently it has been discovered that the human immune system reacts doses dependantly on the total amount and diversity of specific bacteria in the gut (Hapfelmeier et al. 2010) which might give an explanation for bowel diseases when the total microflora is disturbed by external influences such as a phage or antibiosis.

In man species of *Firmicutes, Bacteroidetes, Actinobacteria, Fusobacteria, Proteobacteria, Verrucomicrobia,* and *Cyanobacteria* dominate the gut. The first two genera represent 90–98% of the genetic pool (Qin et al. 2010), of at least 395 different species (Ley et al. 2006a, b) but most probably up to 1,150 different species (Qin et al. 2010).

Qin et al. were able to prove that healthy people and patients suffering from *Colitis ulcerosa* or *Crohn's disease* could clearly be distinguished by their microflora (Qin et al. 2010).

You are what you eat gets a different interpretation with respect of the said. Our diet always contains bacteria. Although the high concentration of acid in our stomach will kill 99.99% of them, some will still enter our gut. On raw meat or beef up to 10,000 bacteria were found per cm^2 and this is best quality food (bfr 2005). If something like *steak tartare* is eaten about 10^6–10^8 bacteria will enter our stomach and 0.01% = 100–10,000 living bacterial cells will finally reach our gut. Similar conditions are found on other nontreated, raw food.

Raw food is good for us – that's what grandmother said. Here we learn why. And we get a rough idea what it may mean for our ecological system, when we eat only "sterile", highly processed food. Today one can buy a (raw) mixed salad broken into pieces which looks and tastes fresh and can be stored "fresh" for nearly a week. That salad most properly has been disinfected by a technology which is called *Catallix* (Bordeau 2005) using lactoperoxidase and hypothiocyanite. The chemistry used in Catallix is part of the passive immune system of many organisms and destroys itself after a short period so that foodstuff must not be marked for containment of any potential residues. Ozone is an alternative to obtain a longer preservation.

Taking all that into consideration, nobody will be surprised when pharmaceutical drugs are available containing living bacteria. Alfred Nißle started treating patients with *Escherichia coli* in 1917. His product was named *Mutaflor* which has been on the market since then (Irrgang and Sonnborn 1988). *Mutaflor* is taken orally like all other products in use for our gut. Other products joined in later, for example *Symbioflor 1 (Enterococcus* spec., *Enterococcus faecalis), Symbioflor 2* (different *E. coli), Utilin, Utilin N* and *Bactisubtil (Bacillus subtilis, Mycobacterium phlei).* These pharmaceutical drugs claim to heal diarrhoea or *Colitis ulcerosa,* chronic obstipation, *Morbus Crohn* (Jungmayr 2009), or obesity (Turnbaugh et al. 2009).

Some other products including living bacteria are marketed as food additives to overcome the rough rules of pharmaceutical drugs, for example *UK Darmflora 10*

Mega (*Lactobacillus acidophilus, Lactobacillus rhamnosus, Lactobacillus casei, Bifidobacterium bifidum*). Help is claimed for chronic dermatosis, diarrhoea, enteritis, enterocolitis and other intestinal dysfunctions or dyspepsia. Decrease of neurodermatitis is claimed and proven using a clinical study on the oral application of *Lactobacillus* GG in newborns (Kalliomäki et al. 2001). *Lactobacillus gasseri* and *Bifidobacterium longum* are used to stabilise the gut microflora (*Omniflora N*).

Others use the same idea but are part of the usual diet, for example yoghurts such as *Actimel, LC1, Yakult, Activa, Activ Bio, SymbioLact* (*Lactobacillus johnsonii LA, B. bifidum, Bifidobacterium lactis, Lactobacillus salivarius, L. acidophilus, B. bifidum, B. lactis, L. casei, L. casei Defensis, L. casei Shirota, Lactobacillus lactis, Streptococcus thermophilus*). Food enriched with special bacteria are named probiotics (Tennyson and Friedman 2008). The application of the following bacteria intends to influence human health, for example *Bacteroides thetaiotaomicron, Bifidobacterium animali, B. bifidum, Bifidobacterium breve, Bifidobacterium infantis, B. lactis, B. longum, E. faecalis, Enterococcus faecium, L. acidophilus, Lactobacillus breve, Lactobacillus brevis, L. casei, L. casei Shirota, Lactobacillus delbrueckii subsp. Bulgaricus, Lactobacillus fermentum, L. johnsonii, Lactobacillus paracasei, Lactobacillus plantarum 299v, Lactobacillus reuteri, L. rhamnosus, L. rhamnosus GG, L. salivarius, Lactobacillus plantarum, Megaspaera elsdenii*, non-pathogenic *Salmonella* strains, *Streptococcus bovis, Streptococcus salivarius, S. thermophilus, S. thermophilus subsp. thermophilus, VSL#3* (Fedorak and Madsen 2004). All these bacteria including orally ingested materials have been proven to be useful for us. In case they claim to heal, for example obesity, they must be considered as a pharmaceutical drug (Heuer et al. 2010; FAZ 2010). In addition there are so many true food stuffs containing living bacteria, for example yoghurt, kefir, cheese, sauerkraut, and all raw vegetables or even raw beef.

Old literature describes dubious methods for the same intention (Paullini 1696). Meant as a help for the poor it is recommended to use faeces of different mammals including *Homo sapiens* to cure an unnumbered amount of human maladies. Whoever believes that these recommendations are somewhat too strange in this context, might be surprised when learning that recently the Bayer AG has claimed something very similar – to cure chicken from the most dangerous *Salmonella typhimurium* (Zündorf 1997). Bayer reported that chicken treated with antibiotics become more easily infected by this very common and dangerous germ. But in cases where chickens were sprayed with "healthy" faeces from their own species, infections of *S. typhimurium* had been avoided in most cases.

Even today, science cannot answer the question, of the origin of the useful and symbiotic microorganism. Young elephants, pandas, hippos and koala eat the faecal pap produced by their mother thus inoculating there intestines. Without that, for example the adult koala is incapable of eating the leaves of *Eucalyptus*, the main food source for koalas. Dogs eat deliberately the faeces of other animals and gorillas and chimpanzees are known to eat their own faeces or that from others of their group. Many other examples of coprophagia, the consumption of faeces, are known (Wikipedia Coprophagia 2010; Lewin 2001). In the case of man *ulcerative colitis* has been observed to slow down after *faecal bacteriotherapy*

(Borody et al. 2003). The same was observed in other bowel diseases (Floch 2010). Since *faecal bacteriotherapy* will infect man with many living microorganisms, this medication is classified as a pharmaceutical drug.

On this point it might be useful to remind ourselves of the fact of abstinence of alimentation during fasting. Fasting is known in many religions and it is used for healing and "clearing" the gut besides as a "soul wash". This is an extreme time for our gut bacteria, which will select the most adapted ones and shift the population to their favour. Ecologically speaking fasting is quite the opposite of an antibacterial medication.

The vagina is populated by *Lactobacillus* spec. and other bacteria. When these are lost for some reason, help might come from *L. fermentum LN 99*, *L. gasseri LN 40*, *L. casei LN 113–2* when included in a tampon (*ellen Tampon*). Healthy women most probably will not consider a tampon that is much more expensive than standard ones. One must presume that these tampons are for healing purposes. *ellen Tampon* is classified as a medical device. *Gynoflor* and *Vagiflor* (different *Lactobacillus* species, e.g. *L. acidophilus*) both are pharmaceutical drugs with full market authorisation and are used for the same purpose as *ellen Tampon* but they claim a healing benefit and fulfil the GMP and other regulations for drugs. A similar purpose is claimed for freeze-dried *L. gasseri* (EB01^{TM}) and *L. rhamnosus* (PB01^{TM}), which are used in *Vagisan* tablets claiming regulation of the pH-value of the vagina. They are also marketed as a medicinal device.

On the basis of the same idea a nose spray was created, which uses *Streptococcus* spec. for infections in the ear (Skovbjerg 2010). The use of *Bdellovibrio bacteriovorus* as a living antibiotic in this field is an outsider since these bacteria kill other bacteria by direct attack (Mielordt 2004). With respect to their use in humans both are pharmaceutical drugs.

15.3.2 Kingdom Animalia

Many of the used living pharmaceutical drugs are found in the kingdom of Animalia. Of the two subkingdoms Parazoa (sponges) and Eumetazoa, only the latter is of interest when it comes to uses in medication. The first pharmaceutically used and only single-cell organisms are *Plasmodium vivax* and *Plasmodium ovale*, two of a roughly 200 different species of parasitic organism from which at least 11 can infect *Homo sapiens*. *Plasmodium* sporozoides (see Fig. 15.1) are transferred by female mosquitoes of the genus *Anopheles*, for example *Anopheles plumbeus*, *Anopheles algeriensis*, *Anopheles claviger*, *Anopheles maculipennis*, *Anopheles messeae*, *Anopheles atroparvus*, as examples of European species from the ca. 100 out of ca. 450 *Anopheles* species known, which can transmit *Plasmodium*. Humans suffer from malaria when infected and will die or suffer for the rest of their life if left untreated. 350–500 Million patients are infected with malaria, most are poor and live in tropical regions of Africa, Asia or South America (Mehlhorn 2008). The observation of *Plasmodium* in the gut of *Anopheles* and the hope for a cure was the reason for the award of the Nobel Prize in Medicine to Ronald Ross in 1902. In 1927 another Nobel Prize on malaria was awarded to Julius Wagner Ritter von Jauregg for the so-called *"malaria-therapy"*. At that time many patients suffered from syphilis and no cure was known until Alexander Fleming (Fleming 1929) invented penicillin some years later. The malaria-therapy was a successful therapy and in regular use until the 1960s (Sama et al. 2006). Patients were transferred to specialised clinics. They were deliberately infected in special rooms by sporozoides of *Plasmodium vivax* or *Plasmodium ovale* using *Anopheles* as a vector (Mehlhorn and Piekarski 2002). Malaria will induce pyrexia to body temperatures as high as 42°C in some stages. Syphilis is the illness produced on infection by the gram-negative bacterium *Treponema pallidum,* which will die at temperatures of 41°C or higher. Patients suffering from syphilis will live longer if infected by malaria. This therapy should be obsolete now but it is still in use for other maladies as long as they cannot be treated properly. So cancer, AIDS and Lyme disease (Aronoff 2002) were claimed to be cured by deliberately induced malaria until recently (Heimlich 2006).

In Europe, United States, Canada, Japan and Australia "malaria-therapy" did not have a market authorisation which would be needed were patients to be treated using malarial stages as a pharmaceutical drug.

In the group of Bilateria all animals from worms to mammals are included. Let's start with roundworms or Nematoda which are interesting animals. *Trichutis suis* (see Fig. 15.2) is one of these pharmaceutically used worms. The orally swallowed eggs of this parasite of swine (*TSO*, *T. suis* ova) are claimed to cure *Morbus Crohn*, *Colitis ulcerosa* (Summers et al. 2005), rheumatoid arthritis, autoimmune diseases, autism, and many allergic reactions. This will be discussed in detail later in this book (Fried and Reddy, helminth therapy). In contrast to *T. suis Necator americanus* and *Ancylostoma duodenale* (see Fig. 15.3) are parasites of man (Mehlhorn and Köhler 2003), about 900,000,000 patients exist, and up to 60,000 dead are claimed by these parasites (Mehlhorn 2008) which penetrate through unprotected skin. *N. americanus* is claimed to have more or less the same positive influence on patients as *T. suis*. The general argument is the observation of the increase in autoimmune diseases within the first world and the strong decrease in worm infections and that both are connected to each other. When the human immune system is triggered by these worms, autoimmune diseases will be cured (Reddy and Fried 2009).

Still in the group of Bilateria the earthworm *Lumbricus terrestris* was once recommended to be swallowed alive for patients suffering from toothache or convulsions (Paullini 1696). All these therapies using worms make pharmaceutical claims, but have not been approved by medical agencies.

If imported from Turkey or bred by specialised companies using a defined environment the leach *Hirudo medicinalis* (see Fig. 15.4) is used for many purposes. Also *Hirudo verbana*, *Hirudo troctina* and *Hirudo orientalis* are used sometimes. Leaches will bite through the dermis and inject a great number of different active compounds into the blood and the tissue of the patient. During the sucking period of 20 min to 2 h, the influence of these compounds is increased. Depending on dose and location, the bite introduces anaesthetic or anticoagulant effects, especially in cases of poor blood flow in extremities or on joints for example after knee surgery. In cases of rheumatism, herpes zoster, varicose, tinnitus, thromboses, furuncle, tonsilla abscess, adnexitis, inflammation of sinus, of mammary gland, of gall bladder, of testicles, phlebitis, hypertona, arthrosis, apoplexy, angina pectoris, and thrombophlebitis the use of leaches is claimed to have positive effects (Moser and Moser 2002; Michalsena et al. 2008; Aurich and Koeppen 2009).

Some special companies sell this animal as pharmaceutically graded leaches. They have to fulfil GMP rules and obtain full market authorisation as a pharmaceutical drug in Europe and Canada. In USA leaches are devices, for which market authorisation can be achieved more easily than for other natural remedies. In some countries importation of this animal is difficult, since *H. medicinalis* is a neozoic organism (= not present before) for some habitats.

The next interesting animals belong to the phylum Anthropoda, covering Arachnida and Insecta. In the Arachnida spiders and scorpions have been used in former times for the treatment of malaria or 3-day-fever (Paullini 1696). Spiders were swallowed alive or on buttered bread.

Man has been interested in the Order of Insecta ever since. Some of the insects followed man and he used them when possible, for example *Cimex lectularius* (see Fig. 15.6). The freshly crushed bed bug was used against fever and as a wound treatment (Hering 2003) or placed alive in the urethra to empty the bladder when adults suffered from bladder obstruction or prostate gland cancer (Keferstein 1827) (the latter intervention is a medical device, all others are pharmaceutical drugs). Children were treated by the smaller body louse *Pediculus humanus corporis* as a medical device. Cockroaches are swallowed in Asia when suffering from convulsion (NN 2010) or *Pulex irritans*, fleas, were recommended to be swallowed in warm water on listlessness or paralysis (Paullini 1696). Swallowing insects alive has been quite common in Europe and although not recommended by modern literature it still occurs when a patient is doomed to death. *Melophagus ovinus* (sheep ket) is an insect that looks like a tick and is used against liver cancer and gout (Paullini 1696). It was swallowed frequently as a medication in the 1970s.

Other insects are bred or cultivated by man, for example the honey bee *Apis mellifera* and other species of *Apis*. Apitherapy meant deliberate stinging of a patient by a bee injecting its venom. Apitherapy is used for multiple sclerosis (MS), rheumatoid arthritis, hay fever, ALS (Lou Gehrig's Disease), shingles, scar pain, arthritis, gout, tendonitis, bursitis, spinal pain, bacterial, viral, and fungal illnesses, acute and chronic wounds, sprains, fractures, benign, malignant cancer, inflammatory or rheumatic pain (http://www.apitherapy.org/what-is-apitherapy/conditions-treated/). The latter can be cured by the "stinging" of ants as well (NN 2010) where formic acid is most probably the active ingredient. Other ants are used in Africa and Australia as medical devices. Individuals of *Ponerinae* spec, *Myrmicinae* spec, *Leptanilloidinae* spec, *Leptanillinae* spec, *Ecitoninae* spec. are allowed to bore their mandibles in such way into a wound, that the wound edges cling together (medical device). When done, the body of the ant is removed and the ant head holds the wound edges like a clamp as good as a suture. *Polyrachis vicina*, also an ant, is swallowed for stimulation of the immune system and to increase virility. In England a drink named ANT is available – who can guess why?

Freshly crushed beetles like *Curculio* (*Rhynchites*) *bacchus Linnaeus, 1758*, *Curculio betulae, Curculio iaceae Fabricius 1775* or *Carabus* (*Poecilus*) *cupreus* were used for toothache in the time before local anaesthetics were invented (Keferstein 1827). The Chinese use ladybirds still for that purpose (NN 2010). Although highly toxic (Schofield 2010) crushed beetles like *Paederus sabaeus*, *Paederus eximius = P. crebrepunctatus* or *Paederus fuscipes* are tested for their potential anti-cancer effects (Krumm 2010). The toxic Spanish fly *Lytta vesicatoria* has been used since ancient times for the removal of verrucas, as an aphrodisiac, abortifacient or as a stimulant producing insomnia (Wikipedia Spanish fly 2010).

The biting action of the mosquito *Anopheles stephensi* (Yoshida and Watanabe 2006; Enserink 2010) has been tested as a way to vaccinate man through stinging. Also the juice of *Anopheles* should expand arteries (NN 2010).

The African mole cricket *Gryllotalpa africana* BEAUVOIS, and *Gryllotalpa gryllotalpa* LINNÉ, the European mole cricket, have been used for wound healing (Zimmer 1997). Both did not show conclusive activity. Also fly larvae from different species of *Lucilia*, for example *Lucilia sericata* (Heuer et al., this book), *Lucilia cuprina* (Paul et al. 2009), and *Lucilia exima* (Wolff Echeverri et al. 2010) have been tested for wound healing. Good clinical results are given for debridement only (Dumville et al. 2009).

All these interesting applications of insects are pharmaceutical drugs and thus must follow the strict rules to produce and use them.

The next group of interest are the *Chordata* which include fish, amphibia, reptilia, birds and mammals; many are used as pets and in that way indirectly as a psychological cure. In some countries horse therapy is becoming regular. But in most countries only guide dogs for the blind are closely connected to patients as a medical device (Heuer et al. 2010). The use of a dog's sense of smell for detecting several diseases at an early stage gives physician's reason for a closer look. Some diseases are claimed to be detectable (e.g. diabetes and lung cancer). The patient sends a tube to an Austrian company into which he/she has exhaled –trained dogs smell these tubes, which are medical devices (Darwin 2010).

Again classified as a medical device is *Garra rufa*, Kangal fish, used for psoriasis, in wound healing and for skin clearance. The patient places his diseased body portion into the water and many little fish feed on the sick skin (Grassberger and Hoch 2006). This therapy is called ichthyotherapy and was first established in Turkey but can be found in many big cities, mostly for foot care.

15.3.3 Kingdom Fungi

Although consumed in high numbers in beer, cheese and in many other foodstuffs, the number of fungi used directly in pharmaceutical applications is low. However, as in bacteria many antibiotics are produced as extracts from fungi such as penicillin from *Penicillium notatum*, discovered by Alexander Fleming (1929). He might not be the first who noticed the antibacterial action of these fungi – but he was the first who acted consequently and developed a pharmaceutical drug from this observation. The use of bread with a blue mould (presumably *Penicillium*) for treating suppurating wounds was a staple of folk medicine in Europe since the Middle Ages (Wikipedia Penicillin 2010).

Living fungi are the active ingredient of *Perenterol*, which is a medication used for gut regulation. Perenterol includes *Saccharomyces boulardii* and different strains of *Saccharomyces cerevisiae*. Perenterol is a pharmaceutical drug.

15.3.4 Kingdom Plantae

Since uncooked plants are a significant contribution to the food of man in all countries, the influence and importance of living plants cells cannot be underestimated. However, up to now none have been used for pharmaceutical applications. This means that no pharmaceutical company claims the healing of patients by eating raw plants. Nevertheless, some clearly pharmacological effects are used by humans when eating/chewing parts of raw plants, for example leaves of *Catha edulis* or *Erythroxylum coca*. Both are used as psychedelic drugs in the countries of origin. Many leaves, roots, fruits or other parts of plants for different typical healing procedures are known but none are officially marketed for medicinal purposes.

15.3.5 Kingdoms Chromista and Protozoa

Both kingdoms are very closely related to each other (Cavalier-Smith 2004). These single-cell organisms might attack man, for example *Trichomonas vaginalis* or *Toxoplasma gondii* (Mehlhorn 2008). Since many protozoa feed on bacteria, they are proposed for use as antibacterial agents (Nacara and Nacarb 2007). However, up to now there is no official use and should they ever be used for healing they will have to be classified as pharmaceutical drugs.

15.3.6 Viruses and Phage

Almost all official pharmaceutical drugs are single chemical entities with a rather "simple" structure and which can be synthesised by chemists in a laboratory and produced following the GMP rules in a pharmaceutical company. Although the action of one "simple" chemical drug can be most complicated for the human intellect, it is the easiest case of all possible healing interventions. Nevertheless, bad outcomes are possible as was unfortunately the case for the remedy *Contagan* taken by pregnant women, leading to the effect that thousands of newborns were crippled for a lifetime. After that incident, in many countries the rules for pharmaceuticals were toughened.

As an active virus or a phage (virus specifically acting on bacteria) is neither living nor dead, it is not surprising that these had been the first "organisms" used for pharmaceutical applications. Vaccination is the name of the process by which small amounts or partly deactivated viruses are inoculated into humans or animals. The inoculum must be able to trigger the immune system but so weakly as not to overwhelm the patient. This has been known for nearly 3,000 years (Wikipedia Pocken 2010), but has been optimised within the last 300 years in Europe (Wikipedia Impfung 2010). Vaccination has been tested or used for *Variola*-virus since 1714 and 1796, *Lyssa*-virus since 1885, *Flavi*-virus since 1935, *Influenza*-virus since 1936, *Polio*-virus since 1955/1960, *Paramyxo*-virus since 1967, *Morbilli*-virus since 1968, Togaviridae since 1969, Flaviviridae since 1973, *Varicella*-virus since 1974, Hepadnaviridae since 1981, *Rota*-virus since 1998 and *Papillom*-virus since 2006. All of these pharmaceutical interventions saved many lives and may be the most successful drugs of all known pharmaceuticals.

Phages are not so impressive but have also been long studied as antibacterial agents (Wikipedia Phage Therapy 2010). Phages have been used against diarrheal diseases caused by *E. coli, Shigella* spec., or *Vibrio* spec. and against wound infections caused by facultative pathogens of the skin like *Staphylococcus* spec., *Proteus* spec, *Klebsiella* spec. and *Streptococcus* spec. Some of the interest can be traced back to 1994, when Soothill demonstrated (in an animal model) that the use of phages could improve the success of skin grafts by reducing the underlying *Pseudomonas aeruginosa* infection. Phages are a very specific agent to fight one bacteria species only. With chemical antibiotics available as a standard therapy, phages have never been submitted to all the needed tests for a pharmaceutical drug. In the future they might receive more interest, because more and more bacteria are becoming resistant to chemical antibiotics.

15.4 Conclusion

In the case of untreatable diseases, many if not all interventions are taken into consideration. If these interventions are part of an individual attempt to achieve a cure for a single patient and undertaken with the full responsibility of the supervising physician, these interventions are allowed in nearly all societies on Earth. If the intervention needs medication, even if this medication uses living organisms, in most cases these pharmaceutical drugs need the approval of the local administration when produced by somebody other than by the physician himself for his patient.

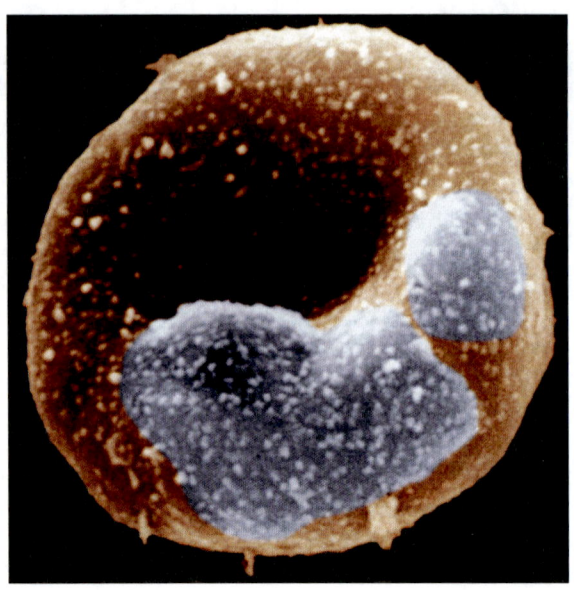

Fig. 15.1 *Plasmodium falciparum* infected red blood cell (photo: Mehlhorn)

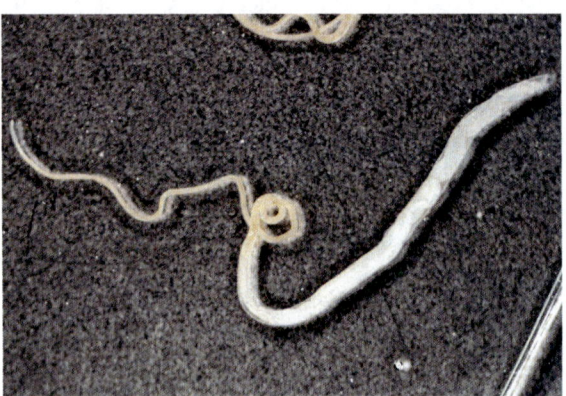

Fig. 15.2 Adult *Trichuris suis* worm (photo: Mehlhorn)

Fig. 15.3 Head of *Ancylostoma duodenale* (photo: Mehlhorn)

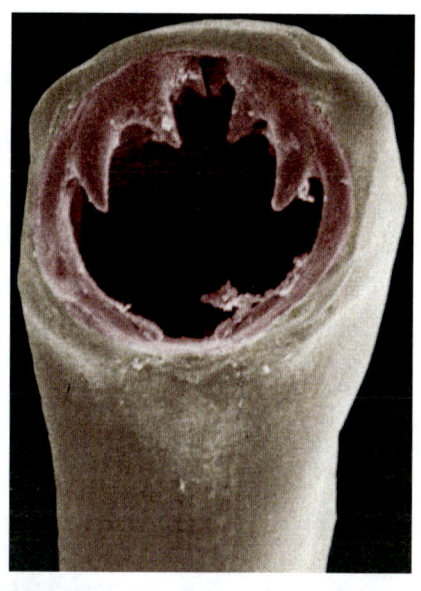

Fig. 15.4 Leech (*Hirudo medicinale*) (photo: Mehlhorn)

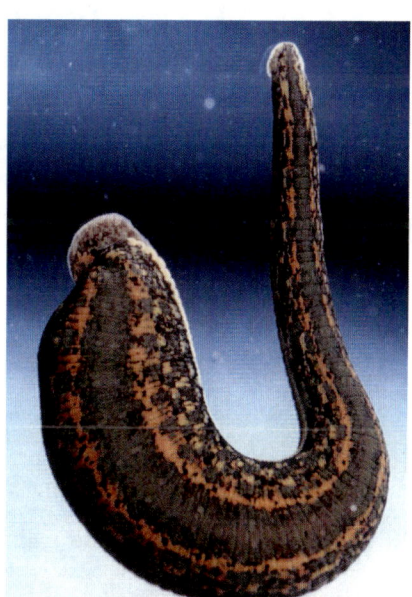

Fig. 15.5 Head louse (*Pediculus humanus capitis*) (photo: Mehlhorn)

Fig. 15.6 Bed bug (*Cimex lectularius*) (photo: Mehlhorn)

References

Aronoff DM (2002) Using live pathogens to treat infectious diseases: a historical perspective on the relationship between GB virus C and HIV. Antivir Ther 7:73–80

Aurich M, Koeppen D (2009) Eine Anwenderumfrage zur Blutegeltherapie – Auswertung aus 171 Falldokumentationen. zkm Zeitschrift für Komplementärmedizin 5:12–18

bfr (2005) http://www.bfr.bund.de/cm/276/ausgewaehlte_fragen_und_antworten_zu_verdorbenem_fleisch.pdf. Accessed 25 Sept 2010. BfR (2005) Bundesinstitut für Risikobewertung, Ausgewählte Fragen und Antworten zu verdorbenem Fleisch, FAQ vom 29. November 2005, http://www.bfr.bund.de/cd/7042

Bik EM, Eckburg PB, Gill SR, Nelson KE, Purdom EA, Francois F, Perez-Perez G, Blaser MJ, Relman DA (2006) Molecular analysis of the bacterial microbiota in the human stomach. PNAS 103:732–737. www.pnas.org_cgi_doi_10.1073_pnas.0506655103. Accessed 18 Sept 2010

Boergen X, Jäckel C, Spiegel JP (2008) Lebende Arzneimittel. Pharma Recht, Germany, pp 357–400

Bordeau P (2005) A biological technique to ensure food safety. Specialities Chemicals Magazine, pp 34–36

Borody TJ, Warren EF, Leis S, Surace R, Ashman O (2003) Treatment of ulcerative colitis using fecal bacteriotherapy. J Clin Gastorenterol 37(1):42–47

Cavalier-Smith T (2004) Only six kingdoms of life. Proc R Soc Lond B 271:1251–1262. http://www.cladocera.de/protozoa/cavalier-smith_2004_prs.pdf. Accessed 18 Sept 2010

Czichos J (2010) Antibiotische Therapie schädigt Darmbakterien stärker als gedacht, Wissenschaft aktuell, 2010-09-14. http://www.wissenschaft-aktuell.de/artikel/Antibiotische_Therapie_schaedigt_Darmbakterien_staerker_als_gedacht1771015587084.html. Accessed 18 Sept 2010

Darwin (2010) Lung cancer http://www.lungenkrebsfinder.at/. Accessed 19 Sept 2010

Dethlefsena L, Relman DA (2010) Incomplete recovery and individualized responses of the human distal gut microbiota to repeated antibiotic perturbation. www.pnas.org/cgi/doi/10.1073/pnas.1000087107

DiBaise JK, Zhang H, Crowell MD, Krajmalnik-Brown R, Decker GA, Rittmann BE (2008) Gut microbiota and its possible relationship with obesity. Mayo Clin Proc 83(4):460–9

Dumville JC, Worthy G, Soares MO, Bland JM, Cullum N, Dowson C, Iglesias C, McCaughan D, Mitchell JL, Nelson EA and Torgerson DJ on behalf of the VenUS II team (2009) VenUS II: a randomised controlled trial of larval therapy in the management of leg ulcers. Health Technol Assess 13(55). doi: 10.3310/hta13550, http://www.hta.ac.uk/pdfexecs/summ1355.pdf. Accessed 02 Sept 2010

Ehrenberg R (2010) Baby's first bacteria depend on birth route. Science News 21 June 2010. http://www.sciencenews.org/view/generic/id/60461/title/Baby%E2%80%99s_first_bacteria_depend_on_birth_route. Accessed 18 Sept 2010

Enserink M (2010) Researchers turn mosquitoes into flying vaccinators, 18 March 2010. http://news.sciencemag.org/sciencenow/2010/03/researchers-turn-mosquitoes-into.html. Accessed 19 Sept 2010

FAZ (2010) Nestlé will Kranke mit Lebensmitteln heilen, Frankfurter Allgemeine Zeitung, p 13, 28 Sept 2010

Fedorak RN, Madsen KL (2004) Probiotics and the management of inflammatory bowel disease. Inflamm Bowel Dis 10(3):286–299

Fleming A (1929) On the antibacterial action of cultures of a penicillium, with special reference to their use in the isolation of B. influenzae. Br J Exp Pathol 10(31):226 36

Floch MH (2010) Fecal bacteriotherapy, fecal transplant, and the microbiome. Clin Gastroenterol 44(8):529

Garrett WS, Gallini CA, Yatsunenko T, Michaud M, DuBois A, Delaney ML, Punit S, Karlsson M, Lynn Bry L, Glickman JN, Gordon JI, Onderdonk AB, Glimcher LH (2010) Enterobacteriaceae act in concert with the gut microbiota to induce spontaneous and maternally transmitted colitis. Cell Host Microbe 8(3):292–300

Goho A (2007) Our microbes, ourselves: how bacterial communities in the body influence human health, society for science & the public. Science News 171(20):314–316. http://www.jstor.org/stable/20055664. Accessed 18 Sept 2010

Grassberger M, Hoch W (2006) Ichthyotherapy as alternative treatment for patients with psoriasis: a pilot study. eCAM 3(4):483–488. doi:10.1093/ecam/nel033, http://ecam.oxfordjournals.org/cgi/content/full/3/4/483. Accessed 19 Sept 2010

Hapfelmeier S, Lawson MAE, Slack E, Kirundi JK, Stoel M, Heikenwalder M, Cahenzli J, Velykoredko Y, Balmer ML, Endt K, Geuking MB, Curtiss R, McCoy KD, Macpherson AJ (2010) Reversible microbial colonization of germ-free mice reveals the dynamics of IgA immune responses. Science 328(5986):1705–1709. doi:10.1126/science.1188454

Heimlich H (2006), http://www.circare.org/malariotherapy.htm. Accessed 19 Sept 2010

Hering (2003) Hering's guiding symptoms of our Materia Medica, vol 2. B. Jain, New Delhi, ISBN 81-8056-316-2

Heuer H, Heuer L, Saalfrank V (2010) Lebende Arzneimittel. Deutsche Apotheker Zeitung 150:46–54

Irrgang K, Sonnborn U (1988) The historical Development of Mutaflor therapy. Ardeypharm GmbH. http://www.ardeypharm.de/pdfs/en/mutaflor_historical_e.pdf. Accessed 18 Sept 2010

Jungmayr P (2009) Deutsche Apotheker Zeitung 48:5450

Kalliomäki M, Salminen S, Arvilommi H, Kero P, Koskinen P, Isolauri E (2001) Probiotics in primary prevention of atopic disease: a randomised placebo-controlled trial. Lancet 357 (9262):1076–1079

Keferstein KG (1827) Über den unmittelbaren Nutzen der Insekten. Maring'sche Buchhandlung, Erfurt

Kroes I, Lepp PW, Relman DA (1999) Bacterial diversity within the human subgingival crevice. PNAS 96:14547–14552

Krumm C (2010) Apotheken Umschau 15 Oct 2010, pp 62–63

Lehrer WP, Pharris ER, Harvey WR, Keith TB (1953) Growth Effects of some Antibiotics on Suckling, Growing, and Fattening Pigs, Journal of Animal Science 12:304–309

Lewin RA (2001) More on Merde. Perspect Biol Med 44(4):594–607

Ley RE, Turnbaugh PJ, Klein S, Gordon JI (2006a) Microbial ecology: human gut microbes associated with obesity Nature 444:1022–1023. doi: 10.1038/4441022a

Ley RE, Peterson DA, Gordon JI (2006b) Ecological and evolutionary forces shaping microbial diversity in the human intestine. Cell 124:837–848. doi:10.1016/j.cell.2006.02.017

Mehlhorn H (ed) (2008) Encyclopedia of parasitology, 2 vols, 3rd edn. Springer, Heidelberg

Mehlhorn H, Köhler S (2003) Reisesratgeber, Parasiten und Schädlinge im Heim und auf Reisen. Alphabiocare Medien Service, Germany

Mehlhorn H, Piekarski G (2002) Grudriß der Parasitenkunde, Spektrum Akademischer Verlag Gustav Fischer, Heidelberg 6te Auflage

Michalsena A, Lüdtkeb R, Cesura Ö, Afraa D, Musiala F, Baeckera M, Finkc M, Dobosa GJ (2008) Effectiveness of leech therapy in women with symptomatic arthrosis of the first carpometacarpal joint: a randomized controlled trial. Pain 137(2):452–459. doi:10.1016/j.pain.2008.03.012

Mielordt S (2004) Bdellovibrio bacteriovorus – ein lebendes Antibiotikum? 04 June 2004. http://www.charite.de/klinphysio/bioinfo/3_k-pathophy-fromm/s_pphy_04ss_05_bdellovibrio-antbiot_mielord.pdf. Accessed 18 Sept 2010

Moser C, Moser K (2002) So hilft Ihnen die Blutegel-Therapie. Haug Verlag, Stuttgart, ISBN 3-8304-2072-2

Nacara A, Nacarb E (2007) Phagotrophic protozoa: a new weapon against pathogens? Med Hypotheses 70(1):141–142. doi:10.1016/j.mehy.2007.03.037

Nelson ML, Dinardo A, Hochberg J, Armelagos GJ (2010) Brief communication: Mass spectroscopic characterization of tetracycline in the skeletal remains of an ancient population from Sudanese Nubia 350–550 CE. Am J Phys Anthropol 143(1):151–154. doi:10.1002/ajpa.21340

NN (2010) http://www.20min.ch/diashow/diashow.tmpl?showid=29081. Accessed 19 Sept 2010

Paul AG, Ahmad NW, Lee HL, Ariff AM, Saranum M, Naicker AS, Osman Z (2009) Maggot debridement therapy with Lucilia cuprina: a comparison with conventional debridement in diabetic foot ulcers, International Wound Journal, Vol 6 No 1. http://onlinelibrary.wiley.com/doi/10.1111/j.1742-481X.2008.00564.x/pdf. Accessed 2010-09-19

Paullini KF (1696) Heilsame Dreck-Apotheke. Frankfurt am Main. http://diglib.hab.de/drucke/xb-3174/start.htm. Accessed 12 Mar 2010

Qin J, Li R, Raes J, Arumugam M, Burgdorf KS, Manichanh C, Nielsen T, Pons N, Levenez F, Yamada T, Mende DR, Li J, Xu J, Li S, Li D, Cao J, Wang B, Liang H, Zheng H, Xie Y, Tap J, Lepage P, Bertalan M, Batto JM, Hansen T, Paslier DL, Linneberg A, Nielsen HB, Pelletier E, Renault P, Sicheritz-Ponten T, Turner K, Zhu H, Yu C, Li S, Jian M, Zhou Y, Li Y, Zhang X, Li S, Qin N, Yang H, Wang J, Brunak S, Dore´ S, Guarner F, Kristiansen K, Pedersen O, Parkhill J, Weissenbach J, MetaHIT Consortium, Bork P, Ehrlich SD, Wang J (2010) Human gut microbial gene catalogue established by metagenomic sequencing. Nature 464:59–65. doi:10.1038/nature08821

Reddy A, Fried B (2009) An update on the use of helminths to treat Crohn's and other autoimmunune diseases. Parasitol Res 104(2):217. doi:10.1007/s00436-008-1297-5

Robinson KL, Coey WE, Burnett GS (1954) The use of antibiotics in the food of fattening pigs, Journal of the Science of Food and Agriculture, Vol. 5(11):541–549

Sagan D, Margulis L (1993) Garden of microbial delights: a practical guide to the subvisible world. Kendall/Hunt, Dubuque, Iowa

Sama W, Dietz K, Smith T (2006) Distribution of survival times of deliberate *Plasmodium falciparum* infections in tertiary syphilis patients. Trans R Soc Trop Med Hyg 100(9):811–816

Schmidt M (2010) Arzneimittel von Tieren. PTAheute 24:30–33

Schofield S (2010) Bugs don't have to bite to do damage: the tale of the Paederus beetle. http://www.forces.gc.ca/health-sante/wn-qn/adv-avi/Paederus-eng.asp. Accessed 19 Sept 2010

Skovbjerg S (2010) Thesis, Inflammatory mediator response to Gram-positive and Gram-negative bacteria in vitro and in middle ear infections, University of Gothenburg 04 Mar 2010. http://gupea.ub.gu.se/bitstream/2077/21533/2/gupea_2077_21533_2.pdf. Accessed 18 Sept 2010

Summers RW, Elliott DE, Urban JF Jr, Thompson RA, Weinstock JV (2005) Trichuris suis therapy for active ulcerative colitis: a randomized controlled trial. Gastroenterology 128(4):1117–1119; 129(2):768–769; author reply 769

Tennyson CA, Friedman G (2008) Microecology, obesity, and probiotics. Curr Opin Endocrinol Diabetes Obes 15(5):422–427

Turnbaugh PJ, Ley RE, Mahowald MA, Magrini V, Mardis ER, Gordon JI (2006) An obesity-associated gut microbiome with increased capacity for energy harvest. Nature 444:1027–1031. doi:10.1038/nature05414

Turnbaugh PJ, Ridaura VK, Faith JJ, Rey FE, Knight R, Gordon JI (2009) The effect of diet on the human gut microbiome: a metagenomic analysis in humanized gnotobiotic mice. Sci Transl Med 1(6):6–14. doi:10.1126/scitranslmed.3000322

Wikipedia Coprophagia (2010) http://en.wikipedia.org/wiki/Coprophagia. Accessed 03 Oct 2010

Wikipedia Hakenwürmer (2010) http://de.wikipedia.org/w/index.php?oldid=78508237. Accessed 19 Sept 2010

Wikipedia Impfung (2010) http://de.wikipedia.org/wiki/Impfung. Accessed 19 Sept 2010

Wikipedia Penicillin (2010) http://en.wikipedia.org/wiki/Penicillin. Accessed 26 Sept 2010

Wikipedia Phage Therapy (2010) http://en.wikipedia.org/wiki/Phage_therapy. Accessed 19 Sept 2010

Wikipedia Pocken (2010) http://de.wikipedia.org/wiki/Pocken. Accessed 19 Sept 2010

Wikipedia Spanish fly (2010) http://en.wikipedia.org/wiki/Lytta_vesicatoria. Accessed 19 Sept 2010

Wolff Echeverri MI, Rivera Álvarez C, Herrera Higuita SE, Wolff Idárraga JC, Escobar Franco MM, Lucilia eximia (2010) (Diptera: Calliphoridae), una nueva alternativa para la terapia larvaly reporte de casos en Colombia. IATREIA 23(2)

World Medical Association (2008) WMA Declaration of Helsinki – Ethical Principles for Medical Research Involving Human Subjects (22.10.2008). http://www.wma.net/en/30publications/10policies/b3/index.html. Accessed 18 Sept 2010

Yoshida S, Watanabe H (2006) Robust salivary gland-specific transgene expression in Anopheles stephensi, mosquito. Insect Mol Biol 15(4):403–410

Zimmer M (1997) Die chinesische Maulwurfsgrille Gryllotapa africana BEAUVOIS und die saarländische Maulwurfsgrille Gryllotalpa gryllotalpa LINNÉ in der Wundheilkunde. http://www.carstens-stiftung.de/nachwuchs/promotionsfoerderung/abstracts/nhk/zimmer.pdf. Accessed 19 Sept 2010

Zündorf U (1997) Salmonellen-Vorbeugender Schutz für Küken. Research Bayer-Forschungsmagazin 9:53–55

Index

A

Active ingredients, 22, 154, 314–317
Active plant compounds, 22, 80–83, 112
Adulticidal activity, 29
Aedes
  A. aegypti, 21, 23–27, 30, 32–34, 36, 38, 41
  A. vexans, 60, 66
Aga toads, 191, 194
AgilPad™, 308, 309
Allergic rhinitis, 221
Amebic infections, 4
Ancylostoma, 363
Angiostrongylus, 134
Anopheles
  An. gambiae, 64, 70
  An. stephensi, 20, 24, 27, 29, 32, 34, 38, 41
Anthelmintics, 110–113, 116, 128, 129
Antibiotic, 353–356, 360, 361
Anti-tick products, 91
Asthma, 212, 220–221
Autoimmune diseases, 211–221
Azadirachta indica, 13, 79, 97–98
AZT, 158

B

Bacillus
  B. sphaericus, 56, 61, 65, 70–71
  B. thuringiensis, 56–61
  B. thuringiensis israelensis, 61–69
  B. thuringiensis kurstaki, 57, 59, 60
  B. thuringiensis tenebrionis, 57, 69
Bacteria, 1, 6, 302, 303, 353
Bacterio therapy, 355, 356
Bed bugs, 89, 95 103, 364
Beetles, 92, 93, 103
Binary toxin, 71
BioBag, 308, 310, 311
BioFOAM, 309–311

Biological control, 78, 83, 86–88, 110, 117, 118, 121–123, 125, 126, 128, 191–193, 195–199, 201
Biosurgical debridement, 311
Blood samplers, 243–265
Blackflies, 57, 61–63, 67–68, 70
Blow flies, 235–236, 301
Brugia malayi, 20
Bufo species, 203

C

Cane toad, 203
Cestodes, 110, 114–127, 213, 216, 220
Chagas disease, 274, 278–292
Chironomus thummi, 69
Chronic wound, 303, 304, 307, 317
Cimex lectularius, 89, 95, 105, 364
Clinical trial phases, 156, 163–176
Cockroaches, 89, 90, 93, 104
Coconut, 109–114, 131
Coconut extracts, 109–136
Coleoptera control, 69
Cost effectiveness, 63–64
Crohn's disease, 211–221
Cry toxin, 57, 60, 61, 69
Culex pipiens, 64, 66, 67, 70
Culex p. quinquefasciatus, 70, 71
Culex quinquefasciatus, 20, 24–30, 32–34, 36, 38, 40, 41
Curcumin, 141–147

D

Damalinia caprae, 29, 38, 39
Debridement, 303–307, 310–317
Debridement of wounds, 303, 304
Decomposition, 236
Dehiscent wound, 304
δ-endotoxin, 56, 57

Dermanyssus gallinae, 86, 100
Dermatophagoides, 89, 100
Detection of drugs, 236
Diagnosis, 273, 274, 277, 279–281, 285, 291–292
Diarrhea, 1–13
Dipetalogaster maxima, 243–246, 248–261, 263–265, 275, 276, 282, 283, 286–290
DNA detection, 238

E
Ease of handling, 59, 63
Echinostoma, 123, 132
Entomotoxicology, 236–237
Environmental damage, 194
Environmental safety, 59, 62, 63
Ethnopharmacology, 1, 2, 7–10, 13
Evidence-based medicine, 350, 351
Extracts, 115, 121

F
Fasciola hepatica, 132
Fever, 21
Fight against pests, 191
Filariae, 20
Fleas, 92, 102
Flies, 90, 303, 327, 350
Forensic entomology, 228, 229, 232, 237–240
Formula of potency, 59
Formulations, 58, 59, 64–67, 70, 71, 81
Fungi, 359

G
γ-radiation, 64
Giardia, 5
Growth curves (flies), 233, 234

H
Haemaphysalis bispinosa, 26, 29, 31, 37–39
Harmonia, 205, 206
Helminth, 142–143
Helminth therapy, 211–221. *See also* Cestodes, Nematodes, Trematodes
Hippobosca maculata, 29, 37–39
Hirudo medicinalis, 357, 363
Hirudo species, 357
Hookworm, 363
Hymenolepis sp., 123, 133, 215, 216

I
IBD. *See* Inflammatory bowel disease (IBD)
Indigenous traditional medicine, 1–13
Inflammatory bowel disease (IBD), 212–217
Insecticidal effects, 91, 99

Insects, 78, 81–85, 89, 92, 96
Integrated biological control (IBC), 65, 66
Integrated mosquito and vector management, 65–67
International toxic unit, 58, 65
Intestinal bacteria, 6–7
Intestinal parasites, 3–5, 10
Invasion of new species, 191–207
Ixodes species, 101

K
KABS, 60, 64, 66, 67

L
Lady beetle, 205, 206
Larval therapy, 302, 303, 305–306, 312, 314, 317
Larveel®, 339–340
Larvicidal activity, 26–29, 32–34, 38, 41
Larvicidal effects, 23–41
Leech, 357, 363
LeFlap™, 308, 310
Lice, 98, 99, 364
Living medication, 349–364
Living syringes, 243–265
Lucilia sericata, 104, 106, 305, 306, 312–317

M
Maggot therapy, 301–318
Maggot treatment, 301–318
Malaria control programs, 71–72
Malaria-therapy, 357
Mallophages, 91, 102
Marine-derived therapeutics, 156, 162
Marine natural products, 155, 156, 176–178
Marine pharmaceutical pipeline, 153
Market overview, 307–311
Medical device, 352, 356, 358, 359
Medicinal plant extracts, 19–42
Medicinal plants, 8, 10–12, 19, 21, 23–41, 112–114, 118–124
Microbial control agent, 56, 59, 62, 65–67, 71–72
Mites, 82, 84–89, 91, 94, 100
MiteStop, 91–96
Mode of action, 143–147
Mosquitocidal toxin, 71
Mosquitoes, 20, 66, 70
Mosquito-larval control, 72
Moths, 203
Multiple sclerosis (MS), 212, 213, 217–218
Myiasis, 232, 302, 305–307

Myxomatosis, 197
Myxoma virus, 197

N
Natural acaricides, 96
Natural insecticides, 96
Natural remedies, 55
Necrophagous insects, 228, 229, 232, 236, 237
Neem, 79, 97–102
Neem extracts, 77–105
Nematoceran flies, 57, 68–69
Nematodes, 3, 112, 126–135, 147, 213–220, 363
Non-healing wounds, 325, 331–341, 345

O
Ochlerotatus sp., 66
Onion, 109–136
Onion extracts, 109–136
Opuntia species, 207, 208
Osmo-regulatory mechanisms, 58
Ovicidal activity, 31, 41
Oviposition-deterrent activity, 31

P
Pain, 307, 310, 312–314
Paramphistomum cervi, 26, 29, 31, 37–39
Parasites, 142–147
Parasitism, 349
Parasitoids, 191, 192, 195, 198, 199
Parasporal protein inclusion, 71
Pathotype A, 57, 60–61
Pathotype B, 57, 60, 61
Pathotype C, 57, 69–70
Pathotypes, 60–63
Pediculus, 98, 99, 364
Pests, 191–208
Phage, 300
Pharmaceutical drug, 350–361
Pharmaceutical law, 349–364
Plant extracts, 84, 85, 94, 96, 109–136
Plasmid, 57, 62
Post mortal changes, 231
Probiotic, 355
Protein crystal, 56–58, 60–62, 69, 70
Protozoa, 143–146, 360
Psychoda alternata, 69

R
RA. *See* Rheumatoid arthritis
Rabbit Haemorrhagic Disease, 127

Raptor bugs, 93, 243–265, 276
Repellent activity, 30, 31, 33, 34, 40, 41
Resistance, 59–63, 71, 72
Retrovir, 158
Rheumatoid arthritis (RA), 212, 213, 218–219
Rhipicephalus (Boophilus) microplus, 26, 29, 31, 37–39
Rhipicephalus sanguineus, 101
Rhodnius prolixus, 244–252, 256, 258–261
Roll back Malaria (RBM), 71

S
Schistosoma species, 216, 220
Sciarid, 69
Side effects, 311, 312
Species identification, 238
Spores, 62, 70
Standardization, 58–59
Struggle for life, 193, 194
Superorganism, 353
Susceptibility, 65
Symbiosis, 353

T
Tapeworms, 109–136
Tenebrio molitor, 103
T1D. *See* Type 1 diabetes (T1D)
Ticks, 84, 89, 91, 94, 101
Time since death, 228, 232, 236, 239
Tipula paludosa, 69
Toxins, 57, 70
Toxocara, 135
Traditional medicine, 9
Treatment, 125, 141, 142
Trematodes, 109–136, 213, 216
Triatoma infestans, 103, 244–252, 258–260
Triatomines, 244–254, 256–258, 260, 261, 264, 265, 274–277, 280–291
Trichinella species, 216, 217, 220
Trichuris species, 123–125, 134, 214
Trypanosoma cruzi, 274, 276–281, 283–292
Type 1 diabetes (T1D), 212, 213, 219–220

V
Vector-derived diseases, 328
Viruses, 168, 360

W
Weaver ants, 202
WHO Pesticide Evaluation Scheme, 62

World population, 136
Worm control, 110, 117–119, 123, 126
Wound healing, 303–304, 312, 314
Wound treatment, 302, 317
Wuchereria bancrofti, 20

X
Xenodiagnosis, 244, 250, 251, 254, 259, 273–292

Z
Zoo animals, 253, 261–264